MOBILITY *and* PROXIMITY *in* BIOLOGICAL MEMBRANES

Sándor Damjanovich
Professor and Director
Department of Biophysics
University School of Medicine
Debrecen, Hungary

János Szöllósi
Associate Professor
Department of Biophysics
University School of Medicine
Debrecen, Hungary

Lajos Trón
Chairman
PET Center
University School of Medicine
Debrecen, Hungary

Michael Edidin
Professor
Department of Biology
Johns Hopkins University
Baltimore, Maryland

CRC Press
Taylor & Francis Group
Boca Raton London New York

CRC Press is an imprint of the
Taylor & Francis Group, an informa business

First published 1994 by CRC Press
Taylor & Francis Group
6000 Broken Sound Parkway NW, Suite 300
Boca Raton, FL 33487-2742

Reissued 2018 by CRC Press

Library of Congress Cataloging-in-Publication Data

Mobility and proximity in biological membranes / Sándor Damjanovich ... [et al.].
p. cm.
Includes bibliographical references and index.
ISBN 0-8493-4931-1
1. Membranes (Biology)—Fluidity. I. Damjanovich. Sándor.
QH601.M59 1994
574.87'5—dc20 93-37831

A Library of Congress record exists under LC control number: 93037831

Publisher's Note
The publisher has gone to great lengths to ensure the quality of this reprint but points out that some imperfections in the original copies may be apparent.

Disclaimer
The publisher has made every effort to trace copyright holders and welcomes correspondence from those they have been unable to contact.

ISBN 13: 978-1-315-89557-4 (hbk)
ISBN 13: 978-1-351-07467-4 (ebk)

Visit the Taylor & Francis Web site at http://www.taylorandfrancis.com and the
CRC Press Web site at http://www.crcpress.com

AUTHORS

Sándor Damjanovich, M.D., D.Sc., is a professor and director of the Department of Biophysics, University School of Medicine, Debrecen, Hungary.

He graduated from the above medical school in 1960, with a degree of M.D. (summa cum laude). Between 1963 and 1967, he studied physics at the L. Kossuth University, Faculty of Sciences. In 1968 he received a Ph.D. and in 1976 a D.Sc. from the Hungarian Academy of Sciences (both in biophysics). He spent several years in Norway (Norsk Hydro's Institute of Cancer Research), Germany (Max Planck Institute of Biophysics Chemie), and the U.S. (National Institute of Health, the Food and Drug Administration, and UCSF) as guest scientist and visiting professor.

He was elected a corresponding member of the Hungarian Academy of Sciences in 1982 and a full member of this distinguished learned society in 1990. He serves on the editorial board of several international journals (*J. Photochem. Photobiol., J. Fluorescence, Neurobiology, Arch. Geront. Geriatr.*).

Dr. Damjanovich received the Went Memorial Medal and Award in 1990, and the Jendrassik Medal in 1993 for his outstanding research activities. He co-authored *The Molecular Enzyme Kinetic Model* with Professors B. Somogyi and G. R. Welch. As a consequence, he was invited to the 16th Solvay Conference on Chemistry.

In the last 12 years he has studied the proximity and dynamic behavior of cell surface molecules with biophysical methodologies.

He is the author of over 170 papers and more than 10 book chapters. Dr. Damjanovich is a member of a number of learned professional societies and serves as vice president of the Hungarian Biophysical Society and member of the Executive Committee of the European Biophysical Societies' Association.

His current research interest is the functional significance of two-dimensional distribution and membrane potential regulation of cell surface molecules in the plasma membrane of lymphocytes.

János Szöllősi, Ph.D., D.Sc., is Associate Professor of the Department of Biophysics, University School of Medicine, Debrecen, Hungary.

Dr. Szöllősi graduated in 1976 from Kossuth Lajos University of Arts and Science, Debrecen, Hungary, with an M.S. degree in chemistry (summa cum laude) and obtained a Ph.D. in 1980 from the Department of Biophysics, University School of Medicine, Debrecen, Hungary. In 1992, he received a D.Sc. degree from the Hungarian Academy of Sciences. While appointed at the Department of Biophysics (University School of Medicine, Debrecen, Hungary) in 1982/1983 he spent 1 year as visiting scientist at the Department of Molecular Biology, Max Planck Institute (Göttingen, Germany), and from 1988 to 1990 3 years as visiting assistant professor at the Department of Laboratory Medicine, University of California, San Francisco.

Dr. Szöllősi, is a member of the Hungarian Biophysical Society, Hungarian Biochemical Society, Hungarian Society of Immunology, and International Society of Analytical Cytology. Currently, he serves as a member of the editorial board of the journal, *Cytometry*.

He has received the Academic Award from the Hungarian Academy of Sciences as well as the Pro Scientia Award for his outstanding teaching activities from the Hungarian Ministry of Education.

Dr. Szöllősi is the author of more than 60 papers. His current major research interests relate to the mapping of cell surface proteins on immunocompetent cells.

Lajos Trón, M.S., Ph.D., D.Sc., is chairman of the PET Center at the University School of Medicine, Debrecen, Hungary.

He obtained his training in physics at the University Debrecen receiving his M.S. degree in 1964. Between 1965 and 1972 he served as a Research Assistant Professor at the Nuclear Research Institute; in 1972 he was appointed an Associate Professor of the Institute of Biophysics at the University Medical School, Debrecen. He became a professor in 1987 and directed the Biomedical Cyclotron Laboratory until 1993.

His Ph.D. was obtained in 1968 in nuclear physics and he received a D.Sc. degree in biophysics in 1986. He spent several years in the U.S.S.R. (Joint Institute for Nuclear Research, Dubna), Germany (Max Planck Institute for Biophysical Chemistry, Göttingen), Norway (Norsk Hydro's Institute for Cancer Research, Oslo), and the U.S. (Food and Drug Administration, Washington, D.C.). He received the Award of the Hungarian Academy of Sciences twice for his research in nuclear physics and biophysics, respectively.

Lajos Trón has published more than 80 papers and has been the author or editor of 4 books. His current major interests include transmembrane signaling in normal and neoplastic cells.

Michael Edidin, Ph.D., is Professor of Biology and Biophysics at The Johns Hopkins University, Baltimore.

Dr. Edidin received a B.Sc. in zoology from the University of Chicago. His doctoral thesis at University College London was supervised by P. B. Medawar.

Dr. Edidin has published over 125 papers in the areas of immunology, cell biology, and cell biophysics. He is a member of the American Association of Immunologists, the American Society for Cell Biology, and The Biophysical Society. He has served on the editorial boards of *Biochemistry* and *Journal of Cell Biology.*

PREFACE

The aim of this book is to outline the major dynamic characteristics of the plasma membrane. Methodical parts summarize some of the currently available methods and approaches that make possible the investigation of such problems as molecular neighborhood of cell surface elements, their proximity, possible two-dimensional localization, and rotational and lateral mobility. These, mostly physical, assets of the cell membrane elements have great importance in biology in general and immunology, neuro-endocrinology, and cell biology in particular. The overview of ion channel activities of the so-called nonexcitable cells helps to understand the strong influence of membrane potential and ionic currents on the above-mentioned dynamic properties of the lipid and protein moieties in the plasma membrane. The plasma membrane is the first theater, at the molecular level, of any encounter between the extra- and intracellular world. Hence, its dynamic properties have strong influence on transmembrane signal processing as well.

The analysis of the above problems was done in collaboration among the four of us. The authors' names are indicated at the individual chapters; thus, we bear our responsibility individually for analyses and views presented there.

We enjoyed the support of many friends who helped us in reading and editing the manuscripts. Especially three of us (S.D., J.Sz., and L.T.) wish to express our thanks to Drs. László Mátyus, György Vereb, and György Panyi for critical reading and help in editing. We are also grateful to Klára Bartók for typing some of the chapters and Dr. Lloyd Sparks for polishing the language for three of us (S.D., J.Sz., and L.T.).

Sándor Damjanovich
János Szöllősi
Lajos Trón
Michael Edidin

CONTENTS

Chapter 1

Experimental Methods to Measure Fluorescence Resonance Energy Transfer Processes

Lajos Trón

CONTENTS

0-8493-4931-1/94/$0.00+$.50

I. INTRODUCTION

In a great variety of biochemical and biological systems defined functions assume well-determined structure. Many enzymes consist of a number of individual polypeptides, called subunits. The unimpaired function of the enzyme is frequently conditioned by the complete ensemble of all the intact constituent subunits.

Apoenzyme DNA-dependent RNA polymerase differing from the holoenzyme only in the lack of σ subunit is not capable to bind to specific promoter sites, but once bound the translational activity of the apoenzyme results in RNA synthesis.[1] This enzyme is just one of the representatives of multi-subunit enzymes changing catalytic activity with altered stoichiometry of subunits or more generally with altered supramolecular structure.

Multienzyme complexes offer another set of examples of activity-regulating effect of the intermolecular association-dissociation processes. Certain enzymes are associated forming multienzyme complexes in the intracellular space.[2,3] The advantages of enzyme organization may manifest in at least three different ways. First, the clustering of the component protein molecules may produce entities with intrinsic catalytic properties very much different from those of the separate proteins. Second, the multienzyme complex may increase the efficiency of the overall process even if the individual intrinsic catalytic activities of the components are not affected by the association. Third, the assembly into an organized cluster of enzymes may enable unique modes of regulation.[3,4]

One of the most striking features of living cells is their ability to respond in a very specific way to specific signals coming from the environment. These extracellular triggers can appear as (bio)chemical transmitters, hormones, different regulating factors, lymphokines, neurotransmitters, etc. Cells express specific receptors with high selectivity and avidity for the information carrying messenger molecules.

The same hormone or lymphokine can be bound by different receptors of differing affinity. Sometimes the change in the ligand receptor binding constant is accompanied by a change in the subunit structure of the receptor. Saito et al. documented that the interleukin-2 (IL-2) receptor consists of at least three subunits.[5] The α polypeptide chain itself can bind the lymphokine with low affinity (10 mM) while β or γ subunits alone cannot form a stable complex with IL-2. Simultaneous expression of two or all the three of the subunits allows binding of the ligand with intermediate or high affinity: $K_d(\beta\&\gamma) = 1$ nM; $K_d(\alpha\&\beta) = 0.15$ nM; $K_d(\alpha\&\beta\&\gamma) = 10$ pM.[5-8]

The process of translocation of information from the extracellular side of the cytoplasm membrane into the interior of the cell is called transmembrane signaling. It comprises a very complex set of "elementary steps" in most if not all of the cases.[9,10] In the chain of these individual molecular events composing the specific cellular response, membrane protein associations are relatively frequent.[11-15] Formation of new molecular complexes can result in the appearance of characteristic three-dimensional structures which in turn can display specific functions or can serve as binding sites of effector molecules. Numerous studies proved that polypeptide- (e.g., growth factors, differentiation factors, and hormones) induced dimerization of the appropriate receptors results in tyrosine kinase activation.[16-18] Receptor oligomerization may permit the transmission of a conformational change from the extracellular domain of the polypeptides constituting the receptor to the cytoplasmic domain without requiring alterations in the positioning of amino acid residues within the transmembrane domain. In other cases interactions in the newly formed supramolecular structures may bring about subtle conformational changes in the interacting partners in a way that bound effector molecules (e.g., specific inhibitors) dissociate resulting in an activation of a specific process.

Facts and arguments mentioned above unequivocally underline the importance and necessity to investigate the supramolecular structure of the membrane proteins in order to explore the inherent relationships between structure and function in these systems. Detailed investigation of this kind is definitely instrumental in giving a better understanding of and deeper insight into the precise process of signal transduction through biological membranes.

II. THEORY OF FLUORESCENCE RESONANCE ENERGY TRANSFER

Fluorescence methods are widely applied to investigate the molecular details of life processes because of the inherent sensitivity, specificity, and temporal resolution of fluorescence spectroscopy. In addition these methods are capable to yield, among others, spatial information as well.

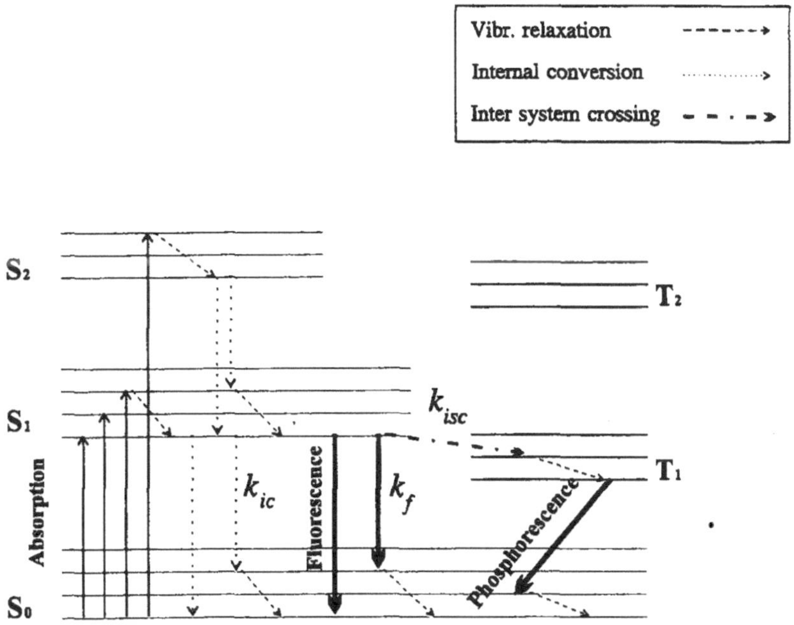

Figure 1. Energy level diagram of a two-electron fluorescent molecule. S (singlet) and T (triplet) refer to the multiplicity of the states and 0, 1, 2, ... designate electronic states of increasing energy. Radiative transitions (fluorescence and phosphorescence) are represented by thick lines, while nonradiative deexcitation processes are explained in the inset. First-order rate constants are denoted by k with subscripts referring to the kind of transition.

Excellent reviews[19-23] and handbooks[24-32] are available describing the fundamental aspects of fluorescence including both theoretical basis and practical considerations. Instead of going into detail we review here just the most basic relations and concepts to facilitate the detailed treatment of fluorescence resonance energy transfer processes.

A. INTRODUCTION TO FLUORESCENCE

Fluorescence emission is a radiative deexcitation process depopulating the lowest energy excited singulet state (S_1) of a polyatomic molecule following optical excitation of the fluorophore.[33] The excitation and deexcitation processes are illustrated by the energy level diagram frequently referred to as the Jablonski diagram (Figure 1). The ground, first, and second electronic levels in a paired two-electron system are designated by S_0, S_1, and S_2. Depending on the vibrational state of the fluorophore following light absorption different vibrational levels may be populated in the same electronic state. Light absorption is very fast, taking place within 10^{-16} s. Excited molecules rapidly (10^{-13} to 10^{-11} s) relax to S_1, the ground vibrational level of the first excited singulet electronic state. The nonradiative vibrational relaxation and internal conversion processes involved are complete before any transition from S_1 can occur. This regularity is referred to as the Kasha-rule and explains that the fluorescence spectrum is generally red shifted relative to the absorption. It is the consequence of these fast radiationless processes that the spectral distribution of the fluorescence emission and the fluorescence lifetime (that of the lowest energy excited singulet state) are independent of the excitation wavelength.

Deexcitation of molecules in the S_1 state can occur via several processes. Fluorescence emission and nonradiative internal conversion populate ground state (S_0). Transition in a fluorophore from state S_1 to T_1 (first excited triplet state) is called intersystem crossing. This nonradiative transition is forbidden because of the change in multiplicity and it results in a low triplet yield. Energy of the first triplet state is lower than that of S_1 so phosphorescence (the radiative transition from T_1 to S_0) is generally shifted to longer wavelengths relative to fluorescence. All the three relaxation processes (internal conversion, fluorescence, and intersystem crossing) are first order processes in the population of the S_1 singulet state with first-order rate constant, k_{ic}, k_f, and k_{isc}, respectively, as indicated in Figure 1.

After a brief excitation pulse the time dependence of N, the population in S_1, can be described by a single decay constant* assuming a homogeneous sample of fluorophores

$$N = N_0 e^{-k^o t} \quad (1)$$

where N_0 is the number of excited fluorophores at t = 0 (end of the excitation), and

$$k^o = k_f + k_{ic} + k_{isc} = k_f + k_d \quad (2)$$

is the rate constant of the decay of the population in S_1. The nonradiative deexcitation processes are characterized by

$$k_d = k_{ic} + k_{isc} \quad (3)$$

The total decay rate constant, k^o, is the inverse of τ^o, the fluorescence lifetime

$$\tau^o = \frac{1}{k^o} \quad (4)$$

An additional characterization of the S_1 state is provided by the quantum efficiency, Q^o, defined as the ratio of the rate of the fluorescence emission to the sum of the rates of all deexcitation processes

$$Q^o = \frac{k_f}{k_f + k_{ic} + k_{isc}} = k_f \tau^o \quad (5)$$

There are numerous cases when the most simple basic scheme, as outlined above, does not describe the situation. Modifications of the standard scheme are required in case of quenching effects adding terms to the decay rate depending on the concentration of quencher; excimer formation, where excited fluorophores interact with ground-state molecules of the same kind; energy transfer processes, where the excitation energy of S_1 state (of a donor fluorophore) can be transferred via dipole-dipole interaction to an appropriate energy accepting (acceptor) molecule; and excited state reactions transforming the initially excited species into different species that may or may not be fluorescent. Energy transfer processes and a special kind of excited state process will be dealt in detail in subsequent chapters.

B. SINGULET-SINGULET RESONANCE ENERGY TRANSFER

A number of mechanisms allow transfer of excitation energy from one molecule (donor) to another (acceptor) one. The energy transfer process results in a diminished donor emission intensity (or quantum efficiency) accompanied by an enhanced acceptor fluorescence (sensitized emission), as compared to the reference value of these parameters in the absence of energy transfer (Figure 2). The trivial case of energy transfer is an emission-reabsorption process. The efficiency of this radiative transfer process depends, among others, on the volume of the system in which donor and acceptor molecules are distributed. It does not change with viscosity of the environment and donor lifetime. Viscosity and donor lifetime both affect the collisional type of energy transfer reducing diffusion constant or the time during which sensitizer and acceptor has to be within a short distance. Formation of a complex between donor and acceptor molecules may also lead to energy transfer processes under favorable conditions.

The long-range resonance energy transfer (fluorescence resonance energy transfer — FRET) can occur if the energy of a deexcitation process of the donor residing in S_1 state corresponds to that of a possible absorption transition in a nearby acceptor molecule. If this condition is fulfilled there is a spectral overlap between donor emission and acceptor absorption spectra (Figure 3). The transfer process requires

* In this chapter we use a special notation to avoid double indexing for the distinction of like parameters of systems in the absence and presence of acceptor. Parameters with superscript o (k^o, t^o, Q^o) refer to systems with no acceptor, and parameters without superscript are the appropriate parameters of the system labeled with both donor and acceptor.

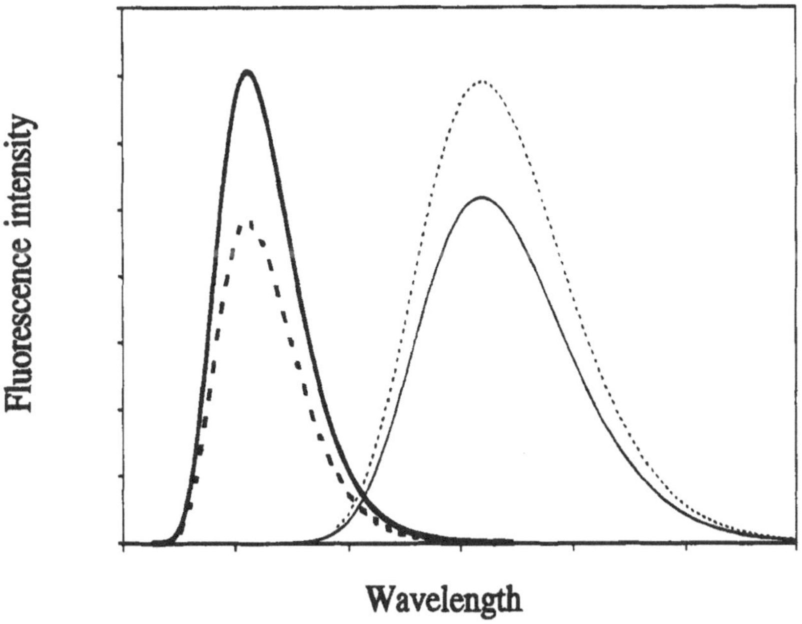

Figure 2. Energy transfer results in changes in the emission intensity (dotted lines) of the donor (thick lines) and acceptor (thin lines) relative to the case with no interaction between donor and acceptor fluorophores (continuous lines).

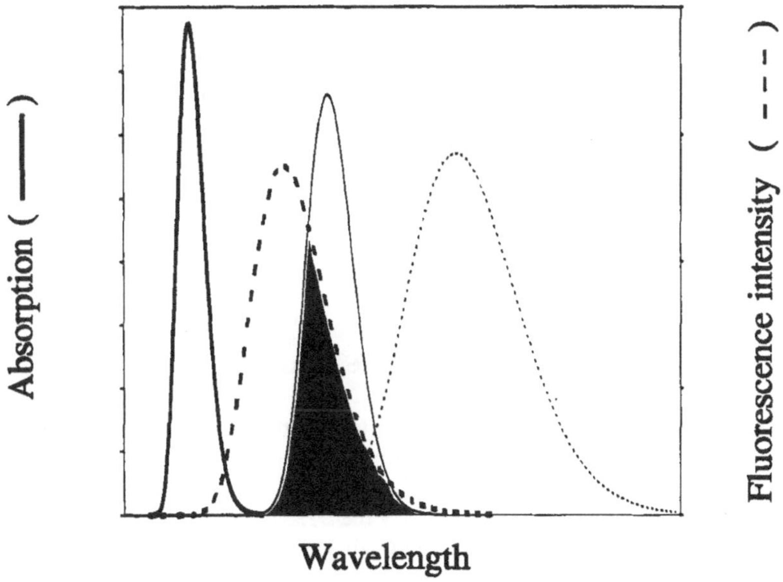

Figure 3. Optical spectra of a donor-acceptor dye pair. Absorption (continuous lines) and emission spectra (dotted lines) are displaced for both donor (thick lines) and acceptor (thin lines). For efficient FRET a substantial overlap is needed between donor emission and acceptor absorption spectra (shaded area).

coupling between the electronic systems of both molecules; consequently, it can take place only over limited distances.[34,35] For dipole-dipole coupling the interaction energy is proportional to the inverse of the third power of the separation distance (R) giving rise to the probability of the energy transfer proportional to R^{-6}. For a donor-acceptor pair separated by a distance, R, the rate constant of the transfer process (k_T) can be described as

$$k_T = \frac{R_o^6}{R^6} k^o \tag{6}$$

with

$$R_o^6 = 8.8 \cdot 10^{-25} \kappa^2 n^{-4} Q^o J \text{ (in cm}^6\text{)} \tag{7}$$

and

$$J = \int_0^\infty q(\lambda)\varepsilon(\lambda)\lambda^4 \, d\lambda \text{ (in cm}^6 \cdot \text{mmol}^{-1}\text{)} \tag{8}$$

In the above expressions R_o is a constant called Förster critical distance, giving a separation distance between donor and acceptor for the case of 50% transfer efficiency (see Equation 14). The orientation factor, κ^2, defines the geometric relationship between the emission dipole of the donor and absorption dipole of the acceptor. The refractive index of the medium is denoted by n. Parameters Q^o and $k^o = 1/\tau^o$ are quantum efficiency and total deexcitation rate of S_1 state of the donor, respectively, (τ^o is the lifetime) in the absence of acceptor (energy transfer). In the expression for the overlap integral, J, expressing the degree of spectral overlap between the donor emission and acceptor absorption spectrum, $q(\lambda)$ and $\varepsilon(\lambda)$, are the relative corrected emission spectrum of the donor (in units of quanta per unit wavelength interval) and the molar decadic extinction coefficient in units of M^{-1} cm^{-1} of the acceptor, respectively.

In the relation between $q(\lambda)$ and $F(\lambda)$, the corrected emission spectrum of the donor is

$$q(\lambda) = \frac{F(\lambda)}{\int_0^\infty F(\lambda) d\lambda} \tag{9}$$

and the orientation factor is defined as

$$\kappa^2 = \left(\cos\Theta_{DA} - 3\cos\Theta_D \cos\Theta_A\right)^2 \tag{10}$$

where Θ_D and Θ_A are the angles defined by the vector joining the centers of the two fluorophores with the emission transition dipole vector of the donor and the absorption transition dipole vector of the acceptor, respectively, and Θ_{DA} is the angle defined by the two transition dipoles.

As the energy transfer process is an additional way the S_1 state can be depopulated, it naturally will reduce the singulet lifetime from τ^o (Equations 2 through 4) to τ

$$\tau = \frac{1}{k_f + k_{ic} + k_{isc} + k_T} \tag{11}$$

Similarly, appearance of energy transfer among the competing processes from S_1 state results in a decreased relative abundance of the fluorescence transitions reducing the quantum efficiency from Q^o (in the absence of acceptor) to Q (in the presence of acceptor):

$$Q = \frac{k_f}{k_f + k_{ic} + k_{isc} + k_T} \tag{12}$$

The lower value of the fluorescence quantum yield manifests in a lower fluorescence intensity of the donor in the presence of the acceptor ($F_{D,A}$) relative to that in the absence of the acceptor (F_D). Owing to the proportionality between donor quantum yield and donor fluorescence intensity we can write

$$\frac{F_{D,A}}{F_D} = \frac{Q}{Q^o} \tag{13}$$

To characterize the relative frequency of the energy transfer processes the efficiency of the energy transfer, E, is defined as

$$E = \frac{k_T}{k_f + k_{ic} + k_{isc} + k_T} = \frac{R_o^6}{R^6 + R_o^6} \tag{14}$$

in analogy to the definition of the fluorescence quantum yield (Equation 5). To determine experimentally the numerical value of E Equation 14 is not instrumental as the various rate constants are not directly measurable in contrast to lifetime, fluorescence quantum yield, and fluorescence intensity. Thus, it is advantageous to derive the interrelationships of these parameters with E using Equations 11 to 14.[24]

$$E = 1 - \frac{\tau}{\tau^o} \tag{15}$$

$$E = 1 - \frac{Q}{Q^o} \tag{16}$$

$$E = 1 - \frac{F_{D,A}}{F_D} \tag{17}$$

Evaluation of FRET efficiency according to Equations 15 to 17 relies on the attenuation of donor emission due to the acceptor's presence. Simultaneously, with the decrease of donor emission intensity, donor lifetime, and donor quantum efficiency, energy transfer leads to an enhancement of acceptor fluorescence. Denoting the measured emission intensity of the acceptor by F_A and $F_{A,D}$ as excited in the donor absorption band in the absence and presence of the donor, respectively, the sensitization of the acceptor emission is described by

$$E = \frac{\varepsilon_A c_A}{\varepsilon_D c_D}\left(\frac{F_{A,D}}{F_A} - 1\right) \tag{18}$$

where c_D and c_A are the molar concentrations and ε_D and ε_A the molar extinction coefficients of the donor and acceptor at the excitation wavelength.[24,36] Any of these formulas (Equations 15 to 18) is equally applicable and widely used for energy transfer determinations.

The significance of transfer efficiency measurements is explained by Equation 14, which describes an explicit interrelationship between transfer efficiency, E, and donor acceptor separation distance, R. This possibility can be exploited in the determination of intramolecular or intermolecular distances in a homogeneous (in terms of this separation distance) molecular system.

The distinction between intrinsic and extrinsic probes has to be emphasized. In the former case the fluorophore is an integral component of the system facilitating a rather direct characterization of the structure. In contrast to intrinsic chromophores (like cofactors of certain enzymes) extrinsic probes with known spectroscopic properties are frequently introduced, although covalent or noncovalent binding of these probes may perturb the system. In these cases an additional difficulty arises in the interpretation of the derived distances especially with large extrinsic probes.

The interpretation of transfer efficiency data is straightforward in a pure "one donor-one acceptor" system. In more complex systems (e.g., those with several defined, but different, donor-acceptor separation distances within the range of energy transfer or associating-dissociating donor and acceptor conjugated molecules) the conversion of measured transfer efficiency values into direct structural parameters is extremely difficult.

Determination of E based on Equations 16 to 18 results in population-averaged values providing no information about homogeneity of the samples or about the number and size of subpopulations differing in the probability of FRET. Lifetime measurements may allow detection of several components by careful analysis of the decay curve. The detection limit of inhomogeneities using lifetime data (Equation 15) strongly depends on numerous factors, such as quality of data, differences in lifetime, etc.

Even more difficulty is encountered when studying the topology of cell surfaces. The measured efficiency of energy transfer between donor and acceptor labeled antigens as determined from spectrofluorimetric emission intensity measurements on cell suspensions is a result of repeated averaging. The first averaging follows from the random conjugation of the labeling ligands. In a substantial fraction of experiments fluorophore-conjugated monoclonal antibodies are used to label cell surface antigens. According to a common practice the covalent coupling reaction occurs frequently between the isothiocyanate (ITC) reactive group of the ITC-derivative of an appropriate fluorophore and the ε-amino groups of the lysin side chains of the immunoglobulin. The number of the exposed lysin side chains of comparable reactivity on an antibody molecule usually exceeds one. Thus, the conjugation of the ligand with fluorescence label itself results in a deviation from the one donor-one acceptor situation of fixed geometry between the dye pairs; therefore the averaging is twofold. Each antibody molecule may carry several fluorophores and the labeled lysine side chains of the individual antibody molecules may be different. An additional averaging is brought about by the eventual distribution of the separation distances between the pertinent antigens (e.g., an association-dissociation equilibrium can be maintained between these polypeptides). The multiple averaging, an inevitable consequence of the nonuniform stoichiometry, explains that FRET measurements are usually carried out with different goals on purified molecular systems, on the one side, and on the surface of cytoplasmic membrane, on the other side. In the former case efficiency data can be converted into absolute distances; thus, mapping of the structure and investigation of steric relations of specific sites can be targeted. In contrast to this, distances in cellular systems derived from FRET measurements can be regarded as approximate values allowing rough estimation. Proximity data, however, even if not expressed as precise distances on the nanometer scale can give appreciable contribution to the characterization of the structure of the cell membrane. The range of the Förster-type FRET processes is between 2 and 10 nm, a range matching the size of most membrane proteins. According to this a nonzero transfer efficiency can be regarded as strong evidence for an existing complex between the respective antigens unless random, noncorrelated distribution of the molecules is giving rise to a comparably efficient FRET process. Calculations of transfer efficiency between membrane antigens assuming relatively high level of expression report on E values very close to zero with an abundance of antigens: 10^6 per cell for lymphocytes. In the majority of experimental systems the level of expression of antigens is lower. In these systems a significant energy transfer efficiency strongly supports macromolecular association.

Aside from the drawback of the approximate feature of proximity data determined on cell surfaces by spectrofluorimetric FRET measurements, this method cannot provide information on homogeneity or, in the case of inhomogeneity, distribution of the population. Protocols for experimental cell isolation rarely result in perfectly homogeneous samples. We only refer to the numerous subsets present in lymphocyte preparations. The different subpopulations may display completely different function, antigens, and interactions between them. This diversity cannot be explored by measurements on cell suspension. In separate studies of defined subsets, investigation of the distribution of proximity parameters within the same subpopulation clearly seeks cell-by-cell type FRET measurements. Flow cytometric energy transfer method is a tool matching these requirements with the additional advantage of being capable to measure large numbers of individual cells at a remarkable rate (200 to 1500 cells per second). Numerous further advantages are summarized in Section V.

III. FLOW CYTOMETRIC DETERMINATION OF ENERGY TRANSFER EFFICIENCY ON A SINGLE CELL BASIS

An energy transfer-related signal can be measured by detecting the acceptor emission intensity with excitation in the donor absorption band. In the case of advantageous spectroscopic features of the donor acceptor pair, this signal is related only to the sensitized fluorescence of the acceptor provided it cannot be excited directly (in the absorption band of the donor) and the directly excited donor emission does not

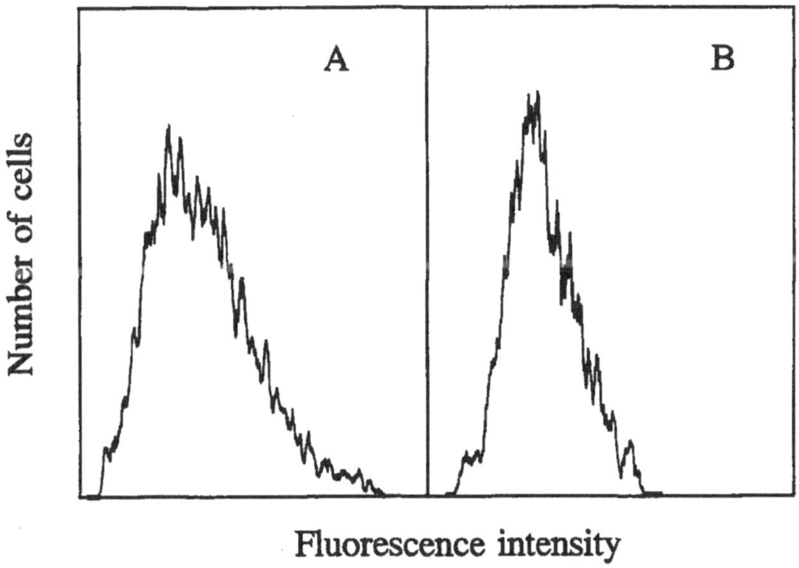

Figure 4. Frequency distribution of fluorescence intensities measured from JY cells labeled with F-W6/32 (A) and T-41 cells labeled with F-Con A (B). Ligands were applied at saturating concentrations.

overlap with the spectral band of the detection either. Unfortunately, in most practical cases these conditions are not guaranteed; consequently, the directly excited contributions of donor and acceptor fluorescence to the detected transfer signal must be corrected. To be able to accomplish these corrections one must determine the increment of the transfer signal due to the directly excited donor and acceptor emission measured from a cell sample stained with both donor and acceptor dye. For this purpose single-labeled samples (two separate samples stained with either donor or acceptor alone) can be used advantageously. These additional samples and the double-stained sample are labeled under identical conditions to maintain exactly the same level of labeling of the appropriate pairs of samples (identical donor staining in the double-labeled and donor-stained, single-labeled samples, on the one side, and the same level of acceptor labeling in the double-labeled sample and acceptor-stained, single-labeled sample, on the other side).

Although this protocol for the correction of the transfer signal is simple and attractive, the procedure usually does not result in data of good quality. This is the consequence of the frequently occurring remarkable biological variance of the cells of the studied population. As examples, Figure 4 displays frequency histograms of fluorescence intensities measured from cells stained with fluorescein-conjugated ligands. The broad distributions (high CV values) fully explain why this type of correction is not accurate.

A significantly better correction can be achieved if all necessary measurements are carried out on the same double-labeled cell. Instead of trying to measure the individual contributions of the directly excited donor and acceptor fluorescence to the distorted transfer signal, three independent fluorescence readings can be taken from each individual double-labeled cell differing from each other in the wavelength of excitation or the spectral range of the detection.[20,37,38] Because of the different wavelength dependence in the absorption and emission spectra of the donor and acceptor, the corrected transfer signal can be evaluated from the three independent measured parameters. In this case biological variance (e.g., in the level of expression of the investigated antigen or receptor) has no effect on the accuracy of the measurements.

A. EVALUATION OF ENERGY TRANSFER EFFICIENCY FROM FLOW CYTOMETRIC DATA

Energy transfer data determined with flow methods are generated on a cell-by-cell basis; thus, flow cytometry allows to assess subpopulations separately. This is a remarkable advantage over microscopic or spectrofluorimetric methods delivering data of poor statistics or averaged over the entire population, respectively. In spite of this great advantage, the first flow methods gave only relative energy transfer efficiencies rather than absolute values.[20,37,38]

Below we give the analytical formulas for the three fluorescence intensities detected from the same cell and their relation to the sensitized emission of the acceptor and to the directly excited donor and acceptor fluorescence. Fluorescence intensities going into the equations have been indexed according to the wavelengths of excitation and emission. Indexing refers to both the dye (donor or acceptor) and the kind of wavelength (excitation or emission). To avoid multiple and complex labeling such as λ_D^{ex} or λ_A^{em} we deliberately choose a given donor-acceptor pair and refer to wavelengths as numbers. In our illustrative case (which does not mean the limitation of the general validity and applicability) the donor is fluorescein and the acceptor is rhodamine. We consider a flow cytometer equipped with an argon ion laser working in all line mode and assume an input optics containing appropriate beam stops obscurating all lines but the 488-nm and the 514-nm ones.[39,40]

The three fluorescence intensities detected are denoted by I_1, I_2, and I_3 followed by two numbers in parenthesis indicating the wavelengths of excitation and detection of emission. In the absence of energy transfer we denote the theoretical, unperturbed, directly excited fluorescence intensity of fluorescein (excited at 488 nm, measured at 535 nm) as I_F and that of rhodamine with excitation at 514 nm and emission >590 nm as I_R. The energy transfer efficiency is E. In Equations 22 to 24 describing I_1, I_2, and I_3 we use three correction factors:

$$S_1 = \frac{I_2}{I_1} \tag{19}$$

determined using cells labeled only with donor,

$$S_2 = \frac{I_2}{I_3} \tag{20}$$

determined using cells labeled only with acceptor, and

$$S_3 = \frac{I_3}{I_1} \tag{21}$$

determined using cells labeled only with donor.

The last parameter, α, going into Equations 22 to 24 describing the three measured fluorescence intensities gives the ratio of the measured rhodamine emission detected (I_2) to the detected emission intensity of fluorescein (I_1) from identical amounts of excited fluorophores. In essence α relates to the different detection efficiencies of the excited states of the two dyes using the given experimental layout of the two detection channels.

With these parameters the three fluorescence intensities from a double-labeled cell are

$$I_1(488,535) = I_F(1-E) \tag{22}$$

$$I_2(488,>590) = I_F(1-E)S_1 + I_R S_2 + I_F E\alpha \tag{23}$$

$$I_3(514,>590) = I_F(1-E)S_3 + I_R + S_3/S_1 \cdot I_F E\alpha \tag{24}$$

According to Equation 22 the detected fluorescein emission intensity (I_1) from a double-labeled cell is smaller than that which could be detected from the same cell labeled only with donor (I_F). This is because a fraction (E) of donor excitation energy is transferred to the acceptor.

The measured signal I_2 is the sum of (1) the fraction of the quenched fluorescein emission overlapping the second detection channel, (2) the fraction of directly excited rhodamine emission detected in the same detection channel, and (3) the contribution of the sensitized emission of the acceptor due to the energy transfer. Sensitized acceptor fluorescence intensity (in I_2) is proportional to the decrease in donor emission, $I_F E$ (in I_1), with a proportionality factor α. The numerical value of this factor is defined by the spectroscopic and detection characteristic of the experimental arrangement and those of the applied pair of donor and acceptor conjugated ligands.

The intensity I_3 has three additive terms: (1) the fraction of the decreased (by energy transfer) donor emission overlapping the acceptor detecting channel, (2) the directly excited rhodamine fluorescence, and (3) the sensitized acceptor emission. This last term can be calculated from the last additive component of the measured I_2 intensity by making the necessary correction for the different excitability of fluorescein at 514 and 488 nm. Because of the smaller amount of excited donor molecules at the longer wavelength the contribution of the sensitized emission to the measured I_3 signal is less than that to the I_2 signal.

From Equations 22 to 24 the interrelationship of E and the three measured fluorescence intensities can be obtained:

$$\frac{E}{1-E} = \frac{1}{\alpha}\left[\frac{I_2 - S_2 I_3}{\left(1 - \frac{S_3}{S_1} S_2\right) I_1} - S_1\right] \tag{25}$$

Denoting the right side of Equation 25 by A, the numerical value of the transfer efficiency can be calculated as

$$E = \frac{A}{1+A} \tag{26}$$

All the parameters going into the right side of Equation 25 as well as involved in Equation 26, an explicit expression for E, are either measured directly (I_1; I_2; I_3) using double-labeled cells or are easily obtained (S_1; S_2; S_3) from recording fluorescence intensities of single-labeled cells as it was explained before. In addition to these six parameters the determination of α is required to calculate the energy transfer efficiency. The detailed procedure for this last parameter is elaborated in Section III.C.

Knowing the actual values of S_1, S_2, S_3, and α and storing the set of intensities I_1, I_2, and I_3 in the memory of a computer in a correlated manner for a large number of cells transfer efficiency can be evaluated for each individual cell. From the calculated E values an energy transfer distribution can be generated.

For a given pair of donor acceptor dyes S_1, S_2, and S_3 are not subject to change due to biological variance of the individual cells. Using the same optical system (including the spectral response of the photodetectors) the magnitude of these spillage factors is constant and their value can be evaluated with great precision measuring fluorescence intensities of appropriately stained cells and averaging the readings over relatively large (e.g., 10,000 cells) cell populations. Similarly, the numerical value of α is determined by the optics and spectroscopic peculiarities of the donor and acceptor; thus, it is constant for the same kind of measurements. However, using different fluorophores S_1, S_2, S_3 as well as α change their numerical values. Furthermore, with respect to the last parameter, changing the ligands (monoclonal antibodies, hormones, etc.) the spectral characteristics of the conjugated dyes may change which manifests in an altered value of α.

Conjugation of the same ligand with the same fluorophore can result in different labeling (dye to ligand molecules) ratios effecting the fluorescence quantum efficiency of the conjugated form of the dye. Accordingly, the constancy of α is guaranteed only for the same batch of labeled ligands. Finally, day-to-day alignment of the flow cytometer may have a poor reproducibility giving rise to slight changes in the steric angles of detection in the two detection channels resulting in altered values of all the four parameters discussed. Each complete set of measurement should be supplemented by all the measurements necessary for the determination of the actual numerical values of S_1, S_2, S_3, and α.

B. INSTRUMENTATION

A flow cytometer capable of dual wavelength excitation is needed for the quantitative evaluation of the transfer efficiency on a cell-by-cell basis with the method discussed in the previous section. Commercially available flow cytometers equipped with two lasers can be used for this purpose. Single laser equipments can be applied, provided the laser has a multiline (all line) working mode. In this case a dispersive element is introduced into the input optics spatially separating the lines of different wavelengths. Displacement of the 488- and 514-nm lines of an argon ion laser can be achieved using either

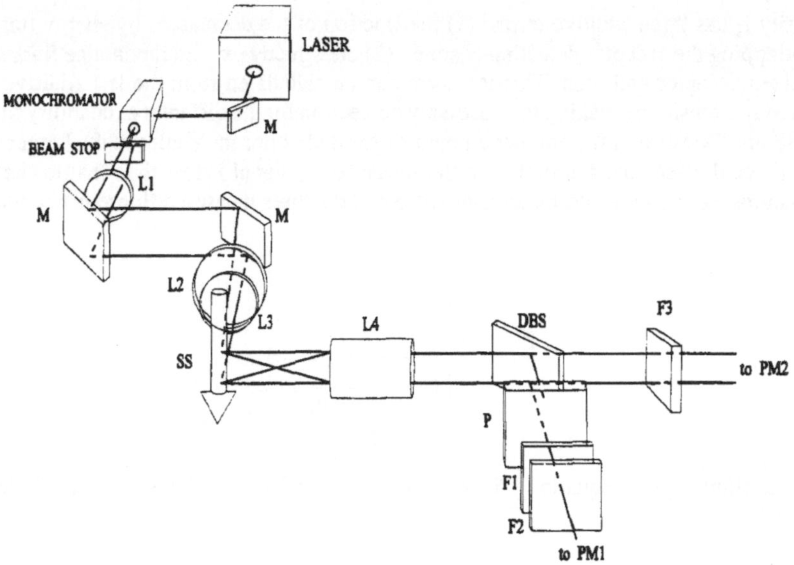

A

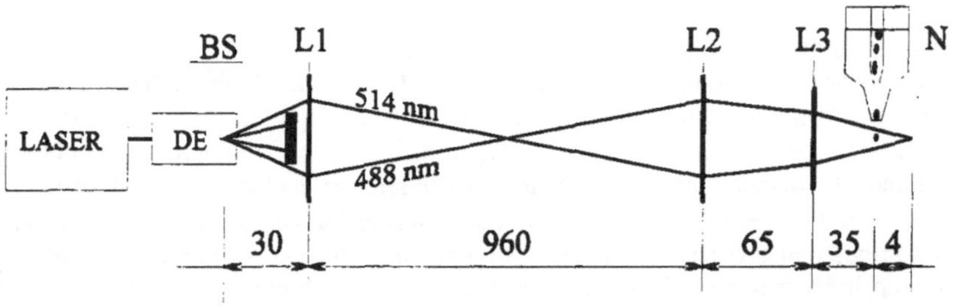

B

Figure 5. The optical arrangement of a flow cytometer with dual wavelength excitation. Different wavelength components of the light of the argon laser working in all line mode were dispersed by a monochromator. The 488- and 514-nm beams were focused onto the sample stream (SS) by three lenses (L) and mirrors (M); all the other lines were blocked by the use of a beam stop. Cellular fluorescence was collected by a condensor (L_4) lens and detected with two photomultiplier (PM) tubes. A dichroic beam splitter (DBS) and three optical filters (F) defined the spectral ranges in the two detection channels. PM2 detected emission excited at both 488 nm and 514 nm, but PM1 detected only that excited at 488 nm since fluorescence excited at 514 nm was blocked out using a beam-stop-pinhole (P). (A) The optical arrangement in perspective. Signals excited at 488 nm coincided in time; that excited at 514 nm appeared in the second detection channel with 29-μs delay. (B) Vertical section of the input optics along the optical axis with indication of distances between optical elements in millimeters. The vertical separation distance between the intersection of SS with the two excitation lines was 0.6 mm. Point of excitation of the cellular fluorescence at 488 nm was 0.5 mm below the orifice of the nozzle tip. DE: dispersive element; BS:beam stop; N: nozzle.

a commercial monochromator[39] or simply a prism. Naturally, all other (undesired) lines must be blocked out by an appropriately positioned beam stop.

In a set of our measurements[39-41] the 488- and 514-nm laser lines were focused on the sample stream carrying a suspension of cells labeled with fluorescent dye-conjugated ligands. Excitation of the fluorophores of the same cell at the selected wavelengths was separated both temporally and spatially (Figure 5). The emission excited at 488 nm was detected in both detection channels, but emission excited at 514 nm was detected only in the rhodamine detection channel and was blocked before reaching the detector (PM1) of the fluorescein channel. This way the detector in the rhodamine channel (PM2) delivered two signals from each individual cell. Separation in time of these signals is determined by the distance between the intersections of the sample stream with the two displaced laser lines and the velocity of the sample stream. In the system used in experiments by Trón et al.[39] the intersection points were displaced by 0.6 mm giving rise to 29-μs separation time between the two signals recorded by PM2. Small-angle forward scattered light was collected by fiber optics and measured by PM3 (not shown). Detection of the signals by PM1 and PM2 was triggered by light scattering signal. Data belonging to the same cells were stored in the memory of a computer in a correlated way.

C. DETERMINATION OF α

As it was already mentioned α is the ratio of detected rhodamine (I_2) to fluorescein (I_1) emission intensity measured from the same number of excited dye molecules. In order to assure the identical number of excited states several kinds of normalization have to be accomplished. Measured averaged rhodamine and fluorescein emission intensities from two stained cell suspensions have to be corrected for the average number of dye molecules (since they may differ from each other even if the same type of ligand is used for donor and acceptor labeling). When accomplishing this type of normalization differences in both the average binding site of the donor and acceptor labeled ligands and in the dye-to-ligand labeling ratio of the two conjugated ligands have to be considered. Another critical point is the same amount of excited fluorophores (calculated as the average number of excited rhodamine and fluorescein molecules per cell), a condition which is not automatically fulfilled even in the case of equal average number of dye molecules because of the different absorption of donor and acceptor molecules. Aiming at a generally applicable equation we present a mathematical treatment of the problem including all the relevant parameters.

Let us consider a homogeneous cell suspension labeled with fluorescein-conjugated ligand. Let B^F denote the number of binding sites per cell occupied by fluorescein-conjugated ligand molecules. Care should be taken to reach saturating occupation during the labeling incubation. L^F means the fluorophore-to-ligand labeling ratio for the donor conjugated ligand; σ^F_λ stands for the absorption cross section of a fluorescein molecule at the excitation wavelength λ (in cm^2); ρ^F for the irradiance of the excitation beam, divided by the energy of the photons (in photons $cm^{-2}s^{-1}$); t for the equivalent time a cell resides in the radiation field of ρ^F density (in seconds). Q^{oF} denotes the fluorescence quantum efficiency of fluorescein, γ^F is the effective steric angle of detection in the fluorescein detecting channel measured in parts of 4π; δ^F is the combined transmission of all optical elements between the places of excitation and detection; ξ^F is the average voltage on the output of the detector per photon reaching the phototube; and g_1 is the electronic gain factor in the first signal channel. The product of the last two factors includes the detection efficiency of the phototube for the fluorescein emission in the first detection channel, the gain of the phototube (in millivolts per photon), and electronic gain in this signal channel. Superscript F refers to fluorescein.

With these parameters the fluorescein-related emission intensity recorded in the first detection channel can be expressed as

$$\left[B^F L^F \sigma^F_\lambda \rho^F t\right]\left[Q^{oF}\gamma^F\delta^F\xi^F g_1\right] = N^F C^F \tag{27}$$

In the above expression, the product of the first five factors yields N^F, the number of the excited fluorescein molecules. Let C^F denote the product of the last five terms, which contains only the spectral characteristics of the fluorophore and instrumental parameters. This factor gives the ratio of the fluorescence signal measured in the first detection channel to the number of excited fluorescein chromophores.

A similar expression can be written for the fluorescence signal measured in the second detection channel from a cell labeled with TRITC conjugated ligand:

$$\left[B^R L^R \sigma^R_\lambda \rho^R t\right]\left[Q^R\gamma^R\delta^R\xi^R g_2\right] = N^R C^R \tag{28}$$

where R refers to rhodamine.

Having measured I_1 and I_2 intensities from a donor- and acceptor-labeled cell suspension, respectively, the mean values of the frequency histograms, M^F and M^R, can be evaluated

$$M^F = \overline{B^F} L^F \sigma^F \rho^F t C^F \tag{29}$$

$$M^R = \overline{B^R} L^R \sigma^R \rho^R t C^R \tag{30}$$

where $\overline{B^F}$ and $\overline{B^R}$ denote values averaged over the analyzed populations.

According to the definition, the numerical value of α can be obtained as the ratio of I_2 and I_1 measured from acceptor and donor stained cells, respectively, having the same amount of excited molecules. These intensities can be expressed as C^R and C^F if the given amount of excited fluorophores is one. With this explanation one can write for α

$$\alpha = \frac{C^R}{C^F} = \frac{M^R \overline{B^F} L^F \varepsilon^F \rho^F}{M^F \overline{B^R} L^R \varepsilon^R \rho^R} \tag{31}$$

where ε^F and ε^R are molar absorption coefficients measured at wavelengths used to measure M^F and M^R, respectively.

Determination of parameters in Equation 31 is straightforward. Measurement of the mean value of fluorescence intensities belonging to the samples stained with fluorescein- and rhodamine-conjugated ligands is a rutin procedure in flow cytometry: dye-to-ligand labeling ratios belong to the characterization of the ligands applied and absorption coefficients of the fluorophores are known or easily measurable. The average number of specific binding sites for donor and acceptor conjugated ligands can be either known from the literature or may be provided by results of appropriate measurements. In the case when the same ligand is conjugated to both donor and acceptor dye the average number of binding sites is identical for fluorescein- and rhodamine-conjugated ligands. If the donor and acceptor labeled ligands are different (e.g., TRITC-Con A and FITC-anti-H2 antibody), one can no longer assume the equality of $\overline{B^F}$ and $\overline{B^R}$ and their ratio must be determined. The irradiance of the excitation beams used to determine M^F and M^R has to be measured by calibrated flux meters; however, it is advantageous to use the same laser line to measure donor- and acceptor-labeled cells making the determination of ρ^F and ρ^R unnecessary.

D. EXPERIMENTAL VERIFICATION OF THE FLOW CYTOMETRIC ENERGY TRANSFER (FCET) MEASURING PROTOCOL

In the experiments discussed below, cells labeled with FITC-Con A (F-Con A) and/or TRITC-Con A (R-Con A) were used to demonstrate the applicability of this calculation method for determination of energy transfer efficiency on a cell-by-cell basis.

1. Labeling of the Cells with F-Con A, R-Con A, and with Dye-Conjugated Anti-HLA Antibodies

Cells of the HK22(C3H × DBA/2)f1) murine lymphoma line[42] (formerly designated T-41 cells) were cultured in Dulbecco's minimum essential medium containing 10% fetal calf serum (FCS) and antibiotics.

Con A was conjugated with FITC and TRITC as previously described.[37,39] The average labeling ratios for fluorophore to protein were 2.2 for F-Con A and 2.6 for R-Con A. Cells were collected from medium and washed two times with phosphate buffered saline (PBS) and incubated (at 2×10^7 cells per milliliter) with 0.2 mg/ml total lectin concentration at 4°C for 60 min with gentle agitation every 5 min. Cells were washed by centrifugation through a layer of 3 ml 50% FCS onto a cushion of Lymphoprep (Flow Laboratories) to avoid aggregation. After this step the cells were either fixed in 1% formaldehyde (called "fixed cells") and measured in the flow cytometer or resuspended in PBS and measured without fixation (called "living cells").

Anti-HLA antibodies were labeled as described earlier by Damjanovich et al.[43] Cells were labeled in PBS (100 μl; 10^7 cells per milliliter) for 60 min at 0°C in the presence of 0.1 μM antibody. The cells were washed and resuspended twice in PBS and used for measurements.

2. Distributions of Measured Flow Cytometric Parameters and the Calculated Energy Transfer Efficiency

The frequency distributions of the signals for cells labeled only with fluorescein or rhodamine-conjugated Con A are displayed in Figure 6. Instrumental factors were kept constant for the experiments.

All the displayed distributions have a rather large standard deviation. Coefficients of variation (CV) are 32 to 38%. In order to demonstrate that the large CV values frequently occur and are not the mere consequence of the "non-physiological" nature of the applied ligand, Figure 7 displays similar frequency histograms measured from JY cells stained with donor- and acceptor-conjugated anti-HLA antibodies.[44]

According to the distributions shown on Figure 7, HLA class I antigens are expressed with an even larger biological variance (CV values are 47 to 51%) than Con A receptors. At least a fraction of the CV is explained by the change in the HLA expression during the cell cycle (unpublished observations). Width

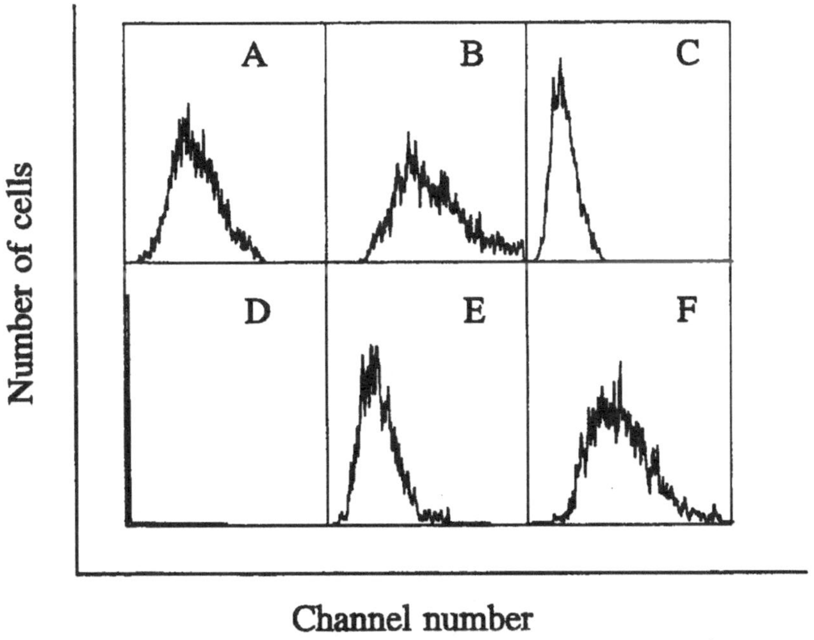

Figure 6. Fluorescence intensity histograms measured from F-Con A (panels A, B, and C) and R-Con A (panels D, E, and F) labeled T-41 cells. Cells were labeled at 4°C in the presence of 0.2 mg/ml lectin concentration. All the histograms of I_1 (panels A and D) I_2 (panels B and E), and I_3 (panels C and F) were constructed from data of 5000 cells.

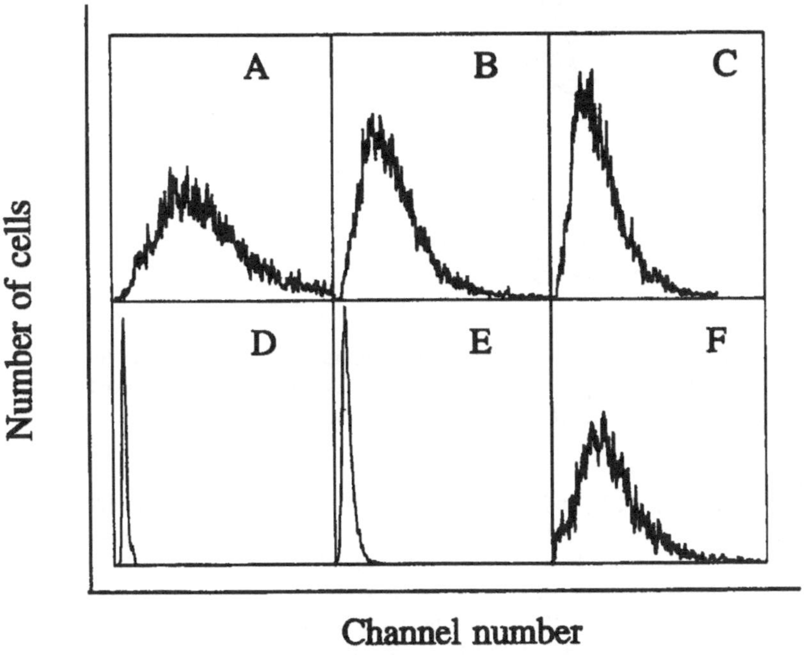

Figure 7. Fluorescence intensity histograms measured from F-W6/32 (panels A, B, and C) and R-L368 (panels D, E, and F) labeled JY cells. Cells were labeled at 4°C in the presence of 0.02 mg/ml antibody concentration. All the histograms of I_1 (panels A and D), I_2 (panels B and E), and I_3 (panels C and F) were constructed from data of 5000 cells.

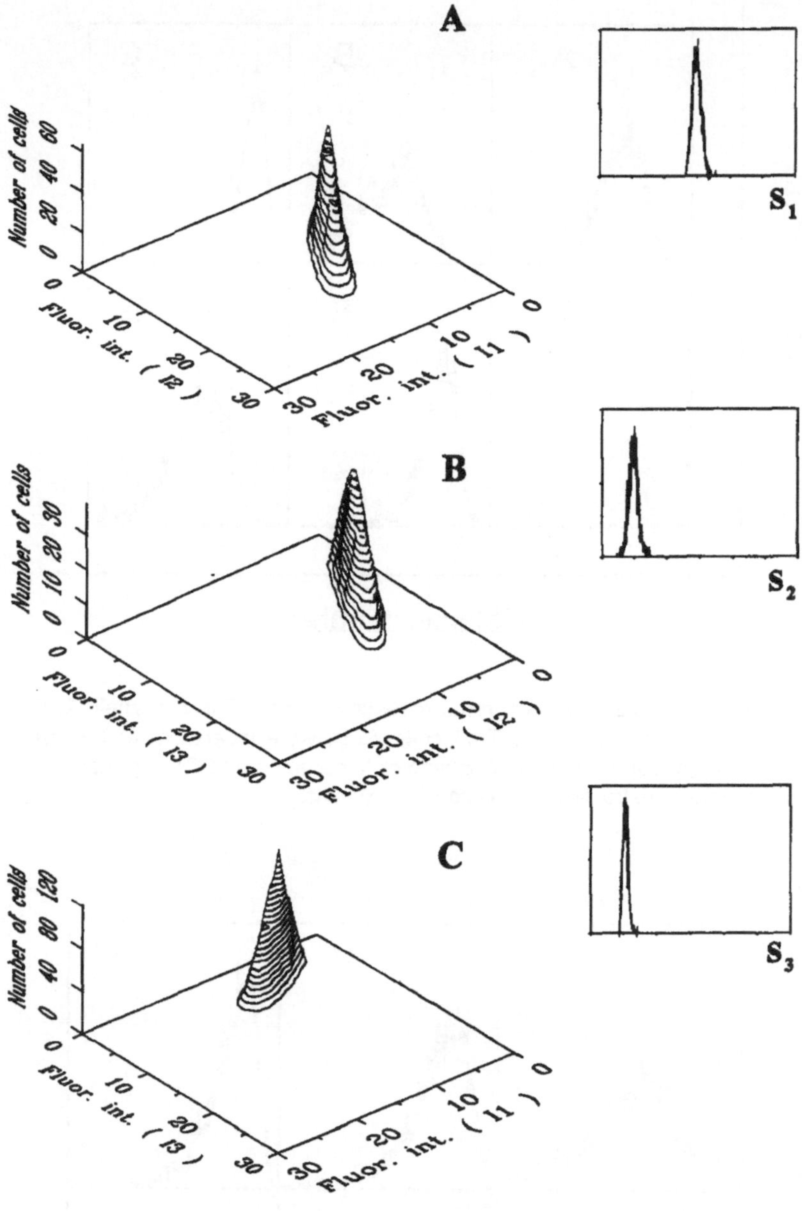

Figure 8. Three-dimensional representation of the correlated distribution of fluorescence intensities (I_1, I_2, I_3) measured from F-Con A (panel A and C) and R-Con A (panel B) labeled T-41 cells. Insets: frequency distributions of the ratios of fluorescence intensities: (A) $S_1 = I_2/I_1$; (B) $S_2 = I_3/I_2$; and (C) $S_3 = I_3/I_1$.

of distributions is different for I_1, I_2, and I_3; however, values are scattered in an interval of similar size for both Con A-labeled T-41 and anti-HLA-labeled JY cells. This observation argues that the inherent biological variance is complemented by a contribution of statistical fluctuation of the signal I_i ($i = 1, 2, 3$) and the amplitude of this fluctuation is different for different values of i. In a subsequent paragraph we describe a protocol for the evaluation of the biological variance.

With the experimental arrangement used for the FCET measurements three fluorescence intensities can be measured from each individual cell. The relationship between any two of the three measured fluorescence signals may be represented either as the distribution of the ratio of the two chosen entities or as a three-dimensional display of the data points (Figures 8 and 9). The S_i parameters used in Equations 22 to 26 to calculate the numerical value of E can be obtained from either representation.

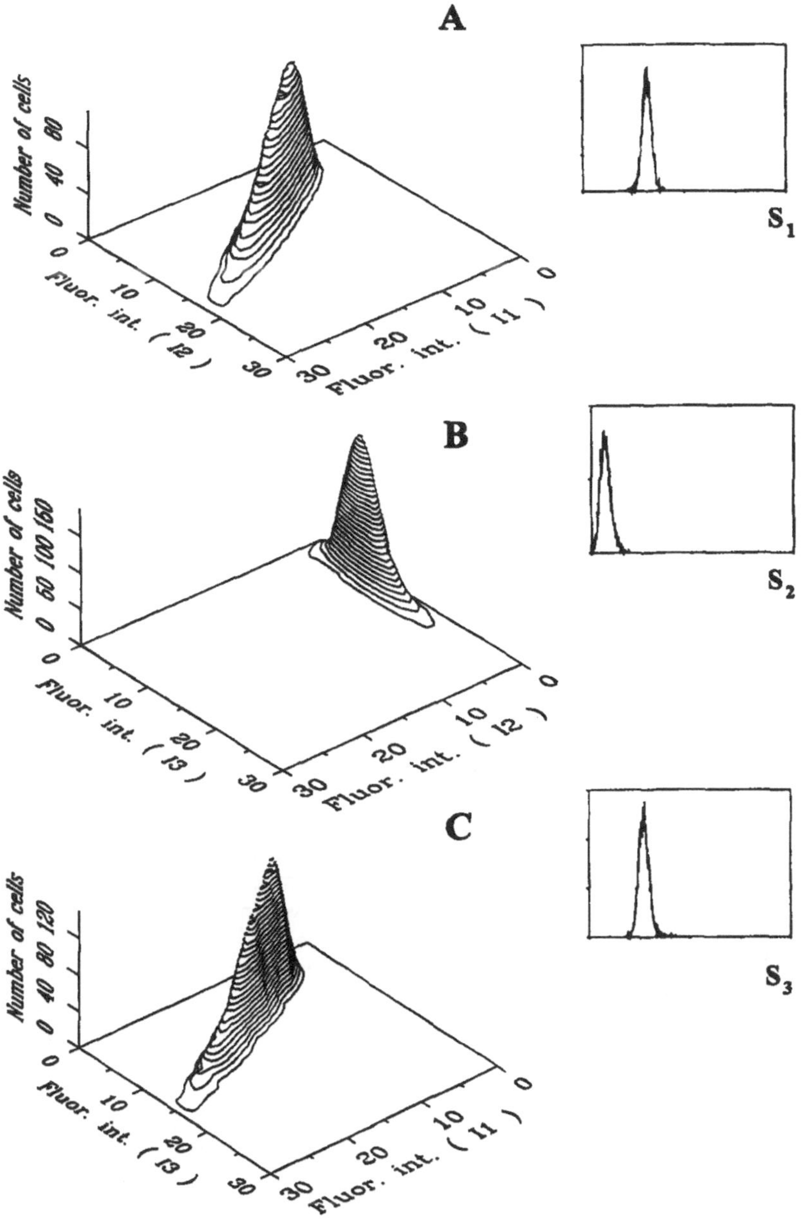

Figure 9. Three-dimensional representation of the correlated distribution of fluorescence intensities (I_1, I_2, I_3) measured from F-W6/32 (panel A and C) and R-L368 (panel B) labeled JY cells. Insets: frequency distributions of the ratios of fluorescence intensities: (A) $S_1 = I_2/I_1$; (B) $S_2 = I_3/I_2$; and (C) $S_3 = I_3/I_1$.

After the determination of correction factors the energy transfer histograms of cells labeled with both F-Con A and R-Con A can easily be generated (Figure 10). Values of the energy transfer efficiency increased together with the relative concentration of the acceptor R-Con A (see also Table 1) maintaining the total concentration of Con A at 0.2 mg/ml saturation levels. The T-41 cells seemed to be homogeneous according to the energy transfer distribution curves indicating no significant differences in the surface density of the Con A binding sites on different cells. A similar energy transfer efficiency histogram is shown in Figure 11 measured from JY cells labeled with F-W6/32 and R-L368. Although all the energy transfer efficiency histograms displayed in Figures 10 and 11 are relatively broad distributions, their CV values are significantly below that of the related fluorescence intensity distributions.

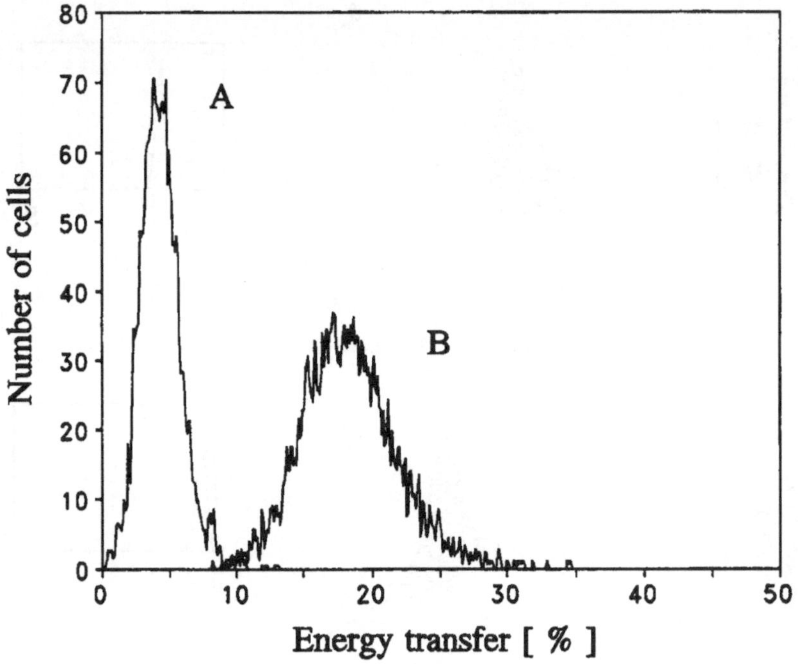

Figure 10. Frequency distribution of energy transfer efficiency from T-41 cells labeled with both F-Con A and R-Con A. Ratio of TRITC to FITC bound to the cells was (A) 0.3 and (B) 2.0.

3. Independence of Fluorescence Energy Transfer Efficiency Data on Detected Spectral Region of Acceptor Emission

Obtaining a constant value of E with different optical filters in the first and the second detection channel is a simple check of the procedure. Upon altering the optical characteristics of the F3 filter (Figure 3), the overlap factors (S_1 to S_3) and the numerical value of the α parameter change. These alterations, however, should not affect the absolute value of the fluorescence energy transfer efficiency. In Table 1 we list the mean values of the transfer efficiency distributions for a number of different doubly labeled samples analyzed successively using three different cutoff filters in the second detection channel. For each spectral range of detection the parameter set S_1, S_2, S_3, and α was determined individually. Good overall agreement was observed between the three energy transfer efficiency values calculated for each of the samples. Similar experiments demonstrated that E values evaluated from fluorescence intensity measurements using different instrumental setting (changing the gain in the individual signal channels) were also very close to each other (data not shown). Results of these measurements provide strong evidence for the validity and correctness of the described energy transfer measuring protocol.

Table 1. **Independence of Calculated Transfer Efficiency Data from the Detected Spectral Region of Acceptor Emission**

Acceptor-to-donor ratio	Energy transfer efficiency (%) OG 570	OG 590	OG 610
0.2	6.0	5.2	6.0
0.29	8.5	7.3	8.5
0.60	14	15	13
1.25	19	18	20
2.54	23	25	24
5.29	27	26	28
8.47	28	28	30

Note: Acceptor-to-donor ratio is rhodamine-to-fluorescein ratio of the labeling mixture of R-Con A and F-Con A. The total concentration of the lectin was 0.2 mg/ml in all cases. The longpass filters in the rhodamine detection channel were OG 570, OG 590, and OG 610. Numerical values of the overlap parameters S_1-S_3 and α with each longpass filter were individually determined and used in the evaluation of the energy transfer efficiency.

From Trón, L., Szöllősi, J., Damjanovich, S., Helliwell, S. H., Arndt-Jovin, D. J., and Jovin, T. M., *Biophys. J.*, 45, 939, 1984, Reprinted by permission of Biophysical Society, Copyright © 1984.

E. ERROR ANALYSIS

It is important to distinguish the relative contributions to the observed distributions of the

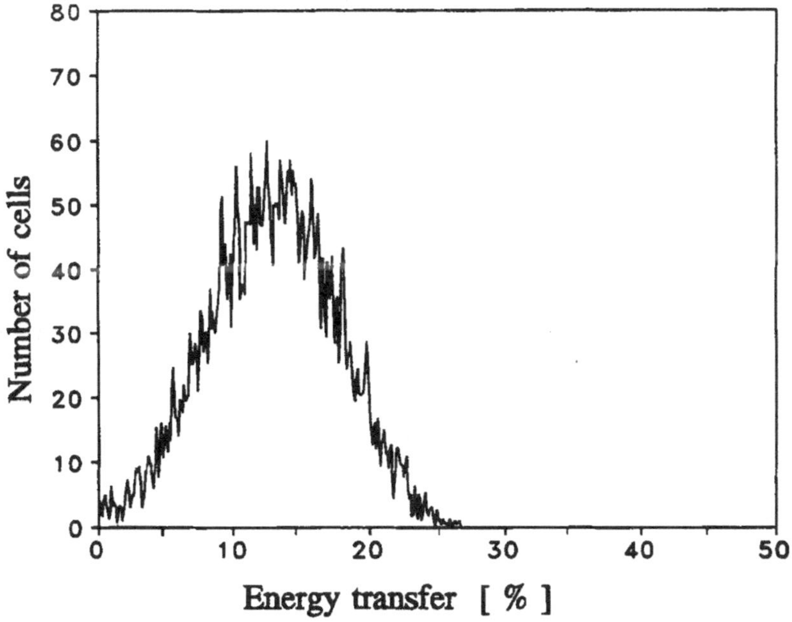

Figure 11. Calculated energy transfer efficiency histogram of JY cells labeled with both F-W6/32 and R-L368.

energy transfer efficiency (E) and A, as calculated according to Equations 25 and 26 arising from instrumental measuring uncertainties and true biological variation. The standard deviation of the resulting value of E can be calculated from the individual standard deviations for S_i, I_i ($i = 1, 2, 3$) and α according to the standard formula[45] for propagation of error assuming orthogonality of the parameters (i.e., negligible covariance). This assumption is generally fulfilled since the measured signals reflect predominant contributions from different sources (fluorescein, rhodamine, energy transfer, etc.). A first-order error propagation analysis of Equations 25 and 26 yields the following expressions for the variances, σ^2, corresponding to the calculated values of A and E in terms of those measured or estimated for the parameters and signals S_1, S_2, S_3, I_1, I_2, I_3, and α. Not included are the possible influences of autofluorescence.

$$\sigma_A^2 \cong \left(\frac{A}{\alpha}\right)^2 \sigma_\alpha^2 + \frac{1}{\left(1 - \frac{S_2 S_3}{S_1}\right)^2} \left\{ \left(\frac{S_2 S_3}{S_1^2} A + \frac{1}{\alpha}\right)^2 \sigma_{S_1}^2 + \left[\frac{S_3}{S_1} A + \frac{\left(S_3 - \frac{I_3}{I_1}\right)}{\alpha}\right]^2 \sigma_{S_2}^2 + \right.$$

$$\left. S_2^2 \left(\frac{A}{S_1} + \frac{1}{\alpha}\right)^2 \sigma_{S_3}^2 + \frac{1}{(\alpha I_1)^2} \left[\frac{(I_2 - S_2 I_3)^2}{I_1^2} \sigma_{I_1}^2 + \sigma_{I_2}^2 + S_2^2 \sigma_{I_3}^2 \right] \right\} \quad (32)$$

$$\sigma_E^2 \cong \frac{1}{(1 + A)^4} \sigma_A^2 \quad (33)$$

Equations 32 and 33 describe the way errors in the individual parameters affect the precision of the energy transfer efficiency resulting from error propagation. In addition to the instrumental uncertainties there is another source of statistical fluctuation, namely, the inherent biological variance of the individual cells for the measured parameter. Denoting the CV of the experimentally determined E distribution by

CV_E and the contribution of the fluctuation due to biological variance to it by CV_{Eb} and assuming that the statistical fluctuation of the transfer efficiency due to biological variability and random instrumental noise are independent we get

$$CV_E^2 = CV_{Eb}^2 + \frac{1}{\left[A(1+A)\alpha I_1\left(1-\frac{S_2S_3}{S_1}\right)\right]^2} \cdot \left[\frac{\left(I_2 - S_2I_3\right)^2}{I_1^2}\sigma_{I_1}^2 + \sigma_{I_2}^2 + S_2^2\sigma_{I_3}^2\right] \quad (34)$$

As it was reasonably assumed in the error propagation analysis parameters S_1, S_2, S_3, and α do not contribute to the statistical fluctuation of the efficiency data (e.g., to the width of the evaluated E distribution) since these factors have constant values for a given system. It must be noted, however, that the actual values of the aforementioned four parameters may also be distorted since their numerical values are derived from experimentally measured data but they do not affect CV_E directly.

The biological variance (or CV_{Eb}) can be obtained from the determined CV_E of a measured energy transfer efficiency distribution using Equation 34. For this purpose the CVs for I_i (i = 1, 2, 3) attributed to only instrumental statistical fluctuations (V_i) have to be known. These parameters can be calculated from the CV values of the derived S_i distributions (CV_S) using the following interrelationships:

$$\left(CV_{S1}\right)^2 = \left(V_1\right)^2 + \left(V_2\right)^2 \quad (35)$$

$$\left(CV_{S2}\right)^2 = \left(V_2\right)^2 + \left(V_3\right)^2 \quad (36)$$

$$\left(CV_{S3}\right)^2 = \left(V_3\right)^2 + \left(V_1\right)^2 \quad (37)$$

provided that V_i values are independent of the measured intensities arising from the individual cells. This assumption is justified only within a narrow dynamic range. (This way the numerical values of V_i for small intensities are underestimated and those for larger intensities overestimated for the real case.) CV values for the S_i distributions (CV_S) can be determined from data like those displayed in Figures 8 and 9 which in turn allow to solve Equations 35 through 37 for V_i. Values of CV_{Si} have to be determined analyzing donor and acceptor single-labeled cells with detected I_i signal levels close to those of the energy transfer sample. Standard deviations of I_i as input parameters for CV_{Eb} (Equation 34) can be obtained from V_i, the solution of Equations 35 through 37, and the calculated mean values of the I_i distributions. We reported results of calculation for the biological variance of Con A-labeled T-41 cells.[40] From a measured E distribution with $CV_E = 11.9\%$ we arrived at a value of $CV_{Eb} = 7.3\%$.

F. FRET EFFICIENCY DETERMINATIONS WITH INCREASED ACCURACY

It is clearly seen from the error analysis that the energy transfer efficiency is biased by errors of quite a number of parameters. Flow instability (e.g., partial clogging of the nozzle and minor displacement of the excitation laser beams) may change the effective steric angle of the detectors applied. Effects of this source of error on the signals measured in the three fluorescence channels are different. This implies that knowledge of the actual numerical values of the S_1, S_2, S_3 spillage factors is critical to gain accurate and precise E data. Although parameters S_1, S_2, and S_3 are directly related to spectroscopic features of the donor acceptor pair, their numerical value is a sensitive function of the actual geometry as well. In an experiment involving the measurement of a whole set of samples, the determination of the three fluorescence intensities from double-labeled "energy transfer" samples and the measurement of the appropriate single-labeled samples (to yield parameters S_1, S_2, S_3) are necessarily displaced in time. Results of these non-coincident (in-time) measurements can be affected to different extent by instrumental instabilities and drifts making measured parameters of Equations 22 to 24 and 31 not completely and exactly related. Hence, the simultaneous measurement of all required variables (intensities) could remarkably reduce the error in the E determination.

1. Mixed Samples

To demonstrate the effect of changing geometry on calculated transfer efficiency data we carried out energy transfer measurements between F-Con A and R-Con A bound to surface glycoproteins of T-41 cells. Cells were labeled separately by the fluorescein- and rhodamine-tagged lectin alone and with both

Table 2. **Dependence of Energy Transfer Efficiency on Instrumental Stability**

Experiment no.	Ligand	Cell	I_1	I_2	I_3	S_1	S_2	S_3	α	E (%)
1.		Chicken RBC	231	308	611					
2.	F-Con A	T-41	320	842	295	8.63		0.921	4.8	
3.	R-Con A	T-41	−11	306	700		0.427			
4.	F-Con A R-Con A *A/D* = 0.3	T-41	237	772	614	2.65	0.427	0.921	4.8	4.02 3.8 3.9
5.	F-Con A R-Con A *A/D* = 2	T-41	81	599	856					14.8 18.2 17.8
6.		Chicken RBC	215	260	608					
7.	F-Con A R-Con A *A/D* = 0.3	T-41	221	517	617	2.63	0.43	0.921	4.8	−2.2
8.	F-Con A R-Con A *A/D* = 2	T-41	76	409	864					8.0
7. corr.	F-Con A R-Con A *A/D* = 0.3	T-41				2.37	0.36	0.10	5.3	3.8
8. corr.	F-Con A R-Con A *A/D* = 2	T-41								16.7

Note: Results of energy transfer measurements between F-Con A and R-Con A bound to surface glycoproteins of T-41 cells. All the samples (single labeled fluorescein stained, single labeled rhodamine stained, and double labeled) used were incubated with the ligands in batch in order to rule out sample-to-sample variations. Labeling was done in the presence of 0.2 mg/ml Con A (total concentration). Experiments 1 and 6 served as calibration or control using glutaraldehyde-fixed chicken red blood cells (RBC). In experiments 1 to 5 instrumental parameters were identical. Between experiments 5 and 6 the laser was slightly retuned and realigned with the new setting unchanged in experiments 7 and 8. In the second column acceptor-to-donor (A/D) ratios are also given for double-labeled samples. For further details see text.

kinds of conjugated ligand as usual. Double-labeled samples were prepared with two different donor-to-acceptor ratios.

Table 2 displays the results of these measurements. Numerical values of S_1 and S_3 were evaluated using fluorescence intensities from donor-labeled cells and S_2 was calculated from those of acceptor-labeled cells. From experiment No. 2 and 3 factor α was also determined. With this set of data transfer efficiency values were obtained running double-labeled cells made up with two different acceptor to donor (A/D) ratios. Results of triplicate samples yielded similar E values for samples of identical (both lower and higher) acceptor-to-donor ratio. Double-labeled samples were rerun after a slight change in the alignment of the system retuning the laser and realigning the excitation beams. To characterize the magnitude of changes, glutaraldehyde-fixed chicken red blood cells were measured as control before (experiment No. 1) and after (experiment No. 6) the change in the setting of the instrument. Transfer efficiency data of the double-labeled cells calculated from fluorescence intensities I_1, I_2, and I_3 and parameter values S_1, S_2, S_3, and α belonging to slightly different instrumental setting (experiments 7 and 8) were astonishingly different from those calculated using parameter values belonging to the same "stable" alignment and setting (experiments 4 and 5). Differences by a factor of two (sample 4 vs. 7 or sample 5 vs. 8) are surprisingly high keeping in mind that realignment affected I_1 and I_2 only by 7 and 16%, respectively, while I_3 was constant within 0.5%. Parameter values S_1, S_2, S_3, and α were corrected for the "forced" instrumental instability using readings of the control sample in both states of the system (experiments 1 and 6). With the corrected set of these parameters transfer efficiency data of experiments 7 and 8 were reevaluated (experiments 7 corrected and 8 corrected). Corrected data compare favorably with those belonging to a "stable" system setting (experiments 4 and 5).

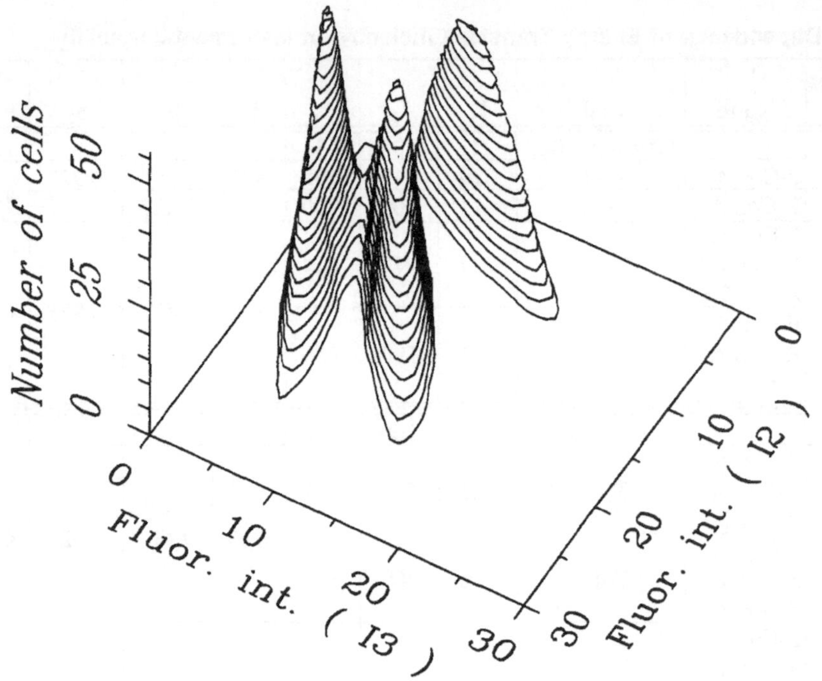

Figure 12. Flow cytometric data of a mixed sample containing T-41 cells. The left and right peaks represent cells incubated with 0.2 mg/ml F-Con A and R-Con A, respectively. Double-labeled (middle peak) cells were incubated with a mixture of both conjugated lectins at 0.2 mg/ml total lectin concentration (acceptor-to-donor ratio was 0.3). Equal amounts of single (both donor- and acceptor-labeled) and double-labeled cells were mixed after labeling.

It is seen from the displayed data (Table 2) that the accurate transfer efficiency determination critically depends on the stability of instrumental parameters including geometry and electronic setting. However, in samples of cells capable to aggregate, partial clogging up is always a potential source of error due to consequent slight changes in the flow; thus, the simultaneous determination of all measured parameters would greatly facilitate obtaining reliable data of high precision.

For this purpose donor and acceptor single-labeled cells were mixed with double labeled cells and the mixed sample was run. When evaluating the transfer efficiency from the stored fluorescence intensities it is essential that data belonging to double-labeled cells are distinguished from those of single-labeled cells. Two-dimensional gating (e.g., gating according to I_3 and I_2 simultaneously) can be applied to assure separate treatment of data from differently labeled cells. Figures 12 and 13 demonstrate that single- and double-labeled cells can be resolved using correlated distributions as a representation of the data. Simultaneous gating according to two (out of the three) measured intensity parameters can be realized in three different ways. In the overwhelming majority of the cases one of the three possibilities allows to treat separately data of all the three different kinds (donor, acceptor, and double labeled) of cells present in the mixed sample.

Table 3 summarizes overlap and energy transfer efficiency parameters vs. instrumental setting for a number of experiments using the same set of samples labeled in batch. Changes in the power of the laser or gain in the different detection channels affected the individual overlap and α parameters significantly more then the intentional change in working parameters of the flow cytometer in experiments summarized in Table 2. In contrast to the remarkable changes in the values of S_1, S_2, S_3, and α parameters energy transfer efficiency data were surprisingly constant arguing for the advantage of using mixed samples. Simultaneous detection of all required measured parameters in a single run abolishes the effect of instrumental instability on the numerical value of the energy transfer efficiency.

2. Overlap Parameters Define Conserving Entities

Below we show that, for a given donor acceptor pair used for labeling the same ligand, a conservation law holds for $P_1 = \alpha/S_1$ and $P_2 = S_1/S_2S_3$. Numerical values of these two expressions are determined by

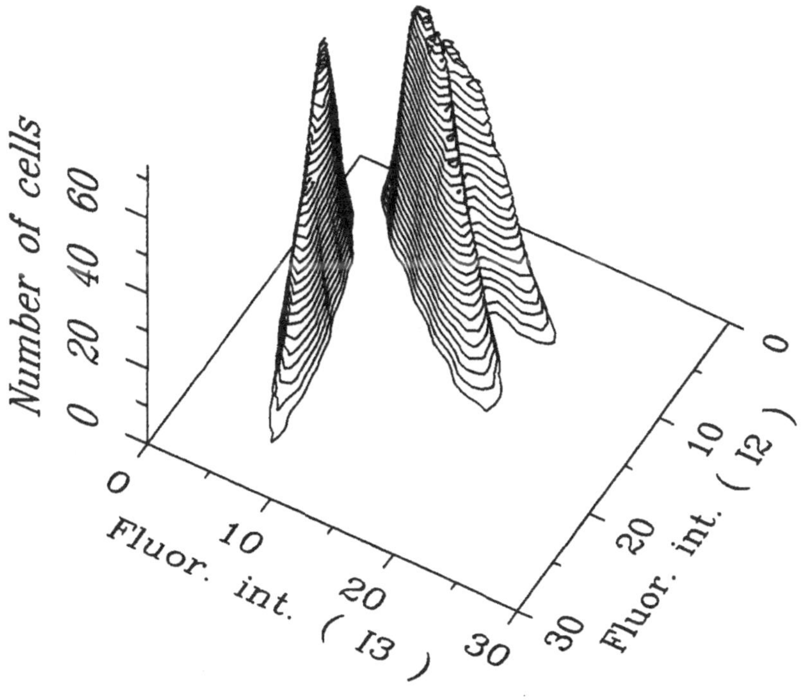

Figure 13. Flow cytometric data of a mixed sample containing T-41 cells. The left and right peaks represent cells incubated with 0.2 mg/ml F-Con A and R-Con A, respectively. Double-labeled (middle peak) cells were incubated with a mixture of both conjugated lectins at 0.2 mg/ml total lectin concentration (acceptor-to-donor ratio was 2.0). Equal amounts of single (both donor- and acceptor-labeled) and double-labeled cells were mixed after labeling.

Table 3. **Energy Transfer Measurements on Mixed Samples Result in Transfer Efficiency Values Independent from Instrumental Instability**

Experiment	Laser power (mW)	Gain channel 1	Gain channel 3	S_1	S_2	S_3	α	E (%)	$P_1 = \alpha/S_1$	$P_2 = S_1/(S_2S_3)$
1	250	1	1	0.70	0.201	0.58	1.33	8.9	1.9	6.03
2	250	0.6	1	1.09	0.203	0.883	1.93	8.7	1.78	6.08
3	250	1	0.75	0.71	0.26	0.45	1.38	8.0	1.95	6.01
4	300	1	1	0.713	0.177	0.66	1.33	8.4	1.87	6.1

Note: T-41 cells were stained with F-Con A and R-Con A (single-labeled cells) and with F-Con A and R-Con A simultaneously (double-labeled cells). The incubations were done in batch and an equal number of the donor and acceptor single-labeled cells were mixed with double-labeled cells and the mixed sample was run under different instrumental conditions. Data displayed in the last two columns demonstrate the conservation laws for S_1/S_2S_3 and α/S_1. For further explanation see text.

parameters involved in the evaluation of E. The inevitable calculation of S_1, S_2, S_3, and α in energy transfer measurements allows an easy recognition of any obvious error in carrying out the experiments or the data evaluation. This control does not require too much effort as the calculation of the actual values of P_1 and P_2, the mentioned conserving entities involves simple operations using parameters determined anyhow.

We regard two samples of cells single labeled with fluorescein and rhodamine-conjugated Con A. Overlap parameters S_1 and S_2 were evaluated from the measurement of the donor-labeled sample while S_2 was calculated from fluorescence intensities of the acceptor labeled sample as

$$S_1 = \frac{I_2^F}{I_1^F} \tag{38}$$

$$S_2 = \frac{I_2^R}{I_3^R} \tag{39}$$

$$S_3 = \frac{I_3^F}{I_1^F} \tag{40}$$

We elaborate the expressions for these five intensities similarly to those in Section III.C while discussing the determination of α factor. We use the same notation: superscripts refer to donor (F) and acceptor (R), subscripts indicate the number of signal channels as defined in Equations 22 to 24.

$$I_1^F = [B^F L^F \sigma_1^F \rho_1 t][Q^{oF}\gamma_1 \delta_1^F \xi_1^F g_1] = N_1^F C_1^F \tag{41}$$

$$I_2^F = [B^F L^F \sigma_2^F \rho_2 t][Q^{oF}\gamma_2 \delta_2^F \xi_2^F g_2] = N_2^F C_2^F \tag{42}$$

$$I_3^F = [B^F L^F \sigma_3^F \rho_3 t][Q^{oF}\gamma_3 \delta_3^F \xi_3^F g_3] = N_3^F C_3^F \tag{43}$$

$$I_2^R = [B^R L^R \sigma_2^R \rho_2 t][Q^{R}\gamma_2 \delta_2^R \xi_2^R g_2] = N_2^R C_2^R \tag{44}$$

$$I_3^R = [B^R L^R \sigma_3^R \rho_3 t][Q^{R}\gamma_3 \delta_3^R \xi_3^R g_3] = N_3^R C_3^R \tag{45}$$

Out of these parameters σ, δ, and ξ are fluorophore and signal channel dependent. Parameters B, L, and Q^o depend only on fluorophore while ρ, γ, and g on signal channel. For the sake of simplicity we assume (although this is not a necessary assumption) that the two excitation laser beams have the same profile giving rise to a unique t value. It is quite evident that the following equalities apply

$$\begin{aligned} &\sigma_1^F = \sigma_2^F, \quad \rho_1 = \rho_2, \quad \delta_2^F = \delta_3^F \\ &\xi_2^F = \xi_3^F, \quad \xi_2^R = \xi_3^R, \quad N_1^F = N_2^F \end{aligned} \tag{46}$$

Using these equalities (Equation 46) and the expressions for the individual intensities (Equations 41 to 45) parameters S_1, S_2, S_3, and α can be written as

$$S_1 = \frac{\gamma_2 \delta_2^F \xi_2^F g_2}{\gamma_1 \delta_1^F \xi_1^F g_1} \tag{47}$$

$$S_2 = \frac{\sigma_2^R \rho_2 \gamma_2 \xi_2^R g_2}{\sigma_3^R \rho_3 \gamma_3 \xi_3^R g_3} \tag{48}$$

$$S_3 = \frac{\sigma_3^F \rho_3 \gamma_3 \delta_3^F \xi_3^F g_3}{\sigma_1^F \rho_1 \gamma_1 \delta_1^F \xi_1^F g_1} \tag{49}$$

$$\alpha = \frac{C_2^R}{C_1^F} = \frac{Q^R \gamma_2 \delta_2^R \xi_2^R g_2}{Q^{oF} \gamma_1 \delta_1^F \xi_1^F g_1} \tag{50}$$

Conservation laws for S_1/S_2S_3 and α/S_1 (Equations 51 and 52) are easy to obtain from Equations 47 to 50:

$$P_1 = \frac{S_1}{S_2 S_3} = \frac{\sigma_3^R \sigma_1^F}{\sigma_2^R \sigma_3^F} \tag{51}$$

$$P_2 = \frac{\alpha}{S_1} = \frac{Q^R \delta_2^R \xi_2^R}{Q^{oF} \delta_2^F \xi_2^F} \tag{52}$$

Both P_1 and P_2 are independent from a number of instrumental setting parameters. P_1 varies only with changing absorption characteristics of the donor and acceptor dyes. These features, however, are very conservative even if different ligands are conjugated with fluorescein and rhodamine. During experimentation the checkup of the constancy of P_1 might indicate systematic errors in the measuring and data evaluation protocol.

P_2 is a function of quantum efficiency of both fluorophores and the response of the photodetector in the second detection channel for fluorescein and rhodamine emission; it changes with altered transmission of the optical elements. It is well known that fluorescence quantum yield is subject to change even in the case of the same conjugation reaction if it is allowed to proceed for a different length of time. For this reason normalization to the same quantum efficiencies is required if one intends to take advantage of the P_2 conservation law while using different batches of conjugated ligands. When applying P_2 conservation law care should be taken to correct for differences, if any, in the combined transmission of the optical elements in the second signal channel if P_2 values from different sets of measurements are to be compared. At the same time P_2 is not susceptible to changes in the laser power, steric angle, optical filter characteristics, number of binding sites, and labeling ratio. Deviation of the actual value of P_2 (corrected for quantum efficiency and gain differences) is an error-indicating sign unless elements of the output optics are changed.

Application of P_1 and P_2 conservation laws has its greatest importance in critical cases of low signals. At a moderate number of binding sites and a high level of autofluorescence, the accurate determination of the overlap parameters and α requires much skill and expertise. Under these conditions verification of P_1 and P_2 conservation is recommended.

G. FRET MEASUREMENTS AND CELLULAR AUTOFLUORESCENCE

In the analysis of the interrelationship between FRET efficiency and the detected three fluorescence intensities measured from double-labeled cells, it was anticipated that I_1, I_2, and I_3 are true, donor, and/or acceptor related parameters. However, mammalian cells often possess a high level of autofluorescence and it is experimentally documented that this phenomenon is a normal characteristic of all viable mammalian cells. Autofluorescence, at least in special cases, can severely influence the detected emission intensities. Precision of transfer efficiency data is effected to different degrees, depending on the relative amplitude of the applied fluorophore's signal and nonspecific autofluorescence. In systems with a large number of occupied receptor (antigen) sites the distortion of the recorded data by the nonspecific background emission is not significant; however, low binding site expression and/or low binding site occupancy may require correction.

Cellular autofluorescence has been noted in a variety of cell types.[46-49] It originates from discrete cytoplasmic vesicle-like regions and is absent from the nucleus;[50] thus, its contribution to the measured fluorescence readings is usually larger for cells with high cytoplasmic-to-nuclear ratios.

It is not only the level of autofluorescence which changes with the type of the cells. Aubin[50] investigated the frequency distributions of autofluorescence using different cell lines and observed remarkable differences in the widths of the histograms. Certain cells (e.g., SC1 clone of Chinese hamster ovary cells) display higher autofluorescence when growing in suspension than the same cells growing in monolayer.

Cellular autofluorescence originates probably from a variety of cellular components. It can be divided into at least two spectral regions, one exciting at 365 nm and emitting at 445 nm and the longer wavelength (green) component exciting at 440 nm and emitting at 520 nm. The green emission of the fluorophores involved interferes especially with the fluorescein emission.

One further feature of autofluorescence is that its level increases with time of cells in culture. Reintroducing frozen cells to culture the initially detected relatively low-level cellular fluorescence increases with time until a plateau level is reached after several days.[50,51] To our best knowledge no procedure has been published so far that results in a decrease of autofluorescence without effecting proliferation rate.

Comparing the distribution of autofluorescence signals of cells with that of fluorescent beads with known number of fluorescein molecules, Roederer and Murphy[52] concluded that the background fluorescence of these cells is equivalent to the emission of approximately 30,000 to 35,000 fluorescein molecules. Keeping in mind flow cytometric FRET measurements between antigens and/or receptors and that the expression of many antigens is below 100,000 per cell, this level of autofluorescence compares unfavorably with immunofluorescence signals measured from such cells. In cases when the contribution of autofluorescence to the measured intensities cannot be neglected an appropriate correction must be carried out.

Because of the cell-by-cell variation in autofluorescence, procedures using its population averaged value are not accurate. Investigation of low signal-to-noise systems requires correction for the background on a cell-by-cell basis. If the autofluorescence of stained cells can be estimated on a single-cell basis, it can be subtracted from the original readings resulting in a resultant signal proportional to the specific extrinsic fluorophore bound.

Alberti et al.[53] developed a method capable of correction on a cell-by-cell basis using a single laser line flow cytometer. Their method relies on the experimental observation that, although the autofluorescence spectrum very often overlaps with that of the applied extrinsic fluorescence label, they are sufficiently different.

To illustrate this Alberti et al.[53] reported that, in the emission above 515 nm of unstained mouse thymidine kinase-deficient L cells,[54] 38% of the signal was between 515 and 555 nm (a band used for fluorescein detection) and 50% was above 580 nm. At the same time the signal from FITC-antibody above 515 nm is basically in the 515- to 555-nm band (77%) and only 7% of the emission is above 580 nm. Since they detected a good correlation between the red (>580 nm) and green (515 to 555 nm) autofluorescence, the ratio (green to red) of these components indicates the fraction of the detected red signal to be subtracted from the green fluorescence emission (measured in the spectral range 515 to 555 nm) of the FITC antibody-labeled cells. This procedure leads to an overcompensation because of the 7% spillover of the fluorescein emission into the autofluorescence (red) detection channel. The same problem arises in any other case too when spillover of the applied fluorochrome into the autofluorescence channel is sufficient to provide a significant portion of the signal in that channel.[52]

To make the compensation more exact Roederer and Murphy[52] introduced a two-way spillover correction. This two-way correction is identical to dual-fluorescence compensation using the autofluorescence parameter as one of the two fluorophores. The actual correction can be executed either by software or by hardware as many of commercial flow cytometers are equipped for dual fluorescence spillover compensation.

An alternative autofluorescence correction method was developed by Steinkammp and Stewart[55] using dual wavelength excitation. Their method relies on the difference of the excitation spectra of the autofluorescence and immunofluorescence rather than on the difference of the appropriate emission spectra. They excite the cells at a wavelength matching the excitation properties of the immunofluorescent dye and use another excitation line of the same laser (working in all line mode) or another laser exciting autofluorescence. It is advantageous to choose the wavelength of the second line not to excite the applied dye efficiently. Both signals are detected in the same spectral range and processed to yield an autofluorescence-corrected immunofluorescence signal.

Both procedures (dual wavelength detection and dual wavelength excitation) assume a close correlation between the two autofluorescence signals. A good correlation could arise either from autofluorescence of essentially a single species emitting (or exciting) in both spectral range or from correlated amounts of different molecules emitting (or exciting) at both wavelength regions.

1. Correction to Autofluorescence in Flow Cytometric Energy Transfer Measurements

To make flow cytometric FRET method applicable in low signal-to-noise systems the mathematical treatment of the phenomenon has to be supplemented to incorporate correction for autofluorescence. As the flow cytometer suitable for this technique is equipped with multi-line excitation and multi-wavelength detection, a combination of the methods by Steinkamp and Stewart[55] and Alberti et al.[53] (multi-wavelength detection and multi-wavelength excitation) seems to be appropriate. As discussed, the standard version of the FRET technique applied dual wavelength excitation and detection in the emission range of both the donor and acceptor. Since autofluorescence of a double-labeled cell can be regarded as the emission of a third dye, and as its spectrum differs from spectra of the donor and acceptor, four independent parameters (energy transfer efficiency, amount of donor and acceptor dye molecules on each cell, and level of autofluorescence) have to be determined. For this purpose the detection of four independent signals is required. This can be achieved using fluorescein and Texas Red as the donor-acceptor pair. For excitation purposes the 488-nm argon ion laser line and the 596-nm dye laser line can be used to have efficient excitation in the donor and acceptor absorption bands, respectively.[56] Emission spectral ranges can be centered according the maxima of fluorescein (525 nm) and Texas Red (630 nm) fluorescence, respectively, with a third detection channel at intermediate wavelength (570 nm).

The four fluorescence intensities to be detected in this case from a particular cell are denoted by I_1, I_2, I_3, and I_4, and the wavelengths of excitation and detection are indicated in parenthesis, as in Section III.A. Equations 53 to 56 describe the way these intensities are composed from the four unknown parameters: (1) the unquenched donor fluorescence intensity and (2) the background autofluorescence intensity (I_F and I_B, respectively) detected in the first signal channel, (3) the nonenhanced acceptor fluorescence intensity (I_T — subscript T refers to the acceptor Texas Red), and (4) the energy transfer efficiency (E):

$$I_1(488, 525) = I_F(1-E) + I_B \tag{53}$$

$$I_2(488, 570) = I_F(1-E)S_1 + I_BS_5 \tag{54}$$

$$I_3(488, 630) = I_F(1-E)S_2 + I_BS_6 + I_TS_4 + I_FE\alpha \tag{55}$$

$$I_4(596, 630) = I_F(1-E)S_3 + I_BS_7 + I_T + \frac{S_3I_FE\alpha}{S_2} \tag{56}$$

Here again the ratios of the individual emissions (I_F, I_R, and I_B) detected in the different signal channels were denoted by the overlap parameters, S_i. The numerical value of S_i can be measured using unstained cells (S_5, S_6, and S_7), cells stained with donor only (S_1, S_2, and S_4), and cells stained with acceptor only (S_3):

$$S_1 = \frac{I_2(488, 570)}{I_1(488, 525)} \tag{57}$$

$$S_2 = \frac{I_3(488, 630)}{I_1(488, 525)} \tag{58}$$

$$S_3 = \frac{I_3(488, 630)}{I_4(596, 630)} \tag{59}$$

$$S_4 = \frac{I_4(596, 630)}{I_1(488, 525)} \tag{60}$$

$$S_5 = \frac{I_2(488, 570)}{I_1(488, 525)} \tag{61}$$

$$S_6 = \frac{I_3(488, 630)}{I_1(488, 525)} \tag{62}$$

$$S_7 = \frac{I_4(596, 630)}{I_1(488, 525)} \tag{63}$$

Instead of giving the relatively complicated analytical form for E, the solution of Equations 53 to 56 we express E/(1–E):

$$\frac{E}{1-E}=\frac{1}{\alpha}\left[\frac{(I_3-I_4S_4)(S_1-S_5)+(I_1S_1-I_2)(S_4S_7-S_6)}{(I_2-I_1S_5)\left(1-\frac{S_3S_4}{S_2}\right)}-S_2\right] \tag{64}$$

The outlined protocol requires slight modification of the experimental hardware relative to that used in the original FCET measuring method. With FITC and Texas Red as donor acceptor dye pair the flow cytometer has to be equipped with an argon ion and a dye laser. The four independent fluorescence intensities can be detected using three photomultiplier tubes requiring two dichroic mirrors for the separation of the three spectral regions of detection. Apart from these alterations the entire data collection procedure remains the same as that for the case with no autofluorescence measurements.

The modification of the original FCET measuring method to include autofluorescence background correction on a cell-by-cell basis extends the applicability of the technique to cells with a significantly lower receptor number. Unfortunately, until now this more sensitive method has not been applied to the experimental investigation of any cellular system.

H. FLOW CYTOMETRIC FRET MEASUREMENTS USING SINGLE LASER EXCITATION

In order to determine the fluorescence energy transfer efficiency between donor and acceptor molecules attached to the cell surface, three independent measurements have to be completed on each individual cell because three unknowns affect the amplitude of signals which carry information about the numerical value of E (see Section III.A). A flow cytometer equipped with an argon ion laser dual laser line excitation (at 488 and 514 nm) can be used for this purpose with the emission intensities measured in the emission bands of the donor fluorescein and acceptor rhodamine. Since many laboratories have flow cytometers with single laser excitation the adaptation of the method to flow cytometers having only single wavelength excitation would potentiate a more widespread application of the technique.

Missing the possibility to use two different wavelengths for excitation one could try to take emission readings in three different spectral regions. In this case, however, the individual spectral detection regions should be defined with relatively narrow band width, otherwise the changes in the contributions of donor and acceptor to the measured intensities in the different detection bands would not be sufficiently high to assure accurate efficiency data. Decrease of detection bandwidth would also give rise to a diminished signal further reducing the precision of the measurements.

Having limited the number of measured fluorescence signals to two (single wavelength excitation and two spectral ranges for detection) the third fluorescence intensity could be substituted by *a/d*, the acceptor-to-donor ratio.[36] This ratio, however, can be measured generally as a population-averaged parameter rather than on a cell-by-cell basis. This is a remarkable drawback of the modified method compared to the more sophisticated double beam procedure, but the simplified technique is still superior to any non-FCET measurements on cell surfaces. In the particular case of homologous ligands (e.g., F-Con A and R-Con A) or ligands that bind to the same membrane protein albeit to different epitopes, adaptation of the original flow cytometric FRET method to a single beam flow cytometer does not reduce the information obtained. In individual cells, labeled with donor and acceptor derivatives of the same ligand, the ratio *a/d* is constant and its exact numerical value can be determined as a population-averaged parameter. This statement does not hold for the general case of two different membrane proteins because the biological variance in the expression of any cell surface antigen is great (see Figures 6 and 7 and Section III.D.2) and usually there is no close correlation between the level of expression of different antigens on individual cells. Further advantages of the single beam FCET measurements over nonflow methods will be summarized in Section V.B.

1. Evaluation of the Energy Transfer Efficiency

Separate cell samples are labeled either with donor- or acceptor-conjugated ligands alone (single-labeled samples) or with both donor- and acceptor-conjugated ligands (double-labeled sample) the same way as when using the double-beam FRET technique. As before we assume that fluorescein (FITC) and rhodamine (TRITC) are used for donor acceptor pair of dyes. Superscripts F and R of the parameters refer to donor and acceptor, respectively. Notation $I(\lambda_1; \lambda_2)$ indicates that the fluorescence intensity is measured at λ_2 nm with excitation at λ_1 nm. With a single excitation line (either 488 or 514 nm using an argon ion laser) and two photodetectors (detecting in two wavelength regions centered around 540 and 580 nm, respectively) the measured I_1 and I_2 fluorescence intensities from a double-labeled cell can be decomposed:

$$I_1(488 \text{ or } 514,\ 540) = I_F(1-E) + I_R S \tag{65}$$

$$I_2(488 \text{ or } 514,\ 580) = I_F(1-E)S_1 + I_R + I_F E\alpha \tag{66}$$

$$\frac{I_R}{I_F} = \frac{B^R L^R \varepsilon^R}{B^F L^F \varepsilon^F}\alpha = \frac{a\varepsilon^R}{d\varepsilon^F}\alpha \tag{67}$$

where I_F and I_R are the theoretical, directly excited fluorescence intensities of fluorescein and rhodamine when no energy transfer takes place. S_1 and S are the corrections for spectral overlap associated with direct excitation of the respective fluorophores. S_1 can be determined as I_2/I_1 using cells labeled with only donor-conjugated ligands and S is I_1/I_2 using cells labeled with only acceptor-conjugated ligands. With 488-nm excitation the value of S is negligibly small. In Equation 65 the effect of energy transfer has been neglected because the contribution of the process to the I_1 signal is small (it is less than 2% using T-41 cells stained with F-Con A and R-Con A at an acceptor-to-donor ratio of 3, yielding about 30% energy transfer efficiency). Definition of factor α is the same as in the double beam method (Section III.C and Equation 31) and it takes care of the fact that an excited fluorescein molecule can be detected with a sensitivity differing from that of an excited rhodamine molecule. Its numerical value can be determined as

$$\alpha = \frac{C^R}{C^F} = \frac{M^R \overline{B^F} L^F \varepsilon^F}{M^F \overline{B^R} L^R \varepsilon^R} \tag{68}$$

Here M^F and M^R are the mean values of fluorescence intensity distributions of I_1 and I_2 determined using cell suspensions saturated either with fluorescein- or rhodamine-conjugated ligands; B^F and B^R denote the number of binding sites per cell occupied by fluorescein- and rhodamine-conjugated ligands, respectively; $\overline{B^F}$ and $\overline{B^R}$ are the same parameters averaged over the analyzed populations of these single-labeled samples. The fluorophore-to-ligand labeling ratio for the fluorescein- and rhodamine-conjugated ligands are designated by L^F and L^R, respectively, and ε^F and ε^R are molar absorption coefficients measured at 488 nm (or at 514 nm).

Equation 67 describes that the ratio I_R/I_F can be calculated for each individual double-labeled cell as the product of the ratio of the excited rhodamine to fluorescein molecules on the particular cell and the ratio of the detectability of an excited rhodamine to fluorescein molecule (α).

Equations 65 to 67 can be solved for E:

$$E = \frac{\frac{I_2}{I_1}\left(1 + S\frac{a\varepsilon^R}{d\varepsilon^F}\alpha\right) - S_1 - \frac{a\varepsilon^R}{d\varepsilon^F}\alpha}{\frac{I_2}{I_1} - S_1 + \alpha} \tag{69}$$

This expression allows the determination of the energy transfer efficiency on a cell-by-cell basis if the fluorescein- and rhodamine-conjugated ligands bind to the same membrane protein (either to the same or different epitopes). Under these conditions the actual rhodamine to fluorescein molecular ratio, *a/d*, does not change from cell to cell and can be measured as the ratio of the total number of rhodamine to fluorescein molecules in an analyzed population. This ratio can be determined by dissolving the labeled cell samples using sodium-dodecyl-sulfate.[57] In the case of homologous ligands accurate *a/d* can be calculated from the initial composition of the incubation mixture of rhodamine- and fluorescein-conjugated ligands, provided that the relative binding affinity of the applied labeled ligand is known. The latter can be determined from competition experiments in the presence of unlabeled ligand.

2. Error Estimation

To demonstrate the propagation of errors, we rewrite Equation 69 neglecting parameter S (it is justified in case of 488-nm excitation):

$$E = \frac{r - S_1 - C\alpha}{r - S_1 + \alpha} \tag{70}$$

where $r = I_2/I_1$ and $C = (a\varepsilon^R)/(d\varepsilon^F)$. The standard deviation of E can be calculated according to Equation 71:

$$\sigma_E^2 = \frac{(1+C)^2}{(r - S_1 + \alpha)^4} \cdot \left[\alpha^2 \left(\sigma_r^2 + \sigma_{S_1}^2 \right) + (S_1 - r)^2 \sigma_\alpha^2 + \frac{\alpha^2 (r - S_1 + \alpha)^2}{(1+C)^2} \sigma_C^2 \right] \tag{71}$$

In Equation 71 σ_r^2 can be evaluated as

$$\sigma_r^2 = \left[\frac{\sigma_{I_2}}{I_1} \right]^2 + \left[\frac{I_2 \sigma_{I_1}}{I_1^2} \right]^2 \tag{72}$$

In order to compare the results of energy transfer efficiency measurements carried out on a single laser excitation flow cytometer with those by double beam equipment, T-41 cells were stained with F-Con A, R-Con A, and with both fluoresceinated and rhodaminated lectins. Transfer efficiency data obtained by the two methods were identical within experimental error in the a/d range 0.4 to 3.0 (transfer efficiency data were between 10 and 26%). For measurements with 488-nm excitation (single laser excitation method) the standard deviation of E values was also determined using the error propagation function described by Equation 71.

The resulting absolute values of standard deviations for energy transfer efficiencies were 2, 2.5, and 3% at acceptor-to-donor ratios 0.4, 0.96, and 2.88, respectively. All these data were in good agreement with the standard deviations determined from the experimentally determined frequency distributions of E (estimated relative errors of individual parameters calculated from experimental data and used in the evaluation of σ_E were 2 and 6% for I_1 and I_2, respectively, and 6, 6, and 10% for parameters S_1, α, and C, respectively).

IV. TIME-RESOLVED FLUORESCENCE AND SPECTROFLUORIMETRIC METHODS TO STUDY FRET

A. FLUORESCENCE LIFETIME MEASUREMENTS

As detailed in Section II.B the transfer of excitation energy from a donor to an acceptor fluorophore molecule brings about changes in the spectroscopic parameters of both molecules. Equation 15 describes how time-resolved fluorescence measurements[58-63] can be used for the determination of the energy transfer efficiency by measuring the shortening of donor fluorescence lifetime induced by the acceptor.

Donor lifetime measurements are infrequently used for FRET measurements on cells. Here we refer to just two publications using time-resolved fluorescence technique. Papp et al.[64] observed efficient energy transfer between donor- and acceptor-labeled Ca^{2+}-ATPase molecules in intact sarcoplasmic reticulum vesicles indicating interaction between these molecules. Kusumi et al.[65] applied a newly developed streak-camera-based time-resolved microscope fluorimetric method to detect fusion of liposomes *in vitro* and that of endosomes in single cells. Fusion can be also followed by macroscopic measurements using the method of time-correlated single photon counting. The latter technique, quite naturally, would allow more accurate measurement of the fluorescence decay than analogue methods, such as those using a streak camera.[65]

Fluorescence microscopic methods,[66,67] however, when combined with imaging techniques[68-72] allow the simultaneous determination of lifetimes from many regions of the interior of the investigated system. Instead of resolving a mean lifetime or a set of lifetimes from millions of cells in suspension, 10^5 to 10^6

pixel lifetimes can be resolved in the image. These decay times can then be displayed as a lifetime image.[70,73] This makes possible to distinguish particular local environments within the cells on the basis of lifetime of labeled species or energy transfer between donor- and acceptor-stained loci.

B. SPECTROFLUORIMETRIC TECHNIQUES

In a large number of experiments the efficiency of FRET was measured using spectrofluorimeters.[22,23,36,40,76] These measurements are most commonly based on the determination of the attenuation of the donor emission (Section II.B, Equation 17) or the sensitization of acceptor fluorescence due to the transfer of excitation energy (Section II.B, Equation 18). Besides these routinely applied methods a spectrofluorimetric assay is suggested by the flow cytometric FRET measuring procedure (Section III).

1. Method I

Similar to the flow cytometric transfer efficiency measurements, the analogue spectrophotometric measurements also require three individual samples besides the double labeled ones (for which E has to be determined): unlabeled cells and cells labeled with saturating concentrations of either donor- or acceptor-conjugated ligands. Assuming fluorescein as donor and rhodamine as acceptor wavelengths for excitation are set to 493 and 553 nm, respectively. Detection spectral ranges are centered at 520 and 580 nm, respectively. All measured intensity readings are designated by I and are further specified by two numbers in parentheses referring to the wavelengths of excitation and emission. To calculate contributions of different origin to the measured fluorescence intensities two proportionality factors (Z_1 and Z_2) have to be introduced:

$$Z_1 = \frac{I_D(493,580)}{I_D(493,520)} \tag{73}$$

$$Z_2 = \frac{I_A(493,580)}{I_A(553,580)} \tag{74}$$

Here subscripts D and A indicate that the appropriate intensities belong to samples of cells labeled with donor only and acceptor only, respectively. I with no subscript (see below) stands for measured fluorescence readings from a double-labeled sample.

The energy transfer efficiency is related to the enhancement of acceptor fluorescence intensity. Denoting "theoretical" measured contributions of the donor and acceptor to the fluorescence readings of the double-labeled sample by F_D and F_A, the increase of detected acceptor emission intensity can be expressed as

$$I(493,580) - F_D(493,580) - F_A(493,580) = KE \tag{75}$$

where K is a constant depending on instrumental and acceptor spectroscopic parameters and donor total concentration. The second and third terms in this expression are the contributions of the directly excited (at 493 nm) donor and acceptor emissions (as detected at 580 nm) to the measured appropriate fluorescence intensity of a double-labeled sample. These terms can be calculated using Equations 73 and 74 which obviously hold also for F_D and F_A as well.

$$F_D(493,580) = Z_1 F_D(493,520) \tag{76}$$

$$F_A(493,580) = Z_2 F_A(553,580) \tag{77}$$

Under our conditions we have not detected any fluorescence intensity above the background from rhodamine samples with 493-nm excitation and 520-nm emission. Thus, the emission of a double-labeled sample under these conditions is entirely due to fluorescein. Detected emission intensity of fluorescein at 580 nm with 553-nm excitation was of background level from donor-stained sample rendering the measured fluorescence of a double-labeled sample under these conditions entirely due to rhodamine. Taking these considerations into account Equations 76 and 77 can be reformulated:

$$F_D(493,580) = Z_1 I(493,520) \tag{78}$$

$$F_A(493,580) = Z_2 I(553,580) \tag{79}$$

Combining Equations 78 and 79 with Equation 75 we arrive at

$$I(493,580) - Z_1 I(493,520) - Z_2 I(553,580) = KE \tag{80}$$

The quenched donor fluorescence measured with 493-nm excitation at 520-nm emission band can be expressed as

$$I_D(493,520) = K'(1 - E) \tag{81}$$

where K′ is a constant similar to K. The ratio of the two constants

$$\frac{K}{K'} = \alpha' \tag{82}$$

α′ has a meaning analogous to that of α as it was defined and used in the two laser line excitation FRET calculation (Section III.A). It was mentioned earlier that the detected donor fluorescence with 493-nm excitation and 520-nm emission is equal to the detected signal from a double-labeled sample under the same conditions; thus, Equation 81 can be written as

$$I(493,520) = K'(1 - E) \tag{83}$$

Dividing Equation 80 by Equation 83 we get

$$\frac{E}{1-E} = \frac{I(493,580) - Z_1 I(493,520) - Z_2 I(553,580)}{\alpha' I(493,520)} = X \tag{84}$$

$$E = \frac{X}{1+X} \tag{85}$$

The determination of α′ may be accomplished analogously to that of α:

$$\alpha' = \frac{I_A(493,580) L^F \varepsilon^F}{I_D(493,520) L^R \varepsilon^R} \tag{86}$$

2. Method II

The energy transfer efficiency, E, can also be calculated by the most commonly used energy transfer efficiency method (see Reference 24):

$$E = \left[\frac{I(493,580)}{I_A(493,580)} - 1\right]\frac{\varepsilon^R a}{\varepsilon^F d} \tag{87}$$

where I and I_A are measured fluorescence intensities from a double-labeled and an only acceptor-stained sample, ε stands for the absorption coefficient (measured at 493 nm), and *a*/*d* is the donor-to-acceptor concentration ratio in the double-labeled sample. It is assumed that the energy transfer sample and the only acceptor-stained sample contain the same rhodamine concentration.

V. CRITICAL COMPARISON OF FLOW CYTOMETRIC AND SPECTROFLUORIMETRIC FRET MEASUREMENTS

Prior to the development of the quantitative FCET measuring technique the steady-state fluorescence resonance energy transfer method has been widely applied to study membrane surface phenomena. The flow method is capable of providing a frequency distribution of E values measured on single cells while spectrofluorimetric measurements yield a simple mean for the entire cell population. Owing to the differences in the two techniques, a critical comparison of results on identical samples by the two methods is unavoidable.

A. DOUBLE LASER EXCITATION EXPERIMENTS

To compare results by steady-state and dual wavelength flow cytometric (FCET) E determination, samples of T-41 cells were stained with F-Con A and R-Con A. The double-labeled energy transfer samples were incubated with both F-Con A and R-Con A at different *a/d* ratios. Sets of identical samples were measured with a flow cytometer and spectrofluorimeter. The steady-state determination of the energy transfer efficiency was carried out for all double-labeled samples with detection of the acceptor emission at 570, 580, and 590 nm. Data displayed in Table 4 represent the mean value of these parallel determinations. For the calculation of the transfer efficiency from steady-state measurements both method I and method II were used (see Section IV.B). Flow cytometric determination of E was based on the protocol discussed in Section III.

For both steady-state methods three different E values are given for each of the five double-labeled samples (Table 4). There is a remarkable difference between uncorrected steady-state E values and flow cytometric results. This discrepancy is explained probably by the presence of unbound ligands due to incomplete washing, since the analysis of samples after repeated washes leads to smaller (but still significant) differences. A better agreement was achieved when correction was made for the fluorescence contribution of the free ligands using fluorescence intensities measured on the supernatant of the centrifuged samples. The best agreement between steady-state and flow cytometric data was achieved using fixed cells. Comparing the formulas to calculate E from spectrofluorimetric data by methods I and II, the former (Equations 84 to 86) are more promising to yield more accurate results. Application of Equation 87 (method II) requires the determination of *a/d* involving digestion of the samples. This is a relatively lengthy procedure and the results strongly depend on the cell concentrations in the appropriate samples. In contrast, results by method I are less sensitive (if at all) to concentration differences in the samples involved. Using this approach cell concentration in the sample stained with both donor- and acceptor-conjugated ligands does not affect the value of E at all. This is achieved by normalizing the energy transfer signal (the sensitized acceptor emission) to the quenched donor fluorescence (method I) rather than to the unquenched acceptor emission (method II). Experimentally determined values of Z_1 and Z_2, as ratios of two fluorescence intensities belonging to the same cell, are naturally also independent of the cell concentration of the samples, similarly to factor α' being a ratio of molecular parameters of the fluorophores. Assuming that binding of the fluorescent ligands to their cell surface receptors does not

Table 4. **Energy Transfer Efficiency (E) between F-Con A and R-Con A Bound to T-41 Cells**

	Steady-state measurements						
	Method I[a]			Method II[b]			FCET[c]
Acceptor-to-donor ratio	E_{liv} (%)	E_{liv}^{corr} (%)	E_{fix} (%)	E_{liv} (%)	E_{liv}^{corr} (%)	E_{fix} (%)	E (%)
4.3	24.3	31.2	31.1	28.9	34.7	29.4	31.8
2.3	20.1	23.3	29.0	20.9	23.1	28.7	28.1
1.2	15.8	19.3	22.6	14.2	17.6	23.2	25.8
0.6	13.3	16.6	17.3	13.1	16.6	15.2	18.0
0.3	6.2	8.1	11.5	6.1	7.3	10.5	11.4

[a]Description of calculation of E values based on method I; see Section IV.B.1.

[b]Evaluation of transfer efficiency data according to method II; see Section IV.B.2.

[c]FCET, flow cytometric energy transfer method; see Sections III.A and C.

From Szöllősi, J., Trón, L., Damjanovich, S., Helliwell, S. H., Arndt-Jovin, D. J., and Jovin, T. M., *Cytometry*, 5, 210, 1984, Reprinted by permission of Wiley-Liss, a division of John Wiley & Sons, Inc., Copyright © 1984.

bring about spectroscopic changes in the optical properties of the fluorophores, α′ can be measured in cell-free systems using the labeled ligands in high-enough concentrations to provide accurate results. The above-outlined differences explain that method I is devoid of several potential sources of additional experimental error compared to method II. Thus, it seems understandable that, although results by method I and method II are similar, those by the former are in a slightly better agreement with the FCET results.

In contrast to the effect of incomplete washing on the spectrofluorimetric E values, flow cytometric protocol resulted in essentially identical data without washing, after repeated washes, and after fixation. This important advantage of the FCET method over steady-state procedures is explained by the small size of the volume illuminated in the flow cytometer containing negligible amounts of unbound ligands. Thus, FCET measurements can be carried out without separation procedures, i.e., without perturbation of the dynamic receptor-ligand binding equilibrium. This feature allows to study systems involving ligands of low binding affinities.

Determination of energy transfer efficiency on a cell-by-cell basis employing dual excitation flow cytometers offers additional advantages over steady-state methods.[19,20,22,23,36-41,43,44,76-88] The FCET method allows you to measure E profiles with different gating according to other parameters. This can be achieved either by collecting data from cells falling into a defined region of a measured cellular parameter or by restricting similarly the evaluation of E from list mode data. Multiple gating (according to two or more parameters) allows the assessment of energy transfer efficiency for individual subpopulations present in a heterogeneous sample. Since each cell is analyzed individually, inhomogeneities can be resolved even without gated data collection. Correlations between E values and other cellular parameters can also be investigated.

With FCET technique it is easy to restrict data collection or data evaluation to undamaged cells using gating with light scatter signals of appropriate amplitudes, thereby suppressing contributions from dead cells or debris.

Another advantage of flow methods is that they are not disturbed by light scatter, a severe problem in spectrofluorimetry at high cell concentrations.

Dynamic processes can also be measured even in cases when cells respond nonsynchronously to ligands. Since the duration of illumination typically is a few microseconds, considerably high speed can be achieved in data collection (several thousands of cells can be analyzed in a second). With this repetition frequency and sequential data acquisition, the time resolution of the studied dynamic processes is limited only by the time necessary to apply the triggered sample to the flow cytometer.

If the measurements are done on a flow cytometer with a high-speed computer, the efficiency of energy transfer can be calculated in real time and populations with different E values can be sorted.

Another flow cytometric method was developed by Chuck et al. to measure energy transfer between fluorescein- and rhodamine-conjugated monoclonal antibodies bound to cell surface antigens.[89] They were able to detect emission from donor (spectral range of detection: 530 ± 30 nm) and acceptor (spectral range: 630 ± 22 nm) without detectable spillover from the other fluorophore using 488-nm (argon laser) and 568-nm (krypton laser) wavelengths to excite fluorescein and rhodamine emission. A third fluorescence signal (excited at 488 nm and detected above 590 nm) was also measured to characterize sensitized acceptor emission. These three intensities registered from each individual cell, and population averages of similar emission intensities belonging to single-labeled cells allowed the evaluation of absolute transfer efficiency data on a cell-by-cell basis. Although the method of calculation is different from that described in Section III, it was not compared with spectrofluorimetric or with other published flow methods.

B. SINGLE LASER EXCITATION EXPERIMENTS

In order to demonstrate the versatility of the single wavelength excitation energy transfer measuring method, T-41 murine lymphoma cells were labeled with fluorescent Con A derivatives. Energy transfer efficiency data measured by different techniques on identical sets of single- and double-labeled cells are displayed in Table 5. The calculation methods were based upon the quenching of donor and sensitization of acceptor (Section II.B, Equations 17 and 18, respectively) for the spectrofluorimetric measurements. In these cases F_A and F_D values were determined from the appropriate intensity readings belonging to the appropriate single-labeled samples of the same concentration of cells labeled with saturating amounts of R-Con A and F-Con A taking into account the percentages of Con A binding sites occupied by the same labeled ligand in the double-labeled sample. E data by single beam and double beam flow cytometric methods allowed the determination of E on a cell-by-cell basis. Results by the two latter techniques were in excellent agreement. The efficiency of FRET increased with higher acceptor-to-donor molecular ratios on the cell surface (Table 5). All the data displayed in Table 5 showed good agreement regardless of the

Table 5. **Energy Transfer Efficiency (E) between F-Con A and R-Con A Bound to T-41 Cells**

Method	Excitation wavelength (nm)	Acceptor-to-donor ratio 0.41	0.96	2.88
Quenching[a]	488	12 ± 2	19 ± 2	27 ± 3
	514	12 ± 2	20 ± 2	28 ± 3
Sensitization[b]	488	8 ± 2	13 ± 2	22 ± 3
	514	8 ± 2	18 ± 2	26 ± 3
Flow cytometric	488	10 ± 2	16 ± 2	25 ± 3
single beam[c]	514	11 ± 2	18 ± 2	26 ± 3
FCET[d]	488/514	10 ± 2	18 ± 2	26 ± 3

[a]Calculation of E values was carried out according to Equation 17 (Section II.B).

[b]Calculation of E values was carried out according to Equation 18 (Section II.B).

[c]E values were evaluated using Equations 68 and 69 (Section III.H).

[d]FCET, flow cytometric energy transfer method; see Section III.A and C.

From Szöllősi, J., Matyus, L., Trón, L., Balázs, M., Ember, I., Fulwyler, M. J., and Damjanovich, S., *Cytometry*, 8, 120, 1987, Reprinted by permission of Wiley-Liss, a division of John Wiley & Sons, Inc., Copyright © 1987.

calculation method used; however, the steady-state sensitization method with 488-nm excitation resulted in systematically lower FRET efficiencies. Cell suspensions prepared from the spleen of Gross virus infected mouse contain normal and leukemic cells which can easily be distinguished according to their size or light scattering signal.[57,90] The modified, single laser excitation FCET method enabled us to study energy transfer efficiency in this heterogeneous population and determine average cell surface density of Con A binding sites on both normal and leukemic cells individually.[36]

VI. MICROSCOPIC TECHNIQUE TO MEASURE ABSOLUTE FRET EFFICIENCY

FRET measurements in a microscopic system are limited to the investigation of a low number of cells unless fluorescence digital imaging microscopy is used. Conventional microscopic procedures[91-96] provided primarily qualitative information about energy transfer processes; however, spectrofluorimetric procedures are readily applicable to microscopic measurements. The measurement of FRET in the microscope offers the advantage of spatial resolution in contrast to both steady-state and flow methods. Thus, one can localize energy transfer efficiency data to different regions inside of a cell or on its surface or study spatial relationships between antigens or receptors of the cytoplasmic membrane in contact regions between cells. Technological advances in image detectors and computer hardware have led to dramatic improvements in quantitative image acquisition and analysis. Combining any spectrofluorimetric procedure (e.g., methods I and II described in Section IV.B) with fluorescence digital imaging microscopy provides a possibility to carry out quantitative energy transfer measurements on a pixel-by-pixel basis. To our best knowledge these kind of measurements have not been carried out on cellular systems so far.

Jovin and Arndt-Jovin developed a microscopic FRET method based on measurement of donor emission and sensitized acceptor emission on a pixel-by-pixel basis.[97] Their procedure resembles the FCET method (outlined in Section III) with some modifications. Below we give a short account of their special microscopic technique and formalism of data analysis.

In this approach three images are recorded from an entire viewing field containing double-labeled cells: I(D/D), I(D/A), and I(A/A). Here D and A refer to the wavelength region of donor and acceptor excitation (first letter) and emission (second letter). Using a high-pressure mercury or xenon arc lamp as a light source one is not limited to discrete wavelengths of excitation as in the case of an argon ion laser with the flow cytometric method. This has the advantage of setting the excitation wavelength of donor (e.g., fluorescein) to 485 nm and that of acceptor (e.g., rhodamine) to 546 nm in a way that the emission of any of these dyes is hardly excited in the spectral region of excitation of the other one eliminating the necessity of corrections inevitable when using argon ion lasers. From the stored intensities determination of transfer efficiency can be obtained for each individual pixel.

Apart from corrections for camera bias and dark current, field inhomogeneities, etc. (see details in Reference 97), the images are processed based on the following. Transfer efficiency, E, can be calculated as the ratio of the amount of the donor intensity transferred to the acceptor ($I_{D,tr}$) and the donor total intensity ($I_{D,tot}$).

$$E = \frac{I_{D,tr}}{I_{D,tot}} \tag{88}$$

In this expression $I_{D,tr}$ means the missing donor fluorescence intensity due to the quenching effect of the acceptor. This decrement is related to the sensitized acceptor emission ($I_{A,s}$)

$$I_{D,tr} = I_{A,s}\zeta \tag{89}$$

where factor ζ is the ratio of emission intensities detected from a given number of excited donor molecules to that measured from the same number of excited acceptor molecules. The numerical value of ζ can be determined in a similar way as factor α for the FCET method (Section III.C) measuring dilute calibration solutions of donor and acceptor and making correction for differences in absorption cross sections, excitation intensities, and detector efficiencies.[97]

Sensitized acceptor emission intensity can be calculated from the registered three intensities for each pixel:

$$I_{A,s} = I(D/A) - b_1 I(D/D) - b_2 I(A/A) \tag{90}$$

where the second and third terms account for overlap of donor emission into the acceptor channel and for the direct excitation of the acceptor emission at the donor wavelength, respectively. Parameters b_1 and b_2 can be derived with appropriate solutions in microcuvettes measured in the microscope[97] or like the spillage coefficients S_i in the FCET method.[39] With known values of b_1, b_2, and ζ one can evaluate $I_{D,tr}$ from the registered three images for each pixel.

The theoretical intensity (not detected in any of the three recorded images), $I_{D,tot}$, the denominator in Equation 88 can be calculated as the sum of $I_{D,tr}$, the missing part of donor emission intensity (due to energy transfer derived using Equation 89) and $I_{D,r}$, the residual, quenched donor emission as detected in the I(D/D) image:

$$I_{D,tot} = I_{D,r} + I_{D,tr} \tag{91}$$

Jovin and Arndt-Jovin claim that the above procedure bypasses some difficulties encountered in the flow cytometric measurements.[97] Using the FCET method a function of $E/(1 - E)$ was defined and calculated from experimentally measured quantities. This function increases without limit in the case of very high energy transfer. Until now, however, there have been no comparative studies conducted and, judging from the reported values of energy transfer efficiencies, this problem does not seem to be a frequently occurring difficulty in energy transfer measurements on the cell surface.

VII. DETERMINATION OF ENERGY TRANSFER EFFICIENCY BY PHOTOBLEACHING MEASUREMENTS

Under appropriate conditions, some fluorophores may undergo irreversible photodecomposition from their excited state. The bleached product of this irreversible process is a nonfluorescent entity.[98] Hirschfeld realized that the integrated fluorescence emission obtained on complete bleaching (I_{int}) is independent from numerous spectroscopic variables.[99] He showed that it does not change with fluorescent quantum yield, absorption cross section, and illumination intensity. This is a very special feature of the integrated fluorescence intensity and it can serve as a basis for application in studying fluorescence energy transfer processes.

In order to measure the efficiency of these processes the number of transitions with transfer of energy from donor to acceptor has to be normalized to the number of excitations of donor molecules per unit time. As I_{int} is naturally linearly related to the size of the donor population and depends on the same parameters characterizing instrumental setting (steric angle, transmission of optical filters, gain of electronics, etc.) as other measured fluorescence intensities of interest, it can readily be used for normalization of the latter

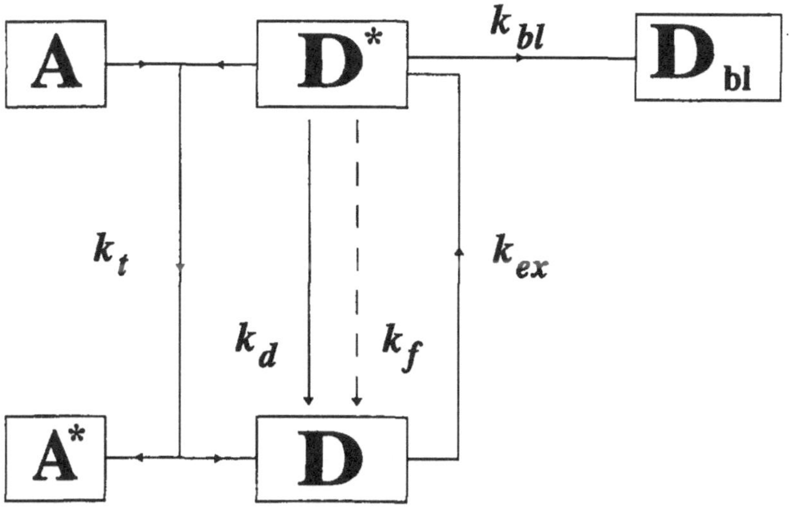

Figure 14. Scheme of processes populating or depopulating the excited state (S_1) of the donor.

related to transfer process. Below we give an account of the theory based on publications by Hirschfeld[99] and Jovin and Arndt-Jovin[97] and arrive at a protocol for transfer efficiency determination using donor photobleaching.

The rate of the production of the bleached donor (D_{bl}) can be calculated as the product of the kinetic constant of the bleaching, k_{bl}, and the amount of excited donor molecules D^*

$$\frac{dD_{bl}}{dt} = k_{bl}D^* \tag{92}$$

However, elimination of the donor molecules from the excited singulet state can also be the result of other deexcitation processes such as decay of the excited state (including both radiationless and radiative transitions to the ground state) and energy transfer. These processes compete with the bleaching. Thus, bleaching occurs with a decreased efficiency in cases when any of the competing processes has higher probability. Consequently, a comparison of the bleaching kinetics of samples labeled only with donor (no energy transfer) and that labeled with both donor and acceptor (characterized by a nonzero energy transfer efficiency) allows the calculation of E.

Figure 14 displays a scheme of possible processes depopulating the S_1 state taking the photobleaching into account. Kinetic constants of specific processes are displayed as well. The total donor population (D_o) in the system studied can be calculated as the sum of the populations in ground state, D, excited state, D^*, and bleached state D_{bl}:

$$D_o = D + D^* + D_{bl} \tag{93}$$

All the processes indicated, but bleaching are transitions between D and D^* in one of the two possible directions. If the rate constant of the bleaching is significantly smaller than all other rate constants a fast equilibrium is maintained between D and D^* before any substantial amount of the donor would be depleted by bleaching. For this quasi-steady state:

$$Dk_{ex} = D^*\left(k_d + k_f + k_T\right) \tag{94}$$

and

$$\frac{D^*}{D} = \beta = \frac{k_{ex}}{k_d + k_f + k_T} \tag{95}$$

where k_{ex}, k_d, k_f, and k_T denote rate constants of excitation, nonradiative deexcitation, fluorescence, and energy transfer processes, respectively.

Assuming a first-order process the probability of the bleaching of the donor per unit time is

$$k_{ex}\frac{k_{bl}}{k_d + k_f + k_T} = \beta k_{bl} \tag{96}$$

So that donor is depleted exponentially according to

$$D = D_o e^{-\beta k_{bl} t} \tag{97}$$

with time constant

$$\tau_{bl} = \frac{1}{\beta k_{bl}} = \frac{1}{k_{ex} k_{bl} \tau} \tag{98}$$

and with the singulet lifetime of the donor

$$\tau = \frac{1}{k_d + k_f + k_T} \tag{99}$$

The time constant of the bleaching evidently shows a strong dependence on k_{ex} and k_{bl}. However, using a comparable pair of samples containing cells labeled only with donor-conjugated ligands, on the one side, and those labeled with both donor and acceptor conjugated ligands, on the other side, k_{ex} does not change and k_{bl} is also constant unless intensity of the bleaching beam is altered. Thus, the determination of the bleaching time constant from dynamic measurements yields parameters the ratio of which gives the ratio of singulet lifetime of the donor in the double- (donor and acceptor) labeled system to that in the single- (donor) labeled system. Using Equations 15 and 98

$$1 - \frac{\tau_{bl}^{o}}{\tau_{bl}} = 1 - \frac{\tau}{\tau^{o}} = E \tag{100}$$

where τ_{bl}^{o} and τ_{bl} are the bleaching time constants measured from the single-labeled and double-labeled samples, respectively.

As another possibility the energy transfer efficiency is routinely calculated also from the decrease of quantum yield of the donor due to energy transfer process induced by the acceptor situated in the immediate proximity of the donor (see Equation 16),

$$E = \frac{Q^{o} - Q}{Q^{o}} = 1 - \frac{Q}{Q^{o}} \tag{101}$$

where Q^o and Q denote quantum yields of the donor in the absence and presence of acceptor, respectively.

Quantum efficiency data can be derived from fluorescence intensities.

$$I_d = \gamma k_f D^{*} \tag{102}$$

where the constant γ incorporates all instrumental and spectroscopic factors, such as intensity of the excitation light, solid angle and response of the detector, transmission of elements of the output optics, and electronic gain. Using Equation 95 we obtain

$$I_d = \gamma k_f \beta D \tag{103}$$

and, accordingly, for the prebleach donor intensity

$$I_{d,o} = \gamma k_f \beta D_o \tag{104}$$

In order to eliminate D_o from this expression one needs the analytical form of the integrated donor emission obtained during a complete bleaching.

The rate of the donor depletion (that of bleaching) is determined by

$$-\frac{dD}{dt} = k_{bl} D^* = k_{bl} \beta D \tag{105}$$

as it was mentioned before since deexcitation and transfer processes do not result in a loss of fluorescent donor molecules. Derivation of an explicit expression for the invariant integrated donor emission is straightforward. Using Equations 97 and 103 we get

$$I_{d,int} = \int_0^\infty I_d dt = \int_0^\infty \gamma k_f \beta D dt = \gamma k_f \beta D_o \int_0^\infty e^{-\beta k_{bl} t} dt = \frac{\gamma k_f D_o}{k_{bl}} \tag{106}$$

Denoting the ratio of prebleach intensity of the donor to the integrated fluorescence intensity by ϕ and using Equations 12, 95, 96, 104, and 106,

$$\Phi = \frac{I_{d,o}}{I_{d,int}} = \beta k_{bl} = \frac{k_{ex}}{k_d + k_f + k_T} k_{bl} = \frac{k_f}{k_d + k_f + k_T} \frac{k_{ex} k_{bl}}{k_f} = Q \frac{k_{ex} k_{bl}}{k_f} \tag{107}$$

Similarly, for the single labeled case one can write

$$\Phi^o = Q^o \frac{k_{ex} k_{bl}}{k_f} \tag{108}$$

With Equations 107 and 108 transfer efficiency can be obtained according to Equation 16

$$E = 1 - \frac{\Phi}{\Phi^o} \tag{109}$$

A. EXPERIMENTAL ASPECTS OF FRET MEASUREMENTS BY PHOTOBLEACHING

Application of the method of photobleaching to obtain FRET efficiency data offers two different possibilities according to Equations 100 and 109. The first type of experiment requires determination of the kinetics of photobleaching[70,97] and like many other kinetic methods it possesses several advantages. Assuming monoexponential bleaching (equivalent to homogenous population in terms of donor-acceptor relative distribution) there is no need to follow the entire bleaching process, since time constant of an exponential decay can be analyzed from any part of the complete curve using appropriate fitting routines. This can be instrumental in the case of a limited dynamic range of the photodetector precluding bleaching beams of very high irradiance. An alternative possibility can be the attenuation of the emitted fluorescence by neutral density filters. One can also apply successive bleaching pulses of high power for shorter time and monitor unbleached donor emission intensity in between using low excitation intensity.

The second type of experiment involves emission intensity recording (followed by integration) during the whole bleaching process. High bleaching intensity is necessary to avoid lengthy measurements (large bleaching time constant) which in turn involves a high level of fluorescence. To avoid saturation of the detector the emission intensity has to be attenuated. The prebleach fluorescence has to be recorded before the bleaching process using reduced excitation intensity. A similar measurement after the bleaching can report on the completeness of the process.

The microscope is the ideal instrument for these type of bleaching experiments. Measurements can be carried out by monitoring a single cell without spacial resolution.[100-102] The viewing field of the instrument has to be defined by an aperture of appropriate size to keep background interference to minimum. Unfortunately, fitting the size (and shape) of the aperture to that of the cell under study can be achieved only approximately. This shortcoming, however, can be bypassed by digital imaging using a CCD camera.

Scanning can also be achieved by moving the microscope stage, although this solution is time consuming; it results in inconveniences in stage control and relatively larger uncertainties in the localization due to the moving mechanical mechanism of the arrangement. Spatial resolution allows to investigate either more cells of the same field of the microscope or to carry out FRET measurements on a pixel-by-pixel basis with a possibility to explore differences between particular cellular components and subcellular assemblies. Possible applications of photobleaching FRET digital imaging microscopy include studies of receptor occupancy, aggregation, and receptor processing; mobility of lipid domains; trafficking of macromolecules between subcellular compartments; chromatin assembly, condensation, and redistribution throughout the cell cycle; and mapping of gene sequences, etc.[97] or binding equilibrium,[101] receptor or binding site topology,[102] and conformational changes of cell surface antigens.[103]

Photobleaching energy transfer measurements can be carried out using different light intensities for the photodestruction process. With relatively low intensities the decay of the fluorescence intensity was monoexponential in all the experimentally studied systems;[70,97,101] however, applying an ACAS 570 stage scanning laser microscope[100,102,103] with a power density of 1.4×10^3 W/cm^2 in the plane of the specimen (a power density exceeding that used by Jovin et al.[70,97,101] by an order of magnitude) the bleaching kinetics was double exponential:

$$I = A + Be^{-\frac{t}{T_1}} + Ce^{-\frac{t}{T_2}} \quad \textbf{(110)}$$

where A is background fluorescence and B and C are amplitudes of the two components decaying with time constants T_1 and T_2, respectively. Although systematic investigations have not been carried out, it seems probable that the number of components in the photobleaching kinetics depends on the applied light intensity. Szabó et al. observed that when the laser power was decreased the B/C ratio increased considerably, i.e., the bleaching process was approximating monoexponential kinetics.[100,104] They suggest that the double exponentiality of the photobleaching kinetics may be explained by a significant number of triplet state molecules present throughout the illumination. The rate of transitions from excited singulet to excited triplet state depends on the population of the excited singulet energy level and the emerging triplet state species are then quenched by diffusional quencher molecules or radicals. This argument is supported by the finding that parameters describing the double exponential decay change when samples were deaerated before experiments.

Reliability of E values acquired by photobleaching techniques depends on constancy of several critical parameters. Although both methods (based on determinations of either ϕ and τ_{bl} values) are "ratio methods" in which one can avoid the measurement or calculation of the exact numerical value of several parameters, the stability of the system during the measurements is essential. When calculating the ratio of bleaching time constants for two separate samples, the local environmental factors determining the actual rate of photobleaching have to be identical. Any change in temperature, oxygen concentration, or concentration of other reactants playing a role in the photodestruction process will inevitably alter k_{bl}. With more complete bleaching curves (comprised of 2000 points) measured within 100 ms[100,102,103] instrumental drift is of smaller significance and determination of prebleach intensities can be avoided by extrapolation of the measured fluorescence readings to the beginning of the bleaching, assuming accurate timing of the measurement.

Microscopy providing spatial resolution offers remarkable advantages over investigations of cell suspension by spectrophotometric methods or flow systems. The studied objects can be characterized from different aspects, using specific staining protocols. Although principally it is possible also with flow cytometry, microscopy allows sequential administration of dyes and reagents while monitoring the same cell. It also differentiates between locations of the cell in terms of specific distances or structural features. This can be achieved either on the surface of interacting cells or inside the same cell between different intracellular sites.

VIII. FLUORESCENCE RESONANCE ENERGY TRANSFER BETWEEN NONASSOCIATED ANTIGENS RANDOMLY DISTRIBUTED ON THE CELL SURFACE

For dipolar interactions the rate of energy transfer from a donor to an acceptor molecule depends on the separation distance, relative orientation of the dipole moments of the fluorophores, and spectral characteristics of the dyes as it was outlined in Section II.B (Equations 6 to 10). The process brings about a change in the donor singulet lifetime, quantum yield (emission intensity), as well as in the fluorescence intensity of the acceptor. In a purified homogeneous system with identical donor-to-acceptor separation distances, Equations 15 to 18 describe the energy transfer efficiency in terms of the observed changes in the mentioned parameters. The donor-acceptor separation distance can also be calculated (Equation 14).

Cells labeled with donor- and acceptor-conjugated antibodies or lipid analogues do not fulfill the requirements for a homogeneous system, i.e., specifically located single energy donors transferring excitation energy to single (within transfer distance) acceptor molecules at another specific location. In these cases the rate of transfer is the sum of k_T (Equation 6) for all donor acceptor pairs. For systems with randomly distributed multiple donors and acceptors appropriate calculations have to be carried out to allow the evaluation of the transfer efficiency in terms of donor-acceptor distribution parameters.

The theory of energy transfer between donors and acceptors distributed on surfaces is based on the work by Förster.[105] For the three-dimensional case Förster developed a theory with a relatively simple ensemble-averaged formula. Although his formula describing donor fluorescence decay is approximate, it holds for low acceptor concentrations and for times long compared to the most rapid transfer possible (k_T^{-1} for the nearest neighbor donor-acceptor pair with transition dipoles parallel to each other and perpendicular to the straight line connecting the fluorophores).

Fung and Stryer[106] derived equations for the two-dimensional case (infinite plane) similar to those by Förster. They introduced a parameter, z, giving the distance of closest approach of donor and acceptor (in Förster theory the influence of the intermolecular separation was neglected) to approach the real relative distribution of donor and acceptor fluorophores conjugated to macromolecules of finite size. Their formula for the fluorescence intensity, F(t), of the donor in such a system at time t after a brief light flash is given:

$$F(t) = F(0)e^{-\frac{t}{\tau^0}}e^{-qS(t)} \tag{111}$$

with

$$S(t) = \int_z^{\infty}\left[1 - e^{\left(\frac{t}{\tau^0}\right)\left(\frac{R_0}{r}\right)^6}\right]2\pi r dr \tag{112}$$

where τ^0 is the excited state lifetime of the donor in the absence of acceptor and q stands for the surface density of the acceptor. For the energy transfer efficiency they obtained

$$E = 1 - \frac{1}{\tau_0}\int_0^{\infty}\frac{F(t)}{F(0)}dt \tag{113}$$

E is independent of the surface density of the donor and depends only on R_0, q, and z. Numerical integration of Equation 111 showed that with increasing surface density of energy acceptor the average lifetime of the donor decreases and its fluorescence decay deviates increasingly from the monoexponential pattern. The emission kinetics of donor in a pure homogeneous system is single exponential while that of an ensemble is not, because donor molecules are in environments of different acceptor distribution.

Wolber and Hudson[107] presented an analytic solution of the Förster energy transfer problem in two dimensions for randomly distributed donors and acceptors. They derived an expression for the relative quantum yield of the donor:

$$\frac{Q}{Q^o} = \sum_{j=1}^{\infty}\frac{(-4.25409C)^j}{j!}\Gamma\left(\frac{j}{3}+1\right) \tag{114}$$

where Q and Q^0 denote donor quantum yields in the presence and absence of the acceptor, respectively, and C is the concentration of acceptors per R_0^2 and

$$\Gamma(x) = \int_0^\infty y^{x-1} e^{-y} dy \tag{115}$$

In accord with the finding by Fung and Stryer[106] Equation 114 also predicts the increase of transfer efficiency with higher acceptor density. The analytic solution was validated by comparing it with numerical simulations. Simple numerical approximations were also given to reduce labor-consuming computations.

The effect of a minimal intermolecular distance can be taken into account also by Monte-Carlo calculations.[108,109] The results suggest that this method can also be used to examine the effects of distinct lattice structures, e.g., phase-separated protein domains. There are other models using various approximations for very long or very short closest approach of acceptor and donor molecules,[107,110] for parallel infinite planes, and for the surfaces of concentric spheres.[111-115] Kellerer and Blumen[116] presented exact expressions for the ensemble-averaged decay of donor fluorescence due to the energy transfer to energy acceptors randomly distributed on a surface. They took into account the randomness in the positions and also in the orientations of the acceptor molecules. These authors showed that in special arrangements the effect of anisotropic interactions may be used in gaining insight into the microscopic structure. They considered several cases explicitly: random orientation of the acceptors in space, with the donor being perpendicular to the plane, or in the plane as well as randomly oriented donor or both donor and acceptor in the space. For all cases they evaluated analytic, Förster-like expressions. These formulas are valid for long times and low acceptor densities. Their ensemble-averaged decay law is exact[116,117] and is expressed in a closed form. The expression holds for any concentration of acceptors and for all times and allows the derivation of a whole series of approximate formulas.

In the majority of investigated cases fluorescence energy transfer was used to study distances between stationary donors and acceptors. In such a system only donors with at least one acceptor within a distance of about R_0 will have an appreciable probability of energy transfer. Energy transfer can be substantially enhanced by diffusion if distances covered by translational motions are similar to R_0 and the average distance between donor and acceptor (s) during the excited state lifetime of the donor.[118,119] The critical parameter is $D\tau_0/s^2$ (D is the sum of the diffusional constants of donor and acceptor).

Three cases can be distinguished. In the static limit characterized by $D\tau_0/s^2 \ll 1$ transfer rate is not affected by diffusion at all. In the intermediate range with $D\tau_0/s^2$ close to 1 the energy transfer rate can be used to monitor diffusional motion. In the rapid diffusion limit with $D\tau_0/s^2 \gg 1$ the transfer rate is highly responsive to the closest approach of donor and acceptor.

Calculations using the theory of Steinberg and Katchalski[120] show that the fast diffusion limit cannot be attained for small molecules in aqueous medium with donors having excited-state lifetime in the nanosecond range. Using long-lived energy donors with lifetimes in the millisecond range allows the investigation of energy transfer processes at extremely low acceptor concentrations. Closest approach distances up to greater than R_0 can be determined under these conditions. Dipole-dipole coupling is the dominant transfer mechanism until the donor and acceptor molecules are nearly in contact. At a separation of 1 nm or below exchange interactions become important. With donors having millisecond lifetimes, translational motions corresponding to D from 10^{-6} to 10^{-10} cm^2/s can be monitored, a range very important in membrane studies.

REFERENCES

1. **Travers, A. A. and Burgess, R. R.,** Cyclic reuse of the RNA polymerase sigma factor, *Nature*, 222, 537, 1969.
2. **Coleman, L.,** Membrane-bound enzymes and membrane ultrastructure, *Biochim. Biophys. Acta*, 300, 1, 1973.
3. **Welch, G. R.,** *Current Topics in Regulation*, Vol. 16, Horecker, B. L. and Stadtman, E. B., Eds., Academic Press, New York, 1978.
4. **Welch, G. R.,** On the role of organized multienzyme systems in cellular metabolism: a general synthesis, *Prog. Biophys. Mol. Biol.*, 32, 103, 1977.

5. **Saito, Y., Tada, H., Sabe, H., and Honjo, T.,** Biochemical evidence for a third chain of the interleukin-2 receptor, *J. Biol. Chem.*, 266, 22186, 1991.
6. **Saito, Y. and Honjo, T.,** Why are multiple chains required for the interleukin-2 receptor?, *Prog. Growth Factor Res.*, 2, 207, 1990.
7. **Honjo, T.,** Shared partners in receptors, *Curr. Biol.*, 1, 201, 1991.
8. **Hatakeyama, M., Tsuda, M., Minamoto, S., Kono, T., Doi, T., Miyata, T., Miyasaka, M., and Taniguchi, T.,** Interleukin-2 receptor β-chain gene: generation of three receptor forms by cloned human α- and β-chain cDNA's, *Science*, 244, 551, 1987.
9. **Berridge, M. J. and Irvine, R. F.,** Inositol phosphates and cell signalling, *Nature*, 341, 197, 1989.
10. **Damjanovich, S., Szöllősi, J., and Trón, L.,** Transmembrane signalling in T cells, *Immunol. Today*, 13, A12, 1992.
11. **Schlessinger, J., Shechter, J., Willingham, M. C., and Pastan, I.,** Direct visualization of binding, aggregation and internalization of insulin and epidermal growth factor on living fibroblastic cells, *Proc. Natl. Acad. Sci. U.S.A.*, 75, 2659, 1978.
12. **Maxfield, F. R., Willingham, M. C., Davies, P. J. A., and Pastan, I.,** Amines inhibit the clustering of α_2-macroglobulin and EGF on the fibroblast cell surface, *Nature*, 277, 661, 1979.
13. **Maxfield, F. R., Schlessinger, J., Shechter, J., Pastan, I., and Willingham, M. C.,** Collection of insulin, EGF and α_2-macroglobulin in the same patches on the surface of cultured fibroblasts and common internalization, *Cell*, 14, 805, 1978.
14. **Schlessinger, J.,** Receptor aggregation as a mechanism for transmembrane signalling: models for hormone action, in *Physical Chemical Aspects of Cell Surface Events in Cellular Regulation*, Vol. 3, DeLisi, C. and Blumenthal, F., Eds., Elsevier, New York, 1979, 89.
15. **Levi, A., Shechter, J., Neufeld, E. J., and Schlessinger, J.,** Mobility, clustering and transport of nerve growth factor in embryonal sensory cells and in a sympathetic neuronal cell line, *Proc. Natl. Acad. Sci. U.S.A.*, 77, 3469, 1980.
16. **Schlessinger, J.,** Signal transduction by allosteric receptor oligomerization, *Trends Biochem. Sci.*, 13, 443, 1988.
17. **Ullrich, A. and Schlessinger, J.,** Signal transduction by receptors with tyrosine kinase activity, *Cell*, 61, 203, 1990.
18. **Seifert, R. A., Hart, C. E., Phillips, P. E., Forstrom, J. W., Ross, R., Murray, M. J., and Bowen-Poppe, D. F.,** Two different subunits associate to create isoform-specific platelet-derived growth factor receptors, *J. Biol. Chem.*, 264, 8771, 1989.
19. **Mátyus, L. and Edidin, M.,** Flow cytometry and cell sorting, in *Topics in Fluorescence Spectroscopy, Vol. 1, Techniques*, Lakowicz, J. R., Ed., Plenum Press, New York, 1991, chap. 8.
20. **Jovin, T. M.,** Fluorescence polarization and energy transfer: theory and application, in *Flow Cytometry and Sorting*, Melamed, M. R., Mullaney, P. F., and Mendelsohn, M. L., Eds., John Wiley & Sons, New York, 1979, 137.
21. **Stryer, L.,** Fluorescence energy transfer as a spectroscopic ruler, *Annu. Rev. Biochem.*, 47, 819, 1978.
22. **Szöllősi, J., Damjanovich, S., Mulhern, S. A., and Trón, L.,** Fluorescence energy transfer and membrane potential measurements monitor dynamic properties of cell membranes: a critical review, *Progr. Biophys. Mol. Biol.*, 49, 65, 1987.
23. **Matkó, J., Szöllősi, J., Trón, L., and Damjanovich, S.,** Luminescence spectroscopic approaches in studying cell surface dynamics, *Q. Rev. Biophys.*, 21, 479, 1988.
24. **Cantor, C. R. and Schimmel, P. R.,** *Biophysical Chemistry*, W. H. Freeman, San Francisco, 1980.
25. **Cundall, R. B. and Dale, R. E.,** Time resolved fluorescence spectroscopy in *Biochemistry and Biology*, Plenum Press, New York, 1983.
26. **Lakowicz, J. R.,** *Principles of Fluorescence Spectroscopy*, Plenum Press, New York, 1983.
27. **Lakowicz, J. R.,** *Topics in Fluorescence Spectroscopy, Vol. 1*, Plenum Press, New York, 1991.
28. **Lakowicz, J. R.,** *Topics in Fluorescence Spectroscopy, Vol. 2*, Plenum Press, New York, 1991.
29. **Lakowicz, J. R.,** *Topics in Fluorescence Spectroscopy, Vol. 3*, Plenum Press, New York, 1992.
30. **Dewey, D. G.,** *Biophysical and Biochemical Aspects of Fluorescence Spectroscopy*, Dewey, D. G., Ed., Plenum Press, New York, 1991.
31. **Taylor, D. L., Waggoner, A. S., Lanni, F., Murphy, R. F., and Birge, R. R., Eds.,** *Application of Fluorescence in the Biomedical Sciences*, Alan R. Liss, New York, 1986.
32. **Melamed, M. R., Lindmo, T., and Mendelsohn, M. L., Eds,** *Flow Cytometry and Sorting*, 2nd ed., John Wiley & Sons, New York, 1990.

33. **Jablonski, A.**, Über den Mechanismus des Photolumineszenz von Farbstoffphosphoren, *Z. Phys.*, 94, 38, 1935.
34. **Förster, T. H.**, Intramolecular energy migration and fluorescence, *Ann. Phys.*, 2, 55, 1948.
35. **Förster, T. H.**, Mechanism of energy transfer, in *Comprehensive Biochemistry*, Vol. 22, Florkin, M. and Statz, E. H., Eds., Elsevier, New York, 1967, 61.
36. **Szöllősi, J., Mátyus, L., Trón, L., Balázs, M., Ember, I., Fulwyler, M. J., and Damjanovich, S.**, Flow cytometric measurements of fluorescence energy transfer using single laser excitation, *Cytometry*, 8, 120, 1987.
37. **Chan, S. S., Arndt-Jovin, D. D., and Jovin, T. M.**, Proximity of lectin receptors on the cell surface measured by fluorescence energy transfer in a flow system, *J. Histochem. Cytochem.*, 27, 56, 1979.
38. **Jovin, T. M. and Arndt-Jovin, D. D.**, Flow sorting on the basis of morphology and topology, in *Trends in Photobiology*, Helene, C., Charlier, M., Montenay-Garestier, Th., and Laustriat, G., Eds., Plenum Press, New York, 1982, 51.
39. **Trón, L., Szöllősi, J., Damjanovich, S., Helliwell, S. H., Arndt-Jovin, D. J., and Jovin, T. M.**, Flow cytometric measurement of fluorescence resonance energy transfer on cell surfaces. Quantitative evaluation of the transfer efficiency on a cell-by-cell basis, *Biophys. J.*, 45, 939, 1984.
40. **Szöllősi, J., Trón, L., Damjanovich, S., Helliwell, S. H., Arndt-Jovin, D. J., and Jovin, T. M.**, Fluorescence energy transfer measurements on cell surfaces. A critical comparison of steady-state fluorimetric and flow cytometric methods, *Cytometry*, 5, 210, 1984.
41. **Szöllősi, J., Trón, L., Damjanovich, S., Helliwell, S. H., Arndt-Jovin, D. J., and Jovin, T. M.**, Energy transfer measurements on the cell surface. Distribution of H-2K^k antigens, in *Quo Vadis? Flow Cytometry and Monoclonal Antibodies for Therapy Monitoring*, Gros, P., Jansen, F. K., Poncelet, P., and Roncucci, R., Eds., Sanofi Research, Montpellier, France, 1984, 33.
42. **Holtkamp, B., Lindahl, K. F., Segall, M., and Rajewsky, K.**, Spontaneous loss and subsequent stimulation of H-2 expression in clones of a heterozygous lymphoma cell line, *Immunogenetics*, 9, 405, 1979.
43. **Damjanovich, S., Trón, L., Szöllősi, J., Zidovetzki, R., Vaz, W. L. C., Regateiro, F., Arndt-Jovin, D. J., and Jovin, T. M.**, Distribution and mobility of murine histocompatibility H-2K^k antigen in the cytoplasmic membrane, *Proc. Natl. Acad. Sci. U.S.A.*, 80, 5985, 1983.
44. **Szöllősi, J., Damjanovich, S., Balázs, M., Nagy, P., Trón, L., Fulwyler, M. J., and Brodsky, F. M.**, Physical association between MHC class I and class II molecules in cytoplasmic membranes. A flow cytometric energy transfer study, *J. Immunol.*, 143, 208, 1989.
45. **Bevington, P. R.**, *Data Reduction and Error Analysis for the Physical Sciences*, McGraw-Hill, New York, 1969.
46. **Barenboim, G. M., Domanskii, A. N., and Turoverov**, *Luminescence of Biopolymers and Cells*, Plenum Press, New York, 1969, 64.
47. **Haitinger, M.**, *Fluoreszence Mikroskopie*, Akademie Verlag, Leipzig, 1959, 59.
48. **Pearse, A. G. E.**, *Histochemistry: Theoretical and Applied*, Churchill Livingstone, London, 1972, 1177.
49. **Udenfriend, S.**, *Fluorescence Assay in Biology and Medicine*, Academic Press, New York, 1969, 543.
50. **Aubin, J. A.**, Autofluorescence of viable cultured mammalian cells, *J. Histochem. Cytochem.*, 27, 36, 1979.
51. **Benson, R. C., Meyer, R. A., Zaruba, M. E., and McKhann, G. M.**, Cellular autofluorescence — is it due to flavins?, *J. Histochem. Cytochem.*, 27, 44, 1979.
52. **Roederer, M. and Murphy, R. F.**, Cell-by-cell autofluorescence correction for low signal-to-noise system: application to epidermal growth factor endocytosis by 3T3 fibroblast, *Cytometry*, 7, 558, 1986.
53. **Alberti, S., Parks, D. R., and Hercenberg, L. A.**, A single laser method for subtraction of cell autofluorescence in flow cytometry, *Cytometry*, 8, 114, 1987.
54. **Kavathas, P. and Herzenberg, L. A.**, Stable transformation of mouse L cells for human membrane T-cell differentiation antigens, HLA and β2-microglobulin: selection by fluorescence-activated cell sorting, *Proc. Natl. Acad. Sci. U.S.A.*, 80, 524, 1983.
55. **Steinkamp, J. A. and Stewart, C. C.**, Dual-laser, differential fluorescence correction method for reducing cellular background autofluorescence, *Cytometry*, 7, 566, 1986.
56. **Mátyus, L., Hochman, J. H., and Edidin, M.**, Correction of autofluorescence in flow cytometric energy transfer measurements, *Cytometry*, Suppl. 2, 90, 1988.

57. **Szöllősi, J., Kertai, P., Somogyi, B., and Damjanovich, S.**, Characterization of living normal and leukemic mouse lymphocytes by fluorescein diacetate, *J. Histochem. Cytochem.*, 29, 503, 1981.
58. **Badea, M. G. and Brand, L.**, Time-resolved fluorescence measurements, in *Methods in Enzymology*, Vol. 61, Hirs, C. H. W. and Timasheff, S. N., Eds., Academic Press, New York, 1979, 378.
59. **O'Connor, D. V. and Phillips, D.**, *Time-Correlated Single Photon Counting*, Academic Press, New York, 1984.
60. **Gratton, E., Jameson, D. M., and Hall, R. D.**, Multifrequency phase and modulation fluorimetry, *Annu. Rev. Biophys. Bioeng.*, 13, 105, 1984.
61. **Chang, M. C., Courtney, S. H., Cross, A. J., Gulotty, R. J., Petrich, J. W., and Fleming, G. R.**, Time correlated single photon counting with microchannel plate detectors, *Anal. Instrum.*, 14, 433, 1985.
62. **Visser, A. J. W. G.**, Time resolved fluorescence spectroscopy, *Anal. Instrum.*, 14, 193, 1985.
63. **Birch, D. J. S. and Imhof, R. E.**, Time-domain fluorescence spectroscopy using time-correlated single-photon counting, in *Topics in Fluorescence Spectroscopy, Vol. 1*, Lakowicz, J. R., Ed., Plenum Press, New York, 1991, chap. 1.
64. **Papp, S., Pikula, S., and Martonosi, A.**, Fluorescence energy transfer as an indicator of Ca^{2+}-ATPase interactions in sarcoplasmic reticulum, *Biophys. J.*, 51, 205, 1987.
65. **Kusumi, A., Tsuji, A., Murata, M., Sako, Y., Yoshizawa, A. C., Kagiwada, S., Hayakawa, T., and Ohnishi, S.**, Development of streak-camera-based time-resolved microscope fluorimeter and its application to studies of membrane fusion in single cells, *Biochemistry*, 30, 6517, 1991.
66. **Kohen, E. and Hirschberg, J. G.**, *Cell Structure and Function by Microspectrofluorimetry*, Academic Press, New York, 1989.
67. **Kasten, F. H.**, The origins of modern fluorescence microscopy and fluorescence probes, in *Cell Structure and Function by Microspectrofluorimetry*, Kohen, E. and Hirschberg, J. G., Eds., Academic Press, New York, 1989, chap. 3.
68. **Bright, G. R., Fischer, G. W., Rogowska, J., and Taylor, D. L.**, Fluorescence ratio imaging microscopy, in *Methods in Cell Biology*, Vol. 29, Taylor, D. L. and Wang, Y., Eds., Academic Press, New York, 1989.
69. **Hiraoka, Y., Sedat, J. W., and Agard, D. A.**, The use of charge coupled device for quantitative optical microscopy of biological structures, *Science*, 238, 36, 1987.
70. **Jovin, T. M. and Arndt-Jovin, A. D.**, Luminescence digital imaging microscopy, *Annu. Rev. Biophys. Biophys. Chem.*, 18, 271, 1989.
71. **Jovin, T. M., Arndt-Jovin, A. D., Mariott, G., Clegg, R. M., Robert-Nicoud, M., and Schorman, T.**, Distance, wavelength and time: the versatile 3rd dimension in light emission microscopy, in *Optical Microscopy for Biology*, Herman, B. and Jacobson, K., Eds., John Wiley & Sons, New York, 1990, 575.
72. **Wampler, J. E. and Kutz, K.**, Quantitative fluorescence microscopy using photomultiplier tubes and imaging detectors, in *Methods in Cell Biology*, Vol. 29, Taylor, D. L. and Wang, Y., Eds., Academic Press, New York, 1989, 239.
73. **Aikens, R. S., Agard, D. A., and Sedat, J. W.**, Solid-state imagers for microscopy, in *Methods in Cell Biology*, Vol. 29, Wang, Y. and Taylor, D. L., Eds., Academic Press, New York, 1989, 292.
74. **Mariott, G., Clegg, R. M., Arndt-Jovin, A. D., and Jovin, T. M.**, Time resolved imaging microscopy. Phosphorescence and delayed fluorescence imaging, *Biophys. J.*, 60, 1374, 1991.
75. **Dale, R. E., Novros, J., Roth, S., Edidin, M., and Brand, L.**, Application of long-range excitation energy transfer to the determination of distributions of fluorescently-labeled concanavalin A receptor complexes at the surfaces of yeast and of normal and malignant fibroblasts, in *Fluorescent Probes*, Beddard, G. S. and West, M. A., Eds., Academic Press, New York, 1981, 159.
76. **Mátyus, L.**, Fluorescence energy transfer measurements on cell surfaces. A spectroscopic tool for determining protein interactions, *J. Photochem. Photobiol. B*, 12, 323, 1992.
77. **Damjanovich, S., Trón, L., Szöllősi, J., Mátyus, L., and Szabó, G., Jr.**, Dynamic properties of the murine histocompatibility H-$2K^k$ antigen in the cytoplasmic membrane, *Mol. Immunol.*, 21, 1151, 1984.
78. **Trón, L., Szöllősi, J., Szabó, G., Jr., Mátyus, L., and Damjanovich, S.**, Cell surface dynamics and distance relationship of integral membrane proteins, in *Membrane Dynamics and Transport of Normal and Tumor Cells, Symposia Biologica Hungarica 1984*, Trón, L., Damjanovich, S., Fonyó, A., and Somogyi, J., Eds., Academic Press, Budapest, 1984, 307.

79. **Trón, L., Szöllősi, J., and Damjanovich, S.,** New trends in studying slow diffusion processes of the cytoplasmic membrane components, *Acta Biochem. Biophys. Acad. Sci. Hung.*, 19, 247, 1984.
80. **Trón, L., Damjanovich, S., Aszalós, A., Szöllősi, J., Mulhern S. A., and Fulwyler, M. J.,** On the role of cell surface dynamics and transmembrane information transfer: cyclosporin A changes physical properties of cell membranes, in *Dynamics of Biochemical Systems, Symposia Biologica Hungarica*, Damjanovich, S., Keleti, T., and Trón, L., Eds., Elsevier, Amsterdam, 1986, 417.
81. **Szöllősi, J., Damjanovich, S., Goldman, C. K., Fulwyler, M. J., Aszalós, A., Goldstein, G., and Waldman, T. A.,** Flow cytometric resonance energy transfer measurements support the association of a 95-kD peptide termed T27 with the 55 kD Tac peptide, *Proc. Natl. Acad. Sci. U.S.A.*, 84, 7246, 1987.
82. **Trón, L., Szöllősi, J., and Damjanovich, S.,** Proximity measurements of cell surface proteins by fluorescence energy transfer, *Immun. Lett.*, 16, 1, 1987.
83. **Trón, L., Aszalós, A., Balázs, M., Mulhern, S. A., Szöllősi, J., and Damjanovich, S.,** On the biophysics of transmembrane signalling, *Mol. Immunol.*, 25, 1075, 1988.
84. **Szöllősi, J., Damjanovich, S., Balázs, M., Nagy, P., Trón, L., Fulwyler, M. J., and Brodsky, F. M.,** Physical association between MHC class I and class II molecules in cytoplasmic membranes. A flow cytometric energy transfer study, *J. Immunol.*, 143, 208, 1989.
85. **Balázs, M., Mátyus, L., Szöllősi, J., Trón, L., and Damjanovich, S.,** A biophysical approach of cell surface phenomena and transmembrane signalling, in *The Physical Aspects of the Living Cell, Proceedings of the Eugene (Jenö) Ernst Memorial Symposium*, Tigyi, J., Hazlewood, C. F., and Kellermayer, M., Eds., Akademiai Kiado, Budapest, 1991, 271.
86. **Liegler, T., Szöllősi, J., Hyun, W., and Goodenow, R. S.,** Proximity measurements between H-2 antigens and the insulin receptor by fluorescence energy transfer: evidence that a close association does not influence insulin binding, *Proc. Natl. Acad. Sci. U.S.A.*, 88, 6755, 1991.
87. **Mátyus, L., Takács, L., Pohubi, L., Balázs, M., Szöllősi, J., Matkó, J., Trón, L., and Damjanovich, S.,** Dynamic behaviour of cell surface receptors as revealed by laser excited fluorescence spectroscopy, in *Light in Biology and Medicine*, Vol. 2, Douglas, R. H., Moan, J., and Rontó, Gy., Eds., Plenum Press, New York, 1991, 383.
88. **Chakrabarti, A., Matkó, J., Rahaman, N., Barisas, G. B., and Edidin, M.,** Self association of class I major histocompatibility complex (MHC) molecules in liposomes, *Biochemistry*, submitted.
89. **Chuck, R. S., Cantor, C. R., and Tse, D. B.,** CD4-T-cell antigen receptor complexes on human leukemia T cells, *Proc. Natl. Acad. Sci. U.S.A.*, 87, 5021, 1990.
90. **Szöllősi, J., Ember, I., Trón, L., and Damjanovich, S.,** A follow up study of the progression or spontaneous regression of leukemia in mouse strains, *Cytometry*, 3, 110, 1982.
91. **Fernandez, S. M. and Berlin, R. D.,** Cell surface distribution of lectin receptors determined by resonance energy transfer, *Nature*, 264, 411, 1976.
92. **Herman, B. A. and Fernandez, S. M.,** Dynamics and topographical distribution of surface glycoproteins during myoblast fusion: a resonance energy transfer study, *Biochemistry*, 21, 3275, 1982.
93. **Nakache, M., Gaub, H. E., Schreiber, A. B., and McConnell, H. M.,** Topological and modulated distribution of surface markers on endothelial cells, *Proc. Natl. Acad. Sci. U.S.A.*, 83, 2874, 1986.
94. **Uster, P. S. and Pagano, R. E.,** Resonance energy transfer microscopy: observations of membrane-bound fluorescent probes in model membranes and in living cells, *J. Cell. Biol.*, 103, 1221, 1986.
95. **Watts, T. H. and McConnell, H. M.,** High-affinity fluorescent peptide binding to I-A^d in lipid membranes, *Proc. Natl. Acad. Sci. U.S.A.*, 83, 9660, 1986.
96. **Watts, T. H., Gaub, H. E., and McConnell, H. M.,** T-cell-mediated association of peptide antigen and major histocompatibility complex protein detected by energy transfer in an evanescent wave-field, *Nature*, 320, 179, 1986.
97. **Jovin, T. M. and Arndt-Jovin, D. J.,** FRET microscopy: digital imaging of fluorescence resonance energy transfer. Application in cell biology, in *Microspectrofluorometry of Single Living Cells*, Kohen, E., Ploem, J. S., and Hirschberg, J. G., Eds., Academic Press, Orlando, FL, 1989, 99.
98. **Peters, R., Peters, J., Tews, K. H., and Bahr, W.,** A microfluorimetric study of translational diffusion in erythrocyte membranes, *Biochim. Biophys. Acta*, 367, 282, 1974.
99. **Hirschfeld, T.,** Quantum efficiency independence of the time integrated emission from a fluorescent molecule, *Appl. Optics*, 15, 3135, 1976.
100. **Szabó, G., Jr., Pine, P. S., Weaver, J. L., Kasari, M., and Aszalos, A.,** Epitope mapping by photobleaching fluorescence resonance energy transfer measurements using a laser scanning microscope system, *Biophys. J.*, 61, 661, 1992.

101. **Kubitscheck, U., Kircheis, M., Schwitzer-Stenner, R., Dreybrodt, W., Jovin, T. M., and Pecht, I.**, Fluorescence resonance energy transfer on single living cells. Application to binding of monovalent haptens to cell-bound immunoglobulin E, *Biophys. J.*, 60, 307, 1991.
102. **Szabó, G., Jr., Weaver, J. L., Pine, P. S., and Aszalós, A.**, Specific disengagement of cell-bound αLAM-1 antibodies by aurintricarboxylic acid, *J. Immunol.*, submitted.
103. **Szabó, G., Jr., Pine, P. S., Weaver, J. L., Rao, P. E., and Aszalós, A.**, CD4 changes conformation upon ligand binding, *J. Immunol.*, submitted.
104. **Szabó, G., Jr.**, personal communication, 1992.
105. **Förster, T.**, Experimentelle und theoretische Untersuchung des zwischenmolekularen Übergangs von Elektronenanregungsenergie, *Z. Naturforsch. A*, 4, 321, 1949.
106. **Fung, B. K. K. and Stryer L.**, Surface density determination in membranes by fluorescence energy transfer, *Biochemistry*, 17, 5241, 1978.
107. **Wolber, P. K. and Hudson, B. S.**, An analytic solution to the Förster energy transfer problem in two dimensions, *Biophys. J.*, 28, 197, 1979.
108. **Freire, E. and Snyder, B.**, Quantitative characterization of the lateral distribution of membrane proteins within the lipid bilayer, *Biophys. J.*, 37, 617, 1982.
109. **Snyder, B. and Freire, E.**, Fluorescence energy transfer in two dimensions. A numeric solution for random and nonrandom distributions, *Biophys. J.*, 40, 137, 1982.
110. **Shaklai, N., Yguerabide, J., and Ranney, H. H.**, Interaction of hemoglobin with red blood cell membranes as shown by a fluorescent chromophore, *Biochemistry*, 16, 5585, 1977.
111. **Dewey, T. G. and Hammes, G. G.**, Calculation of fluorescence resonance energy transfer on surfaces, *Biophys. J.*, 32, 1023, 1980.
112. **Kampman L.**, Application of radiationless energy transfer for distance measurements across membranes. Transition probabilities for radiationless energy transfer within several assemblies of donors and acceptors at stationary distances, *Biophys. Struct. Mech.*, 3, 239, 1977.
113. **Estep, T. N. and Thompson, T. E.**, Energy transfer in lipid bilayers, *Biophys. J.*, 26, 195, 1979.
114. **Koppel, D. E., Fleming, P. J., and Strittmatter, P.**, Intramembrane positions of membrane-bound chromophores determined by excitation energy transfer, *Biochemistry*, 18, 5450, 1979.
115. **Doody, M. C., Sklar, R. A., Pownall, H. J., Sparrow, J. T., Gotto, A. M., Jr., and Smith, L. C.**, A simplified approach to resonance energy transfer in membranes, lipoproteins and spatially restricted systems, *Biophys. Chem.*, 17, 139, 1983.
116. **Kellerer, H. and Blumen, A.**, Anisotropic excitation transfer to acceptors randomly distributed on surfaces, *Biophys. J.*, 46, 1, 1984.
117. **Blumen, A.**, Excitation transfer from a donor to acceptors in condensed media, *Nuovo Cimento B*, 63, 50, 1981.
118. **Thomas, D. D., Carlsen, W. F., and Stryer, L.**, Fluorescent energy transfer in the rapid diffusion limit, *Proc. Natl. Acad. Sci. U.S.A.*, 75, 5746, 1978.
119. **Stryer, L., Thomas, D. D., and Meares, C. F.**, Diffusion-enhanced fluorescence energy transfer, *Annu. Rev. Biophys. Bioeng.*, 11, 203, 1982.
120. **Steinberg, I. Z. and Katchalski**, Theoretical analysis and the role of diffusion in chemical reaction, fluorescence quenching, and nonradiative energy transfer, *J. Chem. Phys.*, 48, 2404, 1968.

Chapter 2

Mapping of Membrane Structures by Energy Transfer Measurements

János Szöllősi and Sándor Damjanovich

CONTENTS

I. INTRODUCTION

The work of Theodor Förster established the potentialities of dipole-dipole interactions between fluorescent molecules more than 42 years ago. The distance between two closely localized dipoles influences their interaction proportionally to its negative third power. The ingenious discovery that a fluorescence dipole-dipole interaction, besides orientational and other spectroscopic parameters, which can be kept under control, depends on the negative sixth power of their distance provided one of the most sensitive methods to measure molecular or atomic distance relations at the nanometer level. The utilization of this possibility in chemistry and biochemistry had a pinnacle in the 1970s. Cell biological application also started in the 1970s, but the real widespread application began only a decade later and is still flourishing. This chapter provides a brief account on this part of proximity measurements. A special emphasis is on the flow cytometric fluorescence energy transfer, since this is the only method so far that can reveal the

0-8493-4931-1/94/$0.00+$.50

biological variations in cell populations. The focus is on the membrane surface, which is understandable because of its easy accessibility and the high number of available target-specific carrier molecules such as monoclonal antibodies. Nevertheless, this account on the state of art is not entirely restricted to membrane protein elements. Lipid moieties, or lipid merging dyes, monitoring proximities in the depth of the membrane are also discussed. The previous chapter provided a concise treatment of the methodology behind the measurements. Thus, in this chapter, we tried to avoid methodological discussions and focused our interest on liposomes, specialized receptor systems of sensory organs, major receptor elements of the muscle and nerve receptor families, and cytoskeletal and blood clotting factors. Data obtained with erythrocyte membranes are also included, but we do not make efforts to hide that investigating the topographical distribution of major plasma membrane elements of the immune system, monitored on living cells, is closest to our heart. The proximity of molecules has an overwhelming dominance in the accessibility by functionally specific ligands and the receptor site forming ability of subunits. Some of them, like lectins, do not differentiate between protein binding sites but have specificity for certain sugar-moieties, thereby identifying a family of molecules. The immune receptors in the plasma membrane of the effector cells, leukocytes, and particularly lymphocytes have an enormous complexity, which has a close correlation to their functions and interactions. The cell activation processes are closely linked to transmembrane signal transducing pathways. The pathways are selected by specific receptor systems, which have a pattern of nonrandom codistribution with other molecules, which do not bind ligands, but are essential to process the signals. The interaction of the major histocompatibility system (MHC) with the lymphokine receptors is still an enigmatic phenomenon, although we know that it is indispensable for the intercellular recognition processes. The colocalization of intercellular adhesion molecules with activation triggering receptor systems was a surprise in our measurements and has yet to be understood. Nevertheless, there can be little doubt about the notion that this may be connected to metastasis problems or basic phenomena of inflammation processes at tissue level. The nonrandom codistributions, pattern formations of receptor assemblies, and other protein molecules may have important biological consequences and give orientation even to drug design. Decoding the physiological and pathological meaning of receptor and antigen patterns will very likely contribute to the final understanding of transmembrane signaling, intercellular communication and of the interaction between cell interior, and the theater of the communication between the exterior and interior of every cell, i.e., the function of the cell membrane. Besides the potentially great practical importance of the two-dimensional structuredness of the membrane surface, it can be recognized as an encoded message that reflects the metabolic state of the cells. Thus, we believe that those measurements we discuss here, and also those which may be initiated by challenging the reader, can provide a field of investigation for interested scientists with a prospective utilization for the benefit of a better life.

II. FRET MEASUREMENTS ON LIPOSOMES

Liposomes serve as a good model system in investigation of the structure of membranes and membrane proteins. In this artificial membrane system the complex effect of cytoplasm on membranes is completely missing and the interpretation of experimental data is much easier. The disadvantage of this simplification is that the influence of the cytoskeleton on membrane protein organization cannot be studied in this type of model systems. Nevertheless, the experiments performed on model systems provide very useful information about the organization of membrane lipids and proteins. Application of FRET in model membrane systems provided very useful information about various aspects of membrane structures. The problems investigated with this technique include lipid fusion phenomena, lipid structures, location of lipid constituents and various drugs in lipid bilayer, structure of membrane proteins, and the organization of erythrocyte membrane.

A. FUSION OF LIPOSOMES

Membrane fusion constitutes an essential, intermediate step in numerous cell biological processes, occurring, for example, during endocytosis, oocyte fertilization, muscle embryogenesis, membrane recycling, and exocytosis. Also, less desirable events, such as the infection of cells by viruses, are mediated by membrane fusion. In this case the viral envelope merges with the cellular membrane causing the expulsion of the viral nucleocapsid into the cytoplasm of the cell as an initial step in virus replication. Much of our current knowledge concerning the mechanism of membrane fusion had been derived from studies using simple artificial membranes such as liposomes or phospholipid vesicles as model systems.

Various assays have been developed for monitoring liposome fusion. In one approach, the result of the membrane fusion is detected in the aqueous phase. In this case, the assay utilizes the mixing of two aqueous probes entrapped in separate liposomes. Wilschut et al.[1] developed an assay where the aqueous phase probes, Tb^{3+} and dipicolinic acid, formed a fluorescent complex upon internal compartment mixing. In the other approach, the mixing of membrane markers is used to determine that the fusion has occurred. In this case, the mixing of membrane components during vesicle fusion can be monitored by differential scanning calorimetry,[2] by electron spin resonance,[3] by freeze-fracture,[4] or by modulation of enzymatic activity of integral membrane proteins.[5]

One of the most useful techniques for monitoring mixing of membrane components is based on the application of FRET. This method is very sensitive and capable of registering membrane fusion in a continuous fashion. For monitoring certain aspects of membrane fusion, mixing of fluorescent lipid probes can be a more practical subject of measurements than that of aqueous contents, as lipid mixing assays are not complicated by the effects of membrane leakiness, and they do not require prior manipulation of the internal aqueous compartments bounded by both membranes.

1. FRET as a Fusion Assay

The first fusion assay system that utilized FRET between two fatty acid probes was developed by Keller et al.[6] It employed a donor fluorescent probe incorporated into the membrane of one population of cells and an acceptor fluorescent probe incorporated into the membrane of another population. Then the cells were infected with a virus that caused fusion, mixed in equal proportions, and the fluorescence of the acceptor probe measured as a function of time after infection. Following the fusion there was an increase in the acceptor probe fluorescence due to FRET from the donor probe. The fluorescent probes used in this assay were synthesized by covalent linkage of fluorescein (donor) or rhodamine (acceptor) to a saturated hydrocarbon chain, 18 carbon atoms long. The assay system was successfully applied for monitoring the time-course of HEL cell fusion induced by the herpes simplex virus type I. In order to determine how the increase of fluorescence depended upon the distance separating the probe molecules, the investigators varied the relative number of probe molecules per cell. Since the probes are localized in planar membrane structures, it was expected that their interactions would follow the kinetics of a two-dimensional array. The experimental data, however, showed the best fit with kinetics derived for three-dimensional interactions. One possible explanation is that the probes are spread throughout all of the cellular membranes and they are present on both sides of a given membrane and thus energy transfer may occur between opposite sides of the bilayer. According to the other interpretation, the three-dimensional effect could result from a large contribution from those probes located in intracellular membrane structures where many layers of closely packed membrane are prominent.[6]

This type of fusion assay utilizing FRET has been applied not only for cells[6] but also for liposomes.[7-9] Vanderwerf and Ullman employed *N*-(5-dimethylaminonaphthalene-1-sulfonyl)dipalmitoyl phosphatidylethanolamine as donor and tetramethylrhodamine-dipalmitoyl phosphatidylethanolamine as acceptor in their assay systems, where phospholipid vesicle fusion was mediated by Ca^{2+} or Mg^{2+}.[7] While the above-mentioned assays[6,7] are regarded as qualitative FRET systems, Uster and Deamer[8] developed a quantitative assay for liposome fusion which was also based on FRET. In their system cholesteryl anthracene-9-carboxylate (CA9C) was used as donor probe and dihexylamino-7-nitrobenzo-2-oxa-1,3-diazole (DH-NBD) as acceptor probe. The structure of the probes suggested that the fluorophore moiety of DH-NBD is located at the hydrophilic/hydrophobic interface of the lipid bilayer, and the fluorophore of CA9C in the hydrophobic region. For making the fusion assay quantitative, the maximal donor fluorescence quenching possible was determined for each experiment by preparing "mock-fused" vesicles containing both probes (100% probe intermixing). Separately labeled vesicles were made and mixed (0% probe intermixing). The signal quenching due to the fusion was compared with the maximal quenching possible, and the extent of fusion was then estimated from probe intermixing.[8] In another system, 7-nitro-2,1,3-benzoxadiazol-4yl (NBD) acyl chain labeled lipid served as donor and (lissamine) rhodamine B-labeled phosphatidylethanolamine as acceptor.[9] In addition, the principle for monitoring fusion by FRET was also changed in this experimental system. Both fluorescent lipids were incorporated into the same lipid vesicle bilayer at appropriate surface density to observe efficient energy transfer. Then such vesicles were fused by addition of Ca^{2+} with a population of pure phosphatidylserine vesicles containing no probe molecules. The mixing of lipids reduces the surface density of the energy acceptor resulting in a decreased efficiency of FRET.[9] In contrast to the previous fusion assay system which is called the "probe mixing" assay, this method is called the "probe dilution" assay.

A number of considerations are involved in the choice of donor and acceptor probes for studying membrane fusion. The fluorophores should be attached to hydrophobic chains of sufficient length so that the rate of exchange via diffusion between liposomes is negligible relative to fusogen-induced vesicle fusion. Intervesicle exchange rate of simple lipid-soluble fluorophores such as perylene and *N*-phenyl naphthylamine was too fast to permit their use in monitoring membrane fusion.[7] It was found that two hydrocarbon chains per fluorophore ensured that the fluorophores remained localized in the membrane on the time scale of the membrane fusion.[7] In addition, the fluorophores should function as good donor-acceptor pairs, i.e., the fluorescence emission spectrum of the donor should significantly overlap with the absorption spectrum of the acceptor. As discussed in the previous chapter, according to the Förster theory a distance (R_0) characteristic for a donor-acceptor pair can be calculated from the spectral and orientational parameters of the fluorophores. R_0 is the distance required for 50% FRET efficiency. For good donor-acceptor pairs the R_0 can be as large as 4.9 nm (for the dansyl-rhodamine pair)[7] or 5.6 nm (for the fluorescein-tetramethylrhodamine pair).[10,11] Of course, the larger the R_0, the better the sensitivity of the fusion assay is. The other aspect which should be considered in choosing the donor-acceptor pair is that the absorption characteristics of the fusion system (liposomes, proteins, or DNA) should not interfere with the spectra of the fluorophores.

It has been noted that careful control experiments should be performed when the fusogenic action (such as pH changes) might cause significant changes in spectral characteristics of the fluorophores. While studying the fusion of Sendai virus at pH 4 to 7 with artificial lipid vesicles, about 60% of the vesicles and the virus appeared to fuse at pH 4 and about 100% at pH 5.[12] Among the spectra of the applied donor-acceptor pair [*N*-(7-nitrobenz-2-oxa-1,3-diazol-4-yl)-phosphatidylethanolamine, NBD-PE, and *N*-(lissamine-rhodamine-B-sulfonyl)-phosphatidylethanolamine, Rho-PE pair] the absorption spectra of the acceptor showed pH dependence. Therefore, the apparent fusion at pH 4 was due to the decrease in the absorption of the acceptor probe, instead of the dilution of the acceptor caused by fusing labeled vesicles with unlabeled virus. After correcting for the change in absorption, the extent of fusion was only 4 to 20% at pH 4 but still 80 to 100% at pH 5. The author concluded that the assay of membrane fusion by FRET can be misleading unless correction is made for changes in energy transfer originating from fusion-independent effects.[12]

Comparing the "probe mixing" assay with the "probe dilution" assay demonstrated that the result of fusion assays might be contradictory.[13] The probe mixing assay was found to be very sensitive to aggregation, even when the internal aqueous contents of the vesicles did not intermix, as in the case of large unilamellar vesicles (LUV) composed of phosphatidylserine in the presence of Mg^{2+}. No lipid mixing was detected in this case by the probe dilution assay. In the presence of Ca^{2+}, LUV exhibited real membrane fusion as determined by the intermixing of aqueous contents concomitant to membrane mixing. In this case, the rate of probe mixing was faster than that of probe dilution, which, in turn, was faster than the rate of aqueous content mixing. The investigators warned that experiments utilizing the probe mixing assay should be interpreted with caution, since aggregated vesicles not intermixing their outer monolayers would still result in a decrease of donor fluorescence due to FRET across the interbilayer space. Furthermore, even in systems that do undergo fusion, rate constants of fusion could be overestimated.[13]

To overcome some of the above-mentioned problems, a series of fluorescent phospholipids was synthesized with various fluorescent groups attached to the methyl-terminal half of one acyl chain in an otherwise normal phospholipid structure.[14] Phospholipids labeled with (dialkylamino)coumarin moieties or with bimane groups served as donors and lipids labeled with {[(dimethylamino)phenyl]azo}phenyl (DAB) or with methyl(nitrobenzoxadiazolyl)amino (NBD) groups served as acceptors. Of all the fluorescent lipid probes tested in this study the {[*N*-[4-[7-(diethylamino)-4-methylcoumarin-3-yl]phenyl]carbamoyl]methyl}thio (CPS) and *N*′-[4-[7-(diethylamino)-4-methylcoumarin-3-yl]phenyl]-*N*-methylthioureidyl (CPT) labeled ones proved to be the best probes. The intense blue fluorescence, low environmental sensitivity, and good photostability of the lipid probes with coumarin moiety make them highly suitable as probes for lipid-mixing assays in combination with energy acceptors such as DAB-labeled lipids. The investigators demonstrated that, in two particularly demanding applications (assays of the calcium-mediated fusion of phosphatidylserine vesicles and of the proton-triggered coalescence of phosphatidylethanolamine vesicles), some combinations of acyl chain-labeled probes offered substantial advantages over the commonly used NBD-PE/Rho-PE pair in accurately monitoring the progress of lipid mixing between liposomes.[14] These fusion assays allow one to readily detect and quantify fusion and they also provide the possibility to relate the kinetics of fusion to the rate at which certain molecular changes in membranes take place. Obviously, this insight is of relevance for understanding the mechanism of membrane fusion.

2. Applications of Fusion Assays Based on FRET

Membrane fusion assays based on the application of fluorescence resonance energy transfer have been applied in a wide variety of membrane systems. A large scale of factors mediating membrane coalescence (called fusogens) have been tested in liposomes and cells. Table 1 summarizes some examples of experimental systems and fusogens where the probe mixing or probe dilution assays have been successfully applied.

It is known that elevated calcium ion levels are coincident with the onset of fusion in sea urchin oocyte cortical granules, neurotransmitter release from synaptic vesicles, and myoblast plasma membranes. Therefore, it is obvious that one of the most thoroughly studied fusogens is Ca^{2+}. The effect of calcium concentration on membrane fusion has been investigated in artificial lipid vesicles[7-9,13-17] and in isolated chromaffin granules.[18]

The other physiological parameter which has been thoroughly studied as membrane fusogen is the pH. The pH plays an important role in the fusion of clathrin-coated liposomes,[19,20] of viruses with liposomes,[12] and in the coalescence of liposomes with nuclear membranes.[21,22]

Liposome fusion can be induced with various physical parameters such as temperature,[23] dehydration,[24] and freeze thawing.[25] Several chemicals proved to be very effective fusogens, including polyethylene glycol,[26,27] GTP,[28] polyvalent anions,[29] polylysine,[30] dolichylphosphate,[31] and colicin N.[32] Enzymes, such as lysozyme[33,34] and glyceraldehyde-3-phosphate dehydrogenase,[35] and hormones, such as insulin[36] and the plant hormone abscisic acid,[37] could also promote membrane fusion. The anti-tumor α-sarcin caused fusion of bilayers of phospholipid vesicles at neutral pH.[38] Annexin I significantly enhanced the Ca^{2+}-mediated fusion of liposomes with plasma membranes prepared from human neutrophil lymphocytes, suggesting that it plays a role in Ca^{2+}-dependent degranulation of human neutrophils.[17]

The effects of toxins and viruses are highly dependent on the fusogenic properties of these agents. α-latroxin[39] and β-bungarotoxin,[40] acting on presynaptic membrane, and diphteria toxin[41] displayed fusogenic activity in the liposome systems tested. Similarly, the action of viruses was closely related to their effect on membrane fusion, as in the cases of herpes simplex virus,[6] Sendai virus,[12] simian immunodeficiency virus,[42] arenavirus,[43] influenza virus,[44] and reconstituted influenza virus.[45]

These examples show that the fusion of membranes occurs in a great variety of artificial lipid vesicles and biological membrane systems and it can be mediated by several physical, chemical, and biological parameters.

B. STRUCTURE OF MEMBRANE BILAYERS

Surface densities of different phospholipids in synthetic membrane vesicles were determined by Fung and Stryer[46] using the FRET approach. Various dansyl conjugates with phosphatidylethanolamine were applied as donor fluorophores and eosinphosphatidylethanolamine served as acceptor in their experimental system. Significant energy transfer was observed between fluorophore-conjugated phospholipids in membrane vesicles containing 1 to 50 energy acceptors per 1000 phospholipids. The dependence of FRET efficiency on the acceptor surface density in this range agreed closely with that predicted by theory for a random distribution of acceptors in two dimensions. Calculation showed that FRET efficiency in a spherical membrane (radius of 9 nm) was nearly the same as in a planar membrane, indicating that this degree of curvature had a negligible effect. For probes located at the surface of a lipid bilayer, there was a little energy transfer from donors in one leaflet to acceptors in the opposite one if the R_0 value was less than 4 nm. Significant energy transfer was observed between donors and acceptors located closer to the center of the bilayer or having a larger R_0 value.[46]

FRET efficiency between alloxazines and isoalloxazines in multicompartment liposomes was measured to be higher than in single-walled liposomes, indicating that the average distance between donor and acceptor is smaller when the external surface of a bilayer is in the vicinity of the internal surface of another bilayer.[47] In comparison with dipalmitoyl phosphatidylcholine liposomes, the energy transfer efficiency in the surfactant (dialkylammonium chloride) vesicle was low, in spite of the efficient energy loss. The investigators suggested that the looseness of the membrane structure due to the electrostatic repulsion between the head groups of the surfactant molecules might cause high lateral and/or transversal mobility of the fluorophores, which could be responsible for the collisional quenching of the donor fluorescence.[47]

Table 1. **Application of Membrane Fusion Assays Based on FRET**

Fusogens	Donor fluorophore	Acceptor fluorophore	Mixing assay	Ref.
Ca^{2+}	Dansyl[a]	Rhodamine	M[b]	7
	CA9C[c]	NBD[d]	M	8
	NBD	Rhodamine	D[e]	9
	NBD	Lissamine-rhodamine	M, D	13, 15–18
	Coumarin	DAB[f]	M, D	14
pH	NBD	Lissamine-rhodamine	D	12, 20, 22
	NDB	Rhodamine	M	19
	FITC	Rhodamine	M	21
Temperature	NBD	Lissamine-rhodamine	D	23
Dehydration	CA9C	NBD	M	24
Freeze thawing	Dansyl	Indocarbocyanine	M	25
PEG[g]	CE-DPH[h]	CE-N-DPH[i]	M, D	26
	Pyrene	α-Parinaric acid	M	27
GTP	Aminofluorescein	Rhodamine	M	28
Polyvalent anions	NBD	Lissamine-rhodamine	D	29
Polylysine	NBD	Lissamine-rhodamine	D	30
Dolichyl phosphate	NBD	Lissamine-rhodamine	D	31
Colicin N	NBD	Lissamine-rhodamine	D	32
Lysozyme	Tryptophan	Pyrene	M	33
	NBD	Rhodamine	M	34
Glyceraldehyde-3-phosphate dehydrogenase	CA9C	NBD	M	35
Insulin	NBD	Lissamine-rhodamine	D	36
Abscisic acid	NBD	Lissamine-rhodamine	D	37
α-sarcin	NBD	Lissamine-rhodamine	M	38
α-latrotoxin	Chlorophyll *a*	Chlorophyll *b*	D	39
β-bungarotoxin	CA9C	NBD	M	40
Diphteria toxin	NBD	Lissamine-rhodamine	D	41
SIV[j]	NBD	Lissamine-rhodamine	D	42
Arenaviruses	NBD	Lissamine-rhodamine	M	43
Influenza virus	NBD	Lissamine-rhodamine	D	44
Reconstituted influenza virus	NBD	Lissamine-rhodamine	D	45

[a]Dansyl; *N*-5-dimethylaminonaphthalene-1-sulfonyl.
[b]M; "probe mixing" assay.
[c]CA9C; cholesteryl anthracene-9-carboxylate.
[d]NBD; 7-nitrobenzo-2-oxa-1,3-diazole.
[e]D; "probe dilution" assay.
[f]DAB; 4-{[4-(dimethylamino)phenyl]azo}phenyl.
[g]PEG; polyethylene glycol.
[h]CE-DPH; 1-(4′-carboxyethyl)-6-diphenyl-*trans*-1,3,5,-hexatriene.
[i]CE-N-DPH; 1-(4′-carboxyethyl)-6-(4″-nitro)diphenyl-*trans*-1,3,5,-hexatriene.
[j]SIV; Simian immunodeficiency virus.

FRET has also been applied to address the question: what is the size of a protein-lipid complex in membranes?[48] The tryptophan and tyrosine fluorescence of trypsin was gradually quenched by increasing concentrations of ONS-kephalin (*N*-octadecyl naphthyl-2-amino 6 sulfonic acid kephalin), a fluorescent lipid derivative, in 1,2-dimyristoyl-*sn*-glycerol-3-phosphoric acid methyl ester vesicles. The stoichiometry of lipid to protein was 40 to 50 lipid molecules per protein molecule in the association complex.[48]

The location of the receptor protein, apo-B100, in the low-density lipoprotein (LDL) particle with quasispherical shape was investigated by Bastiaens et al.[49] using the FRET technique. The 37 tryptophan residues of the apo-B100 were surrounded by pyrenylphosphatidylcholine acceptors inserted in the phospholipid monolayer of the human LDL particle. After analyzing the FRET data with various models, taking into the account the geometry of the donor-containing-protein with acceptors embedded in an associated phospholipid mono- or bilayer, the investigators concluded that significant portions of the apo-B100 protein penetrated the phospholipid layer.[49]

Connor and Schroit[50] described the application of a resonance energy transfer assay to determine the transbilayer distribution of NBD-labeled lipid analogs. Distances between fluorophores distributed symmetrically (outer leaflet label) and asymmetrically in lipid vesicles were determined form FRET between NBD-labeled donor lipids and lissamine rhodamine-labeled acceptor lipids; 1.4-nm difference in the average distance was revealed between NBD lipids and lissamine rhodamine lipids localized in the same leaflet and in opposing leaflets, respectively. Application of this technique for the investigation of the transbilayer distribution of NBD lipid in human red blood cells showed that exogenously supplied NBD phosphatidylserine was selectively transported to the inner leaflet, whereas NBD phosphatidylcholine remained in the outer leaflet.[50] A similar approach was applied in determining the transbilayer distribution of fluorescent lipid analogs by Wolf et al.[51] Energy transfer from NBD to rhodamine was measured using liposomes with symmetric and asymmetric distributions of rhodamine-labeled lipid analogs to determine the relative contribution of energy transfer between donor and acceptor probes in the same (*cis*) and opposite (*trans*) leaflets. The dependence of transfer efficiency on acceptor density was fitted to a theoretical treatment of energy transfer to determine the distances of closest approach for *cis* and *trans* transfer. From these calculations the positions of the fluorescent groups relative to the bilayer center were 2.0 to 3.1 nm for NBD and 3.1 to 5.5 nm for rhodamine, indicating that the rhodamine fluorophores are all well above the lipid phosphates while the NBD fluorophores are at or above the phosphates.

C. LOCALIZATION OF DRUGS IN LIPOSOMES BY FRET

The first scenario of drug action is the interaction with the cell membrane either because the plasma membrane constitutes a cellular target for the drug or because the drug should cross the membrane in order to reach its intracellular target. Fluorescent probes are widely applied for the investigation of the structure of biological membranes. In all cases, the location of drugs or probes in the membrane bilayers is very important regarding their action or the information they convey. A few examples, in which the localization of drugs and probes in membranes was determined by FRET are discussed below.

1,6-diphenyl-1,3,5-hexatriene (DPH) is perhaps the most widely applied among all the so-called "microviscosity" probes in membrane studies for accessing fluidity and structural parameters of the fatty acid chain region of both natural and synthetic membranes via the depolarization of its fluorescence. Experimental data about the transverse location of DPH and other fluidity probes are thoroughly discussed in Chapter 4.

The localization of fluorescent membrane probes, anthracene, 4-*p*-(dimethylamino)styryl-*N*-dodecylpyridinium (DSP-12) and 4-dimethylaminochalcone (DMC) was investigated using both steady-state and time-resolved fluorescence measurements.[52] Among other probes embedded in liposomes, anthracene served as donor and DSP-12 or DMC as acceptors. Steady-state fluorescence data permitted the area of the possible probe localization in membranes to be distinguished, while time-resolved measurements narrowed this area significantly. Fitting the experimental data with theoretical models suggested that the anthracene, the DSP-12, and the DMC are located close to the surface of the lipid bilayers at an average distance of 1.5 to 1.8 nm from the bilayer center.[52]

Frolov et al.[53] investigated the intramembrane location of the pigment molecule, chlorin e6, by measuring the luminescence quenching of DPH evoked by incorporated chlorin e6 molecules. On the basis of the energy transfer data the investigators concluded that the incorporation of chlorin e6 molecules into the vesicles from the aqueous phase occurs mainly into the external monolayer.[53]

Immunogenic activity of membrane bound haptens is strongly dependent of the length and nature of the linker chain (spacer) connecting the hapten to the head group of the lipid.[54] The role of location of the lipid hapten in immunogenic response was investigated by FRET measurements between TNP (1,3,5-trinitrophenol, acceptor), assumed to be randomly distributed in a plane parallel to the membrane surface, and pyrene (donor), embedded in the middle of lipid membranes. The measured distances were almost independent of the length of spacers, indicating that TNP residues of the lipid haptens were located in a similar vertical position at the membrane surfaces even if they had spacers of different lengths. The author suggested that the binding affinity of antibody to lipid haptens on the membrane surfaces was not modulated by the differences in the location of haptens but rather by the variation in the affinity of the spacer chains to the bilayer surface.[54]

FRET between a local anesthetic, dibucaine (donor), and a set of functionalized probes [*n*-(9-anthroyloxy)stearic acids, $n = 2, 3, 6, 7, 9$, and 12, and 16-(9-anthroyloxy)palmitic acid] (acceptors) was studied, in order to determine the location of the anesthetic in a model membrane system, in small unilamellar vesicles of dipalmitoylglycerophosphocholine, both above and below the temperature of the gel to the liquid-crystal phase transition of the phospholipid.[55] Application of the anthroyloxy family of probes was advantageous, since these probes are known to be located at graded series of depths in the lipid bilayer,[56] so a precise mapping can be obtained. In addition, the spectral overlap between dibucaine emission and the absorption of the anthroyloxy fluorophore was identical for all the acceptor probes; the same R_0 was valid for all the donor/acceptor pairs, which made the calculations easier. From the sequence of relative quenching efficiencies the investigators concluded that dibucaine intercalated into the vesicles near the glycerol backbone of the lipid. In addition, it was found that dibucaine location was not significantly affected upon crossing the phase-transition temperature of the phospholipid.[55]

Anthracycline antibiotics, such as daunomycin and adriamycin, are potent cytotoxic drugs used as antitumor agents.[57,58] The classical model to explain anthracycline cytotoxicity has been based on intercalation into nuclear DNA and interference with nucleic acid function. Nevertheless, experiments using the polymer-immobilized drug have shown that the drug is able to exert its cytotoxic activity without entering the cells and is thus interacting solely with the plasma membrane.[58] The interaction of adriamycin with lipids in model and natural membranes was monitored by FRET between anthroyloxy fatty acids and anthracene phosphatidylcholine as donors and the adriamycin as acceptor.[57] According to the experiment performed on model membranes, adriamycin molecules appeared to penetrate the lipid bilayer, with the aminoglycosyl group interacting with the lipid phosphate groups and the dihydroanthraquinone residue coming into contact with the lipid fatty acid chains. In contrast, FRET studies on Chinese hamster ovary cells showed that adriamycin penetrated the plasma membrane of these cells to a much more limited extent than in the model membrane systems. The authors concluded that the penetration of adriamycin into lipid bilayers strongly depends on the molecular packing of the lipid.[57] In the case of daunomycin, the DPH, located mainly in the core of the lipid bilayer, and trimethylammonium-DPH (TMA-DPH), partly interacting with the phospholipids' polar head group, were used as reference points in order to determine the relative location of the drug within the membrane bilayer.[58] The fluorescence of either probe was quenched by the energy acceptor, daunomycin, in a concentration-dependent manner resulting from FRET. Based on the dependence of FRET efficiency upon the acceptor concentration, it was suggested that at low or moderate daunomycin/lipid ratios the preferential incorporation of daunomycin occurred at membrane surface domains. At high acceptor/lipid ratios the efficiency of FRET to daunomycin from the surface and core domains of the membranes was very similar, suggesting that the drug may also traverse the membrane to reach internal regions of the bilayer. In fact, two macroscopic membrane-associated drug populations were detected applying iodide quenching experiments: a large fraction (85%) of membrane-bound daunomycin, which was accessible to the iodide quencher and must, therefore, have been located closer to the membrane surface, and a small non-accessible fraction, which was presumably buried within the bilayer core.[58] In addition to the model membranes, the incorporation of daunomycin into lipid bilayers was also studied in membrane fractions from daunomycin-sensitive and -resistant murine leukemia P-388 cells. FRET between tryptophan and daunomycin in the membranes indicated that interaction of protein components with the drug should not be responsible for the observed differences in drug binding exhibited by plasma membranes from drug-sensitive and -resistant cells.[59] Significant differences were found in the partition of daunomycin between "deep" and "surface" membranes domains when isolated plasma membranes from daunomycin-sensitive and -resistant P-388 cells were compared in this respect.[60] While daunomycin was preferentially confined to "surface" domains in membranes from drug-sensitive cells, the drug distributed more homogeneously between "surface" and "deep" domains in membranes from drug-resistant cells. The investigators assumed that the observed accommodation of a significant daunomycin population into "deep" plasma membrane domains in the drug-resistant cells could result in an easier diffusion of the drug through, and out of, these membranes, as compared to that in membranes from drug-sensitive cells.[60]

D. MEMBRANE PROTEINS IN LIPOSOMES

Protein-protein interactions in biological membranes may play a significant role in a variety of biochemical processes. Such interactions are extremely important in assembling of enzymes (such as ATPases) and receptors (acetylcholine esterase), and are thought to have direct impact on the function of rhodopsin. Association of clathrin molecules in coated pits, the quaternary structure of the complement system, and of the blood clotting factors are also good examples of protein-protein interaction in lipid bilayers. In addition, organization of pore-forming proteins such as toxins, antibiotics, and melittin proved to be very good model systems in studying the structure and function of membrane proteins. Selected aspects of structural organization of membrane proteins studied mostly by FRET techniques are discussed below.

1. ATPases

The major role of ATP-dependent Ca^{2+}-ATPases is the maintenance of a calcium concentration gradient across cellular membranes. In sarcoplasmic reticulum membrane, the translocation of Ca^{2+} requires cyclic changes in the structure of the Ca^{2+}-ATPase. In erythrocytes, the Ca^{2+}-transporting ATPase of the plasma membrane extrudes Ca^{2+} out of the cell against its large electrochemical gradient. Structural and conformational changes of functioning Ca^{2+}-ATPases have been thoroughly studied by techniques based on FRET. These studies can be sorted into two groups. FRET has been applied for investigating the quaternary structure of Ca^{2+}-ATPases in membranes: Do they form dimers, tetramers, or higher oligomers during their enzymic action? How does the oligomerization of Ca^{2+}-ATPases relate to their activity? The other application of FRET has been to map the intramolecular distances between various regulatory and active sites of the Ca^{2+}-ATPases. In the latter studies, attempts were made to detect changes in the conformation of Ca^{2+}-ATPase during its pumping cycle.

Significant FRET efficiency was detected in reconstituted vesicles between Ca^{2+}-ATPase purified from sarcoplasmic reticulum and labeled either with *N*-iodoacetyl-*N*′-(5-sulfo-1-naphthyl)ethylenediamine (IAEDANS, donor) or with iodoacetamidofluorescein (IAF, acceptor) fluorophores.[61] Up to tenfold dilution of the lipid phase of reconstituted vesicles with egg lecithin had no measurable effect on the energy transfer efficiency. Addition of unlabeled Ca^{2+}-ATPase in five- to tenfold molar excess over the labeled molecules abolished energy transfer. These observations suggested that ATPase oligomers exist in the membranes with sufficiently long lifetimes for energy transfer to occur.[61] Efficient energy transfer was also observed from covalently bound fluorescein-5′-isothiocyanate (FITC, donor) to eosin-5′-isothiocyanate (EITC, acceptor).[62,63] While the former fluorophores reacted with the SH groups of the enzyme, due to structural similarities, both FITC and EITC competed for the same site on the Ca^{2+}-ATPase, i.e., for the ATP binding site. Under optimal conditions each ATPase molecule is labeled either with a donor or an acceptor fluorophore, but not with both. The FRET efficiency between FITC-ATPase and EITC-ATPase was also independent of the phospholipid:protein ratio. These findings, too, suggested that the Ca^{2+}-ATPase is not randomly distributed in the plane of the membrane but rather forms ordered clusters on the time scale of fluorescence (nanoseconds) even in the presence of a large excess of phospholipids.[62,63]

Another approach was applied by Fagan and Dewey[64] for detecting the oligomerization of Ca^{2+}-ATPase. Temperature dependence of FRET from a fluorescent lipid donor to an acceptor covalently bound to the CA^{2+}-ATPase was monitored in native sarcoplasmic reticulum vesicles. The FRET efficiency in these systems increases as the size of protein aggregates decreases due to the increased exposure of the protein in the lipid domain that result in the shortening of distances between donors and acceptors. Elevating the temperature caused a decrease in the aggregation of ATPases. According to a theoretical model, the average radius of the ATPase aggregate at low (5°C) temperature was consistent with the formation of a tetramer.[64] Similar kinds of FRET measurements were used for identifying the "coastline" of the aggregate, and a fractal dimension of 1.9 was estimated of Ca^{2+}-ATPase aggregates.[65]

A series of studies were performed to determine the relative distance between various sites of Ca^{2+}-ATPase. During the Ca^{2+} transport cycle, the enzyme alternates between two major conformations (E_1 and E_2) with distinct reactivities toward substrates and substrate analogs. To monitor changes in the relative distances between these sites during the conformational change of the enzyme was the aim of many papers using FRET technique to address this question. Labeling the Ca^{2+}-ATPase at the ATP binding site with FITC and the lipid molecules with rhodamine-5′-isothiocyanate made it possible to measure the distance between this site and the lipid/water interface.[66,67] When DPH was used to label the lipid bilayer, the proximity between the nucleotide binding site and lipid core was monitored.[68] Distances from the Ca^{2+}

binding sites on the ATPase to the lipid headgroup region were estimated using Tb^{3+} as fluorescent donor and labeled lipid as acceptor.[66] The distance between the ATP and Ca^{2+} sites was estimated using Tb^{3+} as a fluorescence donor and either FITC or the ATP analog trinitrophenyl-ATP (TNP-ATP) as acceptor.[69] Alternatively, the Ca^{2+} binding sites could be labeled with the fluorescent carbodiimide, NCD4.[70,71] Highly reactive sulfhydryls can be labeled with the fluorescent probe IAEDANS and other SH reagents, without loss of enzymic activity.[61,72-74] The triazine dye, Reactive Red 120, was found to bind tightly to the Ca^{2+}-ATPase and the dye-binding site was different from both the active site and the regulatory site.[75] Tryptophans of the ATPase can also be used as reference sites,[71,76] since 11 of the 13 tryptophan residues of the Ca^{2+}-ATPase are located in the membrane spanning region, near the headgroups of the phospholipids. The results of these FRET studies are nicely summarized in References 77 and 78 with a scheme showing the estimated distances between characteristic sites of the Ca^{2+}-ATPase. Temperature dependence of the normalized FRET efficiency reported about the fluctuation of domains within the ATPase molecule.[78] According to the theory by Somogyi et al.,[79] the normalized energy transfer would be greater with increasing distances between donor and acceptor fluorophores, because the amplitude of oscillation is likely to increase with length. The change in normalized energy transfer with temperature showed no clear relationship with the average distance between donor and acceptor fluorophores, suggesting the existence of flexibility gradients within the ATPase molecule.[77,78] The conformational changes associated with Ca^{2+} transport do not produce readily identifiable changes in the distance between intramolecular markers analyzed by FRET.[73,74,77,78] No significant difference in energy transfer efficiency was observed between the E_1 and E_2 states of the Ca^{2+}-ATPase at low temperatures (10°C) probably because the structural change was too small compared to the relatively large error inherent in energy transfer measurements.[78] The major source of error comes from the fact that conversion of FRET efficiency into distances is based on the assumption that the relative orientation of donor and acceptor fluorophores is random. This assumption is not valid in every case, since the high fluorescence polarization of FITC bound to ATPase indicates a somewhat restricted motion of the fluorophores.[78] In these cases, the range of the orientation factor can be calculated, so that only a range of the distances can be determined instead of an exact proximity value. Nevertheless, at higher temperatures (30°C) discernible differences begin to appear in the normalized energy transfer efficiency between the E_1 and E_2 conformational states.[78] Formation of the phosphoenzyme intermediate resulted in a measurable increase in FRET efficiency between IAEDANS and another fluorescent SH reagent (4-dimethylaminoazobenzene-4′-maleimide) and this increase was reversible upon phosphoenzyme destabilization by subsequent addition of calcium. No corresponding change was observed in FRET efficiency measured between IAEDANS and FITC, indicting that the change in FRET may occur as a result of reorientation of associated ATPase polypeptide chains with respect to one another.[74] These data also suggest that, regardless of whether the change in FRET is caused by changes in relative distance or relative orientation, the FRET technique may provide valuable information about the conformational changes of Ca^{2+}-ATPase. Although the FRET technique has its inherent limits, the extension of FRET studies to new donor-acceptor pairs, together with information about the structure of Ca^{2+}-ATPase, obtained with other methods, such as X-ray crystallography or electron microscopy, may help to refine our overall picture about the structure of the functioning Ca^{2+}-ATPase molecule of the sarcoplasmic reticulum.

Similarly to the above discussed Ca^{2+}-ATPase from sarcoplasmic reticulum, inter- and intramolecular FRET measurements were performed on erythrocyte Ca^{2+}-ATPase in order to monitor changes in the oligomerization and conformation of the enzyme. Kosk-Kosicka et al.[80] measured the FRET efficiency at different enzyme concentrations, from FITC attached on one erythrocyte Ca^{2+}-ATPase molecule to tetramethylrhodmamine-5-isothiocyanate attached on another enzyme molecule. The energy transfer efficiency showed a sigmoid dependence on enzyme concentration, suggesting the oligomerization of the enzyme molecules.[80,81] No change in FRET efficiency was observed upon subsequent addition of calmodulin at different enzyme concentrations. Adding calmodulin before mixing donor and acceptor-labeled enzyme populations prevented the occurrence of energy transfer. These observations suggested that binding of calmodulin to both monomeric and oligomeric Ca^{2+}-ATPases temporarily locked the enzyme molecule in its conformation.[82] Energy transfer studies also demonstrated that enzyme oligomerization required calcium. These studies showed that the Ca^{2+}-ATPase could be fully activated by a Ca^{2+}-dependent oligomerization mechanism, independently from the activation by calmodulin. The investigators proposed two pathways for the activation of the Ca^{2+}-ATPase, taking into account the interdependencies between Ca^{2+}, calmodulin, and enzyme concentration.[80-82] FRET between the tryptophan residues of Ca^{2+}-

ATPase and dansyl derivative of covalently labeled calmodulin occurred in the presence of EGTA. The measured FRET efficiency was further promoted by the addition of free Ca^{2+}. It was concluded that the interaction of calmodulin with Ca^{2+}-ATPase resulted in different conformational states of the enzyme, since the E_1 conformation of the enzyme was stabilized by 10μM Ca^{2+}, while the E_2 conformation was stabilized by 0.1m*M* EGTA.[83] The interaction of various phospholipids with erythrocyte Ca^{2+}-ATPase was monitored by measuring the FRET from tryptophans of the enzyme to pyrene-labeled analogs of phospholipids (e.g., phosphatidylcholine, various phosphatidylinositol phosphates). A positive correlation was found between the number of negative charges on the phospholipids and the potency of their pyrene-labeled analogs to act as acceptors in FRET from the tryptophans of the enzyme. These observations suggested that the stimulation of Ca^{2+}-ATPase activity by negatively charged phospholipids is based on a binding of these lipids to the enzyme and that the negative charges are a major modulatory factor for this interaction.[84]

The oligomeric nature of Na^+,K^+-ATPase was also detected with the help of FRET technique.[85] The enzyme, which is responsible for active transport of Na^+ and K^+ across the plasma membranes of animal cells, consists of a catalytic subunit (α) and a glycoprotein subunit (β) with yet-unknown function. The ATP protectable, active site domain of the α subunit was labeled with either FITC or erythrosin 5'-isothiocyanate (ErITC). FRET from donor (FITC) to acceptor (ErITC) revealed an apparent 5.6-nm distance between the putative ATP binding sites of α subunits, which is consistent with $(\alpha\beta)_2$ dimers rather than randomly spaced $\alpha\beta$ heteromers.[85] Intramolecular FRET was detected between anthroylouabain bound to the cardiac glycoside site on the α subunit and lucifer yellow bound to the β subunit.[86] The distance was the same (4.7 nm) when anthroylouabain was bound with ATP or P_i as phosphorylating ligands, but increased to 4.9 nm in the presence of vanadate. This increase in the average distance indicated a conformational difference between the complexes of cardiac glycosides with Na^+,K^+-ATPase induced by phosphorylating ligands or by vanadate.[86]

These examples clearly demonstrated that the FRET technique provided valuable information, within its limitations, about the structure and conformation of membrane-bound ATPase enzymes.

2. Acetylcholine Receptor

The other membrane bound protein which has been thoroughly investigated, the acetylcholine receptor, is a cation channel, whose activation is elicited by acetylcholine. The pentameric receptor is composed of four different subunits, designated as α_2, β, γ, and δ, which are arranged like barrel staves around a central pit. The arrangement of the subunits around the perimeter of the central pit is $\alpha\gamma\alpha\beta\delta$. In addition to other techniques, the precise structural arrangement of receptor subunits has also been investigated by FRET methods.[87-89] Snake curarimimetic α-toxins bind competitively with acetylcholine to inhibit the activation of acetylcholine receptor. The FRET approach was used to examine the spatial relationship between the two α-toxins bound to the α-subunits of the acetylcholine receptor. Significantly greater energy transfer was observed between FITC or tetramethyl rhodamine isothiocyanate (TRITC) labeled α-toxins bound to membrane-associated receptors than bound to solubilized receptors suggesting that transfer between fluorophores on separate receptor molecules was greater than that occurring intramolecularly between the two sites on the receptor. The magnitude of the distance (5.0 nm) calculated from the intrareceptor FRET efficiency combined with the considerable inter-receptor efficiency showed that cobra α-toxins bind to the apical surface of extracellular domains of the each of the two α-subunits located on opposite sides of the disc-shaped receptor molecule (having a diameter of 9.0 nm).[87] In addition to the location of α-toxin binding sites, the orientation of α-toxins was also investigated. Toward this end four FITC derivatives of α-toxins, labeled at the ε-amino groups of Lys-23, Lys-35, Lys-49, or Lys-69, and a TRITC derivative, labeled at the ε-amino group of Lys-23, were used in FRET experiments in combination with a lipid probe 5-(*N*-dodecanoylamino)eosin (C_{12}-eosin) and with tryptophans of the receptor. Measurements of FRET between the lipid membrane surface and the sites of labeling on receptor bound α-toxin derivatives showed that Lys-23 was the closest to and Lys-69 was the farthest from the lipid surface. The highest FRET efficiency was measured between tryptophans of acetylcholine receptor and the Lys-49 position. Combining these data, the investigator established a model showing the toxin directly interacting with the receptor surface and the major axis of the neurotoxin tilting from a perpendicular projection to the membrane. The surface of the α-toxin that includes Lys-23 projects away from the central ion channel and the surface that includes Lys-69 faces the ion channel.[88] The binding site of quinacrine, a potent fluorescent noncompetitive inhibitor, was also mapped on the acetylcholine receptor by FRET.[89] The closest distance between the receptor-bound quinacrine and the lipid surface (C_{12}-eosin) was less than 1 nm as calculated from the FRET efficiency measured between them. These result suggested that the quinacrine site is too close to the lipid domain to be in the lumen of the receptor, and therefore it may be located on the outer surface of the membrane-spanning domain of the acetylcholine receptor.[89] FRET was also measured between the tryptophans of the acetylcholine receptor and the *N*-(1-pyrene)maleimide reacting with the sulphydryl (SH) groups of the receptor. The FRET data, as well as amino acid analysis of *N*-(1-pyrene)maleimide-alkylated receptor, indicated that a homologous class of cysteines, involved in the reaction with *N*-(1-pyrene)maleimide, is located in each subunit within the acetylcholine receptor domain embedded in the membrane.[90] In addition to the agonist site, the α-subunits of the acetylcholine receptors possess an allosterically coupled competitive inhibitor site. Decidium diiodide, a bisquaternary analog of ethidium, binds to the agonist site at low concentrations and to the noncompetitive inhibitor site at higher concentrations. No energy transfer was detected between FITC-labeled α-toxin and decidium when each was bound to one of the two agonist sites on the receptor. In contrast, energy transfer was measured between decidium nonspecifically associated with the membrane or with nonspecific sites and the FITC-labeled α-toxins at the agonist sites.[91] Significant FRET efficiency was also measured between a dansylated choline analog (donor) bound to the agonist site and the ethidium (acceptor) bound to the noncompetitive inhibitor site, yielding a distance range of 2.2 to 4.0 nm.[92] Considering the molecular dimensions of the receptor and the location of the agonist sites, these data suggested a location for the noncompetitive inhibitor site near the ion channel at the extracellular surface of the membrane bilayer.[92]

3. Rhodopsin

Many membrane proteins contain, besides the amino acids, other functionally important molecules, prosthetic groups. Sometimes these molecules are colored compounds that absorb light in the visible spectral region. The spatial arrangement of these dye molecules and the amino acid chains of the protein is of fundamental interest for understanding the function of the proteins. Good examples for these proteins are the bacteriorhodopsin of the *Halobacterium halobium*, which functions as a light-energy converter, and the rhodopsin of the rod, which functions as sensor. In both cases, the location of the retinal chromophore within bacteriorhodopsin or rhodopsin molecules has been the subject of many studies using FRET as the experimental approach.

DPH was applied paired with retinal for the investigation of bacteriorhodopsin in phospholipid vesicles. From this analysis a distance of closest approach between DPH (donor) and retinal (acceptor) of 1.8 nm was obtained, which is close to the effective radius of bacteriorhodopsin (1.7 nm) and consistent with the current ideas about the location of retinal in the interior of the protein.[93] Rapid-diffusion energy transfer was also applied to estimate the *trans*-membrane depth of retinal in the purple membrane. In this case, FRET from a long-lived energy donor (a terbium chelate) to the retinal chromophore in the membrane was measured. Because of the long lifetime (around 1 ms) of the terbium donor, the rapid-diffusion limit is reached, making the transfer efficiency extremely sensitive to the depth of the chromophore beneath the membrane surface. In cell envelope vesicles from *H. halobium*, the estimated depth of retinal from the extravesicular face of the purple membranes was 1.0 nm, calculated from the rate constant of the rapid-diffusion limit energy transfer.[94] The DPH-retinal donor-acceptor pair was used as a spectroscopic ruler to estimate the protein-induced perturbation in the lipid phase, which was then found to be larger than 4.5 nm.[95] In another experiment, the retinal group of the bacteriorhodopsin served as acceptor and rhodamine or indocarbocyanine-labeled lipid as donor in the investigation of protein domain or patch formation. The aggregation state of bacteriorhodopsin was controlled by adjusting the temperature and followed with FRET measurements. Quantitative analysis showed that the segregated domains of membrane proteins had radii ranging from 3 to 30 nm.[96] Application of two hydrophobic solutes which perturb lipid packing order, permethrin and allethrin, had significant effect on the aggregated state of bacteriorhodopsin in lipid vesicles as determined by FRET measurements. Permethrin and allethrin were found to cause a decrease in the temperature of bacteriorhodopsin disaggregation which was associated with a decrease in transition temperature of the lipid. In gel-phase dipalmitoylphosphatidylcholine at 25°C, the perturbing effects of permethrin on lipid packing order were associated with a decrease in the average patch radius from 13.3 to 3.3 nm.[97] In addition to temperature, the dissociation/association processes of bacteriorhodopsin molecules were also found to be sensitive to the pH of the medium.[98] Renthal et al.[99] used dansyl chloride to label Lys-41 of the bacteriorhodopsin. FRET was used to probe the proximity of the modified site to the retinal chromophore of bacteriorhodopsin. The projected distance from Lys-41 to retinal was 1.6 nm. This measurement, combined with many other labeling experiments, suggested a possible arrangement of the seven transmembrane helices of the bacteriorhodopsin.[99]

Unbleached rhodopsin from rod outer segment disk membranes reduced the mean fluorophore lifetimes of DPH and TMA-DPH in proportion to the amount of protein present in lipid vesicles as a result of probe-to-retinal energy transfer. The decrease in fluorescence lifetime was caused by redistribution of the relative lifetime contributions in favor of the short lifetime population and by reduction of lifetimes of each derived population.[100] Rapid-diffusion energy transfer was used to measure the depth of retinal chromophore in rod outer segment disc membrane. The distance of closest approach between a terbium chelate and the retinal of rod outer segment disc membrane was measured to be 2.8 nm from the intradiscal membrane surface and 2.2 nm from the extradiscal membrane surface.[101] The sensory function of the rhodopsin in the visual transduction is closely associated with the amplified activation of the cyclic-GMP phosphodiesterase. Photoactivation of the phosphodiesterase requires the presence of a GTP binding protein (G-protein).[102] The interaction between rhodopsin and G-protein was thoroughly studied by FRET using many combinations of donor-acceptor pairs. SH groups of the G-protein and bleached rhodopsin were labeled with pyrene, monobromobimane, or coumarin and reconstituted in phospholipid vesicles. FRET was detected in bleached recombinants of pyrene-rhodopsin/coumarin-G-protein, pyrene-rhodopsin/bimane-G-protein, and pyrene-G-protein/coumarin-rhodopsin, reflecting the binding of G-proteins to membranes. The specificity of G-protein binding was characterized by reduction of energy transfer by an excess of nonhydrolyzable analogs of GTP, which promote the dissociation of the rhodopsin-G-protein complex and by addition of unlabeled G-protein, which competes for binding sites in rhodopsin. According to the results of these measurements the proximity of labeled SH groups was found to be in the range of 3.8 to 4.6 nm.[102]

4. Clathrin

Clathrin-coated vesicles mediate selective translocation of receptors from one cellular membrane to another. They contribute to receptor-mediated endocytosis and the sorting of proteins in the *trans*-Golgi network.[103] Clathrin forms a three-legged structure termed a triskelion, which is composed of three heavy and three light chains.[103,104] Formation of a clathrin-coated vesicle requires polymerization of triskelion into a polyhedral network on the cytoplasmic side of cellular membranes. Clathrin induces fusion of phospholipid membranes containing phosphatidylserine when the pH of the medium is reduced below about six.[20,104] The hydrophobicity of clathrin in the pH range inducing membrane fusion significantly increased, as it was monitored by the blue shift in the fluorescence maximum of 1-anilinonaphthalene-8-sulfonate associated with clathrin. Moreover, the efficiency of FRET from tryptophan to anilinonaphtyl residues in the clathrin molecule also increased in this pH range. These results suggested that clathrin-induced fusion of liposome membranes involved both protein binding to the membranes and a conformational change of clathrin accompanied by the exposure of its hydrophobic domains.[20,104] In addition to this pH induced conformational change, the self-aggregation of clathrin under similar conditions may play an important role in the fusion process. Toward this end, the extent of FRET from *N*-[*p*-(2-benzimidazoyl)phenyl]maleimide-labeled clathrin to *N*-(7-dimethylamino-4-methyl-3-coumarinyl) maleimide-labeled clathrin was measured in the presence of liposomes. A high level of FRET efficiency was observed, and the area on the membrane surface occupied by a clathrin molecule was estimated to be much less than that expected from its size, suggesting that clathrin binds to the liposome membranes in an associated form, which may be crucial for the induction of membrane fusion.[105] Investigation of the fusion abilities of clathrin fragments, obtained by limited proteolysis of clathrin cages with thermolysin, showed that the light chains had no role in membrane fusion, similarly to the terminal domain of the heavy chain. Only the proximal part of the heavy chain showed fusion activity. Both the terminal domain and the proximal part induced reversible liposome membrane aggregation. The investigators concluded that the terminal domain and the proximal portion of clathrin have critical roles in the steps of close apposition and fusion of membranes, respectively.[106]

5. Complement

The cytolytic activity of human serum resides in the C5b-9 membrane attack complex composed of complement proteins C5b, C6, C7, C8, and C9. The spontaneous assembly of this complex from these key immunoproteins initiates hydrophilic-to-hydrophobic conformational transitions within the proteins which result in their insertion into membrane lipid. The concomitant increase in membrane permeability leads to the collapse of ionic gradients and consequently to cell death. The mechanism of this association and polymerization processes was investigated with many experimental techniques including FRET measurements.[107-109] Fluorescence self-quenching by energy transfer of FITC-labeled C9 protein was used to monitor the kinetics of C9 polymerization induced by the membrane-associated complex of C5b-8.[107,108] Fluorescence measurements on C5b-8-treated sheep red blood cells ghost membranes revealed that C9 polymerization induced by the C5b-8 proteins exhibits a temperature dependence similar to that of the complement-mediated hemolysis of these cells. The investigators suggested that the intercalation of tubules of polymerized C9 caused the change in the membrane permeability initiating the collapse of transmembrane electrochemical gradients.[107,108] The association state of membrane-bound complexes of complement proteins C5b-c8 was also studied by FRET.[109] There was no energy transfer between fluorescein- (donor) and Texas red- (acceptor) conjugated C8 molecules embedded in sheep erythrocyte ghost membrane. Upon binding of preformed C5b67 complexes, FRET was detected between donor C8 and acceptor C8, suggesting that C5b-8 complexes are aggregated into polymeric clusters. Good correlation was found between the polymeric clustering of C5b-8 complex and its cytolytic activity.[109]

6. Blood Clotting Factors

In the penultimate step of the blood clotting cascade, prothrombin is converted into thrombin by factor Xa, the activated form of blood coagulation factor X. This enzymatic cleavage proceeds most efficiently when factor Xa is associated with a nonenzymatic protein cofactor, designated factor Va, in the presence of calcium ions and on a membrane surface containing acidic phospholipids. This complex, termed the prothrombinase complex has been extensively studied by a variety of biochemical and biophysical techniques, among others by FRET.

Blood clotting factors II, X, and IX bind to phospholipid interfaces in the presence of Ca^{2+} to activate coagulation. The topology of their binding site on the membrane was studied using isothermal energy transfer experiments with natural lipids in their fluid phase. The tryptophan residues of the factors were the energy donors, and pyrene covalently bound to fatty acids was the energy acceptor. FRET data suggested that factor II exhibited some specificity for acidic phospholipids and was surrounded by noninteracting zwitterionic lipids. Factor IX was surrounded by statistically the same amount of charged and zwitterionic lipids. The investigators concluded that these blood clotting proteins bind to charged surfaces by ionic forces reinforced by hydrophobic contributions.[110] Binding of factor Xa to factor Va did not change the composition of the lipid binding site of factor Va, as shown by FRET experiments between tryptophan residues of factor Va and pyrene-labeled phospholipids.[111] Ca^{2+} induced a greater FRET between prothrombin and membranes labeled with pyrene, probably due to a higher affinity of prothrombin for the membrane and/or to a specific conformational change of the molecule in the presence of Ca^{2+}, inducing an efficient FRET from some particular tryptophan residues to the pyrene group.[111]

Factor Va was labeled with iodoacetamido derivatives of fluorescein and titrated with phosphatidylcholine/phosphatidylserine vesicles containing either octadecylrhodamine or 5-(*N*-hexadecanoylamino) eosin. From the FRET observed, the distance of closest approach between the fluorescein on factor Va and the two different acceptors located on the bilayer surface averaged 9.0 nm. Association of fluoresceinated factor Va with factor Xa on the phospholipid surface reduced this distance to 8.3 nm.[112] The efficiency of diffusion-enhanced energy transfer between rhodamine-labeled factor Va and terbium dipicolinate entrapped inside lipid vesicles was less than 1%, indicating that the location of the dye is far above the inner surface of the vesicle.[112] FRET measurements were also applied to determine the location of the active site of membrane-bound factor Xa relative to the phospholipid surface in the presence and absence of factor Va. The active site of Xa was labeled with 5-(dimethylamino)-1-naphthalenesulfonyl-glutamylglycylarginyl chloromethyl ketone (DEGR-CK) and FRET was observed between the dansyl donor dye and the octadecylrhodamine acceptor dyes at the outer surface of the phospholipid bilayers. From the dependence of energy transfer upon acceptor density, the closest approach between the active site membrane bound factor Xa and the surface of the phospholipid bilayer was calculated to be 6.1 nm in the absence of factor Va and 6.9 in the presence of factor Va. These measurements demonstrated that the active site of factor Xa is located far above the membrane surface.[113]

A similar approach was used by the same group in investigating the proximity between the active site of the membrane-bound thrombin-thrombomodulin complex and the membrane surface.[114] Thrombin functions both as a soluble procoagulant protease that activates factor V, factor VIII, and fibrinogen and also as a membrane-bound anticoagulant that activates protein C.[115] The conversion of procoagulant thrombin to anticoagulant thrombin occurs when thrombin is associated with a cofactor termed thrombomodulin, an integral membrane glycoprotein that is found on the surface of endothelial cells.[115] Significant FRET was measured between DEGR-thrombin bound to thrombomodulin reconstituted in phospholipid vesicles and octyldecylrhodamine located at the bilayer surface. The observed energy transfer occurred only when the fluorescent thrombin was bound to thrombomodulin. The average distance of closest approach between the active site of the thrombin and the membrane surface was found to be 6.6 nm.[114] This means that no part of the thrombomodulin-bound thrombin molecule is within 2.5 nm of the membrane, since its diameter is near 4.0 nm. More likely, assuming that the active site is located at about the middle of the enzyme, the thrombin molecule is separated from the membrane surface by at least 4.5 nm. From these data, the investigators concluded that thrombin does not bind alongside thrombomodulin near the membrane surface to form the thrombin-thrombomodulin complex. Instead, thrombin binds "on top" of thrombomodulin or to a domain of thrombomodulin that extends far above the membrane surface.[114]

A special proteolytic cleavage of prothrombin results in meizothrombin which can easily be labeled fluorescently with DEGR at the catalytically active site.[116] Addition of Ca^{2+} significantly reduced the fluorescence lifetime of dansyl groups conjugated to meizothrombin. The authors suggested that binding of Ca^{2+} to meizothrombin and, by extension, perhaps to prothrombin, elicits a conformational change that extends beyond the fragment 1 domains into the distant active site, the protease domain. FRET measurements showed that the active site of meizothrombin was located 7.1 nm above the membrane surface. In the presence of factor Va the distance was 6.7 nm, indicating that association of factor Va with meizothrombin on the phospholipid surface appears to cause a slight movement of the meizothrombin protease domain toward the bilayer surface.[116] All these calculations were based on R_0 values determined using the assumption that the orientation factor (κ^2) was 2/3. However, because the dye in the active site of the molecules and the dye at the phospholipid surface do not rotate with complete freedom, there is an uncertainty in the relative orientation of the acceptor and donor transition dipoles during the excited state lifetime of the donor. From the measured steady-state anisotropies of the dansyl and rhodamine dyes the calculated maximum uncertainty in R_0 was ±25%.[117] This means that all the calculated distances are burdened with this error, although the relative changes are still very reliable. If the only change in the fluorescently labeled molecule upon addition of other factors is the alteration in the orientation of the fluorophore relative to the membrane surface with no change in the distance between the donor acceptor pair, the altered FRET efficiency will still reflect the modification of the conformation of the protein.

The plasma protein factor VIII functions as a cofactor in the activation of factor X by factor IXa. This cofactor increases the rate of factor X activation by at least four orders of magnitude only in the presence of phospholipid membranes. Binding of factor VIII to small and large unilamellar vesicles composed of phosphatidylserine and phosphatidylcholine was demonstrated by resonance energy transfer measured between intrinsic fluorophores in factor VIII and dansyl-phosphatidylethanolamine incorporated into vesicles.[118] Factor VIII binding was rapid and reversible and fit a simple bimolecular equilibrium model. Factor VIII exhibited nonlinear binding relationship to the mole fraction of phosphatidylserine.[118] This observation might have a biological significance, since factor VIII binds preferentially to activated platelets with lipid composition of less than 2% phosphatidylserine and with lower affinity to resting platelets having 4 to 13% phosphatidylserine.[119]

Activation of platelet by thrombin, ADP plus epinephrine, or phorbol ester converts the membrane GP IIb-GP IIIa into a functional receptor for fibrinogen. Domains of plasma membrane GP IIb-IIIa exposed on the cell surface might undergo a conformational change during platelet stimulation. In order to study the spatial relationship among specific epitopes on platelet GP IIb and IIIa, platelets were labeled with anti-GP IIb and IIIa monoclonal antibodies that had been conjugated with either fluorescein isothiocyanate or TRITC.[120] In flow cytometric experiments, 6 to 8% FRET was observed between antibody B1B5, bound to GP IIb, and antibody SSA6, bound to GP IIIa, regardless of which antibody served as energy donor. Cell stimulation by thrombin caused a twofold increase in FRET efficiency suggesting a change in the separation and/or orientation of these epitopes within the GP IIb-IIIa complex. Because of the segmental mobility of the fluorophore-labeled antibodies and the uncertainty in the location of sites within the antibodies that are labeled by FITC and TRITC, the precise spatial separation of the antibody epitopes on the GP IIb-IIIa complex cannot be deduced from these data. The activation-related conformational change detected by FRET was unaffected by the addition of fibrinogen or by the inhibition of fibrinogen binding by an appropriate peptide. The investigators concluded that the conformational change in GP IIb-IIIa complex, converting this molecule into a competent fibrinogen receptor, occurred prior to fibrinogen binding.[120]

7. Pore-Forming Proteins

The cytotoxic action of various toxins and antibiotics is based on their interaction with membranes, resulting in formation of special pores through the plasma membrane. These pores can be nonspecific holes, big enough for permitting complete exchange of intracellular and extracellular aqueous environment, or specific channels permitting the translocation of specific cations. These types of changes in membrane permeability may lead to the collapse of electrochemical gradients and consequently to cell death. Concomitantly to the formation of pores, association and/or conformational changes of the protein components can be observed. These processes can be successfully explored with FRET-based techniques.

Gramicidin A, a linear peptide antibiotic, makes membrane permeable to alkali cations and hydrogen ions by forming transmembrane channels. FRET studies between dansylated gramicidin and *p*-phenylazobenzene sulfonyl derivative of gramicidin were applied to determine the degree of association of gramicidin molecules. FRET data, together with single-channel conductance measurements, showed that there are two molecules of gramicidin in a channel.[121] The FRET method was also used to study the distribution of phospholipids around the gramicidin molecules. The efficiency of FRET from tryptophan residues of gramicidins to different phospholipids labeled with a dansyl moiety was consistent with a random distribution of each phospholipid in the bilayer in the presence of gramicidin. Addition of Ca^{2+} to vesicles made up of phosphatidylcholine and phosphatidic acid resulted in phase separation. FRET measurements from gramicidin to either dansylphosphatidylcholine or dansylphosphatidic acid showed gramicidin preferentially partitioned into the phosphatidylcholine-rich regions.[122]

The α-toxin of *Staphylococcus aureus* forms hexamers when inserted into lipid bilayers. Concomitantly to the toxin-hexamer formation, the fluorescence of the tryptophan residues of the toxin molecule increases.[123] In order to distinguish the toxin-membrane interaction from the hexamer formation, FRET measurements were performed between the tryptophans of the toxin molecule and dansylated phosphatidylethanolamine. Under moderately acidic or basic conditions, the FRET always preceded the increase in the fluorescence the tryptophan residues. The increase in tryptophan fluorescence below 20°C showed a latent period, whereas FRET did not. The investigators concluded that, when α-toxin damages the target membrane, the molecule interacts with the membrane first, and then undergoes oligomerization within the membranes.[124] FRET studies also demonstrated that, while the toxin-membrane interaction occurred in phosphatidylcholine membranes containing saturated and/or unsaturated fatty acids, oligomerization occurred only in the presence of phosphatidylcholine containing unsaturated fatty acids.[125] In addition, the toxin-hexamer formation correlated well with the leakage of carboxyfluorescein from the phospholipid liposomes.[125]

Colicins are plasmid-encoded proteins from enterobacteria that provide favorable conditions for plasmid replication by killing cells not bearing the same plasmid. They form well-defined voltage-gated ion channels in artificial membranes.[126] Colicins are composed of structural domains specialized in the different steps of the toxic activity, i.e., targeting, translocation, and killing. In colicin A, the 21-kDa C-terminal domain, isolated after thermolysin digestion, carries the complete pore-forming activity.[127] This polypeptide of 204 residues is a bundle of 10 α-helices that are arranged in 3 layers. The middle layer consist of a hydrophobic helical hairpin completely buried within amphipathic helices. The way in which this fragment, which is highly water soluble, alters its conformation to become an integral membrane protein and eventually an ion channel was studied applying the FRET approach.[128] To introduce extrinsic fluorescent probes at specific sites of the molecules, three cysteine mutants were generated, each bearing one SH group on helix 1, on helix 2, and between helices 6 and 7, respectively. Energy transfer was measured between the three tryptophan residues of the colicin fragment and the IAEDANS-labeled SH groups. Upon addition of lipids, the distances between tryptophans and IAEDANS increased by 14 to 22% as compared to the distances measured in the soluble form. Mixing of IAEDANS-labeled fragments with nonfluorescent fragments, in which the tryptophan fluorescence was abolished by *N*-bromosuccinimide treatment, resulted in a greater increase in these specific distances (16 to 38%). This showed that intermolecular transfer reduces the distances estimated in samples containing only labeled protein. The FRET data showed that the amphipathic helices of the C-terminal fragment open out on the surface of the lipid bilayers during the initial phase of membrane insertion. These results correlated well with the "umbrella model", which suggests that the complete hydrophobic hairpin of helices 8 and 9 could be inserted into the membrane without coming in contact with the aqueous phase while the other helices fold upward like a fabric on the surface of the membrane.[128]

Pardaxin, a neurotoxic peptide composed of 33 amino acids, is the main component in the secretion of the Red Sea Moses sole fish *Pardashirus marmoratus*, where it acts as a shark repellent. Pardaxin is composed of two α-helices, with proline residues acting as a hinge between the two. At low concentrations ($<10^{-7}$) pardaxin forms single channels in lipid bilayers, while at higher concentrations ($>10^{-6}$) cytolysis is induced.[129] The aggregation and organization of pardaxin in membranes were studied by measuring FRET between peptide molecules labeled with carboxyfluorescein (donor) at their N-terminal and peptides labeled with carboxytetramethylrhodamine (acceptor) at either their N- or C-terminal amino acid. The self-aggregation monitored by FRET correlated with the pore-forming ability of the pardaxin analogs. A comparison of the FRET efficiency measured between N^1-donor-labeled pardaxin and N^1-acceptor-labeled pardaxin with that measured between the same donor and C^1-acceptor-labeled pardaxin suggested that aggregates are formed in an ordered manner, with preferentially parallel orientation of monomers within the aggregate.[130] The FRET data were in good agreement with the mechanism of pore formation of pardaxin described by the "barrel stave" model. According to this model, the monomers aggregate into a barrel-like formation, in which a central aqueous pore surrounded by proteins is formed. This pore increases in diameter through the progressive recruitment of additional monomers.[130,131]

Melittin is a major component of bee venom with 26 amino-acid residues. Melittin displays an amphiphilic property, since the primary sequence is divided into the hydrophobic 1-20 segment and the hydrophilic 21-26 segment.[132] Interaction of melittin with lipid bilayer membrane may result in increased membrane permeability and formation of voltage-dependent ion channels.[133] Conformation and assembly formation of melittin have been studied by FRET-based methods among many other techniques. These studies can be sorted into two groups: papers dealing with melittin lipid interactions and papers studying the self-association of melittin in aqueous solution and in membrane bilayers.

The incorporation of melittin into phosphatidylcholine bilayer vesicles was studied by FRET experiments using 2- and 6-anthroyloxypalmitic acid as acceptor and peptide tryptophan as donor.[134] Above phase transition temperature T_t the melittin incorporated and distributed more randomly in small unilamellar vesicles than below T_t, when melittin incorporation was reduced. Cooling the temperature below T_t after incubation at temperature above T_t resulted in a deeper incorporation of the peptide into the lipid bilayer due to electrostatic interaction between the lipid phosphate groups and the positively charged amino acids in the hydrophilic segment.[134] Lipid-specific penetration of melittin into phospholipid model membranes was observed by Batenburg et al.[135] FRET data measured between tryptophan of the peptide and an acceptor incorporated into the lipid bilayer was compared in egg phosphatidylcholine and bovine heart cardiolipin model membranes. The results were interpreted as a proof of a deeper positioning of the tryptophan containing part of the peptide molecule in the cardiolipin system in comparison with the situation in phosphatidylcholine.[135] Similarly, FRET data measured between tryptophan and phospholipids bearing the anthrylvinyl fluorophores showed that the depth of penetration of melittin into the bilayer depends on the polar headgroup composition of the phospholipid matrix.[136]

In water, melittin can exist as a monomer or, at high concentration and/or at high ionic strength, as tetramers. Association of melittin in lipid membranes was studied by Vogel and Jähnig[137] using melittin chemically modified at Trp-19 by 2-hydroxy-5-nitrobezylbromide in Koshland reaction as energy acceptor and the tryptophan of unmodified melittin as energy donor. On the basis of FRET data between unmodified and modified Trp-19 residues, it was suggested that melittin was aggregated in the membranes predominantly in the form of tetramers. In tetramers, the hydrophilic sides of four bilayer-spanning helices face each other to form a hydrophilic pore through the membrane.[137] Similarly, using *N*-bromosuccinimide oxidized Trp-19 melittin as energy acceptor, self-association of melittin in phospholipid bilayers was detected at high salt concentration by FRET experiments.[138] In contrast to these observations, melittin formed no association when energy transfer was measured between the tryptophan residue of unlabeled melittin and the *N*-methyl anthranoyl residue covalently bound to a basic lysine residue on melittin.[139] Schwarz and Beschiaschvili found also no evidence for aggregation.[140] The controversy between these results has been resolved by John and Jähnig,[141] who repeated these experiments using time-resolved fluorescence energy transfer measurements instead of steady-state measurements. Their carefully performed experiments supported the observation that the melittin in fluid membranes is usually monomeric. Only at the high melittin/lipid molar ratio of 1/200 and high ionic strength was evidence for aggregation obtained, the percentage of aggregated melittin molecules being of the order of 10%. They pointed out that the difference in the observed results may come from the fact that in one case the direct absorption by acceptor molecules has not been taken into account,[137] and in another case the contribution to FRET due to the random distribution of donors and acceptors has not been considered.[138]

8. Other Proteins

FRET measurements between *N*-methylanthranoyl derivative of the potent depolarizing agent of batrachotoxin and dimethylaminocoumarinyl derivative of the allosteric activator α-scorpion toxin showed that the distance separating the receptor sites for these toxins on voltage-gated sodium channels in synaptosomes was 3.7 ± 1.0 nm.[142] It was found that the binding of α-scorpion toxin must involve conformational changes extending over large distances from the batrachotoxin-binding locus.[142] Kinetics of the aggregation of *Clostridium perfringens* θ-toxins (or perfringolysin O) was studied by Harris et al.[143] using FRET measurements. Perfringolysin O (PFO) was labeled with either FITC or TRITC. FRET was only observed when FITC-PFO and TRITC-PFO were incubated in the presence of membranes. While the temperature-dependent rates of change in FRET upon mixing the fluorescently labeled toxin with erythrocyte membranes proceeded without a lag phase, the hemolysis exhibited a distinct lag phase. The aggregation of PFO was virtually complete before the onset of hemolysis. The investigators concluded that the aggregation of membrane-associated PFO is necessary to initiate the hemolytic process, and the lag phase which occurs before the onset of hemolysis reflects the kinetics of PFO monomer to polymer conversion.[143] Addition of the carboxyterminal tryptic fragment of PFO to the above-defined FRET system inhibited membrane-dependent FRET, suggesting that an alternating copolymer of PFO and the tryptic fragment might form, where the distance between donor and acceptor moieties increases significantly enough to diminish FRET.[144]

In order to monitor the interaction of the GTP-γ-S-bound α subunit of transducin (α-T-GTP-γ-S) with the retinal cyclic GMP phosphodiesterase (PDE), the FRET approach has also been used.[145] Significant FRET efficiency was detected between fluoresceinated PDE and eosin-labeled α-T-GTP-γ-S. The quenching of fluorescence of fluorescein-labeled PDE by eosin-labeled α-T-GTP-γ-S was dose dependent, saturable, and specific for the GTP-γ-S-bound form of the α subunit of transducin.[145] A more thorough study was performed by Heithier et al.[146] to investigate subunit interactions of GTP binding proteins. For this purpose, α_o- and βγ-subunits and trimeric G_o purified from bovine brain, βγ-subunits from bovine rod outer segment membranes, and β_1-adrenoceptor from turkey erythrocyte were all labeled either with FITC or tetramethylrhodamine maleimide under conditions which leave the labeled proteins functionally intact. The high affinity and nonspecific interactions between α_o- and βγ-subunits were readily distinguished by comparing FRET before and after dissociation with 10 μ*M* GTP-γ-S. Similar FRET efficiency was measured between α_o- and βγ-subunits regardless of whether they were prepared from bovine brain or from bovine retinal transducin. The βγ-subunits from bovine brain were found to bind with high transfer efficiency to both hormone-activated and nonactivated β_1-receptors reconstituted in lipid vesicles.[146] This observation supported the previous finding[147] that βγ-subunits can bind to the nonactivated β_1-adrenoceptor. These findings attest to the suitability of FRET for studying protein-protein interactions of G-proteins and G-protein-linked receptors.

Energy was transferred from excited tryptophan residues of membrane proteins of a moderately halophilic bacterium, *Pseudomonas halosaccharolytica*, to the calcium ionophore A23187.[148] According to the calculated distances the ionophore cannot be bound to the membrane proteins but is probably located in the lipid bilayer.[148]

FRET measurement was performed to determine the distance between substrate sites on the Na^+-glucose cotransporter from rabbit intestinal brush border. The putative glucose and Na^+ binding sites were labeled with covalent fluorescent probes. Using different donor/acceptor pairs with different FRET efficiencies, the calculated distance between substrate sites was approximately equal to 3.5 nm. The fact that the Na^+ site and glucose site probes bind to a 75-kDa polypeptide, copurify with the same isoelectric point, and retain FRET after purification indicates that the sites are on the same 75-kDa polypeptide.[149]

Burgun et al.[150] studied the pH-induced reorganization of synaptic membrane by measuring FRET between tryptophan residues of membrane proteins and 2-(9-anthroyloxy) stearate or 12-(9-anthroyloxy) stearate. Decrease in pH provoked an increase of the FRET efficiency, which was more pronounced with the surface probe, indicating either a redistribution of probes to protein-rich regions or a redistribution of both the probes and the proteins to the same membrane domain. Both possibilities lead to a closer average packing between proteins and the probes.[150]

Intermembrane and intramembrane protein-protein association was studied during Ca^{2+}-promoted chromaffin granule membrane fusion by applying the FRET approach.[151-153] Proteins of the chromaffin granule membranes were labeled *in situ* with maleimide iodoaminonaphtyl sulfonate serving as donor and fluorescein mercury acetate or fluorescein-5-maleimide serving as acceptor. If the donors and acceptors were placed on the same membrane, an increase in FRET could be detected upon addition of calcium (400 μM), at a rate five to ten times slower than the aggregation itself. The detected FRET was attributed to the patching of proteins on the granule surface, since no significant granule-granule aggregation would occur at this calcium concentration. Placing the donor and acceptor on separate membranes and mixing them in the absence of calcium resulted in a very slow increase in FRET. Addition of calcium significantly increased FRET but the rate of increase was, once again, five to ten times slower than vesicle aggregation itself. The failure to see any FRET component as rapid as dimer formation itself suggested that either the SH groups are removed too far from the contact points in the intermembrane protein-protein interaction or these interactions are not detectable above the background noise of the FRET experiment.[151-153]

E. STRUCTURE OF ERYTHROCYTE MEMBRANE

The thickness of air-dried erythrocyte plasma membrane was determined using FRET measurements in a microscope. The erythrocyte membrane was embedded between a monolayer of a fluorescent dye, *N,N'*-distearyloxacyanine, and a thin gold layer serving as donor and acceptor, respectively. From the efficiency of FRET the thickness of the membrane was found to be 6.5 nm.[154]

Binding of hemoglobin to red cell membrane was studied through an energy transfer mechanism of hemoglobin quenching the fluorescence of 12-(9-anthroyloxy) stearate embedded in the membrane.[155] The binding constant determined by these FRET experiments indicated a high affinity interaction between hemoglobin and the red cell membrane. This interaction proved to be reversible and electrostatic in nature. Important differences were observed between the interaction of hemoglobin with the red cell membrane and the interaction of hemoglobin with models of artificial membranes composed of the different lipids of the red cell membranes. Although the binding of hemoglobin to the artificial membranes was also electrostatic, it was not reversible, suggesting that hemoglobin was incorporated into the lipid bilayer. In the case of red blood cells, the reversibility shows that hemoglobin is not incorporated into the lipid bilayer. This observation was also supported by distance measurements between hemoglobin and the fluorescent lipid probes. The calculated closest distance between heme groups and 12-(9-anthroyloxy) stearate was 5.0 nm. Considering the possible orientation of heme, it was concluded that the minimal distance between acceptor and donor was around 4.0 to 6.0 nm. This range is in good agreement with the assumption concerning the localization of 12-(9-anthroyloxy) stearate in the outer lipid layer and that of hemoglobin on the interior surface of the membrane.[155] Eisinger and Flores[156] further extended this type of measurement using a series of *n*-(9-anthroyloxy) stearic acids ($n = 2, 5, 9, 12$) as donors and the hemoglobins as acceptors. From the measured FRET efficiency for various probes, the distance between the fluorophore and the hemoglobin was found to range from 4.0 nm for 2-(9-anthroyloxy) stearic acid to 3.1 nm for 12-(9-anthroyloxy) stearic acid, which is in good agreement with the data discussed above.[155] The investigators concluded that the cytosolic hemoglobin is in contact with the inner membrane surface, or nearly so.[156]

The anion transport system of the human erythrocyte plasma membrane catalyses bicarbonate for chloride exchange to facilitate CO_2 release in the lungs. The band 3 protein of 95 kDa is responsible for this exchange. The stilbenedisulfonate inhibitory site (anion binding site) of the human erythrocyte anion-exchange system (band 3 dimer) was labeled with four different fluorescent stilbenedisulfonates to serve as donors and the cysteine residues on the 40-kDa amino-terminal cytoplasmic domain of band 3, sensitive to mild trypsin digestion, were labeled with four different fluorescent SH reagents to serve as acceptors.[157] Significant FRET efficiencies were measured between the donor and acceptor sites. Although these sites are approachable from opposite sides of the membrane by impermeant reagents, they were separated by only 3.4 to 4.2 nm, indicating that the anion binding site is located in a protein cleft which extends some distance into the membrane.[157] The relative proximity of subdomains, split by a central hinge, in the cytoplasmic domain of band 3 was also investigated applying the FRET approach.[158] The fluorescence donors were chosen to be the tryptophans in the fragment, since four of the five tryptophans of the fragment are located between residues 76 and 106, i.e., on the N-terminal side on the of the putative hinge. The two cysteines, located on the C-terminal side of the putative hinge, were labeled with IAEDANS to serve as energy acceptors. The FRET efficiency, measured between donor and acceptor sites, showed significant pH dependence. On the basis of this observation, it was concluded that the donor and acceptor are closer to each other at pH 6.5 than they are at pH 10. The investigators suggested that this movement may involve a pH-dependent bending of the two subdomains of band 3 at the putative central hinge. This conformational change could account also for the observed increase in the Stokes radius with pH.[158]

The mechanical properties of mammalian erythrocytes are defined by the specialized network of skeletal proteins located at the cytosolic side of the membrane. This skeleton consists of a two-dimensional spectrin-actin network, which appears to be anchored to the membrane by at least two protein linkages: the ankyrin-band 3 linkage and the protein 4.1-glycophorin linkage. In addition to protein-protein interaction, protein-lipid association may have an important role in the coupling of this skeleton to the membrane. This hypothesis was tested by Sharokh et al.[159] using FRET measurement between protein 4.1 conjugated with IAEDANS or IAF and bilayer interface labeled with 3,3'-ditetradecyloxacarbocyanine perchlorate or tetramethylrhodamine phosphatidylethanolamine. Using both donor-acceptor pairs, an average distance of 7.5 ± 0.5 nm between the probe on the protein and the surface of inside-out vesicles was calculated. In control measurements with protein 4.1 and liposomes with a phospholipid composition similar to that of the inner leaflet of the red cell membranes, a distance of 4.9 ± 0.5 nm was found. On the basis of these data the investigators suggested that protein 4.1 should have an elongated shape for the following reasons. The SH groups, targets for the fluorescent labeling, are located on the 30-kDa N-terminal segment of protein 4.1 of 80 kDa. In addition, the protein 4.1 is known to bind phosphatidyl serine liposomes with its 30-kDa domain facing the lipids.[159] A globular protein of 80 kDa would have a diameter of 5.7 nm, which should localize the donor site outside the 30-kDa SH-containing N-terminal domain. Instead, an elongated protein that extends perpendicularly away from the bilayer accommodates the donor site at the observed distance of 4.9 nm from the liposome membrane while still within the 30-kDa domain. In the case of inside-out vesicles, the observed longer distance between the donor and acceptor was interpreted through the notion that the cytoplasmic tail of glycophorin is interposed between protein 4.1 and the lipids.[159]

III. FRET MEASUREMENTS ON BIOLOGICAL MEMBRANES

Almost all of the examples previously discussed dealt with the application of FRET technique on model membranes, mostly on liposomes, for studying proximity relationships of membrane components. A more realistic picture can be obtained about the organization of elements in biological membranes, when these types of measurements are performed on intact, in many cases, live cells. Application of FRET on live cells utilizes all the advantages provided by fluorescence techniques, e.g., sensitivity, simplicity, specificity, and little or no interference with biological processes of the labeling itself. Since fluorescence labeling is very gentle, FRET experiments can be performed on functioning cells. These experiments can be performed on cell suspensions, in spectrofluorimeters, or on single cells using microscopes or flow cytometers. Other methods, such as electron microscopy, chemical cross-linking, immunoprecipitation, and electron spin resonance can also be used to obtain valuable information on spatial relationships. However, the interference of these techniques with the system studied is much more aggressive than in the case of fluorescent labeling. These methods are beyond the scope of this review. Here we will discuss a few examples of FRET measurements in biological membranes without aiming at completeness.

A. AGGREGATION OF LECTINS

Lectins are plant proteins with specificity for particular carbohydrate moieties. Concanavalin A (Con A), for example, binds α-methylmannoside, a component of several major glycoproteins in mammalian cell membranes. Lectins, which are usually multivalent, can be used to precipitate and isolate surface components ("lectin receptors"). On living cells, aggregation of these components can be induced by addition of these multivalent lectins, and in some cases a mitogenic, i.e., a proliferative response is evoked.

Topographical distribution of different lectins binding to intact cells or platelets was studied applying FRET in spectrofluorimetry (Table 2). FRET efficiency measured between fluoresceinated and rhodaminated Con A bound to fibroblasts was independent of the extent of patching.[160] The results have been interpreted by assuming that Con A binds in more than a single monolayer thickness, thereby producing a three-dimensional distribution. Patching then caused the layer to thicken in particular regions

Table 2. **FRET Measurements on Lectin Receptors**

Instrument	Lectin	Donor	Acceptor	Ref.
Spectroflurimeter	Con A[a]	FITC[b]	TRITC[c]	11, 160
	Lentil lectin	FITC	TRITC	161, 162
	Succinyl-Con A	FITC	TRITC	163
Microscope	Con A	Dansyl[d]	Rhodamine	164
	Con A	FITC	TRITC	168, 169
	LCA[e]	FITC	S-RH[f]	167
Flow cytometer	Con A	FITC	TRITC	10, 11, 170–172

[a]Con A; concanavalin A.
[b]FITC; fluorescein isothiocyanate.
[c]TRITC; tetramethylrhodamine isothiocyanate.
[d]Dansyl; *N*-5-dimethylaminonaphthalene-1-sulfonyl.
[e]LCA; *Lens culinaris* agglutinin.
[f]S-RH; sulforhodamine.

without much effect on FRET efficiency.[160] The same donor-acceptor pair was used for studying the association and dissociation mechanism of lentil lectin receptor microclustering on HeLa cells.[161,162] No FRET occurred when cells were fixed prior to lentil lectin binding. At 37°C, a maximum FRET efficiency was reached even after 30 s, long before visible patching was observed on a parallel sample using a fluorescence microscope. This microclustering was reversible and the dissociation of these microclusters could be monitored by measuring changes in FRET efficiency at 4°C. Irreversibility was reached concomitantly to the appearance of visible patches, reflecting the aggregation of microclusters.[161,162] Significant FRET was detected between thrombin or dimeric (succinylated) Con A and other membrane proteins after addition of the aggregating agents. These ligands were labeled either with FITC (donor) or rhodamine (acceptor). Evidence was obtained for clustering of membrane proteins induced by platelet activators.[163]

The surface topography of Con A bound to normal and transformed murine fibroblasts has been studied by microscopic FRET technique (Table 2).[164] However, as it was difficult to calculate the absolute value of FRET efficiency between dansylated and rhodaminated Con A, the ratio of fluorescence intensity of the sensitized acceptor to that of the quenched donor emission was taken as a measure of FRET. It was found that the distribution of Con A was nonrandom in each case, but the binding sites were in close proximity on the transformed cells.[164]

Herman and Fernandez[165] employed time-resolved FRET microscopy for investigating the topography of Con A on the surface of cultured chick muscle cells during the period of myoblast fusion. Pyrene-labeled Con A was applied as donor, FITC-labeled Con A as acceptor. Significant decrease in FRET efficiency was observed on cells cultured for 24 to 48 h, indicating a marked increase in average Con A receptor separation on the developing muscle cell surface. This interval corresponds to the time period in which these cells ceased to proliferate and underwent fusion. The increase in average receptor proximity was only transient, since after 48 h — when the myoblast fusion was complete — FRET increased rapidly and leveled off close to the values detected in the first 24 h. This transient change in average proximity can be interpreted with various topographical distributions. If the receptors always exist in clusters, the changes in FRET would reflect different degrees of packing of the clusters. Alternatively, the receptors might exist in two extreme states: free or aggregated. In this case, changes in FRET would monitor changes in the relative populations of these two states. The third possibility would be to assume that receptors are uniformly distributed: in this case changes in FRET may simply reflect changes in the uniform density of receptors. The investigators concluded that their data by themselves do not permit distinction between these various possibilities.[165,166]

FRET microscopy was also used to examine the internalization of lipid and lectin probes.[167] Fluorescence of FITC-labeled *Lens culinaris* agglutinin (LCA) was quenched by a sulforhodamine derivative of phosphatidylcholine (SR-PC) at the plasma membrane of GNK cells at 2°C, owing to energy transfer between the donor acceptor pair. There was no significant change in FRET efficiency measured during internalization of both FITC-LCA and SR-PC, suggesting that they were colocalized. Under microscopic investigation, FITC-LCA appeared only in a subset of those internal membrane compartments labeled with SR-PC, suggesting the fluorescent lectin was sorted into certain intracellular locations, while SR-PC appeared to act as a general membrane marker during the internalization process.[167]

Jovin and Arndt-Jovin[168,169] have developed a new microscopic FRET technique. From the rates of donor bleaching in the presence and absence of acceptor, the transfer efficiency between donor and acceptor molecules can be calculated on a pixel-by-pixel basis. The applicability of this technique has been demonstrated by measuring FRET between FITC-Con A and TRITC-Con A and the surface of rat basophilic leukemia cells (line 2H3). The observed high transfer efficiency (40 to 50%) further supported the previous model that the distribution of Con A on the cell surface is uneven owing to microcluster formation.[168,169] Although the microscopic FRET technique provides an excellent tool for determining a map of FRET efficiency within a cell, it is slightly biased by the fact that the number of cells which can be analyzed within a reasonable time interval is limited. Application of flow cytometric measurements can recuperate this shortcoming of the microscopic FRET technique.

The statistical accuracy of the distribution of FRET efficiency in a heterogeneous cell population has been greatly improved by employing flow cytometers where the rate of analysis can be as high as 1000 cells per second. In early flow cytometric FRET measurements only the mean values of fluorescence intensity distribution curves were used for calculating FRET efficiency (Table 2). Although these experiments did not fully exploit the capability of flow cytometry for determining FRET efficiency on a cell-by-cell basis, a great deal of new information was obtained about the cell surface distribution of lectin A receptors.[170,171] High FRET efficiency between FITC-Con A and TRITC-Con A bound to transformed Friend erythroleukemia cells was observed even at 4°C. The investigators estimated the separation between Con A molecules at saturation on the cell surface to be around 4 nm, a range possibly characteristic for microclusters induced by ligand binding.[170] Using the same donor acceptor pair, Jenis et al.[171] investigated the topography of Con A receptors on HL-60 promyelocytic leukemia cells during cell differentiation. Though the number of Con A binding sites present on cells did not change during granulocytic differentiation over time, the average distance between these binding sites was increased causing a decrease in FRET efficiency. During differentiation into macrophages, however, both receptor density and receptor number increased. The authors suggested that changes in membrane structure responsible for the altered Con A topography appear as a result of functional changes that yield the mature state in the HL-60 cell.[171]

In contrast to the above-mentioned flow cytometric FRET measurements, our flow cytometric method, introduced in 1983/1984, is capable of determining the absolute value of FRET efficiency on a single cell basis.[10,11] We call this technique flow cytometric energy transfer (FCET). With FCET, frequency distributions of FRET efficiency can be generated for the cell population and the correlation of FRET efficiency with other cell parameters can also be investigated. The first version of FCET can be applied only on flow cytometers capable of dual wavelength excitation. FCET was modified and simplified, limiting also the range of its applicability, in order to adapt FCET to flow cytometers with single laser excitation.[172] The algorithm and the technical details of FCET techniques are thoroughly covered in chapter 1 of this book. Since the introduction of FCET, the technique has been applied for studying the topography of cell surface elements in numerous experimental systems. Reviews, summarizing various applications of FCET, are also available in the literature.[173-178]

We applied FCET for studying cell surface distribution of Con A on HK 22 murine lymphoma cells.[10,11] Energy transfer distribution histograms of HK 22 cells labeled with both FITC-Con A and TRITC-Con A are shown on Figure 1. The peak position of the energy transfer distribution curves increased as we increased TRITC to FITC molar ratio, while maintaining the total Con A concentration at saturation level. The HK 22 cells seemed to be homogeneous according to the energy transfer distribution curves indicating that there were no significant differences in the surface density of Con A receptors on different cells. The interpretation of energy transfer efficiency data in terms of distance relationships is straightforward only in case of one donor-one acceptor systems, where the localization and the relative orientation of fluorophores are known. The situation is more complicated when the energy transfer process takes place in membranes, because the labeled components are embedded in membranes imposing two-dimensional restriction. Analytical solutions have been derived for random distribution of donors and acceptors[160,179] and numerical solutions have been applied to predict the transfer efficiency for nonrandom distribution.[179-182] To differentiate between random and nonrandom distribution, the energy transfer efficiency has to be determined as a function of acceptor concentration in the membrane. By fitting the mean values of energy transfer distribution curves shown in Figure 1 and the corresponding relative surface densities of TRITC-Con A to theoretical curves,[179] an absolute surface density of

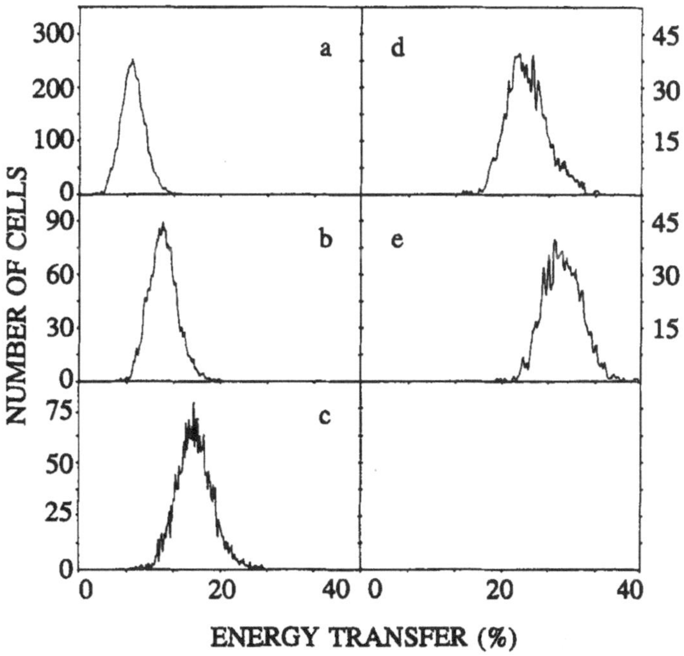

Figure 1. FRET measurements between Con A lectins on cell surface. Frequency distribution histograms of FRET efficiency between FITC-Con A on HK22 mouse lymphoma cells. Rhodamine to fluorescein ratios: (a) 0.21, (b) 0.54, (c) 1.0, (d) 2.3, and (e) 4.6. The total Con A concentration was kept at a saturating level (0.2 mg/ml).

$1.1 \times 10^4/\mu m^2$ was obtained for Con A binding sites.[10,11] The same Con A surface density was obtained for normal mouse lymphocytes and Gross virus leukemia cells prepared from AKR mice, based on energy transfer measurements.[172] In the flow cytometer, normal and leukemic cells were distinguished on the basis of their scattered light intensity which is proportional to their size, so FCET data of normal and leukemic cells prepared from the same animal could be analyzed separately. In addition to FCET measurements, the surface density of Con A can be determined in a trivial way as well. Using a simplified model of the cell surface corresponding to a smooth sphere and the number of Con A binding sites per cell, the surface density of Con A can also be calculated. Surface density values of Con A for various cell types calculated from energy transfer data and by assuming smooth sphere surface are summarized in Table 3. Regardless of the cell type, the effective surface density of Con A receptors calculated from FRET efficiency values was the same. This value was two to three times higher than that calculated for a smooth sphere with a uniform Con A distribution. The ratio between these densities can be even higher if the cell surface is irregular with a much larger area than assumed for the sphere. These data suggest

Table 3. **Cell Surface Density (D_{surf}) of Con A Binding Sites (BS) on Different Cells**

			Leukemic mouse lymphocytes	
Parameters	**HK22 cells**	**Normal mouse lymphocytes**	**SC1[a]**	**SC2[b]**
Diameter (μm)	15	8	8	11.5
Con A (BS/cell)	$3.6 \cdot 10^6$	$7.1 \cdot 10^5$	$7.2 \cdot 10^5$	$1.2 \cdot 10^6$
D_{surf} (BS/μm²)				
Assuming smooth sphere surface	$5.1 \cdot 10^3$	$3.6 \cdot 10^3$	$3.6 \cdot 10^3$	$2.9 \cdot 10^3$
Calculated from ET efficiency[c]	$1.0 \cdot 10^4$	$1.0 \cdot 10^4$	$1.0 \cdot 10^4$	$1.0 \cdot 10^4$

[a]Regions in light scatter histograms (SC1 = small cells).
[b]Regions in light scatter histograms (SC2 = large cells).
[c]Energy transfer (ET) efficiency data were fitted to theoretical curves[179] describing ET processes in two dimensions.

From Szöllősi, J., Mátyus, L., Trón, L., Balázs, M., Ember, I., Fulwyler, M. J., and Damjanovich, S., *Cytometry*, 8, 120, 1987. With permission.

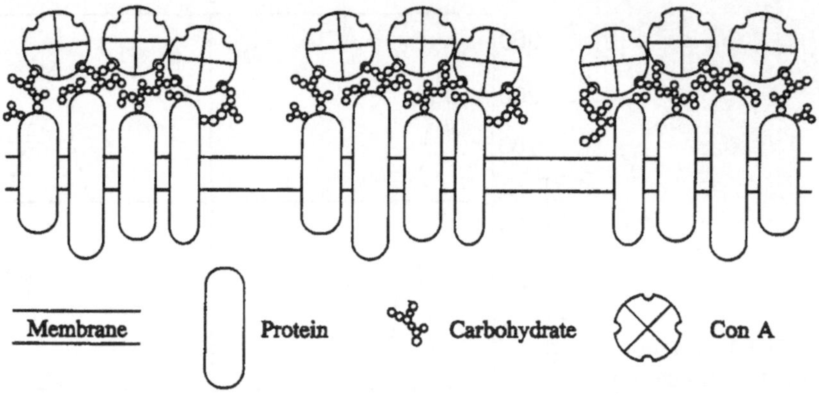

Figure 2. Schematic representation of Con A binding to cell surface receptors. Con A binds to mannoside residues of glycoproteins. According to FRET measurements, Con A receptors form microclusters on the cell surface.

that the Con A receptors could be clustered on the cell surface, a result expected from multiple sites of glycosylation of glycoproteins and glycolipids as shown in Figure 2. According to this model, the tetravalent Con A molecules can bind to any mannoside sugar moiety on the cell surface glycoproteins. The observed variations in surface density of Con A calculated by assuming smooth sphere surface might mean that the average distance between Con A clusters varies among different cell types.[10,11,172] This finding is in accordance with previous studies suggesting the microclustering of Con A receptors.[164-166,170]

B. INTERACTION OF MEMBRANE PROTEINS WITH MHC MOLECULES

Major histocompatibility complex (MHC) antigens, members of the immunoglobulin superfamily, have a central role in immunity. MHC class I proteins on the cell surface consist of a heavy chain (about 45 kDa in size) and a noncovalently associated light chain, the β_2-microglobulin, of about 12 kDa.[183] MHC class II proteins are heterodimers of one α-chain noncovalently attached to one β-chain (about 34 and 29 kDa, respectively).[184] The major function of MHC class I and class II molecules is to bind and present peptide fragments of degraded proteins to T cells.[185,186] Additionally, there is considerable evidence that these molecules have other roles as well.[187,188]

A nonimmunological role has been proposed for MHC products in which selected alleles may influence hormone-receptor binding and postbinding events by inducing conformational changes in receptor structure through direct interaction on the cell surface. Association of MHC gene products with hormone receptors and other cell surface proteins are supported by genetic, biochemical, and biophysical evidence, but the specificity and physiological significance of this interaction remain obscure.[189-194] These types of associations with the ancestral molecules of the immunoglobulin superfamily may well originate from, and could give insight into, the evolutionary process of MHC molecules, and the maintenance of polymorphism in populations.

The most commonly used methods for determining molecular associations in membranes include co-capping, co-immunoprecipitation, chemical crosslinking, modified fluorescence recovery after photobleaching, and FRET experiments. Application of one method alone usually does not give a clear and conclusive result. However, results with two or more methods reinforce each other and lead to a consistent picture of the molecular interactions in the plasma membrane. Although here we will focus mostly on the results gained by FRET technique, some of the supporting data obtained by other biochemical and biophysical methods will also be mentioned briefly.

1. Association of MHC Class I and MHC Class II Molecules

The human MHC class I molecules, HLA-A, -B, and -C, with bound fragments of foreign antigens are recognized by cytotoxic T cells during graft rejection and in the killing of virus-infected cells. The human MHC class II molecules presenting antigens on B cells and macrophages toward T-helper cells are encoded in the HLA-D region, which is comprised by three subregions, HLA-DR, DQ, and DP.

Although class I and class II molecules are structurally and functionally distinct, there is evidence that MHC class I and class II molecules can be more intimately related than previously thought. It has been reported that HLA class I-specific antibodies can cocap HLA class II antigens on B lymphocytes.[195,196] Analyzing pairs of monoclonal antibodies with various specificity to determinant on class I and II molecules, a frequent finding was unidirectional cocapping: here a first antibody cocapped epitopes seen by a second, but this antibody did not cocap epitopes seen by the first.[195,196] Adams et al.[197] have reported that HLA class II-specific monoclonal antibody coprecipitates a protein with an approximate molecular weight of 43 kDa, which appears to be the heavy chain of HLA class I antigen. It has also been observed that class II-restricted immune reactions can be inhibited by antibodies with specificity for class I antigens,[198-200] and that antibodies specific for class II antigens can abrogate recognition of class I determinants.[201]

These observations prompted us to perform studies where a more direct approach, FRET technique, was applied for the investigation of the possible association between HLA class I and class II molecules.[202] Toward this end we selected human B lymphoblastoid homozygous cell lines, PGF (HLA-A2, B7, DR2, DQw1) and JY (HLA-A2, B7, DR4, DQw1,3), which express class I and class II antigens ranging from 0.5 to 3×10^6 per cell. This level of HLA antigen expression is high enough to perform flow cytometric energy transfer measurements in order to map proximity relationship of class I and class II antigens on the cell surface. Using FCET approach we wanted to address the following questions: Is there any association between class I and class II antigens? Is there any homo- or heteroassociation between class I antigens? Is there any homo- or heteroassociation between class II antigens? A panel of monoclonal antibodies specific for various class I and class II antigens were labeled with either FITC or TRITC to serve as donors and acceptors in our FCET experiments. Cell samples were labeled with either FITC-antibody or TRITC-antibody alone to determine correction coefficients necessary for energy transfer calculations. FCET measurements were made on cells labeled with fluoresceinated and rhodaminated antibodies simultaneously. FRET efficiency was calculated on a cell-by-cell basis and energy transfer histograms, similar to those shown on Figure 1, were generated. The mean values of such histograms are summarized in Tables 4, 5, and 6. FRET efficiency of 5% or more can be regarded as significant, and as an indicator of the fact that molecules bearing the FITC- and TRITC-labeled antibodies are located within 10 nm of each other. Table 4 shows mean values of transfer efficiency measurements for pairs of class I- and class II-specific antibodies. Regardless of the monoclonal antibody used, significant transfer efficiencies were observed demonstrating the close proximity of class I and class II antigens. As shown

Table 4. Energy Transfer Efficiency Values (E) between Donor-Acceptor Pairs Bound to HLA Class I and Class II Antigens

Donor	Donor determinant	Acceptor	$E \pm \Delta E^a$ (%)
FITC-L368	β_2-microglobulin	R-P4.1	6.8 ± 1.4
	β_2-microglobulin	R-Leu 10	5.1 ± 1.4
FITC-W6/32	HLA-A,B,C	R-P4.1	8.0 ± 1.5
	HLA-A,B,C	R-Leu 10	4.9 ± 1.2
FITC-BB7.1	HLA-B7	R-P4.1	8.5 ± 1.2
	HLA-B7	R-Leu 10	8.0 ± 1.7
	HLA-B7	R-B7/21	5.0 ± 1.0
FITC-GAP-A3	HLA-A3	R-P4.1	9.2 ± 1.3
	HLA-A3	R-Leu 10	9.5 ± 1.7
	HLA-A3	R-B7/21	5.7 ± 1.0
FITC-L243	HLA-DR	R-W6/32	11.2 ± 1.8
FITC-P4.1	HLA-DR	R-W6/32	8.6 ± 1.1
FITC-Leu 10	HLA-DQw1,3	R-W6/32	12.2 ± 2.1
	HLA-DQw1,3	R-L368	7.1 ± 1.4
FITC-G3.53	HLA-DQw1	R-L368	6.0 ± 1.3
	HLA-DQw1	R-W6/32	11.5 ± 2.1
FITC-B7/21	HLA-DP	R-W6/32	12.8 ± 2.2

[a]Data represent mean ± SD of at least three independent experiments.

From Szöllősi, J., Damjanovich, S., Balázs, M., Nagy, P., Trón, L., Fulwyler, M. J., and Brodsky, F. M., *J. Immunol.*, 143, 208, 1989. With permission.

Table 5. **Energy Transfer Efficiency Values (E) between Donor-Acceptor Pairs Bound to HLA Class I Antigens**

Donor	Determinant	Acceptor[a]	$E \pm \Delta E^{b}$ (%)
FITC-L368	β_2-microglobulin	R-L368	2.1 ± 1.2
FITC-BB7.1	HLA-B7	R-BB7.1	1.9 ± 1.3
FITC-GAP-A3	HLA-A3	R-BB7.1	1.5 ± 0.9
FITC-L368	β_2-microglobulin	R-W6/32	11.3 ± 2.1
FITC-W6/32	HLA-A,B,C	R-L368	8.6 ± 1.8
FITC-BB7.1	HLA-B7	R-W6/32	10.2 ± 1.8
	HLA-B7	R-L368	9.2 ± 1.7
FITC-GAP-A3	HLA-A3	R-W6/32	12.3 ± 2.2
	HLA-A3	R-L368	9.1 ± 1.9

[a]In case of competing monoclonal antibodies care was taken that half of the available binding sites should be occupied by the rhodaminated antibody.
[b]Data represent mean ± SD of at least three independent experiments.

From Szöllősi, J., Damjanovich, S., Balázs, M., Nagy, P., Trón, L., Fulwyler, M. J., and Brodsky, F. M., *J. Immunol.*, 143, 208, 1989. With permission.

Table 6. **Energy Transfer Efficiency Values (E) between Donor-Acceptor Pairs Bound to HLA Class II Antigens**

Donor	Donor determinant	Acceptor	$E \pm \Delta E^{a}$ (%)
FITC-L243	HLA-DR	R-L243[b]	3.2 ± 1.7
FITC-P4.1	HLA-DR	R-P4.1	3.5 ± 1.8
FITC-L243	HLA-DR	R-P4.1	1.2 ± 0.7
FITC-P4.1	HLA-DR	R-L243	2.0 ± 1.0
FITC-Leu 10	HLA-DQw1,3	R-Leu 10	10.4 ± 1.6
FITC-G3.53	HLA-DQw1	R-G3.53	1.6 ± 0.8
	HLA-DQw1	R-Leu 10	13.5 ± 2.1
FITC-Leu 10	HLA-DQw1,3	R-G3.53	7.7 ± 1.7
FITC-DQM	HLA-DQ	R-Leu 10	16.2 ± 2.6
FITC-B7/21	HLA-DP	R-B7/21	3.9 ± 1.9
FITC-L243	HLA-DR	R-Leu 10	9.0 ± 1.9
FITC-Leu 10	HLA-DQw1,3	R-P4.1	10.4 ± 1.8
FITC-G3.53	HLA-DQw1	R-P4.1	6.5 ± 1.2
FITC-DQM	HLA-DQ	R-P4.1	14.4 ± 2.3
FITC-B7/21	HLA-DP	R-P4.1	9.3 ± 1.4
	HLA-DP	R-Leu 10	6.5 ± 1.3

[a]Data represent mean ± SD of at least three independent experiments.
[b]In case of competing monoclonal antibodies care was taken that half of the available binding sites should be occupied by the rhodaminated antibody.

From Szöllősi, J., Damjanovich, S., Balázs, M., Nagy, P., Trón, L., Fulwyler, M. J., and Brodsky, F. M., *J. Immunol.*, 143, 208, 1989. With permission.

in Table 4, any class I molecule can associate with any class II antigen. In accordance with theoretical predictions, higher transfer efficiency was observed when the antibody of higher abundance was rhodaminated and served as acceptor.[173,175] In our experiments intramolecular energy transfer measurements between FITC-L368 and TRITC-W6/32 antibodies served as positive control. These antibodies bind to the same HLA class I antigen complex but to different epitopes (β_2-microglobulin and the heavy chain, respectively). In this case 11.3% transfer efficiency was measured (Table 5). The measured transfer efficiency would have been higher if no competition existed between these antibodies. Due to the slight

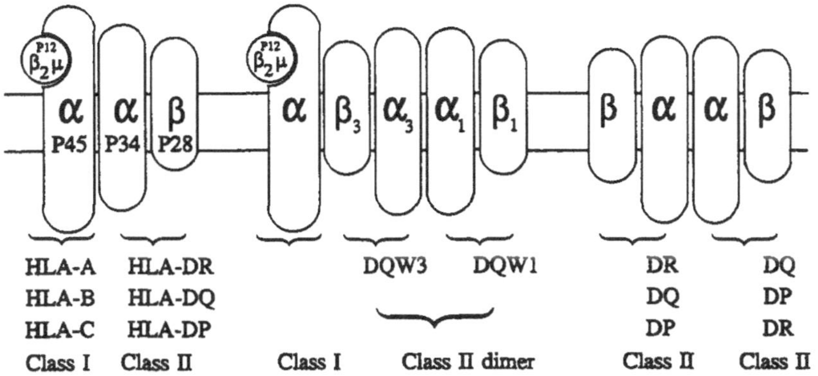

Figure 3. Cell surface distribution of HLA class I and II molecules on the cell surface. On the basis of FRET studies it is proposed that class I and II molecules associate with each other. DP, DQ, and DR molecules may form heteroassociations. In the case of DQ antigens, a DQw_1DQw_3 dimer may associate with HLA class I antigens.

competition (15%), there are HLA class I complexes on the cell surface which are labeled only with FITC-L368 or only with TRITC-W6/32, decreasing the possibility for intramolecular energy transfer. When cells were labeled with FITC-L368 and TRITC-L368 antibodies simultaneously, low FRET efficiency (1.5%) was observed, suggesting that class I antigens are not associated on the cell surface. In this case only intermolecular FRET was observed, since the conjugates block each other's binding. The monomeric nature of class I antigens was further demonstrated by the low transfer efficiency values measured between monoclonal antibodies binding to HLA-A3 and HLA-B7 antigens (Table 5). The anti-HLA-A3 and B7 antibodies bind to separate subpopulations of class I antigens, while the W6/32 and L368 antibodies to all class I antigens. The low transfer efficiency between FITC-labeled anti-HLA-A3 antibodies and TRITC-labeled anti-HLA-B7 antibodies demonstrated that there is no heteroassociation between HLA-A3 and B7 antigens. A similarly low level of homoassociation was observed among class II antigens, indicating that HLA-DR, -DP, and DQw1 antigens are also expressed mainly in monomeric state (Table 6). Interestingly, there was efficient FRET between the anti-DQ, FITC-Leu10, and TRITC-leu10 antibodies on PGF cells, for which no FRET between DQw1 molecules was observed. This suggests that anti-Leu10 recognizes a second population of DQ-related molecules which can associate among themselves or with DQw1. Biochemical evidence for the existence of this second population of antigens recognized by anti-Leu10 antibodies has been reported previously.[203] Significant transfer efficiency was observed between monoclonal antibodies binding to different class II antigens, suggesting that there is an association between DR and DQ, DR and DP, and DQ and DP antigens on the cell surface. This could be explained by the association between two αβ complexes coded by different loci or by the existence of complexes formed from an chain from one locus and a β chain from another locus creating a mixed molecule, e.g., DRαDQβ.[204] Unfortunately, the chain specificity of most of the class II-specific antibodies used here has not been successfully characterized; thus, our results not distinguish between these two possible interpretations. On the basis of our FCET data, we propose a model depicted on Figure 3. Class I and class II antigens are mostly expressed in a monomeric state on the surface of PGF cells. Class II antigens may form heteroassociations among themselves, and there is association between class I and class II antigens. There is an equilibrium between free, nonassociated, class I and class II molecules and the associated class I-class II antigen complexes on the cell surface. The number of class I-class II antigen complexes formed depends upon the available class I and class II molecules expressed in the plasma membrane. These results are also in agreement with cocapping experiments demonstrating class I-class II interaction.[195,196] Our data, however, demonstrated that these complexes are physically associated before cocapping as well.

Most of the data discussed above were obtained on the PGF cell line; the JY cells were used only occasionally. Lately, more experiments were performed on JY cells and, though most of the results were similar to that obtained on PGF cells, on one occasion a slight difference was observed, namely, while there was no homoassociation between HLA class I molecules on PGF cells, homoassociation was detected between class I antigens on JY cells (unpublished observations). The degree of homoassociation of class I antigens highly depends on the culturing conditions, whether the cells are in log phase or in plateau phase. These observations are in accordance with the findings of Chakrabarti et al.,[205] who observed self-association of MHC class I molecules on the surfaces of B lymphoblasts bearing appreciable amounts of β_2-microglobulin-free class I molecules. They used FITC- and Texas red-labeled HLA-A2 antigens reconstituted in liposomes to perform FCET measurements. Significant FRET efficiency was detected between the fluorescently labeled HLA-A2 molecules suggesting the existence of self-association. HLA aggregation was blocked by addition of β_2-microglobulin. This finding was also supported by fluorescence recovery after photobleaching and time-resolve phosphorescence anisotropy measurements.[205] FRET is a sensitive qualitative method for detecting molecular associations, but it does not define the cluster size. The rate of rotational diffusion of erythrosin labeled HLA-A2 molecules is much more sensitive to the size of the associated molecular complex. From measuring this rotational diffusion, it was concluded that aggregates had a radius four to seven times larger than that of a monomer. The aggregation of HLA-A2 molecules is not an artifact of their reconstitution into liposomes. HLA aggregates, revealed by measuring FCET between fluorescently labeled F_{ab} fragments of monoclonal antibodies, were readily detected on the surface of human lymphoblastoid JY cells.[205] The degree of self-association of HLA-A2 molecules is closely related to the amount of β_2-microglobulin-free class I molecules expressed on the surface of JY cells.[206]

In addition to human HLA class I molecules, we have also investigated the distribution and mobility of murine MHC class I antigens, i.e., H-2K^k molecules.[173-176,207-209] Since there is only one determinant on the H-2K^k antigen for a single monoclonal antibody, we were able to study the proximity of the antigens themselves using a mixture of the same monoclonal antibodies labeled with two different dyes. The lack of FRET between fluoresceinated and rhodaminated monoclonal antibodies suggested that the adjacent H-2K^k antigens are beyond the range of FRET (10 nm). Our rotational relaxation measurements on the H-2K^k antigen yielded correlation times consistent with the picture of a monomeric protein embedded in the plasma membrane, freely rotating about its membrane portion. In the case of murine class I antigens, both techniques gave a consistent picture of a monomeric, freely rotating, dispersed antigen.[173-176,207-209]

The self-association of MHC class I molecules is still an open question. Contradictory results have been published in the literature. The contradictions may be apparent or real, coming from experimental errors. Experiments are in progress in our laboratory in order to define physiological conditions where the degree of self-association of MHC class I molecules can be influenced in a controlled way.

2. Effect of Membrane Potential on MHC Conformation

Ion gradients across biological membranes generate membrane potentials, which are ubiquitous in practically all living cells and also in mitochondria.[173,175] Besides the so-called excitable cells of muscles and the nervous system, more and more data are available on the functional role of membrane potential in nonexcitable cells, such as those of the immune or endocrine system.[173,175] Significance of the plasma membrane potential is mainly attributed to its regulatory function in the ion transport between the extra- and intracellular world through gating ion and other transport channels. The membrane potential has an obviously different role in muscle or nerve cells, as compared to that in nonexcitable cells. An additional feature of the 10^7V/m potential gradient is that it can be changed very quickly into both directions, i.e., depolarized or repolarized. The high potential gradient has a remarkable capacity, namely, to alter protein conformation, lipid structure, and intermolecular interactions by generating or eliminating transient dipoles.[210] These effects are well known in the relation of ion channels, but are much less studied for other molecular elements of the cell membranes. Electrical fields have a great variety of functional effects on membrane elements, which have extensively been studied and analyzed in model membranes or processes like cell fusion (electrofusion) and electroporation.[210] A number of studies have been carried out focusing mainly on structural features indicated by either spectroscopic signals or electronmicroscopic observations.[211-214] Accessibility of different protein sites clearly show alterations of protein conformation either upon increasing or decreasing the membrane potential. However, an unambiguous detection of such changes is far from trivial. Wyatt and Cherry[215] presented data on the conformation of the anion transporter band 3 obtained from luminescence quenching experiments, indicating that rotational relaxation parameters were insensitive in detecting such fine alterations. Thus, transient dichroism or phosphorescence anisotropy measurements are not suitable to detect electric field-induced conformational changes.

In our recent experiments, we successfully applied intermolecular FRET between antibodies bound to β_2-microglobulin and the heavy chain of the HLA class I molecules to detect proximity changes between

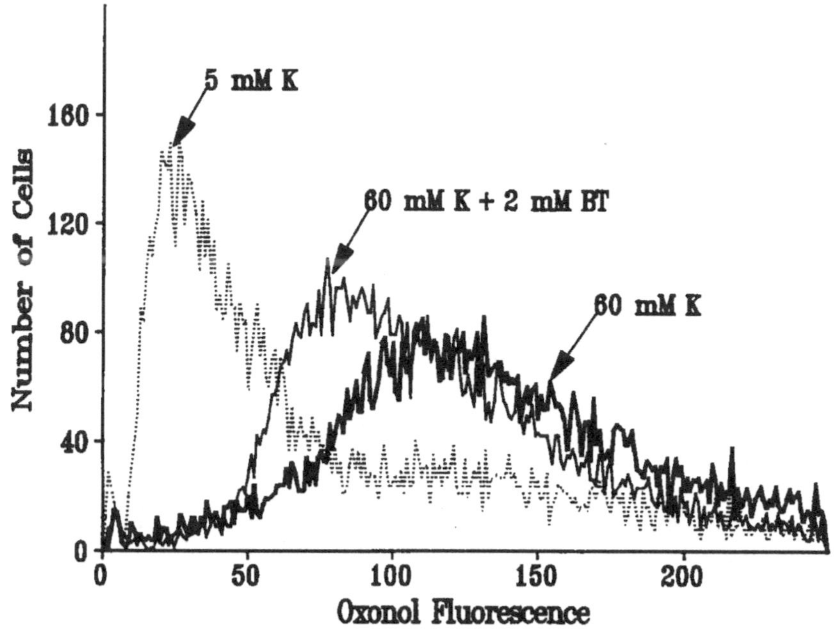

Figure 4. Membrane potential measurements on JY cells. Fluorescence intensity distribution histograms of cells labeled with oxonol, a translocating, potential sensitive dye. Depolarization of cells by high K^+ buffer increases the fluorescence intensity of cells (thick line). Addition of bretylium tosilate partially repolarizes the depolarized cells (thin line).

the two proteins. Due to the localization of β_2-microglobulin in a groove of the heavy chain of HLA class I molecule, proximity changes are supposed to monitor conformational changes as well. The magnitude of the changes was not very large. Thus, in order to determine a causal relationship between the changing potential field and the distance of membrane proteins, energy transfer experiments were first carried out under depolarizing conditions. Subsequently, the same cell sample was hyperpolarized and the energy transfer was determined again between the same donor-acceptor pair. Thus, the membrane potential was changed into both directions and followed by determining the changes of distance between the β_2-microglobulin and the heavy chain of the HLA class I. The observed distance changes pointed toward the opposite direction when the potential was changed from negative to more positive values and vice versa. These data are likely to indicate a conformational change of the heavy chain, which allowed the dislocation of the β_2-microglobulin to a certain extent. From the consistently opposite alteration of proximity between the molecules upon the above pattern of potential changes we could conclude that conformational changes occurred upon depolarization and this was restored by repolarization of the membrane. These experiments had the beauty that changing the potential and FCET measurements were carried out parallel on live samples, providing a perfect self-control of the experiment. The flow cytometric variation of the fluorescence energy transfer measurements yielded data over a large enough cell population to get rid of problems which might be caused by biological variations even among cells of the same cell line.

The actual measurements and data were as follows: JY B lymphoma line was cultured, the cells were harvested in the log-phase and labeled with fluoresceinated and rhodaminated monoclonal antibodies, and directed against the epitopes of β_2-microglobulin or HLA class I heavy chain. The one-to-one molar ratio of the proteins in the plasma membrane was supported by the identical transfer efficiency, regardless of which antibody the rhodamine (acceptor) dye was attached to. The depolarizing conditions were produced by an isotonic increase of the extracellular K^+ ion concentration from 5 to 145 m*M* in a stepwise fashion. Repolarization was ensured adding 2 m*M* bretylium tosylate to the cell suspension, which activated the Na^+-K^+-dependent ATPase, as described in a number of flow cytometric and patch clamp experiments earlier.[217,218] Figure 4 shows the effect of membrane potential changes of the cells upon increasing the extracellular K^+ concentration and subsequent addition of bretylium tosylate. The energy transfer distribution histograms of normo-, de-, and repolarized cells are shown on Figure 5 for aliquots of the same

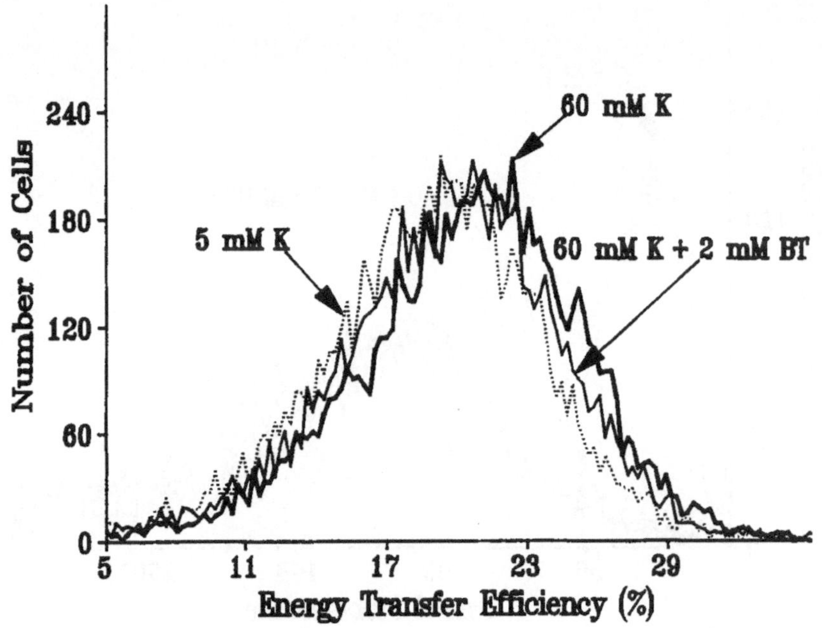

Figure 5. Effect of membrane potential on the conformation of HLA class I molecules. Frequency distribution histograms of FRET measured between FITC-labeled L368 monoclonal antibody, specific for β_2-microglobulin, and TRITC labeled W6/32 antibody, specific for the heavy chain of HLA class I antigens. Depolarization with 60 m*M* K^+ increased FRET efficiency (thick line). Repolarization with bretyllium tosilate (BT) reversed this effect (thin line).

cells properly labeled with the respective antibodies, tagged by donor and acceptor pairs. Table 7 shows data calculated from the means of the histograms. These energy transfer values are obtained as a weighted average over the cell population.

The actual reason for the alteration in the conformation of a membrane-spanning protein can be very different. The likeliest one is the relaxation of the transient dipoles upon decreasing membrane potential and the opposite, when the cells are repolarized. Another possible explanation is the changing structural interaction between proteins and the changing lipid environment. The changing potential level has a well-documented effect on the dynamic behavior of cell membranes and their lipid moieties.[219] The known structural features of the β_2-microglobulin and the heavy chain of HLA class I molecules suggest that proximity changes between these membrane elements are more likely to due to a conformational change in the heavy chain. The β_2-microglobulin is much smaller and does not span the membrane in contrast to the heavy chain of HLA I antigen.

The above paragraph demonstrated the power of combining energy transfer measurements with membrane potential changes, which was able to resolve dynamic conformational changes of particu-

Table 7. **Effect of Membrane Potential Changes on Energy Transfer Efficiency Values between Donor-Acceptor Pairs Bound to HLA Class I Antigens**

	Treatment		
		$\Delta\Psi < 0$	
ET pair determinant)	$\Delta\Psi > 0$ (m*M* K^+)	+ ∅	+ 2 mM BT
F-L368 (β_2-microglobulin)	5	12.2 ± 0.5[a]	10.8 ± 2.4
R-W6/32	60	14.0 ± 2.3	12.5 ± 1.1
(HLA-A,B,C)	145	13.9 ± 0.3	11.8 ± 0.4
	5	13.7 ± 2.0	—
F-W6/32	60	18.4 ± 5.8	17.2 ± 5.7
R-L368	145	15.0	—

[a]Data represent mean (%) ± SEM (%) of at least three independent energy transfer measurements.

lar membrane molecules. A further extension of such an approach over other proteins, especially those which are involved in the transmembrane signal processing, may open a new pathway in understanding the regulatory function of membrane potential.

3. Structure of IL-2 Receptor

The interleukin-2 (IL-2) receptor structure and its "immediate environment" (i.e., the relative two-dimensional localization of the molecular neighborhood, which can be defined also as interleukin-2 receptor-associated proteins) have been thoroughly studied by fluorescence energy transfer measurements applied mostly in flow cytometer.[220] The great interest in this particular type of lymphokine receptors has its explanation in an instrumental physiological role of the structure and function of IL-2 and its receptor (IL-2R) in the lymphocyte maturation and activation. Among lymphokines, i.e., hormone-like molecules which play a decisive role in intercellular communication in the immune system, IL-2 has a pivotal role in controlling the growth and function of T lymphocytes. Beyond this, an emerging new branch of science (the neuro-endocrino-immuno modulation) claims the existence of cross regulation of immune, endocrine, and nervous systems, and IL-2 may participate in this type of regulation as well. In addition, the IL-2 molecule affects the activities of B lymphocytes, monocytes, natural killer, and lymphokine-activated killer cells.[220]

IL-2 production and the expression of its receptor complex in the plasma membrane are "triggered" by a series of steps, which are initiated at the cell surface when, for example, a suitably processed foreign antigen is presented. This latter process is carried out in such a way by the molecules of the MHC that it initiates cell activation in T lymphocytes through the heterodimeric T cell receptor-CD3 complex. Similar programs can be initiated by liganding the T cell receptor-CD3 multisubunit receptor aggregate by appropriate monoclonal antibodies or crosslinking the oligosaccharide moieties of cell surface molecules by lectins.[221] The subsequent IL-2 production and IL-2R expression are consequences of *de novo* protein biosynthesis and processing. The mechanism involves a characteristic positive cascade. The secreted IL-2 binds the newly expressed receptors, and thereby enhances high affinity IL-2R-s, which generally are not present in the membrane of resting T lymphocytes. The IL-2 liganded IL-2R is the key element of the activation of T lymphocytes.

The IL-2R can be classified into three different forms, based on its functional properties.[222] Resting T lymphocytes (and particular T cell lines) may express a "low affinity" binding site, having a dissociation constant of 10^{-6} *M*. This protein is often called Tac antigen, or p55 subunit, based on its enhanced appearance in activated T lymphocyte membranes and its molecular weight, respectively. This p55 subunit has direct IL-2 binding capability, but its molecular structure is clearly insufficient to transmit any signal into the cell interior as it has only a short intracellular tail. Sequencing and cloning of this receptor subunit indicates that it contains 251 amino acid residues and "an NH_2-terminal 21 amino acid signal peptide that is cleaved *in vivo*."[220] 219 terminal amino acids of the molecule form the extracellular, only 13 the intracellular and 19 the membrane spanning region. This subunit is famous for the first monoclonal antibody having been developed against it and this had a remarkable impact on the research of the IL-2 receptor system.[223] The Tac antigen is now called the IL-2Rα subunit. The short, and very likely nonfunctional, intracellular tail of the IL-2Rα subunit raised the possibility of the existence, and lead to the discovery, of the IL-2Rβ, or p75 protein subunit. This subunit generally has a lower density in the cell membrane than the p55 but has a tighter IL-2 binding ability with a dissociation constant of 10^{-9} *M*. This subunit is built from 525 amino acids with a smaller extracellular domain (214 amino acids) than the intracellular tail (286 amino acids). Binding studies with IL-2 lead to the current opinion that the real functional tight binding site of IL-2, i.e., the functional IL-2 receptor is built up from these two different IL-2 binding subunits, with a $K_d = 10^{-11}$. The tyrosine kinase activity, which was found to accompany the IL-2 binding and subsequent cell activation steps, raised the possibility that the IL-2Rβ subunit might be responsible for this. However, every kinase must have an ATP binding site, which is known to have the consensus sequence of amino acids: Gly-X-Gly-X-X-Gly. This is missing from the intracellular part of the IL-2R β chain, although its size could accommodate a kinase activity.

Of course, the "real time" sequence of the research activities did not follow the above logically constructed trend. However, the involvement of flow cytometric resonance energy transfer studies, which significantly contributed to the identification of the molecular "environment" of the multi-subunit IL-2R, focused the interest on the detection and localization of further molecules in the immediate vicinity of the IL-2R. This trend was rewarded very recently, when Sugamura and co-workers[224] cloned a putative further subunit of the IL-2 receptor. They determined that the β chain with the α chain (heterodimers) and the β chain alone are effective in IL-2-mediated signal transduction. The binding of IL-2 to the IL-2Rα chain alone is ineffective. Recently, they obtained several lines of evidence suggesting the implication of a third putative (γ) receptor subunit. Although they found that expression of β chains alone can bind IL-2 in lymphoid cells and transmit growth signals, this was not the case with fibroblasts. The number of intermediate affinity IL-2 binding sites did not correlate with that of the β subunits. Even the coexpression of the α and β chains, which have a higher affinity than the β subunit alone, had not formed a functional IL-2 receptor. Finally, consistent application of molecular biological methods (coprecipitations, transfection, etc.) helped to identify and clone a third, γ subunit or, after its molecular weight, the p64 component of the now really multisubunit IL-2 receptor. The molecular weight varied between 61 and 71 kDa, depending on cell types.[224] Thus, we can claim that in the composition of the human interleukin receptor structure three different molecules participate. The α-β-γ heterotrimers form the high- and the β-γ heterodimers the intermediate-affinity receptors. Murine fibroblastoid cells internalized the IL-2 receptor only in the presence of the γ subunit. The significance of this internalization in signal transduction is yet to be determined. Molecular analysis showed that the β and γ subunits are members of other cytokine receptor families, while the α is not. The three different subunits can be responsible for the unusually high binding affinity. The signal transduction mechanism is not solved yet, but is interesting to note that intracytoplasmic amino acid positions of the γ subunit are very similar to those of the *src* tyrosine kinase family.[222] The hydrophobicity plot predicted a single membrane spanning region between amino acid residues 255 and 283 of the peptide backbone. Normal human peripheral blood leukocytes expressed the γ subunit regardless of their activation state. Of course, a refined numerical analysis has not been done yet. Different cell lines showed variance in their IL-2Rγ expression, but a likely number is around 10,000 or lower. Thus, the new subunit of the already complicated-enough IL-2 receptor is a further challenge for determining its topological relationship to the other receptor subunits and its molecular environment on the cell surface.

In 1985/1986, an international and interdisciplinary team started a vigorous cooperation to look for adjacent molecules around the IL-2R, which may have functional relationship with the IL-2 binding subunits. Waldmann and co-workers at the National Institutes of Health identified a 95-kDa peptide that showed a rather weak coprecipitation with the Tac antigen. Antibodies were raised against the peptide and the method of FCET was introduced to study the question of whether Tac and the antigen that time code named T-27 were neighbors or not.[194] The same technique had already been successfully applied earlier to solve similar problems, as we have seen in preceding sections.[10,11,172,202,207] However, the complexity of IL-2R significantly contributed to the need for elaborating the minimal conditions of measurement, but finally the conclusion was reached that the two membrane-bound molecules are in close proximity, and thus have a nonrandom codistribution.

The T27 antigen was found widely distributed, being expressed on B lymphocytes, monocytes, and adult T-leukemic cells, but not on polymorphonuclear leukocytes or platelets. The level of the expression of T27 was distinctly different on resting and activated T lymphocytes, respectively, and for good reason, since 3 years later the T27 was found to be identical to the intercellular adhesion molecule ICAM-1.[225] The appropriately selected specific monoclonal antibodies, tagged with fluorescein and rhodamine, respectively, immediately revealed a significant energy transfer between their respective binding sites, the Tac and T27 (ICAM-1) antigens. Despite the ample amount of measurements in similar systems,[10,11,172,202,207] a serious problem arose: where is the lower limit of significance, when measuring flow cytometric energy transfer, which provides averaged data on a cell-by-cell basis? It is not possible to decide theoretically on such a problem yet; thus, experiments were carried out with "auxiliary" systems, where available data, obtained with entirely different methodology, supported or excluded close association of membrane-bound protein entities. One easy target was offered by the HLA class I antigen structure.

The β_2-microglobulin and the "frame" (the heavy chain) of the heterodimer HLA class I antigen must be in close proximity, i.e., well within energy transfer distance, when labeled with the fluorescein-rhodamine donor-acceptor pair, carried by monoclonal antibodies with specificity toward the two interlinked subunits. This conclusion was drawn from the known molecular size and other structural parameters of the heterodimeric HLA class I. The approximately 12 ± 2% energy transfer efficiency, obtained with reasonably high accuracy from independent experiments, gave us a proper orientation as to what "close proximity" would be.[194] (As we have seen in preceding sections, the actual energy transfer efficiency depends also on the fluorescence labeling ratio of antibodies and the binding site density of the fluorescently tagged monoclonal antibody carriers.) Luckily enough, the density of transferrin receptors of the particular T cell line (HUT-102B2), which expressed a high density of IL-2α subunit and the T27 antigen, showed a density identical to that of T27. This was a very valuable asset, because the number of the T27 and Tac was not proportional; thus, the possibility of a chance transfer, caused by the mere high density of one component, had to be excluded. The transferrin receptors were already identified having a dimeric structure. Accordingly, the energy transfer efficiency between the rhodaminated and fluoresceinated antibody mixture specific for transferrin receptors was around 9%, corresponding to our expectations. However, transferrin did not show energy transfer, when paired with either T27 (1 to 2%) or Tac (0.1%) antigens. The transfer efficiency between Tac and T27 antigen was found between 4.2 to 9.3%.[194] Based on these experiments, it had been decided that T27 and IL-2 receptor α subunit had a nonrandom codistribution, with a likelihood of functional interaction. This functional interaction has to be demonstrated yet, keeping in mind the very promising and interesting perspective of the new discovery that T27 and ICAM-1 are identical.[225] The HLA class I structure was utilized as a sort of energy transfer efficiency standard to determine well-established close proximity. This was justified by the fact that the proximity of β_2-microglobulin and the heavy chain of the HLA class I are not exposed to variations with changing density of cell surface antigens. Thus, experiments were carried out to determine the localization of HLA class I antigen, relative to Tac and T27 antigens. The energy transfer efficiency was significant both for T27 and Tac. The data suggested that Tac and T27 could be around the HLA class I antigen, but the high density of the MHC antigen in the membrane of the HUT-102B2 cells also gave a warning, and our first publication discussed this finding as a possible chance transfer.[194] Independent experiments, carried out with a different biophysical measuring method, fluorescence recovery after photobleaching, provided additional positive data on the possible nonrandom codistribution of HLA class I and Tac antigens.[226] That was the first time, when the conclusion that HLA class I and the Tac antigens are in a close molecular proximity was clearly drawn. This finding was further supported by Sharon et al.,[227] who applied biochemical techniques to study proximity between the two respective antigens. It is known that spatially distinct epitopes are associated with the binding of IL-2 molecules to the α and β subunits of the IL-2R.[228] Thus, stereochemistry dictates that the vicinity of the receptor subunits can influence the binding capability of IL-2 to its receptor.[228] Using FCET, the low-affinity binding site of IL-2 receptor (p55) was found to be in the vicinity of class I HLA molecules also in human peripheral T cell membranes.[229] This was an important observation, because normal cells were used instead of cell lines, which may display particular properties, not characteristic of normal cells. From the pure technical point of view, we have to note that the simplified form of the FCET method, applied by the authors of this work, using a single laser excitation, has its inherent technical limitations.[172] This could be a reason why the authors applying this simplified technique did not find energy transfer between the β_2-microglobulin-directed antibodies and the anti-Tac. Another obvious problem of using native cells and not cell lines arises from the lower density of the antigens. Efforts to overcome this difficulty by increasing the labeling ratio of the antibodies and dyes is also limited, owing to arising spectroscopic problems. Thus, it is likely that, in the case of native T cells, either not all of the class I HLA "frame" (heavy chain) molecules are associated with β_2-microglobulin or the sensitivity of the single laser line-based method is not enough to resolve the neighborhood of the β_2-microglobulin and the Tac peptide.[194]

The IL-2 receptor in its complexity represents a challenge for scientists. Understanding the reason behind the complex structure and the functional significance of the identified distinct molecules which are in the vicinity of the IL-2 binding structures points beyond this receptor complex proper. Decoding of the message of such a well-ordered two- or three-dimensional molecular assembly may open a new trend in the efforts to reveal the secrets of molecular patterns on the cell surface. The ever-growing choice of different kinds of monoclonal antibodies may provide the right key to such investigations.[228,230]

4. Association of MHC Antigens and ICAM-1 Molecules

The finding of Burton et al.[225] that the former T27-encoded antigen and the ICAM-1 molecules are identical raised a number of questions and suggested the reinterpretation of the possible physical vicinity of this important adhesion antigen and other cell membrane antigens. Experiments revealed that cell surface densities of the ICAM-1, HLA class I and II antigens, or IL-2Rα can at least partially be influenced by cell cycle, activation state, and cell types. Two well known lymphoma lines, the HUT-102B2, a representative line of T cell adult leukemia, and JY, a B-lymphoma line, were studied in a complex way. Here these very recent experiments of ours are presented and discussed in more detail (Bene et al., personal communication).

Table 8. **Relative Surface Density of IL-2Rα, ICAM-1, HLA Class I and II Molecules on the Surface of HUT-102B2 Cells**

mAB	Determinant	F-mAb:F-W6/32[a] intensity ratio in (%)
W-6/32	HLA-A,B,C	100.0
L-368	β_2-microglobulin	71.4
7F7	ICAM-1/T27/CD54	7.5
α-Tac	IL-2Rα	7.7
L-243	HLA-DR	17.2

[a]Antibodies were labeled with fluorescein and intensity ratios were corrected for labeling ratios.

Monoclonal antibodies were applied for site-specific labeling of antigens on the HUT-102B2 T lymphoma line in order to establish the cell surface density of ICAM-1, IL-2Rα, and HLA class I and II. Table 8 provides data on the relative density of the above antigens, normalized to the expression level of HLA class I antigens. Data were calculated from mean values of flow cytometric distribution histograms.

Figure 6 demonstrates energy transfer distributions for ICAM-1, IL-2R, the heavy chain of HLA class I, β_2-microglobulin, and HLA-DR antigens on HUT-102B2 cells. Tables 9 and 10 summarize the mean transfer efficiency values between fluo-rescently labeled donor acceptor pairs on HUT-102B2 and JY lymphoma cells. The energy transfer distribution over the cell populations for β_2-microglobulin and the heavy chain of HLA class I antigen, labeled with FITC-L368 and TRITC-W6/32 monoclonal antibodies, respectively, was narrow (Figure 6A). The mean values of the energy transfer histograms were practically identical when the fluorescence donor acceptor pair was swapped, i.e., for FITC-W6/32 and TRITC-L368. The range of transfer efficiency was also independent of the cell lines, HUT-102B2 or JY, used. Energy transfer between IL-2Rα and β_2-microglobulin was as strong as between ICAM-1 and β_2-microglobulin. The anti-ICAM-1 antibodies labeled either with FITC or TRITC showed significant energy transfer between individual ICAM-1 molecules on the cell surface of both tested lymphoma lines. About two thirds of this value (yet a strong transfer) was found from both IL-2Rα and ICAM-1 to the heavy chain of HLA class I. Both of the FITC-labeled anti-IL-2Rα and the anti-ICAM-1 had low or even negligible energy transfer to rhodamine-labeled L243 on the HUT-102B2 cells. JY cells, which have no IL-2 receptor, but possess both HLA class I and II at high density, showed a strong energy transfer

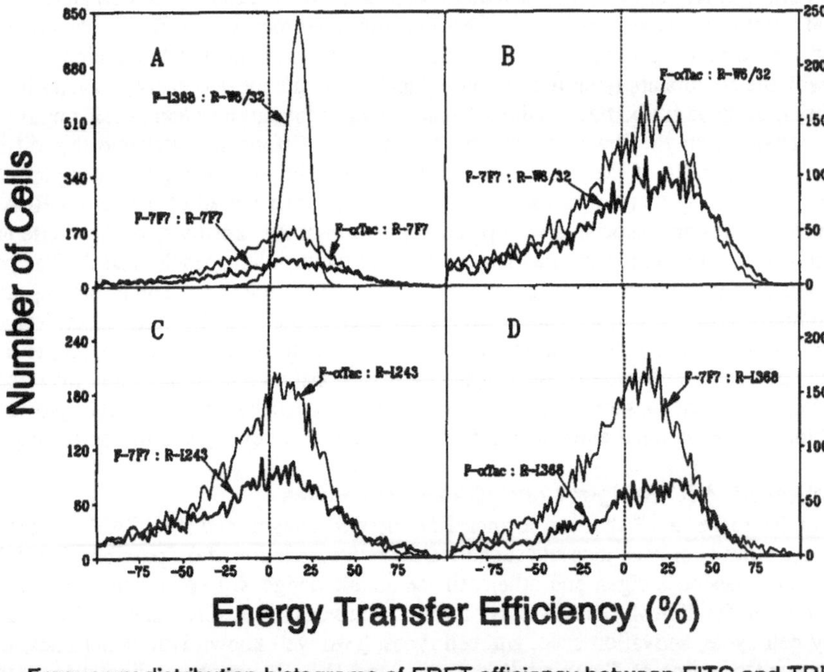

Figure 6. Frequency distribution histograms of FRET efficiency between FITC and TRITC-labeled monoclonal antibodies bound to HUT-102B2 cells. Applied monoclonal antibodies are as follows (with specificities in parentheses): L368 (β_2-microglobulin); W6/32 (HLA- A,B,C); 7F7 (ICAM-1); anti-Tac (IL-2α); L243 (HLA-DR).

Table 9. **Energy Transfer Efficiency Values (E) between Donor-Acceptor Pairs Bound to HLA Class I, Class II, ICAM-1, and IL-2Rα Molecules on the Surface of HUT-102B2 Cells**

Donor	Determinant	Acceptor	E ± ΔE (%)[a]
F-W6/32	HLA-A,B,C	R-L368	17.3 ± 2.6
F-L368	β_2-microglobulin	R-W6/32	17.6 ± 0.2
F-L243	HLA-DR	R-W6/32	18.0 ± 1.8
F-7F7	ICAM-1/T27/CD54	R-W6/32	16.1 ± 4.1
		R-L368	21.9
		R-L243	7.9 ± 1.1
		R-7F7	19.9 ± 0.4
F-αTac	IL-2Rα	R-W6/32	16.2 ± 4.0
		R-L368	24.5
		R-L243	3.1 ± 1.3
		R-7F7	8.7 ± 2.7

[a]Data represent mean ± SEM of at least three independent measurements.

Table 10. **Energy Transfer Efficiency Values (E) between Donor-Acceptor Pairs Bound to HLA Class I, Class II, and ICAM-1 Molecules on the Surface of JY Cells**

Donor	Determinant	Acceptor	E (%)
F-L368	β_2-microglobulin	R-W6/32	16.3
		R-L368	11.2
F-7F7	ICAM-1/T27/CD54	R-W6/32	8.8
		R-L243	12.6
F-L243	HLA-DR	R-W6/32	16.8

between these two classes of MHC molecules. Data from JY cells demonstrate a higher energy transfer between ICAM-1 and HLA-DR than between ICAM-1 and HLA class I heavy chain.

Of course, more than one model can be suggested to explain these findings. However, we are inclined to accept that our earlier unexpected finding that HLA class I and II molecules can be at a molecular distance from each other is true for the JY B lymphoma line,[202] but may not be true for the HUT-102B2 T lymphoma line. Another possibility is that only a fraction of the HLA class I and II molecules will form quasi-oligomeric groups in the cell membrane and the density of these formations varies with different cell lines. The invariable energy transfer efficiency between the β_2-microglobulin-specific L368 and the HLA-A, -B, and -C specific W6/32 antibodies, regardless of the cell type and the direction of the labeling, clearly supports our earlier data.[194,202]

The relatively low energy transfer between the investigated antigens and the HLA class II-specific L243 antibody on HUT-102B2 cells may have a number of different causes. The lower number of HLA class II antigens can be critical when they are labeled with antibodies carrying the acceptor. Another possibility is that in case of the HUT-102B2 cells, where these values were lower than those for the JY cell line, the antigens have a different distribution pattern. Since proximity data determined from energy transfer are averages on a cell-by-cell basis, there is no point in drawing more detailed conclusions. Nevertheless, relative distance can be detected between different pairs, which may serve as a basis for two-dimensional mapping.

The above data provide grounds for a possible model to analyze combined FRET measurements as follows. Relative energy transfer efficiency values between suitably labeled antigens provide a powerful tool even for a two-dimensional mapping of cell surface elements, based on their stepwise FRET measurements on the same cell types, under identical conditions. Individual differences between cell surface measurements of different cell lines can suggest distinct topological distribution of cell surface elements.

As we have shown earlier, the T27 (ICAM-1) and Tac (IL-2Rα) antigens are in close proximity to each other and to the HLA class I heavy chain and the β_2-microglobulin on the HUT-102B2 cell surface.[194,226] As our current data demonstrated, the 7F7 monoclonal antibody directed against a different epitope of the

ICAM-1 antigen showed a similar cell surface distribution pattern for ICAM-1, as was detected with anti-T27 antibody.

Molecular heteroassociations have been detected between ICAM-1, HLA class I heavy chain, β_2-microglobulin, and HLA-DR on the cell surface of JY B lymphoma cells. Similar heteroassociations were found in the plasma membrane of HUT-102B2 T lymphoma cells, but the significantly different quantitative data suggested a somewhat different cell surface distribution pattern of the molecules. Figures 7 and 8 suggest a possible explanation of the quantitative variance of similar data sets from the two cell lines. The strong energy transfer from ICAM-1 and IL-2Rα to β_2-microglobulin and 66 to 72% of this value from the same donors toward the heavy chain of the HLA class I (A, B, C,) molecule suggest a close proximity of the these four molecules. However, the two-dimensional distribution shown in Figure 7 suggests that ICAM-1 or IL-2Rα are closer to the β_2-microglobulin than to the HLA class I. Energy transfer data between ICAM-1 and IL-2α justify the existence of a longer distance between these two molecules. The proximity of β_2-microglobulin and the heavy chain of the HLA class I was independent of the direction of the energy transfer in case of the HUT-102B2 cells. We are reminded that the energy transfer efficiency depends on distance, fluorophore orientation, and acceptor density. However, when topological conclusions are drawn, only those data can be taken into account where the same acceptor density is maintained, e.g., energy transfer from ICAM-1 and IL-2Rα are directed toward the same molecules, labeled with antibodies carrying acceptors.

The heteroassociations involving ICAM-1 molecules may play an important role in antigen presentation, T-cell recognition, cytotoxicity, site-directed lymphokine secretion, and other immunological processes. Furthermore, we think that these data, demonstrating the possibility for a two-dimensional mapping of cell surface elements, call the attention to the existence and role of molecular patterns on the cell surface. Such patterns are likely to depend on cell type and the physiological state of cells. On the other hand, some important patterns might be conserved even during the evolution. This hint may be derived from the co-occurrence of particular cell surface elements, promoting defense functions or physiological intercellular interactions. The existence and variety of cell surface patterns may even suggest the probability of a new code system for intercellular communication, even more so, since the organization of protein molecules into the best accommodating lipid regions depends on their primary structure and occurs on the surface of the rough endoplasmic reticulum.[231] Since the primary structure of proteins is determined by the genetic code, there is little doubt that the molecular organization of the cell surface must be restricted by individual features of proteins and lipid moieties. Of course, the known, although limited, mobilities in lateral, rotational, and perpendicular directions give plenty of possibilities for accommodation and individual behavior pattern of different cell types.

5. Interaction between Insulin Receptor and MHC Class I Molecules

Insulin was one of the first peptide hormones to be purified, sequenced, and to undergo structural analysis. The physiology of insulin was evaluated in the 1950s.[232,233] Studies on the binding of insulin to its cell surface receptor served as a model of hormone-receptor interaction for other peptide hormones. It is generally accepted that the insulin receptor exists as a covalent heterotetrameric disulfide-linked complex composed of two identical α (135-kDa) and two identical β (95-kDa) subunits. The α subunit is thought to encompass the extracellular high-affinity insulin binding site, whereas the β subunits contain the intracellular ATP binding and tyrosine-specific protein kinase domains.[233,234]

Several lines of evidence support the hypothesis that there is interaction between MHC antigens and insulin receptors on the cell surface. Simonsen et al.[235] showed that insulin binding to the human cell line U937 was almost totally blocked by PA 2.2 monoclonal antibody specific to an invariant heavy chain of MHC class I determinants. These experiments were performed at 4°C, when no capping occurred. In addition, capping of MHC class I antigens with anti-β_2-microglobulin antibody or with PA 2.2 antibody virtually precluded insulin binding. In contrast, capping of MHC class II antigens had no influence on either insulin binding or expression of the insulin receptor. The blocking phenomenon was also observed in the reverse case: cells pretreated with insulin at 4°C for 30 min and subsequently washed could only bind PA 2.2 to about 50% of normal.[235] In another set of experiments, photoaffinity labeling technique was used to detect molecular association between insulin receptors and H-2 class I antigens.[189,190,236] In this case, a photoreactive insulin analog was radioactively labeled with ^{125}I and incubated in the dark for 3 h with a purified liver cell membrane preparation form mice of known H-2 genotype. Following photoactivation with ultraviolet light, the membranes were solubilized and incubated with various monoclonal antibodies. Immunoprecipitates were analyzed by polyacrylamide gel electrophoresis fol-

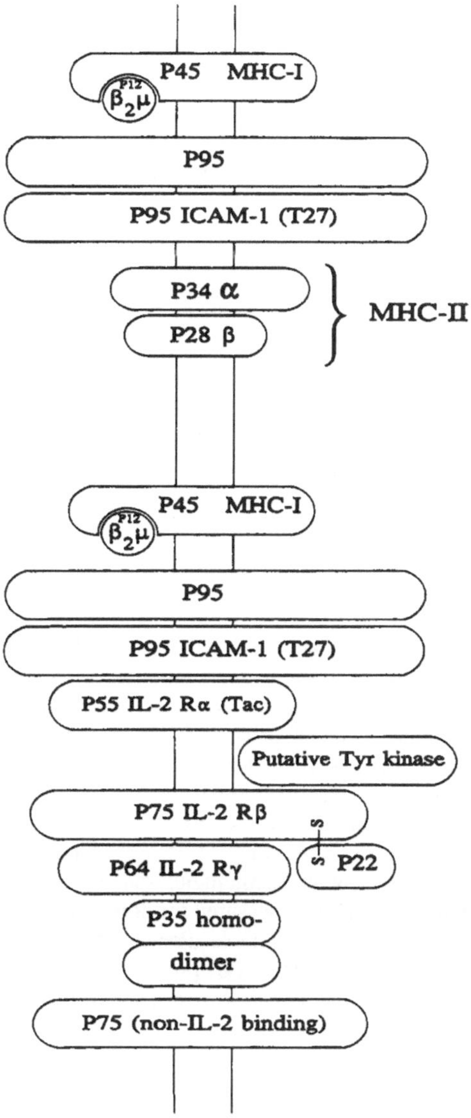

Figure 7. Schematic representation of the structure of IL-2 receptor complex. Our FRET measurements suggested that ICAM-1 and HLA class I molecules are components of the IL-2 receptor complex. HLA class II antigens, however, are not associated with the IL-2α subunit.

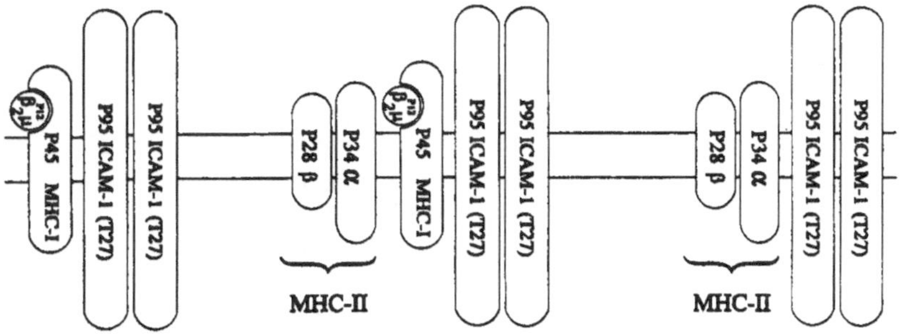

Figure 8. Distribution of HLA class I, class II, and ICAM-1 antigens on JY cell surface. ICAM-1 molecules, which form at least dimers, are associated with HLA class I and HLA class II antigens.

lowed by autoradiography. The results indicated that a significant amount of labeled insulin (linked to insulin receptor) was precipitated by specific monoclonal antibodies to K and D determinants of the liver donor, but not by antibodies of "wrong" H-2 specificity.[189,190,236] Similar results were obtained on the R1 murine thymoma cell line as well.[237] Photoaffinity labeling showed association between insulin receptor and HLA class I antigens on IM-9 and 526 human B lymphoblastoid cell lines.[191] It is very difficult to interpret these data without granting an association between the insulin receptor and MHC class I antigens. Additional evidence for the close relationship between MHC antigens and the insulin receptor was provided by Kittur et al.[238] They measured insulin binding to cells of a HLA-heterozygous B cell line, LCL 721, and to derivative mutants from which all or part of the HLA complex had been deleted. All independently derived mutants of the same HLA phenotype had the same insulin-binding phenotype. Furthermore, transfection of a single HLA-A or HLA-B gene into HLA-A-, B-, or C-negative cells changed their insulin binding to that determined for the same phenotype. This close relationship between the HLA haplotype and insulin binding also suggested that this interaction is likely to involve molecular association between the receptors and MHC antigens rather than some indirect coupling between the two.[238] Edidin and Reiland[239] performed fluorescence recovery after photobleaching experiments in order to detect any dynamic interaction between the insulin receptor and HLA class I antigens. Crosslinking of insulin receptors on LCL 721.1 cells reduced the diffusion coefficient, though not the mobile fraction, of class I HLA antigens about two- to threefold. The effect of crosslinking was specific for MHC class I antigens, since crosslinking of insulin receptors had no effect on the lateral mobility of surface Ig.[239] This measurement showed a good example of how fluorescence recovery after photobleaching technique can detect molecular association in an intact cell membrane.

To directly assess the specificity and significance of class I product-insulin receptor interactions, we estimated the proximity of class I products and insulin receptors on the cell surface by measuring FCET between FITC-(donor) and TRITC-(acceptor) labeled antibodies specific for class I or insulin receptor proteins on intact cells.[240] FCET has advantages over immunocoprecipitation techniques in that the measurements are made *in situ* on intact cells, thereby avoiding introduction of experimental artifacts that may result from incomplete membrane solubilization. NIH 3T3 cells transfected with human insulin receptor gene were chosen for these measurements because of the high insulin receptor expression level. To maximize expression of both products and thus facilitate FCET measurements, cells were cultured with interferon and sodium butyrate, agents known to increase MHC class I antigen and insulin receptor expression. FCET measured for donor/acceptor labeled antibody pairs recognizing (1) different epitopes on the same class I molecules (intramolecular transfer) served as positive control and (2) the same epitopes on different class I products or different epitopes on different class I antigens as negative control. The specificity of antibody pairs and the efficiency of FRET measured between them are summarized in Table 11. Significant intramolecular FRET efficiency (16.1 and 9.5%) was measured between R1-9.6

Table 11. Energy Transfer Efficiency Values (E) between Donor-Acceptor Pairs Bound to Class I H-2 and Insulin Receptor

Donor	Donor determinant	Acceptor	$E \pm \Delta E^a$ (%)
FITC-CP28	H-2K^s	TRITC-CP28	3.5 ± 3.5
FITC-28-14-8	H-2D^q, L^q	TRITC-30-5-7	16.4 ± 3.2
FITC-R1-9.6	H-2D^q	TRITC-28-14-8	16.1 ± 2.3
	H-2D^q	TRITC-30-5-7	9.5 ± 1.7
	H-2D^q	TRITC-CP28	3.1 ± 0.8
FITC-28-14-8	H-2D^q, L^q	TRITC-CP28	3.4 ± 1.4
FITC-30-5-7	H-2D^q, L^q	TRITC-CP28	1.3 ± 1.8
FITC-Ab-1	hu^b IR	TRITC-R1-9.6	18.7 ± 3.7
	hu IR	TRITC-28-14-8	19.4 ± 4.9
	hu IR	TRITC-30-5-7	23.9 ± 6.8
	hu IR	TRITC-CP28	19.5 ± 5.7

[a]Data represent mean ± SD of five independent experiments.

[b]hu = Human.

From Szöllősi, J., Damjanovich, S., Balázs, M., Nagy, P., Trón, L., Fulwyler, M. J., and Brodsky, F. M., *J. Immunol.*, 143, 208, 1989. With permission.

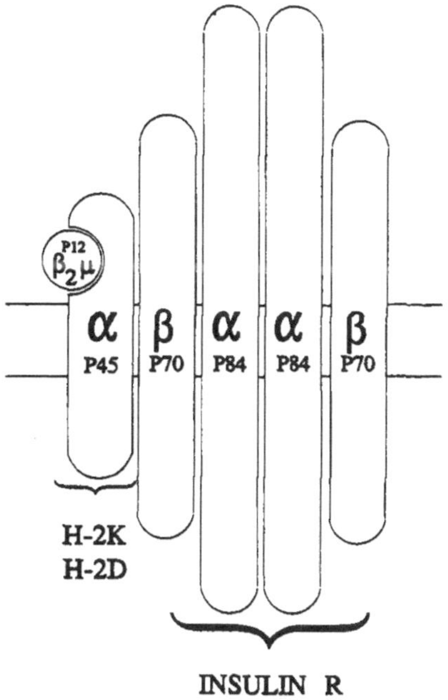

Figure 9. Association of H-2 class I molecules with insulin receptors. Both H-2D or H-2K may associate with the heterotetramer insulin receptor.

antibody recognizing monomorphic regions of H-2 products and either 28-14-8 or 30-5-7 antibodies specific for polymorphic regions on H-2D end products. Low FRET efficiency (3.5%) was found when anti-H-2K reagents were used as both donor and acceptor, indicating that H-2K products appear as monomers on the NIH 3T3 cell surface. This finding is in accordance with our earlier observations that $H\text{-}K^k$ antigens are not self-associated on HK22 murine lymphoma cells.[207] Low FRET efficiency values were also detected between antibodies specific for H-2D end products and CP28 antibody specific for H-2K end products, indicating that H-2K and H-2D antigens do not associate. High FRET efficiency was calculated for the anti-insulin receptor antibody in combination with any of the four H-2 reagents. Our results showed specific association between H-2 products and the human insulin receptor. This finding is not surprising considering the conservative structure and function of these molecules from species to species. Equally close associations were measured between H-2K vs. H-2D end products and insulin receptor, in agreement with reports based on coprecipitation.[189,190]

On the basis of FCET data, we propose a model for the cell surface distribution of insulin receptor and H-2 class I antigens (Figure 9). Since class I/class I dimers were not found, class I H-2 products must associate with insulin receptor as single proteins. A single class I protein could associate with the insulin receptor heterotetramer ($\alpha_2\beta_2$), creating a molecular complex.

We also investigated the physiological significance of the association of insulin receptor with H-2 class I antigens on the cell surface by quantitating insulin receptor number and affinity on a pair of cell lines that are positive and negative for H-2 class I products, respectively. Results from radioreceptor assays demonstrated that binding of insulin to its receptor was independent of MHC expression. However, insulin-stimulated insulin receptor internalization differed among the two lines: the class I null line exhibited both a greater rate and degree of receptor internalization. This is not surprising, since a 25-residue peptide derived form the murine MHC class I molecule also influences the internalization of insulin receptor without affecting insulin binding or dissociation.[241] The nonimmunological role of MHC class I molecules may be to assure regulation of some receptors upon activation by their corresponding ligands. The ability of MHC class I antigens to regulate membrane-bound hormone receptors by association has far-reaching implications for cell growth and metabolism.

6. Association of Peptides with MHC Molecules

The primary and unique property of the vertebrate immune system lies in its ability to distinguish between closely related, but different molecular structures. The immune system should be able to recognize "foreign matter" that may penetrate the body, but at the same time ignore a large variety of "host" molecules with which it is constantly in contact. The adaptive immune system has two arms: humoral and cellular, consisting of B and T lymphocytes. Antigen recognition in humoral immunity has been long understood: the antigen-specific (F_{ab}) portion of the antibody molecule binds directly to the antigen in its native state. The antibodies can be found as antigen receptors on the B cell plasma membrane and also as secreted molecules. In contrast, T cells do not recognize antigens in their native conformation, but only after partial proteolysis within antigen processing or presenting cells (APC) into constituent peptides that are then bound to MHC molecules. The bimolecular complex of antigen peptide and MHC molecule is then displayed on the surface of APC cells and recognized by T cell clones bearing antigen receptor specific for it. This recognition leads to their stimulation and proliferation. Two types of glycoproteins can compromise healthy mammalian cells: those made within the cells (e.g., viral proteins made after viral infection of the cell) and those made elsewhere, which can then enter the cell by endocytosis (e.g., bacterial toxins). MHC class I molecules primarily specialized to present peptides derived form endogenous proteins, while class II molecules present peptides derived from endocytosed antigens.[242,243]

Several techniques have been applied for studying the structure and function of the peptide-MHC antigen complex, including equilibrium dialysis,[244] gel filtration,[245,246] photoaffinity labeling,[247] and crystallography.[248] Very few papers have been published in which the FRET approach was used for studying the interaction of peptides and MHC molecules.

Watts et al.[249] performed an elegant FRET measurement using a microscope capable of evanescent wave excitation in order to detect association between a peptide antigen and an MHC class II protein. In the absence of T cells, no interaction was detected between the fluoresceinated peptide antigen (donor) and the Texas red-labeled class II protein (acceptor) embedded in model membranes. In the presence of T cells, FRET was observed between the peptide antigen and the class II protein showing that T cells stabilized an association between the peptide antigen and the class II protein within a distance of about 4 nm.[249] This finding is consistent with the suggested complex formation between the peptide and the class II protein.[242,243] Unfortunately, this experiment does not answer the question properly, as to whether the peptide antigen and the class II protein are associated prior to T cell binding or the association is induced by the binding of T cells, since the peptide antigens and the class II proteins were embedded in model membranes, not in membranes of intact cells, where the cytoskeleton and the membrane structure may play a significant role in this respect.

The same laboratory applied the FRET approach to address the question whether the class II molecule can bind one or two peptides.[250] They labeled a 17-amino acid peptide from chicken ovalbumin with fluorescein at the amino terminus or with Texas red near the carboxyl terminus. The peptide binding capacity of the two forms of class II molecules, the "floppy" and the "compact", was compared. The floppy conformation is an intermediate in the dissociation of compact into separate α and β chains. FRET was detected between the fluorescein- and Texas red-labeled peptides, suggesting that a pair of peptides is bound to the same MHC molecules. Systematic FRET measurements revealed that two full-length peptides can bind to the floppy $\alpha\beta$ heterodimer, and two truncated peptides can bind to the compact $\alpha\beta$ and also to floppy $\alpha\beta$ heterodimer. No simultaneous binding of the two full-length peptides to the compact $\alpha\beta$ conformation was detected. The biological significance of these findings may be that they suggest a mechanism for peptide exchange in class II antigens, since the floppy conformation can bind two full-length peptides.[250] This possibility is particularly important, since peptide-free $\alpha\beta$ may be unstable in the absence of bound peptide.[242,243]

Cytotoxic T cells recognize the complex of peptide antigen bound to class I MHC protein. The length of the peptide selected for presentation with a given class I molecule is much shorter (only eight to ten amino acids) than that of peptides associated with class II molecules. In addition, these peptides must contain a motif defined by their primary structure. It has been shown that peptides are involved in the assembly of class I molecules through their ability to stabilize the interaction between the class I heavy chain and β_2-microglobulin.[251-253] Addition of specific peptides induced conformational change in the class I heavy chain.[252] On the basis of this observation, we performed preliminary experiments using the FCET technique in order to detect peptide-induced conformational changes of class I antigens in the membrane of live cells (Mátyus L., personal communication).

We used fluoresceinated and rhodaminated, noncompeting monoclonal antibodies, which bind to the same H-2K^k antigen but to different epitopes, to label the surface of HK 22 murine T lymphoma cells. Significant transfer efficiency was detected between FITC-labeled 30/6 and TRITC-labeled 27/55 antibody as it would be expected for intramolecular energy transfer efficiency (Table 12 and Figure 10).[207] Addition of specific peptide to the double-labeled cells increased the FRET efficiency measured between

Table 12. **Effect of N-50-63 Peptide on Transfer Efficiency Values (E) between Donor-Acceptor Pairs Bound to Different Epitopes of the H-2K^k Antigens**

Donor	Determinant	Acceptor	Peptide (n*M*)	E ± ΔE (%)[a]
F-27/55	H-2K^k	R-27/55	0	0.5 ± 1.2
			244	0.9 ± 2.5
F-30/6	H-2K^k		0	11.7 ± 1.2
			2	15.3 ± 0.7
			24	15.4 ± 0.3
			81	18.7 ± 0.4
			244	16.8 ± 1.0

[a]Data represent mean ± SEM of at least three independent measurements.

the fluorescently tagged, noncompeting monoclonal antibodies. The peptide, N-50-63 (a generous gift of Alain Townsend), applied in our experiments, was derived from influenza nuclear protein.[252] The increase in FRET efficiency upon addition of peptide showed concentration dependence (Table 12). As one can see in Figure 10, the energy transfer distribution curves are much wider than in the case of HLA class I molecules on HUT-102B2 cell surfaces (Figure 6A), since the level of expression of the H-K^k antigen in HK 22 cells is lower than that of HLA class I expression in HUT-102B2 cells. The lower level of expression results in lower signal-to-noise ratio, significantly decreasing the precision of FCET measurements.[10,11] Low intermolecular FRET efficiency was observed when 27/55 monoclonal antibody was used both as donor and acceptor. Addition of N-50-63 peptide had no effect on the monomeric state of H-2K^k antigens. In addition, this experiment served also as a control to exclude the direct effect of the peptide on the fluorophores. When similar experiments were performed on human JY cells expressing the HLA system, where this peptide can be regarded as nonspecific, no change in intramolecular energy transfer was observed (data not shown). Figure 11 shows our proposed scheme, how peptide binding to the class I molecule might induce a conformational change sensed by FCET measurements. Since a large portion of MHC class I antigens expressed on the cell surface is already complexed with endogenous peptides, our observation, i.e., the increase in FCET efficiency upon addition of exogenous peptide can be interpreted in the following ways.

Addition of peptide induces a huge conformational change in formerly free class I molecules, which can be 10 to 30% of the total class I molecules, and these changes are reflected in our FCET data. Another possibility is that the exogenously added peptide replaces almost all the endogenous peptides, and due to its different structure induces a slight conformational change on all but class I molecules. At this point, our data cannot discriminate between these possibilities.

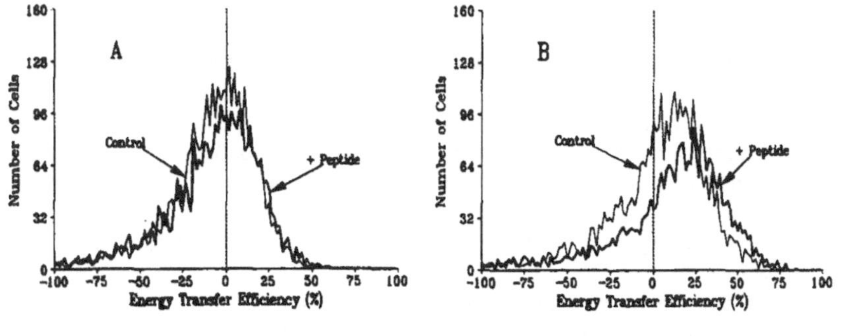

Figure 10. Effect of N-50-63 peptide, prepared from influenza nuclear protein, on the conformation of H-2K^k molecules. (A) Frequency distribution histograms of FRET efficiency between FITC-labeled 27/55 and TRITC-labeled 27/55 monoclonal antibodies bound to different H-2K^k antigens. (Intermolecular transfer). The peptide has no effect. (B) Frequency distribution histograms of FRET efficiency between FITC-labeled 30/6 and TRITC-labeled 27/55 monoclonal antibodies bound to different epitopes of the same H-2K^k antigen. Addition of peptide increased the transfer efficiency, suggesting a change in the conformation of H-2K^k antigen.

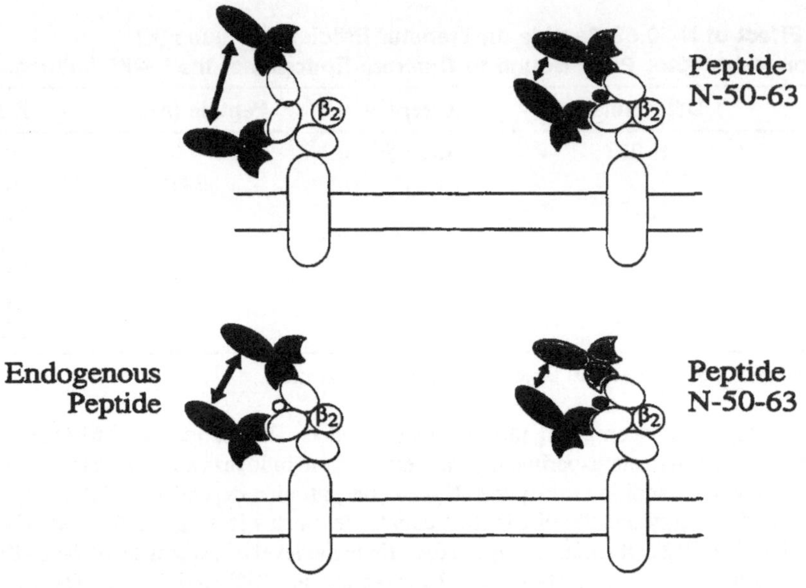

Figure 11. Schematic representation of H-2K^k antigens labeled with noncompeting, fluorescently tagged antibodies bound to the same molecule but to different epitopes. Addition of specific peptide results in a conformational change of H-2K^k antigen bringing the tagged epitopes closer to each other.

Although the FRET approach has only been applied on a few occasions, the investigation of these types of questions warrants significant contribution to our knowledge about the mechanism and function of antigen processing involving class I and class II MHC antigens.

C. COMPLEX FORMATION OF CD MOLECULES ON THE CELL SURFACE

The donor photobleaching method of Jovin and Arndt-Jovin to measure FRET was applied to study the proximity of the CD4 antigen and the TcR-CD3 receptor system.[254] This very versatile method, the utilization of altered photobleaching kinetics of donor dyes in the presence of acceptor, has been adapted to an ACAS 570 laser scanning microscope system by Szabó et al.[255] As we have seen previously, the basic idea behind these methods is that of Theodore Förster's singlet-singlet resonance energy transfer between strategically bound fluorescence donor and acceptor molecules. The flow cytometric application of the method provided valuable data over large cell populations. The data obtained resulted in proximity relationships between target molecules on a cell-by-cell basis, where the donor-acceptor distance was averaged for each cell. The unique advantage of the microscope-based determination of molecular distances by FRET is that, at the expense of losing the distribution data over a larger cell population, the sensitivity is higher. This is mainly due to the greater viewing angle of the microscope, which is very difficult to attain in a flow cytometer. Another obvious advantage of the photobleaching technique is that it determines the transfer efficiency from the bleaching kinetics of the donor, i.e., at the donor side, in contrast to FCET, which is forced to utilize the fluorescence enhancement of the acceptor. The concept behind the method is that exclusively the absorbed energy has photobleaching capacity. In the case of energy transfer, the transferred part of the energy increases the capacity of the donor to absorb more light without further bleaching.

Since the technical and theoretical features of the method have been discussed earlier, here we focus our interest on the data obtained so far. On the other hand data will be included which have been obtained by FCET on the CD4 antigen.

The principal target of the very first epitope mapping in a laser scanning microscope system was the CD4 antigen. This antigen has an important role in the physiological functions of the T helper/inducer lymphocyte subset, and is known as a key molecule in the pathogenesis of acquired immune deficiency syndrome (AIDS). CD4 molecules serve as the actual binding site of the HIV-1 virus and are phosphorylated by its gp120 coat protein. The CD4 is subsequently internalized by the helper cells, which then lose their viability, possibly due to apoptosis.[254-257]

CD4, which is a co-receptor of the TcR-CD3 system, is an approximately 63-kDa molecular weight glycoprotein. Its average plasma membrane protein parameters do not distinguish the CD4 from other protein elements of the membrane. Thus, its connections, actual conformations, and proximity relationships to other surrounding cell surface elements may reveal some new features having a potential importance in the better understanding of its physiological and pathological role. The proximity of CD4 to the T-cell receptor complex has already been investigated by Chuck et al.[258] applying energy transfer measurements in human T leukemia cells and by Mittler et al.[259] in a murine T-cell hybridoma. Both groups applied FCET to determine physical association between different elements of TcR/CD3 complex and CD4. Mittler et al.[259] applied a remarkable approach, when studying the proximity relationships in a TcR/CD3$^+$ and L3T4$^-$ murine T-cell hybridoma line, which was transfected in such a way that it expressed intact or cytoplasmic domain-depleted mutants of the human CD4 gene. According to their data, neither the wild-type hybridoma infectants nor the CD4 cytoplasmic domain-depleted mutants showed any detectable association between CD4 and the TcR/CD3 receptor complex. These experiments were carried out at 4°C. However, when rhodaminated anti-CD3 and fluoresceinated anti-CD4 monoclonal antibodies were added and the cells were kept at higher temperatures (21 to 37°C), energy transfer was observed between the two labeled receptors. This observation indicated that temperature-dependent receptor migration occurred, a process which might also be required for T-cell activation. The rapid association was dependent on the expression of the cytoplasmic domain of the CD4 molecule. Chilling the cells abrogated any observable association of the molecules. Anti-CD3 $F(ab')_2$ failed to produce this phenomenon. It is noteworthy that anti-CD3 $F(ab')_2$ also failed to activate cells or induce IL-2 secretion. Thus, activation may be necessary for the observed translocation of the receptors or those conditions which cause the translocation are also prerequisites of cell activation.

Szabó et al.[254] expanded this work through the higher resolution power of the epitope mapping by photobleaching FRET in an ACAS 570 laser scanning microscope. A panel of anti-TcR-CD3 antibodies was applied in combination with anti-CD4 antibodies, directed against different subunits and epitopes of the complicated multisubunit receptor system. The sometimes low signal-to-noise ratio was overcome by applying labeled second antibodies or the large phycoerythrin (PE), containing 35 fluorophores, or allophycocyanin. Energy transfer between T-cell antigen subunits and their epitopes was studied with OKT4-FITC and Leu-3a-PE, OKT4E-FITC and OKT4-PE, and OKT4-FITC and Leu-5b-PE donor acceptor pairs. The first two pairs showed a substantial energy transfer efficiency (17 to 20%), suggesting a reasonably close proximity, while the last pair showed only a poor transfer efficiency (3 to 4%). This set of data supported the view that different epitopes on the CD4 must be in close proximity, within a 2- to 10-nm scale, while CD4 and CD2 were separated by a greater distance. In this statement the size of the PE is also taken into account. Of course, this relatively low energy transfer does not necessarily mean that CD2 is not associated with the T-cell receptor CD3 antigen complex. On the contrary, data are available based on functional interactions and coprecipitation experiments that CD2 associates with the TcR-CD3 complex on the surface of human T-lymphocytes.[260] Nevertheless, the significant advantage of these data is that they were obtained on human peripheral blood lymphocytes. Another significant asset of the measurements is that in this case a close physical association was directly demonstrated for epitopes within the TcR-CD3 complex, and the CD2 cannot be physically too far from the former at a nm scale. The relatively low CD4 density on the cell surface was compensated by the phyco-dye and the better optical parameters of the instrument, as compared to a flow cytometer.

The sensitivity of the method in an ACAS instrument was checked by comparing the energy transfer data with those gained in a flow cytometer or applying labeled secondary antibodies and the $F(ab')_2$ fragments of the same antibodies. The $F(ab')_2$-labeled samples showed a higher energy transfer than those which were conjugated to whole antibodies. The whole antibodies, having a distribution of the dye molecules over F(ab) and also Fc fragments, resulted in an apparent increase in the donor-acceptor distance. This was expected, but the response of the two systems underlined the sensitivity of the applied method. HPB-ALL cell lines were also probed for energy transfer between different elements of the T-cell receptor complex.[254]

A more recent article of the same authors presented direct physical evidence again with photobleaching energy transfer measurements that CD4 changed its conformation upon ligand binding.[255] Pathological conditions predict that drugs, which can prevent the binding of the gp120 of HIV, may interfere with the development of the diseases. There are known negatively charged agents, such as dextran sulfate, that can inhibit gp120 binding. The CD4 is known to be protected against the binding of gp120 also by a triphenylmethane polymer, the aurintricarboxylic acid, ATA.[255] Szabó et al.[255] carried out a series of elegant experiments, combining photobleaching energy transfer, FCET, fluorescence recovery after photobleaching, and spectrofluorimetry, in order to understand the mechanism of the mode of action of ATA. It has been shown that ATA blocked the binding to CD4 not only of the HIV gp120, but also the monoclonal antibody Leu-3a. Furthermore, ATA caused the disengagement of the antibody OKT4E when it has already been bound to CD4 on human peripheral blood lymphocytes. Another antibody, Leu8, which was specific for a leukocyte homing receptor was also disengaged by ATA, but none of nine other ones, directed against cell surface antigens. The Leu-3a was the only antibody that could prevent the ATA-induced release of OKT4E. A number of nonspecific treatments (SH reagents, chelators, complex carbohydrates, or nonionic detergents) were ineffective in disengaging either OKT4E-FITC or anti-Leu8-FITC. That pointed toward a specific effect of the ATA. The photobleaching FRET revealed that the energy transfer efficiency, i.e., the physical distance relationship between the FITC-OKT4E and OKT4-PE, was significantly changed in presence of ATA. A noteworthy technical innovation was applied in order to overcome the problem that ATA was also disengaging the OKT4E antibody.

The donor concentration does not influence the actual transfer efficiency of a donor acceptor pair. Thus, the bleaching rates of the donor- and acceptor-labeled samples were compared in the presence and absence of ATA treatment. The actual speed of the measurements and the applied drug concentration was such that the OKT4E disengagement was at an intermediate phase. This technical variation did not influence the transfer efficiency of the donor acceptor pair and the difference in the data was derived from pairs binding the OKT4E and the ATA simultaneously, for a short period. Of course, the applicability of such a kinetic approach is highly dependent on the kinetics of binding and disengagement processes. Nevertheless, the significant difference between transfer efficiency values provided strong evidence for a conformational change of the CD4 in the presence of ATA.[255] The importance of the above findings can be further underlined by the fact that enzymes like the $p56^{lck}$ tyrosine kinase of the *src* tyrosine kinase family are involved in the functional and physical association of CD4 and TcR/CD3 complex.[261]

Importantly enough, it has been shown by Mittler et al. that not only kinases but also phosphatases can be associated with CD4 or CD8 antigens.[262] The equilibrium between phosphorylation and dephosphorylation processes is one of the key mechanisms to regulate cell activation processes. The great variety of phosphatases and kinases may suggest that particular types of these enzymes can be regulated by either different regulatory proteins or by their different conformations during intermolecular interactions. CD45 is a protein tyrosine phosphatase on the cell surface. It was established that activation of human peripheral blood lymphocytes with solid phase anti-CD3 antibodies, or during a mixed lymphocyte response (MLR), induced the association of CD4 or CD8 antigens with CD45. While TcR/CD3 receptor-mediated signal processing initiated the above association, mitogen-triggered activation did not induce the formation of the complex. The maximal association occurred in the late phase of activation, i.e., 72 to 96 h after exposure to the anti-CD3 monoclonal antibodies. Kinetic measurements of the association process were made by FCET. The association was confirmed by radiolabeled immunoprecipitates as well. The protein tyrosine phosphatase activity was detectable in the precipitates. The enzyme activity changed parallel with the association process, as observed by FCET. The dephosphorylation process may reset the receptor for the reception of the next signal.[262]

In addition to immunoprecipitation, de la Hera et al.[263] used FCET studies for investigating the structure of T cell receptor (TCR)/CD3 complexes. They wanted to determine the subunit stoichiometry of TCR/CD3 complex, which is supposed to be $\alpha\beta\gamma\delta\zeta_2$ or $\alpha\beta\gamma\delta\zeta\eta$.[264-266] However, previous studies did not accurately quantitate the number of individual chains; therefore, data supporting the above presentation of TCR/CD3 subunits are lacking. In this work, the number of CD3ε subunits was at stake. Toward this end, surfaces of T cells prepared from transgenic mice carrying and expressing the human CD3ε gene were tagged with FITC-labeled anti-mouse CD3ε and TRITC-labeled anti-human CD3ε. The chimeric antigen receptors functioned normally, allowing normal T cell development and selection of repertoires *in vivo* and transducing activation signals *in vitro*. The relative spatial distribution of the binding sites for both conjugated monoclonal antibodies were analyzed by FCET. Significant FRET efficiency was detected indicating that anti-mouse CD3ε and anti-human CD3ε monoclonal antibodies and their respective binding sites are in close proximity (<10 nm). The reciprocal donor-acceptor combination yielded similar results. When similar experiments were carried out using donor/acceptor antibody pairs specific for either human of mouse CD3ε and mouse TCR-β, the measured FRET efficiency values indicated that both the human and mouse CD3ε chains and the TCR-β are in close proximity to each other. These FCET measurements demonstrated that the TCR/CD3 complex contains both human and mouse CD3ε chains, implying that more then one CD3ε subunit occurs per TCR. Antigen comodulation studies were consistent with the stochastic use of human and mouse CD3ε during receptor assembly, and further suggested a structure for the TCR/CD3 complex with two CD3ε chains. On the basis of their data the investigators proposed a minimal model for the structure of TCR/CD3 complex.[263]

D. STRUCTURE OF EGF RECEPTOR

Epidermal growth factor (EGF), a 53 amino acid polypeptide, is a potent mitogen for various epidermal and epithelial cells.[267] Binding of the factor to its cell surface receptor triggers the immediate stimulation of a receptor-associated tyrosine kinase activity and results in a variety of early and late cellular responses. Early responses include the stimulation of inositol lipid metabolism, change in intracellular free calcium concentration, and increases in ion fluxes through the plasma membrane, while late responses include the stimulation of DNA synthesis and the onset of mitosis.[267] The effects of EGF are mediated through the EGF receptor (170 kDa), which is a single chain 1186 amino acid polypeptides comprised of three domains: an extracellular, an intracellular, and a transmembrane domain.[268] The 621 amino acid N-terminal extracellular domain can bind EGF with high affinity; the 541 amino acid intracellular domain possesses tyrosine kinase activity. The extracellular ligand binding domain is connected with the cytoplasmic tyrosine kinase domain by an approximately 23 amino acid transmembrane domain which traverses the plasma membrane once.[267,268] The important question in EGF action is how mitogenic signal, initiated by EGF binding, is transduced across the membrane activating the cytoplasmic tyrosine kinase via a minimal structure that connects these two major domains of the receptor.

Two types of models have been proposed explaining the EGF-induced tyrosine kinase activation. The first type of mechanism (the intramolecular) invokes an EGF-induced intramolecular conformational change which is propagated across the single membrane-spanning helix to the cytoplasmic domain and results in the direct activation of the tyrosine kinase function.[269] Such a mechanism predicts that upon EGF binding the monomeric form of the receptor is a fully active tyrosine kinase. The second type of mechanism (the intermolecular) assumes that a receptor aggregation event is necessary for the signaling process. In this case, binding of EGF causes its receptors to aggregate into a dimeric state and this aggregation event stimulates the receptor kinase activity.[270] EGF-induced oligomerization has been demonstrated with biochemical methods,[270,271] lateral diffusion measurements,[272,273] and rotational diffusion measurements.[274,275]

Carraway et al.[276] applied the FRET approach to monitor EGF receptor-receptor interaction in plasma membranes isolated from human epidermoid A431 cells. EGF molecules labeled at the amino terminus with FITC served as donor molecules in these FRET measurements, while EGF molecules labeled with eosin isothiocyanate at the amino terminus served as acceptors. In the absence of metal ion activators relatively little energy transfer (10%) was found between the fluorescently labeled growth factors. These data suggested the binding of the EGF to the receptor in itself may not be sufficient to fully stimulate interactions between receptor molecules within the plasma membrane. However, divalent metal ion activators of the EGF receptor/kinase caused a significant increase in the extent of energy transfer between the fluorescently tagged EGF molecules. The ability of these metal ions to increase the extent of energy transfer was closely related to their ability to stimulate the autophosphorylation of the receptor. Addition of 20 m*M* $MgCl_2$ enhanced the level of FRET efficiency to about 2.5-fold over its basal level (10%) under conditions where $MgCl_2$ activated the autophosphorylation of the receptor. The addition of $MnCl_2$ (4 m*M*) to the purified membrane preparation increased the FRET efficiency still further, to 40 to 50%, and elicited a corresponding enhancement in the levels of receptor autophosphorylation. Moreover, the dissociation constants for the binding of these metal ions to the EGF receptor, as measured by the stimulation of FRET, are in excellent agreement with their activation constants for the EGF-stimulated phosphorylation of A431 membranes.[276] The low level of energy transfer observed in the absence of divalent metal ions would require that the donor and acceptor EGF molecules within an aggregate (i.e., dimer) were separated by at least 8 nm. The Mn^{2+}-induced increase in FRET would then require a considerable metal-induced translocation of the labeled growth factors with respect to each other, resulting in an approximately 6-nm distance between the donor and acceptor probes. This large distance change of the extracellular domain of the EGF receptor argues against a global conformational change in the receptor molecules and suggests that metal ions play an integral role in promoting receptor-receptor aggregation.[276]

Azevedeo and Johnson[277] studied the temperature-dependent aggregation and extracellular protrusion of EGF receptor in isolated plasma membranes from A431 cells applying the FRET technique. Aggregation of the EGF receptor was measured by determining the increase in FRET between receptor-bound FITC-labeled and TRITC-labeled EGF. FRET between the fluorescently tagged EGF molecules was significantly greater at 37°C than at 4°C. Subsequent cooling of samples from 37 to 4°C produced FRET efficiencies indistinguishable from those for the same sample before warming to 37°C. These data indicated that the temperature-induced receptor aggregation was completely reversible. The distance of closest approach between the N-terminus of receptor-bound EGF and the lipid membrane surface was also determined by FRET technique. In this case, FRET was measured between receptor bound FITC-EGF and 5-(*N*-dodecanoylamino)-eosin (C_{12}-eosin) partitioned into membranes. By titrating A431 membranes with C_{12}-eosin, no significant change in the distance of closest approach between the N-terminus of EGF and the plasma membrane was observed at 4°C (6.9 nm) and 37°C (6.7 nm). Since the observed difference (0.2 nm) is well within the accuracy of the FRET-based distance measurements, it was suggested that the extracellular protrusion of the occupied EGF receptor did not change detectably upon receptor aggregation. The investigators also concluded that upon receptor aggregation membrane lipids are not excluded from the aggregated receptors, since no significant change was observed in the FRET efficiency between receptor-bound EGF and C_{12}-eosin partitioned into the lipid membrane domain. Reduction in FRET efficiency would have been observed if aggregation excluded lipids. Consequently, domains of EGF receptor aggregates did not appear to exclude lipids.[277]

While the above-mentioned two studies were performed on isolated plasma membrane preparations, intact A431 cells were used by Carraway and Cerione[278] for monitoring microaggregation of the EGF receptor within the plasma membrane. The purpose of their study was to determine the extent of EGF receptor aggregation on the surface of intact A431 cells and to see whether or not any observed EGF receptor aggregation could be correlated with the transduction of the signal of early cellular response to EGF. They used the same donor acceptor pair, i.e., fluorescein-labeled EGF and eosin-labeled EGF, as before in their study on isolated membranes,[276] to monitor any microaggregation of the EGF receptor on the cell surface. No energy transfer was observed between receptor-bound, fluorescently labeled EGF on the surface of whole cells under conditions where these growth factors triggered the release of Ca^{2+} from intracellular stores and stimulated Ca^{2+} influx. However, after disrupting cells by homogenization, a significant FRET efficiency (10 to 20%) was observed in membranes, indicating receptor aggregation. The authors suggested that EGF receptor microaggregation is significantly more restricted on the surface of cells than in isolated membranes.[278]

These results argue that receptor aggregation events observed in isolated membranes are not necessary for the transduction of the signal across the plasma membrane. Thus, either a very small population of cell surface receptors aggregate in response to growth factor binding and act to transduce the signal or a conformational change within the receptor monomer is sufficient for the early signaling events.

These observations imply that the first type of mechanism (intramolecular) is valid for the EGF-induced signal transduction processes responsible for the early events. The role of receptor aggregation could be (1) either targeting active receptors for down regulation or (2) after endocytosis, coupling the aggregated from of the receptor to intracellular substrates and stimulate later cellular responses to the growth factor.

E. STRUCTURE OF IGE RECEPTOR

In mast cells and basophils, aggregation of the membrane receptor for immunoglobulin E (IgE) initiates a cascade of reactions culminating in exocytosis and secretion of granules containing mediators of inflammation. IgE binding to its high-affinity receptor ($F_{c\varepsilon}RI$) does not, by itself, stimulate the cell, but subsequent cross-linking of IgE-receptor complexes by multivalent antigens triggers the cellular degranulation which plays central role in the allergic response. Interactions that control this receptor-mediated cell stimulus are binding of IgE to receptor, cross-linking of IgE-receptor complexes by antigen, and coupling cellular components to transduce the signal across the plasma membrane.[279]

Resonance energy transfer has been used to investigate the structure of IgE bound to receptors on isolated plasma membranes of rat basophilic leukemia (RBL) cells.[280] Donor probes of several different sites on the IgE molecule were applied and the distance was measured between each of these sites and amphipathic acceptors probes at the plasma membrane surface when IgE was bound to its receptor. FITC preferentially labeled the F_{ab} segments of IgE, and limited reduction, followed by alkylation with *N*-{4-[7-(diethylamino)-4-methylcoumarin-3-yl]phenyl}maleimide caused selective modification of the inter-ε chain disulfides in the $C_{\varepsilon}2$ domains. For fluorescein donor, 5-(hexadecanoylamino)eosin or octadecylrhodamine were used as energy acceptors; for methylcoumarin donor, 5-(hexadecanoylamino)eosin or didecyloxacarbocyanine were used as acceptors. From the fluorescence quenching of the donor probes measured as a function of acceptor density, the closest distance from the membrane surface was calculated to be 8.6 to 9.1 nm for the fluorescein-labeled F_{ab} segments.[280] The distance to the membrane from the antibody-combining site at the tip was larger than 10 nm.[281] These results indicated that the F_{ab} segments are sticking up, away from the membrane. The distance from the conjugated SHs in the $C_{\varepsilon}2$ domains was only about 4.5 nm, which is considerably less that the estimated distance of 8 nm between sites in this domain and the C-terminal end of the IgE molecule. This result indicated that the C-terminal of the IgE molecule is probably not involved in binding to the receptor.

Using a similar FRET experimental system, the same group studied the possible conformations of IgE upon binding.[282] The donor probes were placed on the IgE F_{ab}, and the acceptor probes on the IgE F_c part. There was a difference in FRET efficiency measured within IgE in solution compared to when IgE was bound to the receptor, suggesting that a change of conformation occurs when IgE binds to its receptor. On the basis of their data the investigators proposed the following model for IgE binding. IgE does not bind to its receptor via its terminal end, rather, in a bent conformation, with the F_{ab} arms sticking up and the C-terminal end of the F_c segment sticking out sideways parallel to the membrane.[279-282]

Kubitscheck et al.[283] used a microscope-based FRET assay to determine the specific binding of 2,4-dinitrophenyl (DNP)-haptens to IgE molecules bound to $F_{c\varepsilon}RI$ on the surface of single, living rat basophilic leukemia cells, subclone 2H3. They used a recently developed technique that permits the quantitative determination of FRET between two fluorophores on single cells in a microscope from the photobleaching kinetics of the donor fluorophore.[168,169] Fluorescein-conjugated IgE specific for DNP served both as donor and binding site for the acceptor molecule. Bound DNP was an effective acceptor for fluorescein, despite the unfavorable spectral overlap, probably due to the close proximity of the antibody combining site to the conjugation site of fluorescein on the employed monoclonal IgE. FRET measurements yielded a transfer efficiency of 18 to 20% for fluorescein-labeled IgE upon saturation with DNP-haptens bound to type I $F_{c\varepsilon}$ receptors on living cells. Using FRET measurement, the investigators were able to determine the equilibrium binding constants for various DNP haptens. Binding constants of DNP haptens to $F_{c\varepsilon}RI$-bound IgE are found to be 4 to 11 times higher that those for binding of these haptens to free IgE in solution. These data indicated that IgE bound to adherent single cells behaves differently from that bound to cells in suspension.[283] Combining this observation with that of Holowka and Baird,[279-282] describing a bending of IgE upon binding to its receptor, one might suggest that the conformational change induced by binding of IgE to its receptor causes a favorable change in the affinity of IgE toward the DNP hapten.

The data described above in a concise form clearly demonstrated the power of FRET in general and the photobleaching variation of the method in particular. We believe that application of microscopic FRET technique will provide a significant contribution to our knowledge about the structure and function of biological membranes.

IV. CONCLUSIONS AND PERSPECTIVES

All the examples discussed in this chapter strongly proved that the FRET technique is a powerful method for studying interactions of molecules in membranes. The method virtually extends the resolving power of optical microscopes and flow cytometers to a molecular level by measuring distances between donor- and acceptor-labeled molecules on the nanometer scale. As it was discussed above, FRET has been successfully applied for monitoring membrane fusion, for localization of probes and drugs in the membrane, and assessing the orientation of proteins relative to the plane of the membranes. In addition, with the help of fluorescently conjugated monoclonal antibodies, specific labeling of various cell surface proteins can be accomplished, thus providing a way for detecting the aggregation state of proteins, i.e., the existence of molecular patterns on the cell surface. Conformational changes of membrane proteins during their function in model and cellular membranes have also been effectively followed by FRET. Of course, FRET alone provides only one-sided information about the structural changes in the molecular assembly of membranes. Therefore, FRET is usually used to reconfirm the putative associations of membrane molecules detected with other, more invasive, biochemical methods, such as immunoprecipition or crosslinking. On the other hand, associations, conformational changes detected by FRET, should be reinforced by other biochemical or dynamic biophysical techniques (e.g., fluorescence recovery after photobleaching or phosphorescence anisotropy measurement). Only this type of combined approach will provide convincing evidence about the possible structural and functional arrangement of membrane components.

Improving the sensitivity of the detection systems in spectrofluorimeters, flow cytometers, and microscopes widens the field where FRET can be applied for studying molecular interactions. Moreover, introduction of new fluorescent dyes (e.g., phyco dyes, Cy3 dye) with better photophysical properties as donor acceptor pairs also promotes the application of FRET. Developments in microscopic techniques, such as the introduction of image analysis systems and confocal arrangements, will also fertilize the FRET approach. These improvements in microscopic techniques open up new areas in biological problems which otherwise could not be investigated with the classical FRET method. For example, cell-cell interactions can be studied in a very special way, using microscopic FRET technique, since with this approach the protein-protein interactions in the area of cell-cell contact can be investigated selectively. This new approach will provide a valuable tool for studying cell-cell interactions and communication, revealing the special orientation and molecular dimension of membrane components participating in establishing this specific cell-cell contact.

More and more data are available on the physical association of cell surface proteins, stressing the functional importance of such physical arrangements. We believe that the combination of various FRET techniques with other biochemical and biophysical methods has the capacity and resolution power to provide a significant contribution to the solution of cell biological problems.

REFERENCES

1. **Wilschut, J., Düzgünes, N., Fraley, R., and Papahadjopoulos, D.,** Studies on the mechanism of membrane fusion: kinetics of calcium ion induced fusion of phosphatidylserine vesicles followed by a new assay for mixing of aqueous vesicle contents, *Biochemistry*, 19, 6011, 1980.
2. **Papahadjopoulos, D., Vail, W. J., Newton, C., Nir, S., Jacobson, K., Poste, G., and Lazo, R.,** Studies on membrane fusion. III. The role of calcium-induced phase changes, *Biochim. Biophys. Acta*, 465, 579, 1977.
3. **Liao, M.-J. and Prestegard, J. H.,** Fusion of phosphatidic acid-phosphatidylcholine mixed lipid vesicles, *Biochim. Biophys. Acta*, 550, 157, 1979.
4. **Verkleij, A. J., Mombers, C., Gerritson, W. J., Leunissen-Bijvelt, L., and Cullis, P. R.,** Fusion of phospholipid vesicles in association with the appearance of lipidic particles as visualized by freeze fracturing, *Biochim. Biophys. Acta*, 555, 358, 1979.
5. **Miller, C. and Racker, E.,** Fusion of phospholipid vesicles reconstituted with cytochrome c oxidase and mitochondrial hydrophobic protein, *J. Membrane Biol.*, 26, 319, 1976.
6. **Keller, P. M., Person, S., and Snipes, W.,** A fluorescence enhancement assay of cell fusion, *J. Cell Sci.*, 28, 167, 1977.
7. **Vanderwerf, P. and Ullman, E. F.,** Monitoring of phospholipid vesicle fusion by fluorescence energy transfer between membrane-bound dye labels, *Biochim. Biophys. Acta*, 596, 302, 1980.
8. **Uster, P. S. and Deamer, D. W.,** Fusion competence of phosphatidylserine-containing liposomes quantitatively measured by a fluorescence resonance energy transfer assay, *Arch. Biochem. Biophys.*, 209, 385, 1981.
9. **Struck, D. K., Hoekstra, D., and Pagano, R. E.,** Use of resonance energy transfer to monitor membrane fusion, *Biochemistry*, 20, 4093, 1981.
10. **Trón, L., Szöllősi, J., Damjanovich, S., Helliwell, S. H., Arndt-Jovin, D. A., and Jovin, T. M.,** Flow cytometric measurement of fluorescence resonance energy transfer on cell surfaces. Quantitative evaluation of the transfer efficiency on a cell-by-cell basis, *Biophys. J.*, 45, 939, 1984.
11. **Szöllősi, J., Trón, L., Damjanovich, S., Helliwell, S. H., Arndt-Jovin, D. A., and Jovin, T. M.,** Fluorescence energy transfer measurements on cell surfaces. A critical comparison of steady state and flow cytometric methods, *Cytometry*, 5, 210, 1984.
12. **MacDonald, R. I.,** Energy transfer measurements of fusion between Sendai virus and vesicles corrected for decreased absorption of acceptor probe, *J. Biol. Chem.*, 262, 10392, 1987.
13. **Düzgünes, N., Allen, T. M., Fedor, J., and Papahadjopoulos, D.,** Lipid mixing during membrane aggregation and fusion: why fusion assays disagree, *Biochemistry*, 26, 8435, 1987.
14. **Silvius, J. R., Leventis, R., Brown, P. M., and Zuckermann, M.,** Novel fluorescent phospholipids for assays of lipid mixing between membranes, *Biochemistry*, 26, 4279, 1987.

15. **Nayar, R., Tilcock, C. P., Hope, M. J., Cullis, P. R., and Schroit, A. J.**, N-succinyldioleoylphosphatidylethanolamine: structural preferences in pure and mixed model membranes, *Biochim. Biophys. Acta*, 937, 31, 1988.
16. **Wilschut, J., Scholma, J., Eastman, S. J., Hope, M. J., and Cullis, P. R.**, Ca^{2+}-induced fusion of phospholipid vesicles containing free fatty acids: modulation by transmembrane pH gradients, *Biochemistry*, 31, 2629, 1992
17. **Oshry, L., Meers, P., Mealy, T., and Tauber, A. I.**, Annexin-mediated membrane fusion of human neutrophil plasma membranes and phospholipid vesicles, *Biochim. Biophys. Acta*, 1066, 239, 1991.
18. **Morris, S. J. and Bradley, D.**, Calcium-promoted fusion of isolated chromaffin granules detected by resonance energy transfer between labeled lipids embedded in membrane bilayer, *Biochemistry*, 23, 4642, 1984.
19. **Lawaczeck, R., Gervais, M., Nandi, P. K., and Nicolau, C.**, Fusion of negatively charged liposomes with clathrin-uncoated vesicles, *Biochim. Biophys. Acta*, 903, 112, 1987.
20. **Maezawa, S., Yoshimura, T., Hong, K., Düzgünes, N., and Papahadjopoulos, D.**, Mechanism of protein-induced membrane fusion: fusion of phospholipid vesicles by clathrin associated with its membrane binding and conformational change, *Biochemistry*, 28, 1422, 1989.
21. **Arvinte, T., Wahl, P., and Nicolau, C.**, Resonance energy-transfer and fluorescence intensity studies of the transport of liposome-encapsulated molecules into isolated mouse liver nuclei, *Biochemistry*, 26, 765, 1987.
22. **Lawaczeck, R., Nandi, P. K., and Nicolau, C.**, Interaction of negatively charged liposomes with nuclear membranes: adsorption, lipid mixing and lysis of the vesicles, *Biochim. Biophys. Acta*, 903, 123, 1987.
23. **Eum, K. M., Riedy, G., Langley, K. H., and Roberts, M. F.**, Temperature-induced fusion of small unilamellar vesicles formed from saturated long-chain lecithins and diheptanoylphosphatidylcholine, *Biochemistry*, 28, 8206, 1989.
24. **Womersley, C., Uster, P. S., Rudolph, A. S., and Crowe, J. H.**, Inhibition of dehydration-induced fusion between liposomal membranes by carbohydrates as measured by fluorescence energy transfer, *Cryobiology*, 23, 245, 1986.
25. **MacDonald, R. I. and MacDonald, R. C.**, Lipid mixing during freeze-thawing of liposomal membranes as monitored by fluorescence energy transfer, *Biochim. Biophys. Acta*, 735, 243, 1983.
26. **Morgan, C. G., Thomas, E. W., and Yianni, Y. P.**, The use of fluorescence energy transfer to distinguish between poly(ethylene glycol)-induced aggregation and fusion of phospholipid vesicles, *Biochim. Biophys. Acta*, 728, 356, 1983.
27. **Seibicke, S., Zimmermann, H. P., and Haeffner, E. W.**, Fusion of lipid vesicles with ascites tumor cells and their lipid-depleted variants. Studies with radioactive- and fluorescent-labeled vesicles, *Biochim. Biophys. Acta*, 944, 487, 1988.
28. **Comerford, J. G. and Dawson, A. P.**, The mechanism of action of GTP on Ca^{2+} efflux from rat liver microsomal vesicles. Measurement of vesicle fusion by fluorescence energy transfer, *Biochem. J.*, 249, 89, 1988.
29. **Düzgünes, N., Goldstein, J. A., Friend, D. S., and Felgner P. L.**, Fusion of liposomes containing a novel cationic lipid, N-[2,3-(dioleyloxy)propyl]-N,N,N-trimethylammonium: induction by multivalent anions and asymmetric fusion with acidic phospholipid vesicles, *Biochemistry*, 28, 9179. 1989.
30. **Walter, A., Steer, C. J., and Blumenthal, R.**, Polylysine induces pH-dependent fusion of acidic phospholipid vesicles: a model for polycation-induced fusion, *Biochim. Biophys. Acta*, 861, 319, 1986.
31. **van Duijn, G., Valtersson, C., Chojnacki, T., Verkleij, A. J., Dallner, G., and de Kruijff, B.**, Dolichyl phosphate induces non-bilayer structures, vesicle fusion and transbilayer movement of lipids: a model membrane study, *Biochim. Biophys. Acta*, 861, 211, 1986.
32. **Massotte, D. and Pattus, F.**, Colicin N and its thermolytic fragment induce phospholipid vesicle fusion, *FEBS Lett.*, 257, 447, 1989.
33. **Arnold, K., Hoekstra, D., and Ohki, S.**, Association of lysozyme to phospholipid surfaces and vesicle fusion, *Biochim. Biophys. Acta*, 1124, 88, 1992.
34. **Arvinte, T., Wahl, P., and Nicolau, C.**, Low pH fusion of mouse liver nuclei with liposomes bearing covalently bound lysozyme, *Biochim. Biophys. Acta*, 899, 143, 1987.
35. **Lopez Vinals, A. E., Farias, R. N., and Morero, R. D.**, Characterization of the fusogenic properties of glyceraldehyde-3-phosphate dehydrogenase — fusion of phospholipid vesicles, *Biochem. Biophys. Res. Commun.*, 143, 403, 1987.

36. **Lai, J. Y., Chow, D. D., and Hwang, K. J.**, Effect of lipid composition on insulin-mediated fusion of small unilamellar liposomes: a kinetic study, *J. Pharm. Sci.*, 77, 432, 1988.
37. **Stillwell, W., Brengle, B., and Wassall, S. R.**, Abscisic acid enhances aggregation and fusion of phospholipid vesicles, *Biochem. Biophys. Res. Commun.*, 156, 511, 1988.
38. **Gasset, M., Thomas, P. G., and Gavilanes, J. G.**, Fusion of phospholipid vesicles produced by the anti-tumor protein α-sarcin, *Biochem. J.*, 265, 815, 1990.
39. **Lishko, V. K., Terletskaya, Ya. T., and Trikash, I. O.**, Fusion of negatively charged phospholipid vesicles by α-latrotoxin, *FEBS Lett.*, 266, 99, 1990.
40. **Rufini, S., Pedersen, J. Z., Desideri, A., and Luly, P.**, β-bungarotoxin-mediated liposome fusion: spectroscopic characterization by fluorescence and ESR, *Biochemistry*, 29, 9644, 1990.
41. **Papini, E., Colonna, R., Cusinato, F., Montecucco, C., Tomasi, M., and Rappuoli, R.**, Lipid interaction of diphtheria toxin and mutants with altered fragment B. I. Liposome aggregation and fusion, *Eur. J. Biochem.*, 169, 629, 1987.
42. **Martin, I., Defrise-Quertain, F., Mandieau, V., Nielsen, N. M., Saermark, T., Burny, A., Brasseur, R., Ruysschaert, J. M., and Vandenbranden, M.**, Fusogenic activity of SIV (simian immunodeficiency virus) peptides located in the GP32 NH2 terminal domain, *Biochem. Biophys. Res. Commun.*, 175, 872, 1991.
43. **Glushakova, S. E., Omel'ianenko, V. G., Lukashevich, I. S., Bogdanov, A. A., Moshnikova, A. B., Kozhich, A. T., and Torchilin, V. P.**, Fusion of artificial lipid membranes induced by synthetic "fusion peptide" of arenaviruses, *Biokhimiia*, 56, 579, 1991.
44. **Nir, S., Stegmann, T., and Wilschut, J.**, Fusion of influenza virus with cardiolipin liposomes at low pH: mass action analysis of kinetics and extent, *Biochemistry*, 25, 257, 1986.
45. **Stegmann, T., Morselt, H. W., Booy, F. P., van Breemen, J. F., Scherphof, G., and Wilschut, J.**, Functional reconstitution of influenza virus envelopes, *EMBO J.*, 6, 2651, 1987.
46. **Fung, B. K.-K. and Stryer, L.**, Surface density determination in membranes by fluorescence energy transfer, *Biochemistry*, 24, 5241, 1978
47. **Aso, Y., Kano, K., and Matsuo, T.**, Energy transfer in artificial membrane systems. Singlet-singlet energy transfer from alloxazines to isoalloxazines in dipalmitoyl phosphatidylcholine liposomes and dialkylammonium chloride vesicles, *Biochim. Biophys. Acta*, 599, 403, 1980.
48. **Vaz, W. L. C., Kaufmann, K., and Nicksch, A.**, Use of energy transfer to assay the association of proteins with lipid membranes, *Anal. Biochem.*, 83, 385, 1977.
49. **Bastiaens, P., de Beus, A., Lacker, M., Somerharju, P., Vauhkonen, M., and Eisinger, J.**, Resonance energy transfer from a cylindrical distribution of donors to a plane of acceptors. Location of apo-B100 protein on the human low-density lipoprotein particle, *Biophys. J.*, 58, 665, 1990.
50. **Connor, J. and Schroit, A. J.**, Determination of lipid asymmetry in human red cells by resonance energy transfer, *Biochemistry*, 26, 5099, 1987.
51. **Wolf, E. D., Winiski, A. P., Ting, A. E., Bocian, K. M., and Pagano, R. E.**, Determination of transbilayer distribution of fluorescent lipid analogues by nonradiative fluorescence resonance energy transfer, *Biochemistry*, 31, 2865, 1992.
52. **Dobretsov, G. E., Kurek, N. K., Machov, V. N., Syrejshchikova, T. I., and Yakimenko, M. N.**, Determination of fluorescent probes localization in membranes by nonradiative energy transfer, *J. Biochem. Biophys. Methods*, 19, 259, 1989.
53. **Frolov, A. A., Zenkevich, E. I., Gurinovich, G. P., and Kochubeyev, G. A.**, Chlorin e6-liposome interaction. Investigation by the methods of fluorescence spectroscopy and inductive resonance energy transfer, *J. Photochem. Photobiol. B*, 7, 43, 1990.
54. **Kimura, K., Arata, Y., Yasuda, T., Kinosita, K., and Nakanishi, M.**, Location of membrane bound hapten with different length spacers, *Immunology*, 69, 323, 1990.
55. **Coutinho, A., Costa, J., Faria, J. L., Berberan-Santos, M. N., and Prieto, M. J. E.**, Dibucaine interaction with phospholipid vesicles. A resonance-energy transfer study, *Eur. J. Biochem.*, 189, 387, 1990.
56. **Blatt, E. and Sawyer, W. H.**, Depth-dependent fluorescent quenching in micelles and membranes, *Biochim. Biophys. Acta*, 822, 43, 1985.
57. **Dupou-Cezanne, L., Sautereau, A. M., and Tocanne, J. F.**, Localization of adriamycin in model and natural membranes. Influence of lipid molecular packing, *Eur. J. Biochem.*, 181, 695, 1989.
58. **Ferrer-Montiel, A. V., Gonzalez-Ros, J. M., and Ferragut, J. A.**, Association of daunomycin to membrane domains studied by fluorescence resonance energy transfer, *Biochim. Biophys. Acta*, 937, 379, 1988.

59. **Escriba, P. V., Ferrer-Montiel, A. V., Ferragut, J. A., and Gonzalez-Ros J. M.,** Role of membrane lipids in the interaction of daunomycin with plasma membranes from tumor cells: implications in drug-resistance phenomena, *Biochemistry*, 29, 7275, 1990,
60. **Ferrer-Montiel, A. V., Gonzalez-Ros, J. M., and Ferragut, J. A.,** Different distribution of daunomycin in plasma membranes from drug-sensitive and drug-resistant P388 leukemia cells, *Biochim. Biophys. Acta*, 1104, 111, 1992.
61. **Vanderkooi, J. M., Ierokomas, A., Nakmura, H., and Martonosi, A.,** Fluorescence energy transfer between Ca^{2+} transport ATPase molecules in artificial membranes, *Biochemistry*, 16, 1262, 1977.
62. **Papp, S., Pikula, S., and Martonosi, A.,** Fluorescence energy transfer as an indicator of Ca^{2+}-ATPase interactions in sarcoplasmic reticulum, *Biophys. J.*, 51, 205, 1987.
63. **Munkonge, F., Michelangeli, F., Rooney, E. K., East, J. M., and Lee, A. G.,** Effect of phospholipid:protein ratio on the state of aggregation of the (Ca^{2+}-Mg^{2+})-ATPase, *Biochemistry*, 27, 6800, 1988.
64. **Fagan, M. H. and Dewey, T. G.,** Resonance energy transfer study of membrane-bound aggregates of the sarcoplasmic reticulum calcium ATPase, *J. Biol. Chem.*, 261, 3654, 1986.
65. **Dewey, T. G. and Datta, M. M.,** Determination of the fractal dimension of membrane protein aggregates using fluorescence energy transfer, *Biophys. J.*, 56, 415, 1989.
66. **Teruel, J. A. and Gomez-Fernandez, J. C.,** Distances between the functional sites of sarcoplasmic reticulum (Ca^{2+} + Mg^{2+})-ATPase and the lipid/water interface, *Biochim. Biophys. Acta*, 863, 178, 1986.
67. **Gutierrez-Merino, C., Munkonge, F., Mata, A. M., East, J. M., Levinson, B. L., Napier, R. M., and Lee, A. G.,** The position of the ATP binding site on the (Ca^{2+} + Mg^{2+})-ATPase, *Biochim. Biophys. Acta*, 897, 207, 1987.
68. **Teruel, J. A. and Gomez-Fernandez, J. C.,** Distances between the FITC-binding site in the (Ca^{2+} + Mg^{2+})-ATPase from sarcoplasmic reticulum and fluorescent probes located in the membrane, *Biochem. Int.*, 14, 409, 1987.
69. **Scott, T. L.,** Distances between the functional sites of the (Ca^{2+} + Mg^{2+})-ATPase of sarcoplasmic reticulum, *J. Biol. Chem.*, 260, 14421, 1985.
70. **Munkonge, F., East, J. M., and Lee, A. G.,** Positions of the sites labeled by N-cyclohexyl-N′-(4-dimethylamino-1-naphthyl)carbodiimide on the (Ca^{2+} + Mg^{2+})-ATPase, *Biochim. Biophys. Acta*, 979, 113, 1989.
71. **Sumbilla, C., Cantilina, T., Collins, J. H., Malak, H., Lakowicz, J. R., and Inesi, G.,** Structural perturbation of the transmembrane region interferes with calcium binding by the Ca^{2+} transport ATPase, *J. Biol. Chem.*, 266, 12682, 1991.
72. **Squier, T. C., Bigelow, D. J., Garcia de Ancos, J., and Inesi, G.,** Localization of site-specific probes on the Ca-ATPase of sarcoplamic reticulum using fluorescence energy transfer, *J. Biol. Chem.*, 262, 4748, 1987.
73. **Bigelow, D. J. and Inesi, G.,** Frequency-domain fluorescence spectroscopy resolves the location of maleimide-directed spectroscopic probes within the tertiary structure of the Ca-ATPase of sarcoplasmic reticulum, *Biochemistry*, 30, 2113, 1991.
74. **Bigelow, D. J., Squier, T. C., and Inesi, G.,** Phosphorylation-dependent changes in the spatial relationship between Ca-ATPase polypeptide chains in sarcoplasmic reticulum membranes, *J. Biol. Chem.*, 267, 6952, 1992.
75. **Coll, R. J. and Murphy, A. J.,** Inhibition of the sarcoplasmic reticulum Ca^{2+}-ATPase by Reactive Red 120, *Biochim. Biophys. Acta*, 904, 227, 1987.
76. **Jona, I. and Martonosi, A.,** Interaction between the $Ca^{2(+)}$-ATPase and the proteolipid in artificial membranes, *Acta Physiol. Hung.*, 74, 215, 1989.
77. **Martonosi, A. N., Jona, I., Molnar, E., Seidler, N. W., Buchet, R., and Varga, S.,** Emerging views on the structure and dynamics of the $Ca^{2(+)}$-ATPase in sarcoplasmic reticulum, *FEBS Lett.*, 268, 365, 1990.
78. **Jona, I., Matko, J., and Martonosi, A.,** Structural dynamics of the Ca^{2+}-ATPase of sarcoplasmic reticulum. Temperature profiles of fluorescence polarization and intramolecular transfer, *Biochim. Biophys. Acta*, 1028, 183, 1990.
79. **Somogyi, B., Matkó, J., Papp, S., Hevessy, J., Welch, G. R., and Damjanovich, S.,** Förster-type energy transfer as a probe for changes in local fluctuations of the protein matrix, *Biochemistry*, 23, 3403, 1984.
80. **Kosk-Kosicka, D., Bzdega, T., and Wawrzynow, A.,** Fluorescence energy transfer studies of purified erythrocyte Ca^{2+}-ATPase. Ca^{2+}-regulated activation by oligomerization, *J. Biol. Chem.*, 264, 19495, 1989.

81. **Kosk-Kosicka, D., Bzdega, T., Wawrzynow, A., Scaillet, S., Nemcek, K., and Johnson, J. D.,** Erythrocyte $Ca^{2(+)}$-ATPase: activation by enzyme oligomerization versus by calmodulin, *Adv. Exp. Med. Biol.*, 269, 169, 1990.
82. **Kosk-Kosicka, D. and Bzdega, T.,** Effects of calmodulin on erythrocyte Ca^{2+}-ATPase activation and oligomerization, *Biochemistry*, 29, 3772, 1990.
83. **Wrzosek, A., Famulski, K. S., Lehotsky, J., and Pikula, S.,** Conformational changes of (Ca^{2+}-Mg^{2+})-ATPase of erythrocyte plasma membrane caused by calmodulin and phosphatidylserine as revealed by circular dichroism and fluorescence studies, *Biochim. Biophys. Acta*, 986, 263, 1989.
84. **Verbist, J., Gadella, T. W., Jr., Raeymaekers, L., Wuytack, F., Wirtz, K. W., and Casteels, R.,** Phosphoinositide-protein interactions of the plasma-membrane $Ca^{2(+)}$-transport ATPase as revealed by fluorescence energy transfer, *Biochim. Biophys. Acta*, 1063, 1, 1991.
85. **Amler, E., Abbott, A., and Ball, W. J., Jr.,** Structural dynamics and oligomeric interactions of Na^{+},$K^{(+)}$-ATPase as monitored using fluorescence energy transfer, *Biophys. J.*, 61, 553, 1992.
86. **Lee, J. A. and Fortes, P. A.,** Spatial relationship and conformational changes between the cardiac glycoside site and beta-subunit oligosaccharides in sodium plus potassium activated adenosinetriphosphatase, *Biochemistry*, 25, 8133, 1986.
87. **Johnson, D. A., Voet, J. G., and Taylor, P.,** Fluorescence energy transfer between cobra α-toxin molecules bound to the acetylcholine receptor, *J. Biol. Chem.*, 259, 5717, 1984.
88. **Johnson, D. A., Cushman, R., and Malekzadeh, R.,** Orientation of cobra α-toxin on the nicotinic acetylcholine receptor. Fluorescence studies, *J. Biol. Chem.*, 265, 7360, 1990.
89. **Valenzuela, C. F., Kerr, J. A., and Johnson, D.A.,** Quinacrine binds to the lipid-protein interface of the *Torpedo* acetylcholine receptor: a fluorescence study, *J. Biol. Chem.*, 267, 8238, 1992.
90. **Clarke, J. H. and Martinez-Carrion, M.,** Labeling of functionally sensitive sulfhydryl-containing domains of acetylcholine receptor from *Torpedo californica* membranes, *J. Biol. Chem.*, 261, 10063, 1986.
91. **Johnson, D. A., Brown, R. D., Herz, J. M., Berman, H. A., Andreasen, G. L., and Taylor, P.,** Decidium. A novel fluorescent probe of the agonist/antagonist and noncompetitive inhibitor sites on the nicotinic acetylcholine receptor, *J. Biol. Chem.*, 262, 14022, 1987.
92. **Herz, J. M., Johnson, D. A., and Taylor, P.,** Distance between the agonist and noncompetitive inhibitor sites on the nicotinic acetylcholine receptor, *J. Biol. Chem.*, 264, 12439, 1989.
93. **Rehorek, M., Dencher, N. A., and Heyn, M. P.,** Fluorescence energy transfer from diphenylhexatiene to bacteriorhodopsin in lipid vesicles, *Biophys. J.*, 43, 39, 1983.
94. **Leder, R. O., Helgerson, S. L., and Thomas, D. D.,** The transverse location of the retinal chromophore in the purple membrane by diffusion-enhanced energy transfer, *J. Mol. Biol.*, 209, 683, 1989.
95. **Rehorek, M., Dencher, N. A., and Heyn, M. P.,** Long-range lipid-protein interactions. Evidence from time-resolved fluorescence depolarization and energy-transfer experiments with bacteriorhodopsin-dimyristoyl-phosphatidylcholine vesicles, *Biochemistry*, 24, 5980, 1985.
96. **Hasselbacher, C. A., Street, T. L., and Dewey, T. G.,** Resonance energy transfer as a monitor of membrane protein domain segregation: application to the aggregation of bacteriorhodopsin reconstituted into phospholipid vesicles, *Biochemistry*, 23, 6445, 1984.
97. **Stelzer, K. J. and Gordon, M. A.,** Effect of alterations in lipid packing order by hydrophobic solutes on the association state of protein assemblies in model membranes, *Biophys. Chem.*, 23, 173, 1986.
98. **Kouyama, T., Kinosita, K., Jr., and Ikegami, A.,** Structure and function of bacteriorhodopsin, *Adv. Biophys.*, 24, 123, 1988.
99. **Renthal, R., Cothran, M., Dawson, N., and Harris, G. J.,** Fluorescent labeling of bacteriorhodopsin: implications for helix connections, *Biochim. Biophys. Acta*, 897, 384, 1987.
100. **Straume, M. and Litman, B. J.,** Equilibrium and dynamic bilayer structural properties of unsaturated acyl chain phosphatidylcholine-cholesterol-rhodopsin recombinant vesicles and rod outer segment disk membranes as determined from higher order analysis of fluorescence anisotropy decay, *Biochemistry*, 27, 7723, 1988.
101. **Thomas, D. and Stryer, L.,** Transverse location of the retinal chromophore of rhodopsin in rod outer segment disc membranes, *J. Mol. Biol.*, 154, 145, 1982.
102. **Borochov-Neori, H. and Montal, M.,** Rhodopsin-G-protein interactions monitored by resonance energy transfer, *Biochemistry*, 28, 1711, 1989.

103. **Brodsky, F. M., Hill, B. L., Acton, S. L., Nathke, I., Wong, D. H., Ponnambalam, S., and Parham, P.,** Clathrin light chains: arrays of protein motifs that regulate coated-vesicle dynamics, *Trends Biochem. Sci.*, 16, 208, 1991.
104. **Yoshimura, T., Maezawa, S., and Hong, K.,** Exposure of hydrophobic domains of clathrin in its membrane fusion-inducible pH region, *J. Biochem.*, 101, 1265, 1987.
105. **Maezawa, S. and Yoshimura, T.,** Assembly of clathrin molecules on liposome membranes: a possible event necessary for induction of membrane fusion, *Biochem. Biophys. Res. Commun.*, 173, 134, 1990.
106. **Maezawa, S. and Yoshimura, T.,** Determination of the regions of the clathrin molecule inducing membrane fusion, *Biochemistry*, 29, 1813, 1990.
107. **Sims, P. J.,** Complement protein C9 labeled with fluorescein isothiocyanate can be used to monitor C9 polymerization and formation of the cytosolic membrane lesion, *Biochemistry*, 23, 3248, 1984.
108. **Sims, P. J. and Wiedmer, T.,** Kinetics of polymerization of a fluoresceinated derivative of complement protein C9 by the membrane-bound complex of complement proteins C5b-8, *Biochemistry*, 23, 3260, 1984.
109. **Cheng, K.-H., Wiedmer, T., and Sims, P. J.,** Fluorescence resonance energy transfer study of the associative state of membrane-bound complexes of complement proteins C5b-8, *J. Biol. Chem.*, 135, 459, 1985.
110. **Prigent-Dachary, J., Faucon, J., Boisseau, M. R., and Dufourcq, J.,** Topology of the binding site of blood-clotting factors in model membranes. A fluorescence study, *Eur. J. Biochem.*, 155, 133, 1986.
111. **Prigent-Dachary, J., Lindhout, T., Boisseau, M. R., and Dufourcq, J.,** A further insight into the binding of blood clotting factors to membranes, *Eur. J. Biochem.*, 181, 675, 1989.
112. **Isaacs, B. S., Husten, E. J., Esmon, C. T., and Johnson, A. E.,** A domain of membrane-bound blood coagulation factor Va is located far from the phospholipid surface. A fluorescence energy transfer measurement, *Biochemistry*, 25, 4958, 1986.
113. **Husten, E. J., Esmon, C. T., and Johnson, A. E.,** The active site of blood coagulation factor Xa. Its distance from the phospholipid surface and its conformational sensitivity to components of the prothrombinase complex, *J. Biol. Chem.*, 262, 12953, 1987.
114. **Lu, R. L., Esmon, N. L., Esmon, C. T., and Johnson, A. E.,** The active site of the thrombin-thrombomodulin complex. A fluorescence energy transfer measurement of its distance above the membrane surface, *J. Biol. Chem.*, 264, 12956, 1989.
115. **Esmon, C. T.,** The roles of protein C and thrombomodulin in the regulation of blood coagulation, *J. Biol. Chem.*, 264, 4743, 1989.
116. **Armstrong, S. A., Husten, E. J., Esmon, C. T., and Johnson, A. E.,** The active site of membrane-bound meizothrombin. A fluorescence determination of its distance from the phospholipid surface and its conformational sensitivity to calcium and factor Va, *J. Biol. Chem.*, 265, 6210, 1990.
117. **Dale, R. E., Eisinger, J., and Blumberg, W. E.,** The orientational freedom of molecular probes. The orientation factor in intramolecular energy transfer, *Biophys. J.*, 26, 161, 1979.
118. **Gilbert, G. E., Furie, B. C., and Furie, B.,** Binding of human factor VIII to phospholipid vesicles, *J. Biol. Chem.*, 265, 815, 1990.
119. **Nesheim, M. E., Pittman, D. D., Wang, J. H., Slonosky, D., Giles, A. R., and Kaufman, R. J.,** The binding of 35S-labeled recombinant factor VIII to activated and unactivated human platelets, *J. Biol. Chem.*, 263, 16467, 1988.
120. **Sims, P. J., Ginsberg, M. H., Plow, E. F., and Shattil, S. J.,** Effect of platelet activation on the conformation of the plasma membrane glycoprotein IIb-IIIa complex, *J. Biol. Chem.*, 266, 7345, 1991.
121. **Veatch, W. and Stryer, L.,** The dimeric nature of the Gramicidin A transmembrane channel: conductance and fluorescence energy transfer studies of hybrid channels, *J. Mol. Biol.*, 113, 89, 1977.
122. **Wang, S., Martin, E., Cimino, J., Omann, G., and Glaser, M.,** Distribution of phospholipids around gramicidin and D-β-hydroxybutyrate dehydrogenase as measured by resonance energy transfer, *Biochemistry*, 27, 2033, 1988.
123. **Ikigai, H. and Nakae, T.,** Conformational alteration in alpha-toxin from *Staphylococcus aureus* concomitant with the transformation of the water-soluble monomer to the membrane oligomer, *Biochem. Biophys. Res. Commun.*, 130, 175, 1985.
124. **Ikigai, H. and Nakae, T.,** Interaction of the α-toxin of *Staphylococcus aureus* with the liposome membrane, *J. Biol. Chem.*, 262, 2150, 1987.

125. **Ikigai, H. and Nakae, T.**, Assembly of the α-toxin-hexamer of *Staphylococcus aureus* in the liposome membrane, *J. Biol. Chem.*, 262, 2156, 1987.
126. **Nogueira, R. A. and Varand, W. A.**, Gating properties of channels formed by colicins Ia in planar lipid bilayer membranes, *J. Membr. Biol.*, 105, 143, 1988.
127. **Martinez, M. C., Lazdunski, C., and Pattus, F.**, Isolation, molecular and functional properties of the C-terminal of colicin A, *EMBO J.*, 2, 1501, 1983.
128. **Lakey, J. H., Baty, D., and Pattus, F.**, Fluorescence energy transfer measurements using site-directed single cysteine mutants. The membrane insertion of colicin A, *J. Mol. Biol.*, 218, 639, 1991.
129. **Shai, Y., Bach, D., and Yanovsky, A.**, Channel formation properties of synthetic pardaxin and analogues, *J. Biol. Chem.*, 265, 20202, 1990.
130. **Rapaport, D. and Shai, Y.**, Aggregation and organization of pardaxin in phospholipid membranes. A fluorescence energy transfer study, *J. Biol. Chem.*, 267, 6502, 1992.
131. **Rapaport, D. and Shai, Y.**, Interaction of fluorescently labeled pardaxin and its analogues with lipid bilayers, *J. Biol. Chem.*, 266, 23769, 1991.
132. **Otoda, J., Kimura, S., and Imanishi, Y.**, Interaction of melittin derivatives in lipid bilayer membrane. Role of basic residues at the C-terminal and their replacement with lactose, *Biochim. Biophys. Acta*, 1112, 1, 1992.
133. **Hanke, W., Methfessel, C., Wilmsen, H., Katz, E., Jung, G., and Boheim, G.**, Melittin and a chemically modified trichotoxin form alamethicin-type multi-state pores, *Biochim. Biophys. Acta*, 727, 108, 1983.
134. **Schulze, J., Mischeck, U., Wigand, S., and Galla, H. J.**, Incorporation of highly purified melittin into phosphatidylcholine bilayer vesicles, *Biochim. Biophys. Acta*, 901, 101, 1987.
135. **Batenburg, A. M., Hibbeln, J. C., and de Kruijff, B.**, Lipid specific penetration of melittin into phospholipid model membranes, *Biochim. Biophys. Acta*, 903, 155, 1987.
136. **Gromova, I. A., Molotkovsky, J. G., and Bergelson, L. D.**, Anthrylvinyl-labeled phospholipids as fluorescent membrane probes. The action of melittin on multilipid systems, *Chem. Phys. Lipids*, 60, 235, 1992.
137. **Vogel, H. and Jähnig, F.**, The structure of melittin in membranes, *Biophys. J.*, 50, 573, 1986.
138. **Talbot, J. C., Faucon, J. F., and Dufourcq, J.**, Different states of self-association of melittin in phospholipid bilayers. A resonance energy transfer approach, *Eur. J. Biophys. J.*, 15, 147, 1987.
139. **Hermetter, A. and Lakowicz, J. R.**, The aggregation state of melittin in lipid bilayers. An energy transfer study, *J. Biol. Chem.*, 261, 8243, 1986.
140. **Schwarz, G. and Beschiaschvili, G.**, Thermodynamic and kinetic studies on the association of melittin with a phospholipid bilayer, *Biochim. Biophys. Acta*, 979, 82, 1989.
141. **John, E. and Jähnig, F.**, Aggregation state of melittin in lipid vesicle membranes, *Biophys. J.*, 60, 319, 1991.
142. **Angelides, K. J. and Brown, G. B.**, Fluorescence resonance energy transfer on the voltage-dependent sodium channel. Spatial relationship and site coupling between the batrachotoxin and *Leiurus quinqustriatus quinquestriatus* α-scorpion toxin receptors, *J. Biol. Chem.*, 259, 6117, 1984.
143. **Harris, R. W., Sims, P. J., and Tweten, R. K.**, Kinetic aspects of the aggregation of *Clostridium perfringens* Θ-toxin on erythrocyte membranes. A fluorescence energy transfer study, *J. Biol. Chem.*, 266, 6936, 1991.
144. **Tweten, R. K., Harris, R. W., and Sims, P. J.**, Isolation of a tryptic fragment from *Clostridium perfringens* θ-toxin that contains sites for membrane binding and self-aggregation, *J. Biol. Chem.*, 266, 12449, 1991.
145. **Erickson, J. W. and Cerione, R. A.**, Resonance energy transfer as a direct monitor of GTP-binding protein-effector interaction: activated α-transducin binding to the cGMP phosphodiesterase in the bovine phototransduction cascade, *Biochemistry*, 30, 7112, 1991.
146. **Heithier, H., Frohlich, M., Dees, C., Baumann, M., Haring, M., Gierschik, P., Schiltz, E., Vaz, W. L., Hekman, M., and Helmreich, E. J.**, Subunit interactions of GTP-binding proteins, *Eur. J. Biochem.*, 204, 1169, 1992.
147. **Kurstjens, N. P., Frohlich, M., Dees, C., Cantrill, R. C., Hekman, M., and Helmreich, E. J. M.**, Binding of α- and βγ-subunits of G0 to β_1-adrenoceptor in sealed unilamellar lipid vesicles, *Eur. J. Biochem.*, 197, 167, 1991.
148. **Hyono, A., Kuriyama, S., and Masui, M.**, Fluorescence energy transfer between ionophore, A23187, and membrane proteins of isolated outer and cytoplasmic membranes of a Gram-negative bacterium, *Biochim. Biophys. Acta*, 813, 111, 1985.

149. **Peerce, B. E. and Wright, E. M.**, Distance between substrate sites on the Na-glucose cotransporter by fluorescence energy transfer, *Proc. Natl. Acad. Sci. U.S.A.*, 83, 8092, 1986.
150. **Burgun, C., Waksman, A., and Cremel, G.**, pH-induced reorganization of synaptic membrane as revealed by fluorescence anisotropy and energy transfer, *Arch. Biochem. Biophys.*, 286, 394, 1991.
151. **Morris, S. J., Südhof, T. C., and Haynes, D. H.**, Lipid and protein interactions in Ca^{2+}-promoted aggregation and fusion of chromaffin granule membranes, *Biophys. J.*, 37, 117, 1982.
152. **Morris, S. J., Südhof, T. C., and Haynes, D. H.**, Calcium-promoted resonance energy transfer between fluorescently labeled proteins during aggregation of chromaffin granule membranes, *Biochim. Biophys. Acta*, 693, 425, 1982.
153. **Morris, S. J., Costello, M. J., Robertson, D., Südhof, T. C., Odenwald, W. F., and Haynes, D. H.**, The chromaffin granule as a model for membrane fusion: implications for excocytosis, *J. Auton. Nerv. Syst.*, 7, 19, 1983.
154. **Peters, R.**, The thickness of air-dried human erythrocyte membranes as determined by energy transfer, *Biochim. Biophys. Acta*, 330, 53, 1973.
155. **Shaklai, N., Yguerabide, J., and Ranney, H. M.**, Interaction of hemoglobin with red cell membranes as shown by a fluorescent chromophore, *Biochemistry*, 16, 5585, 1977.
156. **Eisinger, J. and Flores, J.**, Cytosol-membrane interface of human erythrocytes. A resonance energy transfer study, *Biophys J.*, 41, 367, 1983.
157. **Rao, A., Martin, P., Reithmeier, R. A. F., and Cantley, L. C.**, Location of the stilbenedisulfonate binding site of the human erythrocyte anion-exchange system by resonance energy transfer, *Biochemistry*, 18, 4505, 1979.
158. **Low, P. S., Westfall, M. A., Allen, D. P., and Appell, K. C.**, Characterization of the reversible conformational equilibrium of the cytoplasmic domain of erythrocyte membrane band 3, *J. Biol. Chem.*, 259, 13070, 1984.
159. **Sharokh, Z., Verkman, A. S., and Shohet, S. B.**, Distance between skeletal protein 4.1 and the erythrocyte membrane bilayer measured by resonance energy transfer, *J. Biol. Chem.*, 266, 12082, 1991.
160. **Dale, R. E., Novros, J., Roth, S., Edidin, M., and Brand, L.**, Application of Förster long-range excitation energy transfer to the determination of distribution of fluorescently-labeled concanavalin A receptor complexes at the surfaces of yeast and of normal and malignant fibroblasts, in *Fluorescent Probes*, Beddard, G. S. and West, M. A., Eds., Academic Press, London, 1981, 159.
161. **Schreiber, A. B., Hoebeke, J., Vray, B., and Strossberg, D.**, Evidence for reversible microclustering of lentil lectin membrane receptors on HeLa cells, *FEBS Lett.*, 111, 303, 1980.
162. **Schreiber, A. B., Hoebeke, J., Vray, B., and Strossberg, D.**, Resonance energy transfer studies of the mechanisms of microclustering of lentil lectin membrane receptors on HeLa cells, *Exp. Cell Res.*, 132, 273, 1981
163. **Steiner, M.**, Changes in the distribution of platelet membrane proteins revealed by energy transfer, *Biochim. Biophys. Acta*, 805, 53, 1984.
164. **Fernandez, S. M. and Berlin, R. D.**, Cell surface distribution of lectin receptors determined by resonance energy transfer, *Nature*, 264, 411, 1976.
165. **Herman, B. A. and Fernandez, S. M.**, Dynamics and topographical distribution of surface glycoproteins during myoblast fusion: a resonance energy transfer study, *Biochemistry*, 21, 3275, 1981.
166. **Herman, B. A.**, Resonance energy transfer microscopy, *Methods Cell Biol.*, 30, 219, 1989.
167. **Uster, P. S. and Pagano, R. E.**, Resonance energy transfer microscopy: observations of membrane-bound fluorescent probes in model membranes and living cells, *J. Cell Biol.*, 103, 1221, 1986.
168. **Jovin, T. M. and Arndt-Jovin, D. J.**, FRET microscopy: digital imaging of fluorescence energy transfer. Application in cell biology, in *Cell Structure and Function by Microspectrofluorometry*, Kohen, E. and Hirschberg, J. G., Eds., Academic Press, London, 1989, 99.
169. **Jovin, T. M. and Arndt-Jovin, D. J.**, Luminescence digital imaging microscopy, *Annu. Rev. Biophys. Chem.*, 18, 271, 1989.
170. **Chan, S. S., Arndt-Jovin, D. J., and Jovin, T. M.**, Proximity of lectin receptors on the cell surface measured by fluorescence energy transfer in a flow system, *J. Histochem. Cytochem.*, 27, 56, 1979.
171. **Jenis, D. M., Stepanowski, A. L., Blair, O. C., Burger, D. E., and Sartorelli, A. C.**, Lectin receptor proximity on HL-60 leukemia cells determined by fluorescence energy transfer using flow cytometry, *J. Cell. Physiol.*, 121, 501, 1984.
172. **Szöllősi, J., Mátyus, L., Trón, L., Balázs, M., Ember, I., Fulwyler, M. J., and Damjanovich, S.**, Flow cytometric measurements of fluorescence energy transfer using single laser excitation, *Cytometry*, 8, 120, 1987.

173. **Szöllősi, J., Damjanovich, S., Mulhern, S. A., and Trón, L.**, Fluorescence energy transfer and membrane potential measurements monitor dynamic properties of cell membranes: a critical review, *Prog. Biophys. Mol. Biol.*, 49, 65, 1987.
174. **Trón, L., Szöllősi, J., and Damjanovich, S.**, Proximity measurements of cell surface proteins by fluorescence energy transfer, *Immunol. Lett.*, 16, 1, 1987.
175. **Matkó, J., Szöllősi, J., Trón, L., and Damjanovich, S.**, Luminescence spectroscopic approaches in studying cell surface dynamics, *Q. Rev. Biophys.*, 21, 479, 1988.
176. **Trón, L., Aszalós, A., Balázs, M., Mulhern, S. A., Szöllősi, J., and Damjanovich, S.**, On the biophysics of transmembrane signalling, *Mol. Immunol.*, 25, 1075, 1988.
177. **Mátyus, L.**, Fluorescence energy transfer measurements on cell surfaces. A spectroscopic tool for determining protein interactions, *J. Photochem. Photobiol. B*, 12, 323, 1992.
178. **Damjanovich, S., Mátyus, L., Balázs, M., Gáspár, R., Krasznai, Z., Pieri, C., Szöllősi, J., and Trón, L.**, Dynamic physical interactions of plasma membrane molecules generate cell surface patterns and regulate cell activation processes, *Immunobiology*, 185, 337, 1992.
179. **Wolber, P. K. and Hudson, B. S.**, An analytic solution to the Förster energy transfer problem in two dimensions, *Biophys. J.*, 28, 197, 1979.
180. **Snyder, B. and Freire, E.**, Fluorescence energy transfer in two dimensions. A numeric solution for random and nonrandom distributions, *Biophys. J.*, 40, 137, 1982.
181. **Estep, T. N. and Thompson, T. E.**, Energy transfer in lipid bilayer, *Biophys. J.*, 26, 195, 1979.
182. **Dewey, T. G. and Hammes, G. G.**, Calculation of fluorescence resonance energy transfer on surfaces, *Biophys. J.*, 32, 1023, 1980.
183. **Ploegh, H. L., Orr, H. T., and Strominger, J. L.**, Major histocompatibility antigens: the human (HLA-A, -B, -C) and murine (H-2K, H-2D) class I molecules, *Cell*, 24, 287, 1981.
184. **Kaufman, J. F., Auffray, C., Korman, A. J., Shackelford, D. A., and Strominger, J.**, The class II molecules of human an murine major histocompatibility complex, *Cell*, 36, 1, 1984.
185. **Askonas, B. A. and Openshaw, P. J.**, MHC and antigen presentation, *Immunol. Today*, 10, 396, 1989.
186. **Yewdell, J. W. and Bennink, J. R.**, The binary logic of antigen processing and presentation to T cells, *Cell*, 62, 203, 1990.
187. **Edidin, M.**, Molecular associations of major histocompatibility complex antigens: probing the complexity of cell surface receptors, *Biochem. Soc. Trans.*, 18, 152, 1990.
188. **Edidin, M.**, Function by association? MHC antigens and membrane receptor complexes, *Immunol. Today*, 9, 218, 1988.
189. **Fehlman, M., Peyron, J.-F., Samson, M., Van Obberghen, E., Brandergurg, D., and Brosette, N.**, Molecular association between major histocompatibility complex class I antigens and insulin receptors in mouse liver membranes, *Proc. Natl. Acad. Sci. U.S.A.*, 82, 8634, 1985.
190. **Phillips, M. L., Moule, M., Delovitch, T. L., and Yip, C. C.**, Class I histocompatibility antigens and insulin receptors: evidence for interactions, *Proc. Natl. Acad. Sci. U.S.A.*, 83, 3474, 1986.
191. **Samson, M., Cousin, J.-L., and Fehlmann, M.**, Cross-linking of insulin receptors to MHC antigens in human B lymphocytes: evidence for selective molecular interactions, *J. Immunol.*, 137, 2293, 1986.
192. **Schreiber, A. B., Schlessinger, J., and Edidin, M.**, Interaction between major histocompatibility complex antigens and epidermal growth factor receptors on human cells, *J. Cell Biol.*, 98, 725, 1984.
193. **Claas, F. H. J., van Ree, J. M., Verhoeven, W. M. A., van der Poel, J. J., Verduyn, W., de Wied, D., and van Rood, J. J.**, The interaction between γ-type endorphins and HLA class I antigens, *Hum. Immunol.*, 15, 347, 1986.
194. **Szöllősi, J., Damjanovich, S., Goldman, C. K., Fulwyler, M. J., Aszalós, A. A., Goldstein, G., Rao, P., Talle, M. A., and Waldmann, T. A.**, Flow cytometric energy transfer measurements support the association of a 95 kD peptide termed T27 with the 55 kD Tac peptide, *Proc. Natl. Acad. Sci. U.S.A.*, 84, 7246, 1987.
195. **Neppert, J. and Mueller-Eckhardt, C.**, Monoclonal mouse antibodies to human class I antigens cocap class II antigens, *Tissue Antigens*, 24, 187, 1984.
196. **Neppert, J. and Mueller-Eckhardt, C.**, Interdependent membrane mobility of human MHC coded antigens detected by monoclonal antibodies to various epitopes on class I and class II molecules, *Tissue Antigens*, 26, 51, 1985.
197. **Adams, R. S., Bodmer, J. G., and Bodmer, W. F.**, Production and characterization of monoclonal antibodies recognizing α-subunit of human Ia alloantigens, *Immunology*, 50, 613, 1983.

198. **Bright, S. and Munro, A. J.**, Studies on the role of HLA-DR in macrophage-T cell interactions, *Tissue Antigens*, 18, 217, 1981.
199. **Pawelec, G. P., Shaw, S., Ziegler, A., Müller, C., and Wernet, P.**, Differential inhibition of HLA-D- or SB-directed secondary lymphoproliferative responses with monoclonal antibodies detecting human Ia-like determinants, *J. Immunol.*, 129, 1070, 1982.
200. **Sterkers, G., Henin, Y., Kalil, J., Bagot, M., and Levy, J.-P.**, Influence of HLA class I- and class II-specific monoclonal antibodies on DR-restricted lymphoproliferative responses. I. Unseparated populations of effector cells, *J. Immunol.*, 131, 2735, 1983.
201. **Anichini, A., Strassmann, G., and Bach, F. H.**, Inhibition of anti-class I cytotoxicity by anti-class II monoclonal antibodies (MoAb). II. Blocking of anti-class I CTL clones by anti-DR MoAb, *Hum. Immunol.*, 13, 139, 1985.
202. **Szöllősi, J., Damjanovich, S., Balázs, M., Nagy, P., Trón, L., Fulwyler, M. J., and Brodsky, F. M.**, Physical association between MHC class I and class II molecules detected on the cell surface by flow cytometric energy transfer, *J. Immunol.*, 143, 208, 1989.
203. **Brodsky, F. M.**, A matrix approach to human class II histocompatibility antigens: reactions of four monoclonal antibodies with the products of nine haplotypes, *Immunogenetics*, 19, 179, 1984.
204. **Lotteau, V., Leyton, L., Burroghs, D., and Charron, D.**, A novel HLA class II molecule (DRα-DQβ) created mismatched isotype pairing, *Nature*, 329, 339, 1987.
205. **Chakrabarti, A., Matko, J., Rahman, N. A., Barisas, B. G., and Edidin, M.**, Self-association of class I major histocompatibility complex molecules in liposomes and cell surface membranes, *Biochemistry*, 31, 7182, 1992.
206. **Bushkin, Y., Demaria, S., Le, J., and Schwab, R.**, Physical association between the CD8 and HLA class I molecules on the surface of activated human T lymphocytes, *Proc. Natl. Acad. Sci. U.S.A.*, 85, 3985, 1988.
207. **Damjanovich, S., Trón, L., Szöllősi, J., Zidovetzki, R., Vaz, W. L. C., Regateiro, F., Arndt-Jovin, D. A., and Jovin, T. M.**, Distribution and mobility of murine histocompatibility H-2K^k antigen in the cytoplasmic membrane, *Proc. Natl. Acad. Sci. U.S.A.*, 80, 5985, 1983.
208. **Damjanovich, S., Trón, L., Szöllősi, J., Mátyus, L., and Szabó, G., Jr.**, Dynamic properties of the murine histocompatibility H-2K^k antigen in cytoplasmic membrane, *Mol. Immunol.*, 21, 1151, 1984.
209. **Trón, L., Szöllősi, J., and Damjanovich, S.**, New trends in studying slow diffusion processes of the cytoplasmic membrane components, *Acta Biochim. Biophys. Acad. Sci. Hung.*, 19, 247, 1984.
210. **Tsong, T. Y. and Astumian, R. D.**, Electro-conformational coupling and membrane protein function, *Prog. Biophys. Mol. Biol.*, 50, 1, 1987.
211. **Blumenthal, R., Klausner, R. D., and Weinstein, J. N.**, Voltage dependent translocation of the asialoglycoprotein receptor across lipid membranes, *Nature*, 288, 333, 1980.
212. **Balázs, M., Matkó, J., Szöllősi, J., Mátyus, L., Fulwyler, M. J., and Damjanovich, S.**, Accessibility of cell surface thiols in human lymphocytes is altered by ionophores or OKT3 antibody, *Biochem. Biophys. Res. Commun.*, 140, 999, 1986.
213. **Dunnant, Y., Garcia-Segura, L. M., Muller, D., and Párducz, A.**, Momentary alteration of the postsynaptic membrane during transmission of a single nerve impulse, *Proc. Natl. Acad. Sci. U.S.A.*, 86, 1717, 1989.
214. **Muller, D., Garcia-Segura, L. M., Párducz, A., and Dunnant, Y.**, Brief occurrence of a population of presynaptic intramembrane particles coincides with transmission of a nerve pulse, *Proc. Natl. Acad. Sci. U.S.A.*, 84, 590, 1987.
215. **Wyatt, K. and Cherry, R. J.**, Effect of membrane potential on band 3 conformation in the human erythrocyte membrane detected by triplet state quenching experiments, *Biochemistry*, 31, 4650, 1992.
216. **Pieri, C., Recchioni, R., Moroni, F., Balkay, L., Márián, T., Trón, L., and Damjanovich, S.**, Ligand and voltage gated sodium channels may regulate electrogenic pump activity, in human, mouse and rat lymphocytes, *Biochem. Biophys. Res. Commun.*, 160, 999, 1989.
217. **Pieri, C., Recchioni, R., Moroni, F., Marcheselli, F., Falasca, M., Krasznai, Z., Gáspár, R., Mátyus, L., and Damjanovich, S.**, A sodium channel opener inhibits stimulation of human peripheral blood mononuclear cells, *Mol. Immunol.*, 29, 517, 1992.
218. **Gáspár, R., Jr., Krasznai, Z., Márián, T., Trón, L., Recchioni, R., Falasca, M., Moroni, F., Pieri, C., and Damjanovich, S.**, Bretylium-induced voltage-gated sodium current in human lymphocytes, *Biochim. Biophys. Acta*, 1137, 143, 1992.

219. **Lakos, Zs., Somogyi, B., Balázs, M., Matkó, J., and Damjanovich, S.**, The effect of transmembrane potential on the dynamic behavior of cell membrane, *Biochim. Biophys. Acta*, 1023, 41, 1990.
220. **Waldmann, T. A.**, The interleukin-2 receptor, *J. Biol. Chem.*, 266, 2681, 1991.
221. **Chilson, O. P. and Kelly-Chilson, A. E.**, Mitogenic lectins bind to the antigenic receptors of human lymphocytes, *Eur. J. Immunol.*, 19, 389, 1989.
222. **Minami, Y., Kono, T., Yamada, K., and Taniguchi, T.**, The interleukin-2 receptors: insights into a complex signalling mechanism, *Biochim. Biophys. Acta*, 1114, 163, 1992.
223. **Waldmann, T. A.**, The structure, function and expression of interleukin-2 receptors on normal and malignant lymphocytes, *Science*, 23, 727, 1986.
224. **Takeshita, T., Asao, H., Ohtani, K., Ishii, N., Kumaki, S., Tanaka, N., Munakata, H., Nakamura, M., and Sugamura, K.**, Cloning of the γ chain of the human IL-2 receptor, *Science*, 257, 379, 1992.
225. **Burton, J., Goldman, C. K., Rao, P., Moos, M., and Waldmann, T. A.**, Association of intercellular adhesion molecule 1 with the multichain high affinity interleukin-2 receptor, *Proc. Natl. Acad. Sci. U.S.A.*, 87, 7329, 1990.
226. **Edidin, M., Aszalós, A., Damjanovich, S., and Waldmann, T. A.**, Lateral diffusion measurements give evidence for association of the Tac peptide of IL-2 receptor with T27 peptide in the plasma membrane of HUT-102-B2 cells, *J. Immunol.*, 141, 1206, 1988.
227. **Sharon, M., Gnarra, J. R., Baniyash, M., and Leonard, W. J.**, Possible association between IL-2 receptors and class I HLA molecules on T cells, *J. Immunol.*, 141, 3512, 1988.
228. **Rebollo, A., De Groote, D., Baudrihay, M., Theze, J., and Jankovic, D. L.**, Immunochemical characterization of antigenic domains on human IL-2: spatially distinct epitopes are associated with binding to the p55 and p70 subunits of IL-2 receptor, *Mol. Immunol.*, 29, 119, 1992.
229. **Harel-Bellan, A., Krief, P., Rimsky, L., Farrar, W. L., and Mishal, Z.**, Flow cytometric energy transfer suggests an association between low-affinity interleukin 2 binding sites and HLA class I molecules, *Biochem. J.*, 268, 35, 1990.
230. **Damjanovich, S., Szöllősi, J., and Trón, L.**, Transmembrane signalling in lymphocytes, *Immunol. Today*, 13, A12, 1992.
231. **Hartmann, E., Rapoport, T. A., and Lodish, H. F.**, Predicting the orientation of eukaryotic membrane spanning proteins, *Proc. Natl. Acad. Sci. U.S.A.*, 86, 5786, 1989
232. **Levine, R.**, Insulin: the effects and mode of action of the hormone, *Vit. Horm.*, 39, 145, 1982.
233. **Saltiel, A. R.**, Signal transduction in insulin action, *J. Nutr. Biochem.*, 1, 180, 1990.
234. **Swanson, M. L. and Pessin, J. E.**, High insulin binding in the human placenta insulin receptor requires heterodimeric subunit interactions, *J. Membr. Biol.*, 108, 217, 1989.
235. **Simonsen, M., Skjodt, K., Crone, M., Sanderson, A., Fujita-Yamaguchi, Y., Due, C., Ronne, E., Linnet, K., and Olsson, L.**, Compound receptors in the cell membrane: ruminations from the borderland of immunology and physiology, *Prog. Allergy*, 36, 151, 1985.
236. **Chvatchko, Y., Van Obberghen, E., Kiger, N., and Fehlmann, M.**, Immunoprecipitation of insulin receptors by antibodies against class I antigens of the murine H-2 histocompatibility complex, *FEBS Lett.*, 163, 207, 1983.
237. **Verland, S., Simonsen, M., Gammeltoft, S., Allen, H., Falvell, R. A., and Olsson, L.**, Specific interaction between the insulin receptor and a D product of MHC class I, *J. Immunol.*, 143, 945, 1985.
238. **Kittur, D., Shimizu, Y., DeMars, R., and Edidin, M.**, Insulin binding to human B lymphoblasts is a function of HLA haplotype, *Proc. Natl. Acad. Sci. U.S.A.*, 84, 1351, 1987.
239. **Edidin, M. and Reiland, J.**, Dynamic measurements of the associations between class I MHC antigens and insulin receptors, *Mol. Immunol.*, 27, 1313, 1990.
240. **Liegler, T., Szöllősi, J., Hyun, W., and Goodenow, R. S.**, Proximity measurements between H-2 antigens and insulin receptors by fluorescence energy transfer: evidence that a close association does not influence insulin binding, *Proc. Natl. Acad. Sci. U.S.A.*, 88, 6755, 1991.
241. **Stagsted, J., Reaven, G. M., Hansen, T., Goldstein, A., and Olsson, L.**, Regulation of insulin receptor functions by a peptide derived from a major histocompatibility complex class I antigen, *Cell*, 62, 297, 1990.
242. **Harding, V. C. and Unanue, E. R.**, Cellular mechanisms of antigen processing and the function of class I and II major complex molecules, *Cell Regulation*, 1, 499, 1990.
243. **Levine, T. P. and Chain, B. M.**, The cell biology of antigen processing, *Crit. Rev. Biochem. Mol. Biol.*, 26, 439, 1991.
244. **Rabbit, B. P., Allen, P. M., Matsueda, G., Haber, E., and Unanue, E. R.**, Binding of immunogenic peptides to Ia histocompatibility molecules, *Nature*, 317, 359, 1985.

245. **Sette, A., Buus, S., Colon, S., Miles, C., and Grey, H. M.,** Structural analysis of peptides capable of binding to more than one Ia antigens, *J. Immunol.*, 142, 35, 1989.
246. **Chen, B. P. and Parham, P.,** Direct binding of influenza peptides to class I HLA molecules, *Nature*, 337, 743, 1989.
247. **Luescher, F. L., Loez, J. A., Malissen, B., and Cerottini, J.-C.,** Interaction of antigenic peptides with MHC class I molecules on living cells studied by photoaffinity labeling, *J. Immunol.*, 148, 1003, 1992.
248. **Fremont, D. H., Matsumara, M., Stura, E. A., Peterson, P. A., and Wilson, I. A.,** Crystal structures of two viral peptides in complex with murine MHC class I H-2K^b, *Science*, 257, 919, 1992.
249. **Watts, T. H., Gaub, H. E., and McConnel, H. M.,** T-cell-mediated association of peptide antigen and major histocompatibility complex protein detected by energy transfer in an evanescent wave-field, *Nature*, 320, 179, 1986.
250. **Tampé, R., Clark, B., and McConnel, H. M.,** Energy transfer between two peptides bound to one MHC class II molecule, *Science*, 254, 87, 1991.
251. **Townsend, A., Öhlén, C., Bastin, J., Ljunggren, H.-G., Foster, L., and Klärre, K.,** Association of class I major histocompatibility heavy and light chains induced by viral peptides, *Nature*, 340, 443, 1989.
252. **Elliott, T., Cerundolo, V., Elvin, J., and Townsend, A.,** Peptide induced conformational change of the class I heavy chain, *Nature*, 351, 402, 1991.
253. **Elliott, T., Cerundolo, V., and Townsend, A.,** Short peptides assist to the folding of free class I heavy chains in solutions, *Eur. J. Immunol.*, 22, 3121, 1992.
254. **Szabó, G., Jr., Pine, P. S., Weaver, J. L., Kasari, M., and Aszalós, A.,** Epitope mapping by photobleaching fluorescence resonance energy transfer using a laser scanning microscope system, *Biophys. J.*, 61, 661, 1992.
255. **Szabó, G., Jr., Pine, P. S., Weaver, J. L., Rao, P. E., and Aszalós, A.,** CD4 changes conformation upon ligand binding, *J. Immunol.*, 149, 3596, 1992.
256. **Wang, J., Yan, Y., Garrett, T. P. J., Liu, J., Rodgers, D. W., Garlick, R. L., Tarr, G. E., Husain, Y., Reinherz, E. L., and Harrison, S. C.,** Atomic structure of a fragment of human CD4 containing two immunoglobulin like domains, *Nature*, 348, 411, 1990.
257. **Ryu, S.-E., Kwong, P. D., Trunch, A., Porter, T. G., Arthos, J., Rosenberg, M., Dai, X., Xuong, N., Axel, R., Sweet, R. W., and Endrickson, W. A.,** Crystal structure of an HIV-binding recombinant fragment of human CD4, *Nature*, 348, 419, 1990.
258. **Chuck, R. S., Cantor, C. R., and Tse, D. B.,** CD4-T cell antigen receptor complexes on human leukemia T cells, *Proc. Natl. Acad. Sci. U.S.A.*, 87, 5021, 1990.
259. **Mittler, R. S., Goldman, S. J., Spitalny, G. L., and Burakoff, S. J.,** T-cell receptor physical association in a murine T-cell hybridoma: induction by antigen receptor ligation, *Proc. Natl. Acad. Sci. U.S.A.*, 86, 8531, 1989.
260. **Brown, M. H., Cantrell, D. A., Brattsand, G., Crumpton, M. J., and Gullberg, M.,** The CD2 antigen associates with the T-cell antigen receptor CD3 antigen complex on the surface of human T lymphocytes, *Nature*, 339, 551, 1989.
261. **Collins, T. L., Uniyal, S., Shin, J., Strominger, J. L., Mittler, R. S., and Burakoff, S. J.,** p56^{lck} association with CD4 is required for the interaction between CD4 and the TcR/CD3 complex and for optimal antigen stimulation, *J. Immunol.*, 148, 2159, 1992.
262. **Mittler, R. S., Rankin, B. M., and Kiener, P. A,** Physical associations between CD45 and CD4 or CD8 occur as late events in antigen receptor-stimulated human T-cells, *J. Immunol.*, 147, 3434, 1991.
263. **de la Hera, A., Müller, U., Olsson, C., Isaaz, S., and Tunnacliffe, A.,** Structure of the T cell antigen receptor (TCR): two CD3ε subunits in a functional TCR/CD3 complex, *J. Exp. Med.*, 173, 7, 1991.
264. **Minami, Y., Weissman, A. M., Samelson, L. E., and Klausner, R. D.,** Building a multichain receptor: synthesis, degradation, and assembly of the T-cell antigen receptor, *Proc. Natl. Acad. Sci. U.S.A.*, 84, 2688, 1987.
265. **Meuer, S. O., Acuto, O., Hussey, R. E., Hodgdon, J. C., Fitzgerald, K. A., Schlossman, S. F., and Reinherz, E. L.,** Evidence for the T3-associated 90K heterodimer as the T-cell antigen receptor, *Nature*, 303, 308, 1983.
266. **Krissansen, G. W., Verbi, W., Totty, N. F., and Crumpton, M. J.,** Purification and identification by amino acid sequence analysis of the subunits of the CD3 antigen of human tonsil T-lymphocytes, *Mol. Immunol.*, 24, 1069, 1987.
267. **Carpenter, G. and Cohen, S.,** Epidermal growth factor, *J. Biol. Chem.*, 265, 7709, 1990.

268. **Ullrich, A., Coussens, L., Hayflick, J. S., Dull, T. J., Gray, A., Tam, A. W., Lee, J., Yarden, Y., Libermann, T. A., Schlessinger, J., Downward, J., Mayes, E. L. V., Whittle, N., Waterfiled, M. D., and Seeburg, P. H.,** Human epidermal growth factor receptor cDNA sequence and aberrant expression of the amplified gene in A431 epidermoid carcinoma cells, *Nature*, 309, 418, 1984.
269. **Koland, J. G. and Cerione, R. A.,** Growth factor control of epidermal growth factor receptor kinase activity via an intramolecular mechanism, *J. Biol. Chem.*, 263, 2230, 1988.
270. **Yarden, Y. and Schlessinger, J.,** Epidermal growth factor induces rapid, reversible aggregation of the purified epidermal growth factor receptor, *Biochemistry*, 26, 1443, 1987.
271. **Cochet, C., Kashles, O., Chambaz, E. M., Borello, I., King, C. R., and Schlessinger, J.,** Demonstration of epidermal growth factor-induced receptor dimerization in living cells using a chemical covalent cross-linking agent, *J. Biol. Chem.*, 257, 1523, 1982.
272. **Hillman, G. M. and Schlessinger, J.,** Lateral diffusion of epidermal growth factor complexed to its surface receptors does not account for the thermal sensitivity of patch formation and endocytosis, *Biochemistry*, 21, 1667, 1982.
273. **Benveniste, M., Livneh, E., Schlessinger, J., and Kam, Z.,** Overexpression of epidermal growth factor receptor in NIH-3T3-transfected cells slows its lateral diffusion and rate of endocytosis, *J. Cell Biol.*, 106, 1903, 1988.
274. **Zidovetzki, R., Yarden, Y., Schlessinger, J., and Jovin, T. M.,** Rotational diffusion of epidermal growth factor complexed to cell surface receptors reflects rapid microaggregation and endocytosis of occupied receptors, *Proc. Natl., Acad. Sci. U.S.A.*, 78, 6981, 1981.
275. **Zidovetzki, R., Yarden, Y., Schlessinger, J., and Jovin, T. M.,** Microaggregation of hormone-occupied epidermal growth factor receptors on plasma membrane preparations, *EMBO J.*, 5, 247, 1986.
276. **Carraway, K. L., Koland, J. G., and Cerione, R. A.,** Visualization of epidermal growth factor (EGF) receptor aggregation in plasma membranes by fluorescence resonance energy transfer, *J. Biol. Chem.*, 264, 8699, 1989.
277. **Azavedeo, J. R. and Johnson, D. A.,** Temperature-dependent lateral and transverse distribution of the epidermal growth factor receptor in A431 plasma membranes, *J. Membrane Biol.*, 118, 215, 1990.
278. **Carraway, K. L. and Cerione, R. A.,** Comparison of epidermal growth factor receptor-receptor interactions in intact A431 cells and isolated plasma membrane. Large scale receptor micro-aggregation is not detected during EGF-stimulated early events, *J. Biol. Chem.*, 266, 8899, 1991.
279. **Baird, B., Shopes, R. J., Oi, V. T., Erickson, J., Kane, P., and Holowka, D.,** Interaction of IgE with its high-affinity receptor. Structural basis and requirements for effective cross-linking, *Int. Ach. Allergy Appl. Immunol.*, 88, 23, 1989.
280. **Holowka, D. and Baird, B.,** Structural studies on the membrane-bound immunoglobulin E-receptor complex. II. Mapping of distance between sites on IgE and the membrane surface, *Biochemistry*, 22, 3475, 1983.
281. **Baird, B. and Holowka, D.,** Structural mapping of Fc-receptor-bound immunoglobulin E: proximity to the membrane surface of the antibody combining site and another site in the Fab segments, *Biochemistry*, 24, 6252, 1985.
282. **Baird, B., Conrad, D. A., and Holowka, D.,** Structural mapping of membrane-bound immunoglobulin E-receptor complexes: use of monoclonal anti-IgE antibodies to probe the conformation of receptor bound IgE, *Biochemistry*, 24, 6260, 1985.
283. **Kubitscheck, U., Kircheis, M., Schweitzer-Stenner, R., Dreybrodt, W., Jovin, T. M., and Pecht, I.,** Fluorescence energy transfer on single living cells. Application to binding of monovalent happens to cell-bound immunoglobulin E, *Biophys. J.*, 60, 307, 1991.

Chapter 3

Fluorescence Photobleaching and Recovery, FPR, in the Analysis of Membrane Structure and Dynamics

Michael Edidin

CONTENTS

I. INTRODUCTION

Fluorescence photobleaching and recovery, FPR (also termed fluorescence recovery after photobleaching, FRAP), was developed as a method for quantitating lateral diffusion in cell membranes and synthetic bilayers. At first its main results were values of lateral diffusion coefficients. These were useful in speculating about membrane function, especially about the role of lateral diffusion in coupling reactions in membranes.[1-3] In recent times it has become evident that FPR can be used to investigate a number of other problems in membrane organization and, coupled with techniques of molecular biology, to further analyze membrane function as well. In this review I will describe the evolution of FPR techniques and then discuss their application to three topics in cell surface organization and function:

0-8493-4931-1/94/$0.00+$.50

1. Constraints to lateral diffusion — the physical bases of the diffusion coefficients, D, measured by FPR.
2. FPR as a probe of cell responses to signals and cell differentiation.
3. FPR as a probe for the spatial differentiation of membranes and membrane domains.

Though listed separately, the topics are often connected. For example, the efficiency of diffusional coupling of reactions in membranes is important in signaling at the cell surface. This depends upon D, upon the surface concentration of interacting molecules, and upon the stability of their associations with one another. In turn, the surface concentration of interacting molecules is increased if they are confined to membrane domains rather than diffusing freely over the entire membrane area.[4]

When using FPR to analyze membranes, it must be kept in mind that the technique analyzes the behavior of populations of hundreds to thousands of labeled molecules, on a scale of micrometers. In order to address some questions of membrane organization it may be necessary to resolve the mobility and behavior of one or a few molecules, on a scale of nanometers. Hence, after discussing FPR, I will briefly mention newer techniques of particle tracking that are useful on these small scales.

II. THE ORIGINS OF FPR

The method of FPR has its roots in early work on the dynamics of cell membranes. Part of that work showed that proteins were mobile in the plane of the membrane of heterokaryons, cells formed by fusing cultured fibroblasts of two different species.[5] At first, fluorescent labels for mouse or human antigens were segregated in separate regions of the heterokaryons' surface, but within 5 min of fusion the membranes of some cells contained large regions that were doubly labeled, indicating that the surface antigens of mouse and human membranes had intermixed in this short time.

Controls and later experiments on intermixing of surface antigens in fixed cells[6] showed that this intermixing was not metabolically driven, but was due to diffusion in the plane of the heterokaryon membrane. The experiment provided a vivid qualitative picture of lateral diffusion and it established the scale of protein mobility — from a fraction of a micrometer to tens of micrometers. However, it could not measure transport coefficients and could only characterize lateral diffusion in populations of cells, rather than on single cells.

A second part of the early work showed that one protein, rhodopsin, moved through the plane of a specialized membrane, the discs of retinal rod outer segments, by Brownian motion, diffusion. Two groups made these measurements and evaluated the diffusion coefficient of rhodopsin.[7,8] Each group used a microspectrophotometer to measure the amount of rhodopsin in a small area of the outer segment. This area was then partly bleached by a brief flash of light. Though the absorbance dropped upon bleaching, it rapidly recovered, reaching a maximum within tens of seconds after the bleach and there was a corresponding loss of unbleached rhodopsin molecules from the area of the disks that had not been exposed to the bleaching pulse (Figure 1). Knowing the area bleached and the half-time for recovery a diffusion coefficient, D, could be calculated for rhodopsin. The experimental data also gave an estimate of the fraction of all rhodopsin molecules that were mobile in the time of an experiment, R (the mobile fraction or fractional recovery).

FPR evolved from the two approaches to measuring lateral mobility just described. The sensitivity of fluorescence and the generality of antibody labels were exploited together with the principle of measuring lateral diffusion in terms of the change in concentration of a marker in a defined area of a cell membrane. Instead of measuring changes in absorbance of membrane pigment — something rarely found in animal cell membranes, the method measures changes in fluorescence of exogenous labels. A gradient of fluorescence is created by bleaching a spot in an otherwise uniformly fluorescent membrane and dissipation of the gradient is measured in terms of recovery of fluorescence in the bleached spot. This greatly increases the sensitivity of the method and also allows the tracking of membrane lipids and of as many different membrane proteins as can be labeled by fluorophores.

FPR was first used by Peters and co-workers[9] in an attempt to measure D of fluorescein-labeled erythrocyte membrane proteins. No recovery of fluorescence was detected after bleaching half of a fluorescent erythrocyte ghost using a mercury arc lamp, but the approach pointed the way to a powerful and general method for measuring translational diffusion.

III. PRESENT-DAY FPR METHODS

A. SPOT FPR

The essence of all modern FPR is the use of a focused laser beam to define a region in a fluorescently labeled membrane. The attenuated beam is used to monitor fluorescence in the region before and after

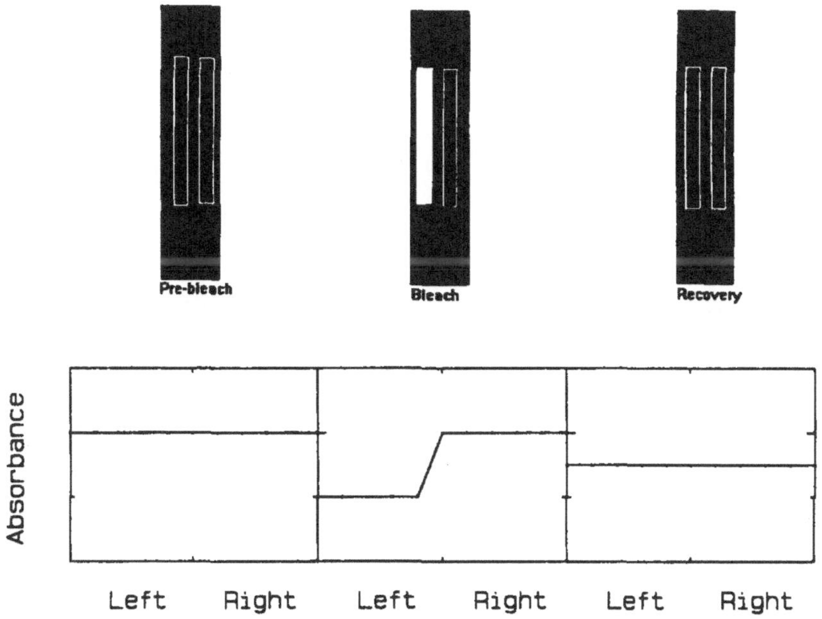

Figure 1. Absorbance recovery after photobleaching of visual rhodopsin. The top panel shows a single rod outer segment (large dark rectangle) with the images of the measuring and bleaching masks projected onto it at three different steps in the experiment. The bottom panel shows the absorbance of unbleached rhodopsin measured through the left and right masks. The total range of the vertical scale is ~0.05 absorbance units. Left — before bleaching the same absorbance is measured through the right and left masks. Center — immediately after a pulse of bleaching light the amount of unbleached rhodopsin under the left mask is lower than that under the right mask. Right — with time (tens of seconds) after bleaching, absorbance measured through the left mask increaes due to lateral diffusion of unbleached rhodopsin from the area under the right mask. The absorbance measured through the right mask is correspondingly decreased.[7,8]

a fraction of the fluorophores is bleached by a millisecond pulse of laser light at full intensity. An idealized FPR experiment is shown in Figure 2. The (asymptotic) extent of recovery of fluorescence after bleaching, F_∞ is a measure of the mobile fraction, R, of label, the fraction of label that is free to diffuse in the time of an FPR measurement, tens of seconds to a few minutes.

The half-time to reach the maximum recovery is proportional to the area bleached and to the diffusion coefficient of the labeled species. D is calculated from

$$t_{1/2} = \frac{\omega^2}{4D}\gamma \tag{1}$$

where ω^2 is the $1/e^2$ radius of the focused laser spot and γ is a function of the fraction of unbleached molecules. For bleaches between 5 and 95% of total fluorescence $1 < \gamma < 2$. The method is useful for diffusion coefficients in the range 10^{-8} cm^2 s^{-1} to ~10^{-12} cm^2 s^{-1}.

FPR origins, methodology, and results have been usefully summarized in two recent reviews.[10,11] A user-friendly and thorough discussion of the design and construction of an FPR instrument is given by Wolf.[12]

The most common FPR instrument couples a laser light source and a photomultiplier detector to a standard microscope. Most FPR machines use 1 W and larger water-cooled argon lasers, but useful FPR work has been done with low-power air-cooled HeCd lasers.[13-15,158] A simple beam-splitting device is used to produce co-axial measuring and bleaching beams and an auxiliary lens is positioned to focus the laser in a secondary image plane, so that the objective brings it into sharp focus at the object plane. An aperture, limiting depth of field, is placed between the object and the photomultiplier. This reduces the amount of fluorescence collected from the cell's interior. In fact the FPR microscope is operated in a confocal mode. The photomultiplier tube (PMT) detector may vary greatly in quality, sensitivity, and noise levels. Many

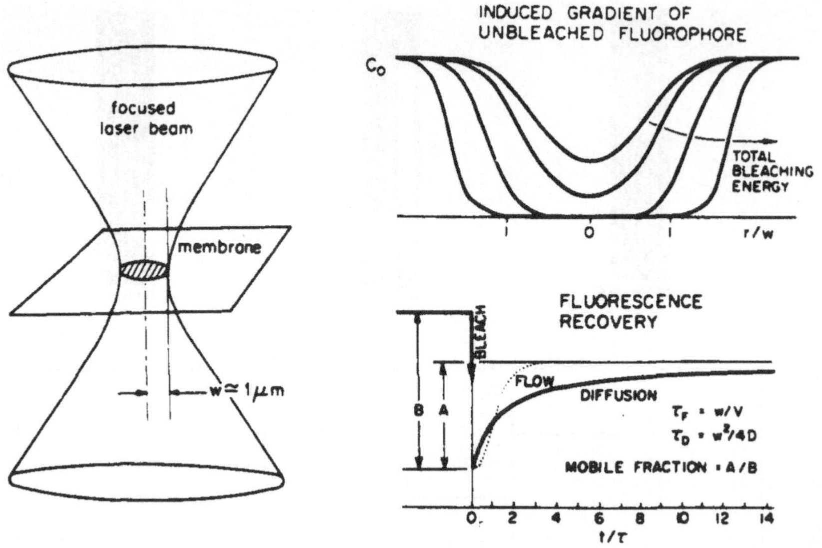

Figure 2. The spot FPR experiment. The left panel of the figure diagrams a fluorescently labeled membrane surface with the laser beam focused to a spot on this surface. The upper right panel shows gradients of fluorophore concentration created by the bleaching pulse. The shape and steepness of the gradient depend upon the depth of bleaching. This is taken into account by the term γ in Equation 1. The lower right panel schematizes an FPR curve. The ordinate indicates fluorescence intensity in arbitrary units.The abscissa in this case is normalized to dimensionless units of time t/τ.

FPR microscopes use cooled, gated, end-window tubes. We find that quite good measurements can be made with a simple and rugged (but selected) side-window PMT. PMT output is usually collected at a computer/controller, but in many early experiments data were taken as chart recorder tracings.

Analysis of FPR recovery curves from the type of instrument just described is complicated by the Gaussian intensity profile of the laser beam which bleaches a corresponding Gaussian profile in the fluorescence in the spot of interest. The series solution to the diffusion equation for this boundary condition converges very slowly.[16] In practice recovery curves have been fit by various approximations,[16] particularly nonlinear least squares methods.[12,17] Though in principle they could be made for more than 1 D per curve, the fits all assume a single recovery. Some linear transformations are good approximations that allow ready dissection of recovery curves into two components.[14,18]

The shape of the recovery curve in a spot FPR measurement and the calculation of D from the curve depend upon the membrane area defined by the focused laser beam. Measurement of the size of the laser spot is critical to the evaluation of D. One convenient way to do this is to translate a small (submicron-size) fluorescent bead through the laser beam while recording the fluorescence at the PMT. The intensity profile should be Gaussian and the $1/e^2$ radius of the beam is measured directly from the scan. Unfortunately, the precision micrometer stage required, translating in 0.1-μm steps, is expensive. It is cheaper and easier to measure the beam radius reflected off a first-surface mirror into a conventional video camera, using some sort of commonly available image analyzer to quantitate the intensity profile. The image of the fluorescence excited from a layer of fluorescent dye may also be used for quantitation with a more sensitive video camera (see, for example, References 19 and 20). The spot size probably cannot be determined to better than an uncertainty of 20%, implying about a 40% uncertainty in D. Hence, most measurements of D are better made as comparisons within an experiment, rather than as absolute values. Having said this, I must also state that we have obtained measurements of D for a given cell type and label made a year apart that are within 10% of one another and have also quantitatively reproduced literature values for D of particular membrane proteins.[21]

B. OTHER FORMS OF THE FPR EXPERIMENT

The spot FPR experiment provides information from a spot on the surface of a single cell and the sampled area in turn is largely defined by the intensity profile and focus of the laser beam. There are three drawbacks to this arrangement. First, only a limited sample is taken of the surface of any given cell.

Second, the geometry of the experiment makes fitting of recovery curves somewhat clumsy. Third, by integrating over the entire spot, a few squared micrometers, the method loses information about the spatial distribution of fluorescence (note that such information was obtained in the pre-FPR experiments of Liebman and Entine[7] and Poo and Cone[8]).

Several methods have been developed that overcome one or more of the drawbacks mentioned.

1. Pattern Photobleaching

In pattern photobleaching a pattern of bright and dark stripes is imposed on the labeled surface, either by illuminating through a mask[22] or by creating a pattern of interference fringes on the surface.[23] Recovery of fluorescence intensity, I, after bleaching the bright regions of the pattern is due to diffusion from the dark regions and is described by a simpler function than that required for spot bleaching:

$$I_t = \pi A - 2Bc^{-Da^2t} - \frac{2}{9}Bc^{-9Da^2t} \tag{2}$$

where coefficients A and B are determined by the contrast of the projected pattern and the extent of bleaching and a is the spatial frequency of the stripe pattern. The third term of the equation can be neglected for $t > 0.1/Da^2$ and so the recovery curve is fit by a single exponential.

The method of making a pattern with a mask can be easily used for large synthetic bilayers; a commercially available Ronchi ruling is used to make the pattern.[22] Creating a pattern on cells requires "contact printing" through masks with submicron rulings that may be difficult to obtain. A greater problem is that due to the finite thickness of the mask the stripe width is expanded at the cell surface from its width in the mask and so the correct value of a is difficult to determine. The method also requires more intense laser light sources than spot photobleaching since the illuminating beam is expanded over many micrometers.

Stripe patterns made by interference fringes overcome the drawbacks of mask-determined patterns. The pattern is oscillated on the cell surface by mechanical vibration of a mirror and so the recovering fluorescence signal, seen as a decay of contrast, can be picked out of the noise using a lock-in amplifier set to the oscillation frequency.

The stripe patterns produced by either mask or fringe method also contain information about the spatial distribution of fluorophores, but this is neglected for the sake of a simple analysis of fluorescence recovery.

2. Scanning Spot Photobleaching

FPR with significant spatial information can indicate if recoveries are isotropic, and if the recoveries are in fact due to flow as well as Brownian motion. Isotropy of recovery cannot be estimated from spot photobleaching. Though in principle the shape of the recovery curve indicates flow, in practice a significant flow component cannot be separated from the diffusional recovery in spot FPR data.

Both the problems raised in the preceding paragraph are addressed by a method of multipoint scanning, in which the cell surface is interrogated with a spot or a line at points to either side of the bleached region (Figure 3).[24-26] Systematic drift produces a phase shift in the pattern which is readily detectable. The size of the measuring and bleaching spot is defined by the scans and so spot size does not have to be determined separately as in FPR. The multipoint scanning method has been used in several biological systems with notable results. The method is also the one used in the only commercially available FPR microscope (ACAS, Meridian Instruments).

3. Video FPR

Some spatial information is lost even in multipoint scanning, since the beam only samples a small fixed number of points on the surface. A much higher spatial resolution can be obtained from video images of the labeled surface.[27] A few years ago this approach required a rather expensive computer and some custom interfacing. Today, interface boards and digital analysis software are available for microcomputers and the greatest problem is storage of the large amount of data in a single image.[12]

The richness of the images raises a problem in sampling cells of a population. Spot FPR allows measurements on hundreds of cells a day; small changes in R & D, correlated with changes in cell physiology, can easily be detected in such large data sets. Video imaging FPR will show anisotropic

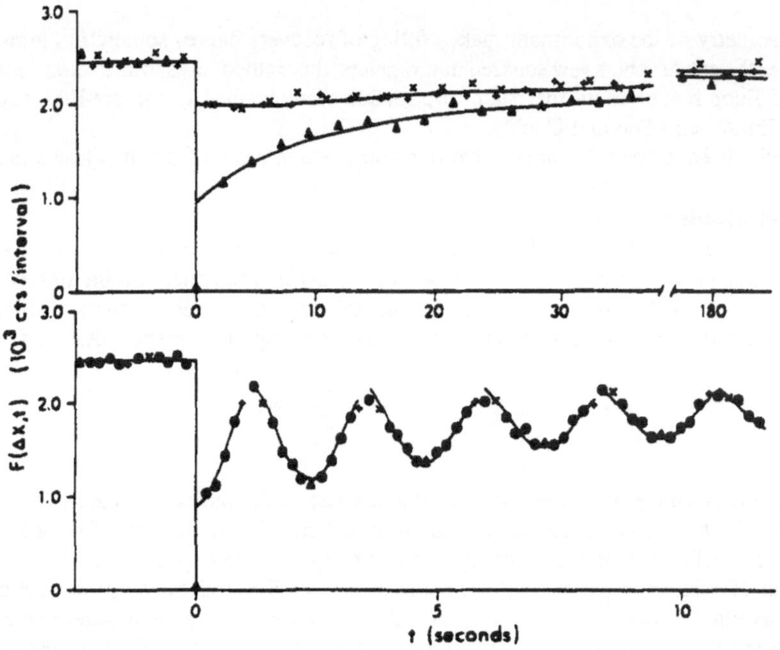

Figure 3. Scanning spot/multipoint FPR. A synthetic multibilayer membrane was labeled with NBD-PE (see Figure 4). A spot was bleached in the usual way, but fluorescence recovery was monitored in a line of 12 points on the membrane, spaced 1 μm apart. Top panel — recovery of fluorescence at three points — the area bleached, ▲, and two other points, 5 μm to each side of the bleached area, +, x. Bottom panel — fluorescence intensity at all 12 points with time after bleaching. The three points of the top panel are marked ▲, +, x. (Reproduced from Koppel, D. E., *Biophys. J.*, 28, 281, 1979. With permission.)

recovery, and other features of the surface of a single cell, but it has limited application for determining R & D when these vary over a wide range. A combination of conventional video and FPR may be useful in such cases.[28]

IV. FLUORESCENT LABELS FOR FPR

A. LIPID LABELS

Different approaches are used for labeling cell membrane lipids for FPR than are used for labeling membrane proteins. Lipid labels are typically fluorescent analogs of native lipids or lipophilic molecules. Since almost all FPR measurements use green or red fluorescence these analogs are often native lipids covalently linked to fairly bulky fluorophores, e.g., NBD-phosphatidylcholine (NBD-PC) or rhodamine-phosphatidylethanolamine (R-PE) (Figure 4). The chemistry and biology of the NBD-lipids have been reviewed[29] and this review can be used as a guide to choosing one of these as a probe for FPR. Alternatives to NBD-lipids are lipid analogs whose fluorophores are not as bulky as those of the first class, but which differ structurally from native phospholipids, e.g., the indocarbocyanine iodides, so-called diIs (Figure 4).[30]

Lipid probes must integrate into the membrane bilayer if they are to report lateral diffusion correctly. However, the image of a labeled membrane does not always reflect the integration of fluorescent label. Micelles of a lipid analog stuck to the cell surface may give as discrete a membrane "ring" stain as molecules of the same analog inserted into the bilayer itself. In this respect, diIs are better probes than NBD-phospholipids. diI molecules self-quench at high concentrations and so micelles of dye stuck to the cell surface will be weakly fluorescent. More important, the diI fluorophore is oriented when the molecule is inserted into membranes and the emitted fluorescence is therefore polarized. Hence, if the fluorescence of diI-labeled cells is viewed through a polarizing filter only two of the four quadrants of the surface appear fluorescent. Their fluorescence is extinguished and that of the other two quadrants appears when the filter is rotated 90°.[31]

The best indication that NBD-PC or other lipid analogs labeled on acyl chains have integrated into the cell membrane is in fact measurable lateral diffusion. Micelles of lipid analog stuck to the membrane will

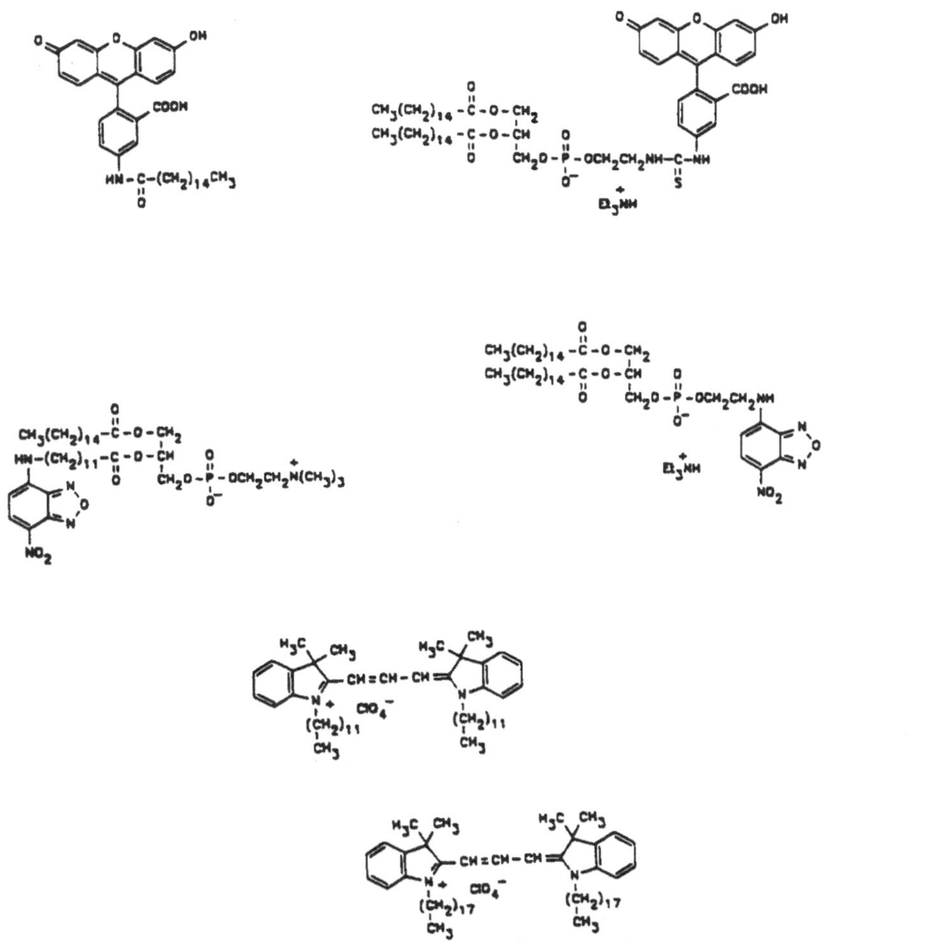

Figure 4. Lipid labels for FPR. Top left — HEDAF, hexadecylaminofluorescein, a fatty acid analog. The bulky, polar, fluorophore is free at the membrane surface, but the single fatty acid chain may act as a detergent to perturb the bilayer. Right — headgroup-labeled phosphatidylethanolamines, PE. Top right — fluorescein-PE. Center right — NBD-PE. The undmodified fatty acyl chains of these derivatives can pack properly with native membrane lipids. However, they may not readily intercalate into cell membranes. Center left — C6-NBD-phosphatidylcholine, NBD-PC. This lipid analog readily intercalates into cell membranes. However, the bulky NBD group is forced toward the surface of the bilayer.[32] Hence, this analog may disrupt native phospholipid packing or may be excluded from some regions of the membrane. Bottom — indocarbocyanines with undmodified alkyl chains, diI C12 and diI C18. I thank Molecular Probes Inc. (Eugene, OR) for help with this figure.

show no measurable diffusion in an FPR experiment. However, this is not always so. Analogs which can dissociate from the surface, or which have appreciable water solubility, may give spurious FPR traces.

Both types of lipid probe must perturb the native bilayer. diI has native, unmodified acyl chains which should not disturb the packing of membrane lipids, but its head cannot associate with other molecules in the membrane in the same manner as native phospholipids. On the other hand, while NBD-PC bears a polar head group, its bulky, polar, NBD fluorophore lies near the surface of the bilayer[32] and disrupts the packing of lipid acyl chains in its vicinity. These differences between the lipid probes should be considered when choosing one or the other type of probe for a particular set of FPR measurements.

B. PROTEIN LABELS

The range of practical fluorescent labels for FPR of membrane proteins is enormous. It extends from small covalent labels, e.g., fluorescein directly coupled to cell surfaces,[33] through antibodies and antibody fragments, to large markers, such as fluorescent low density lipoprotein (LDL).[34,35] There are trade-offs

Table 1. **Effect of Label Size and Valence on Lateral Diffusion**

Cell type	Label	D ($cm^2\ s^{-1} \times 10^{10}$)	R (%)
Normal human fibroblasts	β_2-m	3.3[a]	55[b]
	Fab	3.0	51
	IgG	1.0	60
Transformed human fibroblasts	β_2-m	1.9	60
	Fab	1.5	58
	IgG	0.8	50
Mouse Lymphoblasts	β_2-m	1.7	39
	Fab	2.6	30
	IgG	0.6	32

[a]D are geometric means. Ninety-five percent confidence intervals are approximately ± of the mean.
[b]Standard deviation for sets of 10 to 20 determinations is 20 to 30% of the mean.

to be made in preparing such labels for proteins, among parameters of small size, high signal, affinity, and specificity.

1. Covalent Labels

Direct coupling of fluorescein isothiocyanate and the related (4,6-dichlorotriazinyl) amino fluorescein (DTAF) are generally useful only for membranes containing high concentrations of a few especially reactive species of protein. This approach has therefore been confined largely to work on erythrocytes where band 3 is selectively labeled[9,36-38] and vertebrate rod outer segments, where a sulfhydryl-reactive derivative of tetramethylrhodamine labels rhodopsin in disc membranes.[39]

The protein labeled depends in some detail on the reactive dye and the labeling conditions used. In rods, iodoacetamidotetramethyl rhodamine labeled rhodopsin, but the more polar iodoacetamido fluorescein, though it labeled the disc membranes, did not label rhodopsin[40] (R. A. Cone, personal communication). In red cells, 80% of the proteins reacting with eosin maleimide are band 3, but 75% of all the proteins reacting with fluorescein-5-thiosemicarbazide are glycophorin.[41,42]

2. Lectins

Lectins are readily available noncovalent labels for cell surfaces. The broad specificity of lectins makes them particularly tempting labels, but this temptation should, in most cases, be resisted. Lectins are difficult labels for FPR on two counts. First, most are multivalent and cannot readily be cleaved to monovalent forms. This means that they crosslink membrane proteins as they label, and aggregates of such crosslinked proteins are certainly liable to diffuse differently to the monomeric proteins (see Reference 43 and the comparison in Table 1 of D determined using divalent antibody labels and D determined with monovalent antibody fragments). A second count against lectins is their broad specificity. FPR results using these labels are averages of the diffusion of a collection of membrane glycoproteins and glycolipids. While D depends weakly on molecular weight, it is not reasonable to calculate D for an ensemble ranging over 3 orders of magnitude in molecular weight.

3. Antibodies

Antibodies and antibody fragments, Fab, offer much greater scope for controlled and specific labeling than do lectins. They are easily prepared against many different cell surface proteins, and they are readily prepared, purified, and labeled.[44] Antibodies and Fab may perturb FPR measurements in two ways, by mimicking agonists or otherwise perturbing membrane organization (for example, see References 45 and 46) and by their mass and valence alone. Antibodies that mimic agonist in the effects on membrane receptors should be avoided. If they are used for FPR probes, they should be used at temperatures at which cells are not activated and, even when used in this way, their targets should be checked for the biochemical sequelae of receptor activation.

Aggregated antibodies or Fab can reduce D and R.[47] In my laboratory, solutions of Fab are ultracentrifuged before use to remove aggregates. Once such a solution is prepared it can be stored in ice for a week or more before requiring centrifugation. If the solution is frozen and thawed, it must be centrifuged again before use.

If IgG can crosslink surface proteins, it may affect D and R, even if it has no other biological activity. Zhang and co-workers compared the effects of two different mAbs to the trimeric VSV G protein on D and R of VSV G expressed in COS cells.[48] One IgG gave values slightly different from those measured with labeled Fab. $D = 2.8 \pm 1.3 \times 10^{-10}\ cm^2\ s^{-1}$ and 52% mobile for IgG compared to $D = 3.8 \pm 1.3 \times 10^{-10}\ cm^2\ s^{-1}$ and 50% mobile for Fab. The other IgG, against a different epitope of the VSV G protein, gave $D = 1.7 \pm 1.3 \times 10^{-10}\ cm^2\ s^{-1}$ and 27% mobile compared to $D = 4.2 \pm 1.3 \times 10^{-10}\ cm^2\ s^{-1}$ and 63% mobile for Fab. The differences in both D and R are statistically significant.

In experiments on class I MHC molecules my laboratory has found effects of label on D similar to those just described.[48] Table 1, based on unpublished data, compares D and R for class I MHC molecules, MW ~60 kDa, labeled with one of three molecules conjugated to fluorescein: monoclonal IgG anti-class I MHC, Fab made from the IgG, and β_2-microglobulin, the light chain of the class I MHC molecule. There is no difference in R for the three labels, but D is consistently lower when IgG is used as label than for Fab or the "internal" label β_2-m. These differences are too great to be caused by the difference in molecular weight (mol wt) between D for a monomer MHC molecule + Fab (mol wt ~105 kDa) and D for a complex of IgG and two MHC molecules (mol wt ~270 kDa). They probably reflect the fact that some class I molecules form dimers in the absence of antibody.[49,50] Even though $D \propto \ln mw$,[51] a divalent IgG could create clusters of MHC molecules large enough to reduce D.

4. Hormones, Toxins, and other Labels

Fluorescent ligands for specific cell surface receptors, though more difficult to prepare than Fab, have been used for FPR. Mobility of receptors and channels has been studied using fluorescent epidermal growth factor (EGF), fluorescent toxins (labeling ion channels), and fluorescent agonists and antagonists for the neutrophil N-formyl-peptide receptor.[52-55] Of course, all of these are biologically active and so must be used so that they do no activate the receptors that they mark. A recent study[55] is a nice example of the proper use of such ligands. Fluorescent agonist and antagonist peptides were prepared that bound specifically to the neutrophil N-formyl-peptide receptor and their potency was characterized before using them to study changes in lateral mobility after activation of the receptor. A full cellular response to the compounds was prevented by performing all measurements at 14°C. In contrast to these studies, other peptide ligands used for studies of receptor dynamics, e.g., fluorescent insulin, show little specifity of binding[56] and measurements made using such ligands may not properly report the state of cell receptors.

Soluble proteins that form complexes with integral membrane proteins may also be used as labels. The β_2-m light chain of class I MHC molecules has been used to study mobility of these molecules on epithelial cells[57] as well as in the work summarized in Table 1. Cytochrome *c* is another example of a water-soluble protein that binds to membranes and interacts with their proteins. Fluorescent cytochrome *c* derivatives have been used to investigate the dynamics of mictochondrial inner membranes by FPR[58] though the diffusion of cytochrome *c* on membranes is quite complex.[59]

Extended discussions of fluorescent toxin, peptide, and hormone labels may be found in recent reviews.[60]

V. PROBLEMS AND PITFALLS IN FPR MEASUREMENTS

A. LOCAL HEATING AND CELL PIGMENTS

In an FPR experiment high-intensity light is focused into a small area that is sensitized (by dye labeling) to capture much of this light. Power densities during the bleaching phase of the experiment (tens of milliseconds) may reach 1 MW/cm^2. There is the possiblity of local heating and of photoinduced crosslinking or other damage to the labeled proteins and lipids, especially at high concentrations of lipid labels. Otherwise, if no pigment is present in or beneath the membrane it can be shown that local heating due to the laser flash is well under 1°C.[61]

Local heating has been a problem in pigmented cells, notably in erythrocytes. Intact erythrocytes cannot be probed with the usual labels, e.g., fluorescein, if the usual brief bleaching times are used.

Under these conditions hemoglobin absorbs so much energy that the cells burst.[62] Recently, FPR measurements have been successfully made on intact erythrocytes, using blue light.[42] D and R in this case are close to values measured by others using yellow exciting light.[61] The key to success here appears to be the use of much lower bleaching powers for much longer times than usual. This seems to allow the heat produced by the bleaching photons to dissipate.

B. PHOTODAMAGE

Phototolysis of label during bleaching could produce excited-state species of molecules with high probability of reaction with other excited-state molecules or with other proteins in the spot. Such crosslinked species would likely diffuse more slowly than the native proteins, so that FPR would report lower D than that in fact obtained. Two classes of experiments have been performed to address this question. One class used irradiation of suspensions of labeled cells or membranes. Under these conditions, irradiating for many seconds at power densities much lower than that obtained in an FPR experiment, proteins were crosslinked by irradiation after labeling. The most careful of these experiments concluded that for a constant total light flux production of crosslinked proteins was inversely proportional to the intensity of irradiation.[63] Extrapolating the data of this experiment suggests that no significant crosslinking occurs under conditions of an actual FPR measurement.

The second class of experiment designed to detect photobleaching damage is one in which conditions of an FPR experiment are varied in ways that ought to enhance or reduce photodamage. Variation in D and R for different conditions of the experiment would be strong evidence for such damage. In one such set of experiments multiple bleaches of the same spot did not affect the measured D; with R, the mobile fraction sytematicaly increased with successive bleaches, as expected if incomplete recovery is due to immobilization of a fraction of the labeled molecules.[64] Other experiments in the same series found no effect of quenchers or traps for singlet oxygen or free radicals (suspected mediators of photodamage) on measured D or R.

In a second experiment a surface protein, the FcεR1 receptor for IgE was labeled with a mixture of rhodamine and fluorescein. The receptor was first labeled with either R-IgE or Fl-IgE and then labeled with either Fl-Fab anti-IgE or R-Fab anti-IgE.[65] FPR measurements were made on the doubled-labeled receptors after one color label had been bleached by exposure to light from a mercury arc lamp or from an expanded laser beam. For example, all the fluorescein label on a cell was bleached to 50% of its initial intensity, and this was followed by an FPR measurement using laser excitation of the rhodamine label. Since both fluorescein and rhodamine labels were on each receptor, any crosslinking or other photodamage from the first bleach, of the fluorescein label, would be reflected in the FPR measurement using the second label. This was not seen, even though the first bleach required between 10 s if the expanded laser beam was used and 5 min if the arc lamp was used. Thus, no evidence for photodamage or other artifacts was found even in conditions that approached those used in the experiments on photodamage to cell suspensions.

The only experiments in which there is evidence for photodamage or photoinduced crosslinking are those on FPR of labeled microtubules or microfilaments in which high concentrations of labeled proteins are irradiated in solution for times ranging from milliseconds to minutes.[66,67] Here irradiation results in crosslinking of tubulin or breakage of actin filaments. However, acceptable measurements of D can even be made in these systems if conditions are carefully controlled (see the treatment in References 67 and 68).

VI. RESULTS FROM FPR

A. THE RANGE OF LATERAL DIFFUSION

This section is best introduced by Table 2, a survey of D and R for a number of cell membrane proteins. Most of the special topics to be covered in the pages that follow can be exemplified by the results in the table.

The importance of the first topic to be covered, the constraints to lateral diffusion, is evident when we survey the range of D in Table 2. The largest D, 3 to 5×10^{-9} cm^2 s^{-1} for vertebrate rhodopsin, is about that calculated from

$$D = \frac{kT}{4\pi\eta r} \tag{3}$$

for a sphere with radius, r, of some tens of angstroms immersed in a fluid with a viscosity, η, of about 1 P. This D is also measured for purified cell membrane proteins reconstituted into fluid lipid bilayers.[69-72] The smallest D in Table 2, 5×10^{-11} cm^2 s^{-1}, for voltage-dependent K^+ channels in frog muscle, is 100-fold smaller than that of rhodopsin. Measurements of membrane lipid viscosity by many different methods indicate that the viscosity of cell membrane lipids is in the range 0.5 to 5 P, at physiological temperatures.[73] The difference between the largest D and the smallest D then suggests that the lateral diffusion of membrane proteins is constrained in some manner.

A similar comment can be made on R, the mobile fraction of probe, expressed in Table 2 as the percent recovery of fluorescence. R of proteins in synthetic, fluid-phase, lipid bilayers, is always high, 80 to 100% (see References 69 to 72). Yet, R listed in Table 2 ranges from 0 to 100%, with many between 30 and 40%. Again, assuming that the experiments are technically correct, this implies proteins may be immobile on the minute time scale of an FPR experiment.

Table 2. **Lateral Diffusion of Membrane Proteins Measured by FPR**

Protein	Cell	D[a]	R[b]	Ref.
Vertebrate	Outer segments	30–50	100	7, 8, 39
Rhodopsin	Chicken photoreceptors			74
Na^+,K^+-ATPase	Undifferentiated	20	65	
	Differentiated	20	40	
Voltage-dependent Na^+ channels	Rat neurons			54
	Cell body	10–20	80–90	
	Axon hillock	2	30–40	
	Frog muscle	—	0	75
Voltage-dependent K^+ channels	Frog muscle	0.5	25	75
Class I MHC molecules	Mouse L-cells	~8	30	6
		7	50	77
	L-cell clone	2	30	77
	Spleen			
	Lymphocytes	2–4	30–45	80
	B-cells	2–4	30–45	78
	T-lymphoma	5	15–40	76
FcεR1	Rat basophil	2–3	70–80	65
	Leukemia	3	~85	81
		3	60	21

[a] $cm^2\ s^{-1} \times 10^{-10}$.

[b] Percentage recovery of fluorescence.

Changes in lateral diffusion with cell differentiation are evident when we compare R for the Na^+,K^+-ATPase of chicken photoreceptor cells.[74] These cells, isolated from chick embryos, differentiate into cone cells, with their Na^+,K^+-ATPase localized to one region of the cell membrane. This localization is reflected in a drop in R of nearly half from 65% on average to 40%. Further evidence that cell differentiation is reflected in lateral mobility comes from the comparison of D and R for voltage-dependent Na^+ channels on rat neurons.[54] Nearly all the channels diffuse freely in the cell body of the neuron, while at the specialized region of the axon hillock D is 0.1 of that on the cell body and R is <0.5 of R on the cell body.

Cell-type-specific variations in D and R emerge from comparing a number of the results in Table 2. In contrast to the results for rat neurons, the voltage-dependent Na^+ channels of frog muscle are immobile.[75] Comparing the lateral mobility of class I MHC molecules on different types of mouse cells, and even comparing results with clones, shows a cell-specific variation in D and R.[6,76-80]

A sense that the variation is not due to variation in the FPR instruments or techniques used for measurement can be gained from the results on lateral diffusion of the FcεR1, the high-affinity IgE receptor, on rat basophil leukemia cells. D measured by three different groups over a decade are identical, and even R's are fairly close to one another.[21,65,81]

B. CONSTRAINTS TO LATERAL DIFFUSION

As previously noted, with a few exceptions neither D nor R measured for cell membrane proteins is as large as that measured for the same proteins reconstituted into lipid bilayers. The reasons why this is so and the loci of constraints to lateral diffusion have been discussed for many years (see Reference 82) and have been the subject of numerous experiments.

1. Constraints in the Cell Cytoplasm

Early models for constraints were heavily influenced by work on the erythrocyte cytoskeleton and emphasised the interactions between membrane proteins and proteins of the cytoskeleton. These interactions are certainly important in the erythrocyte. D of band 3, the erythrocyte anion channel, is 50-fold higher in spectrin-deficient mouse erythrocyte ghosts than in normal erythrocytes.[83] It is 10-fold higher than normal in pathological human erythrocytes whose spectrin-to-band 3 ratio is 70% of normal[84] and it is 50-fold higher than control in human erythrocyte ghosts treated with trypsin to destroy spectrin.[85,86]

Cytoplasmic constraints to D may act in other cells, but need not be the only or even the major constraints. The evidence for such constraints comes from three sorts of experiments. One sort shows that membrane proteins are immobilized when they are associated with cytoskeleton in focal adhesions[28] or after ligation by antigen or antibody.[78,81,87] A second sort shows that truncation of the cytoplasmic domain of a class II MHC molecule on B cells reduces and ultimately disrupts its ability to interact with a cytoplasmic enzyme, protein kinase c, and, at the limit of complete trunction of the cytoplasmic domains of its two chains, increases D tenfold.[88] The third sort of experiment shows D for proteins and lipid analogs increases 50- to 100-fold when a cell membrane is induced to form blebs that are pulled away from the underlying cytoskeleton.[89]

Experiments in which D of wild-type molecules is compared to D of mutants lacking all or part of their cytoplasmic domains weigh against the importance of the cell cytoskeleton as strongly affecting D. Truncation of vesicular stomatitis virus G protein, VSV G cytoplasmic domain from 29 amino acid residues to 1 residue,[90] truncation of the class I MHC molecule H-2 L^d cytoplasmic domain from 31 residues to 4 or no residues,[91,92] and truncation of the epidermal growth factor receptor cytoplasmic domain from 542 residues to 9 residues[93] has no effect on D or R for these proteins (Table 3).

Proteins anchored to the bilayer by glycosylphosphatidylinositol (GPI), tails, and GPI proteins do not even penetrate both leaflets of the lipid bilayer. Yet D of such proteins and of hybrids of transmembrane proteins subsituted with GPI tails does not differ significantly from D for homologous proteins with cytoplasmic tails (Table 3).[48]

Table 3 shows clearly that modification of membrane anchors or cytoplasmic domains does not yield D and R similar to those for proteins in synthetic lipid bilayers. However, some of the results summarized in Table 3 do suggest that the cytoplasmic domains of many proteins do influence D and R, though their influence may not dominate the parameters of FPR. The two most striking are those on class II MHC molecules already mentioned[88] and those on truncation of VSV G protein.[90] Those data as presented in Table 3 show a threefold difference between D for wild-type VSV G and D for VSV G with its normal tail substituted with the 11-residue tail of influenza virus HA. This difference is even more striking if we use the data in the paper to calculate a mean D that is not biased by small numbers of outlying values, the geometric, rather than the arithmetic mean. Calculating in this way, we find that substitution of the normal tail of VSV G results in a sixfold increase in D, from $\sim 1 \times 10^{-10}$ cm^2 s^{-1} to $\sim 6 \times 10^{-10}$ cm^2 s^{-1}.

2. Constraints in the Lipid Bilayer

The viscosity, η, of the lipid bilayer certainly affects D for proteins, yet, as we have mentioned earlier, values of η are too small by factors of 10 to 100 to yield measured D. We also note that changing lipid viscosity, e.g., by modifying membrane lipids or simply by increasing temperature, has little effect on D, though, significantly, it may change R to a great extent. D might also depend upon the state of bilayer lipids. If these are a mixture of gel and fluid, and only the fluid lipids can be penetrated by the proteins, then long-range D will be reduced by the obstacles presented by patches of gel lipids.[94-96] Though the study of lipid domains is best developed in synthetic lipid mixtures (reviewed in Reference 97 and see Reference 98), there is evidence, to be reviewed later in this chapter, that cell membranes contain patches of lipid that differ in viscosity and presumably in composition from the average for the membrane in which they occur. Hence it is possible that these patches present obstacles to random walks that reduce D.

High concentrations of membrane proteins will also serve as obstacles to lateral diffusion. In addition, weak interactions between proteins will further reduce D. Diffusion in high solute concentrations in three dimensions, in solution, has been studied for many years,[99] though until recently only empirical corrections, rather than complete theory, have been developed to deal with these conditions. The most useful of these corrections is[100]

$$\eta = \frac{\eta_0}{1 - 2.5\phi} \qquad (4)$$

where ϕ is the volume fraction of solute in solution or the area fraction in a cell membrane. There early suggestions that this sort of relationship might affect D in cell membranes where the volume fraction of protein was high,[101] but models dealing with these effects were not developed until the 1980s.[102-104]

Table 3. **Effect of Truncation or Modification of Cytoplasmic Domains on the Lateral Diffusion of Membrane Proteins**

Protein	Cell	Cytoplasmic tail modification	D[a]	R[b]	Ref.
EGF receptor	COS	None (523 AA)	1.2	61	93
		–63 AA	1.6	59	
		–442 AA	1.5	70	
		–523 AA	1.2	77	
NCAM (180 kDa)	Neuroblastoma	None	1.1	90	175
		GPI-tail	0.9	90	
VSV G protein	COS	None (28 AA)	0.85	75	90
		–27 AA	2.1	62	
		–25 AA	2.1	66	
		–28 + 11 AA from influenza HA	2.9	68	
VSV G protein	COS	None (28 AA)	3.8	50	48
		–28 AA + GPI-anchor	5.9	62	
Thy-1	COS	None(GPI tail)	27	67	48
		+ 29 AA (VSVG)	16	58	
Class II MHC	B-lymphoma M12.C3	None (12α + 18β AA)	1.3	73	88
		–12α –6β	4.3	68	
		–12α –18β(no tail)	8.0	63	
Class I MHC	L-cells	None (31 AA)	12	–	91
		–24 AA	10	30	
		–27 AA	15		
	Hepa-OVA	None (31 AA)	4	40	92
		GPI-tail	5	53	
	Hepa-OVA	None (31 AA)	6	70	153
		GPI-tail	11	70	
		None (31 AA)	13	–	Edidin et al., unpublished
		–24 AA	13	40	
		–27 AA	10		
Influenza HA	CV-1	None (Cys 543)	12	85	176
		Tyr 543	5	80	
NGF receptor	COS	None	6	40	117
		–155 AA	8	40	

[a] $cm^2\ s^{-1} \times 10^{-10}$.
[b] Percentage recovery of fluorescence.

The few measurements of D as a function of protein concentration in synthetic bilayers[70,105] suggest that factors other than the effect of ϕ on lipid viscosity reduce D.

Two sets of experiments on proteins in cell membranes give conflicting results. One[106] shows that D of epidermal growth factor receptor decreases about threefold for every tenfold increase in surface concentration of receptor. This could reflect increased area fraction of receptor proteins in the bilayer or could reflect the increased volume fraction of the receptor protein outside the bilayer. The latter is more likely the case than the former, since the EGF receptor spans the bilayer only once, but has about 500 residues each in its extracellular and cytoplasmic domains. In the second experiment[107] electrophoresis *in situ* was used to raise the local concentration of membrane proteins of RBL cells more than fivefold (ϕ approaching 80% of the accessible membrane area) and D of FcεR1, the high-affinity receptor for IgE, in these areas was compared to D in regions depleted of protein by the electrophoresis. Despite the fact that one of the three subunits of the receptor has multiple intrabilayer spans, there was no effect of protein concentration on D in unperturbed RBL. However, if the cells were swollen to separate surface membrane from cytoskeleton,[89] then D did depend upon the surface concentration (Figure 5). Thus, in RBL cells

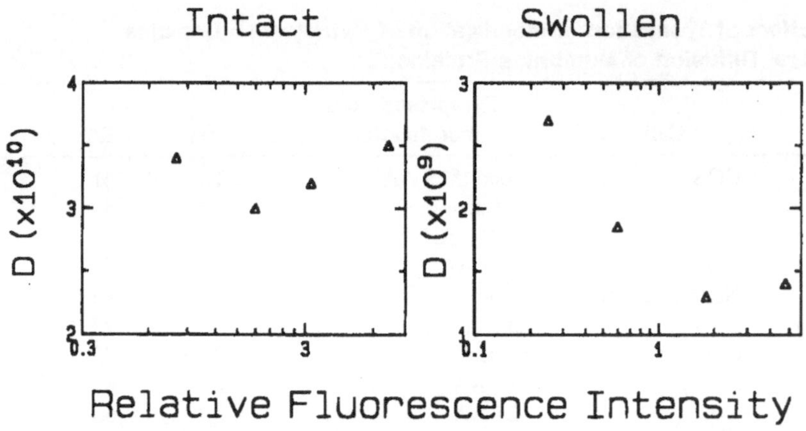

Figure 5. D is insensitive to surface concentration of protein in intact RBL cells, but is sensitive to surface concentration of protein in osmotically swollen cells. A concentration gradient of surface proteins was established by electrophoresis *in situ*. After the field was turned off, diffusion of IgE receptor was measured at various points along the gradient. Left panel — intact cells. D is insensitive to surface concentration of protein, expressed as the ratio of fluorescence intensity of the region probed to the average fluorescence of the entire surface. Right panel — swollen cells. D falls nearly 3× as protein concentration increases. Note that the scale of D is tenfold greater in this panel than in the other.[107]

there seems to be an influence of protein concentration on D, but this effect is not as strong as the influence of the cytoskeleton.

3. Constraints through Exodomains of Membrane Proteins

Many of the membrane glycoproteins for which we have D and R have large fractions of their mass in the extracellular domains. The lateral diffusion of such proteins will be hindered by the bulk and shape of these exodomains. Such proteins are also likely to interact with one another because of their high concentrations. Hence we expect that their exodomains will have strong effects on lateral diffusion. There are experiments that show this to be the case, but again the effects depend upon the protein and the cell in which it is expressed.

Exodomains of type 1 glycoproteins may be modified in one of three ways: truncation, substitution, or directed mutation of sites for N-linked glycosylation. The approach was first used to show that addition of a third glycosylation site to VSV G protein or deletion of one of the existing sites from the protein had little effect (less than 2×) on D when the protein was expressed in COS cells.[90] In contrast, other workers[77] showed that deletion of sites for N-linked glycosylation increased D of class I MHC molecules, H-2L^d expressed in L-cells. The H-2L^d molecule has three such sites and when all three were removed D approached the limit set by membrane lipid viscosity, $\sim 2 \times 10^{-9}$ cm^2 s^{-1}. H-2L^d has a mol wt of about 60 kDa and removing three oligosaccharides reduces this by about 10%. The effects of removing oligosaccharides then appear to reflect the volume occupied by these structures and their interactions with other oligosaccharides on the cell surface. Truncation of two thirds of the exodomain of H-2L^d, resulting in a protein of mol wt ~30 kDa with one glycosylation site, also increased D again, to about the viscosity limit.

The results on glycosylation of H-2L^d and D suggested that D would also be increased if the molecule was expressed in cell mutants defective in N-linked glycosylation. The cell mutants required for these experiments had all been made in CHO cells, by selecting for lectin resistance[108] and so H-2L^d was transfected into these cells. In the event, D varied only twofold between all of the glycosylation mutants. Even D of the unglycosylated H-2L^d was no different from the wild-type proteins when these were expressed in CHO cells.[109] This suggests that either other constraints, e.g., in the cytoplasm, affect D in these cells, or that the higher-order structure of H-2L^d expressed in CHO cells is different to that structure expressed in L-cells. Some support for this suggestion is found in recent studies on the structure of mouse MHC molecules in CHO cells.[110] In any event the result again emphasizes the variation in constraints to D and the way in which D varies from protein to protein and from cell to cell.

The most systematic study of the contributions of each region of a membrane protein to D is one in which exodomains and membrane anchors of a number of conventional and GPI-anchored membrane glycoproteins were incorporated into neoproteins whose lateral diffusion could be compared with that of the parent proteins.[48] The results showed that, in COS cells, the exodomains of the proteins, rather than the membrane anchors or cytoplasmic domains, dominated D. For example, D for Thy-1, a small, heavily glycosylated GPI-anchored protein, was ~3×10^{-9} cm^2 s^{-1}, while D of Thy-G, the Thy-1 exodomain attached to the 49 amino acid transmembrane and cytoplasmic region of VSV G, was 1.6×10^{-9} cm^2 s^{-1}. On the other hand, D for VSV G in this series was $3.8 \pm 2 \times 10^{-10}$ cm^2 s^{-1}, while D for G-Thy, the 463 amino acid exodomain of VSV G attached to the GPI-anchor of Thy-1, was $6 \pm 2 \times 10^{-10}$ cm^2 s^{-1}.

The whole set of results was analyzed in terms of the contribution of each domain of the protein, exodomain, and "anchor" (membrane anchor + cytoplasmic domain if any) to the frictional coefficient, f_{total}, of the protein. Once f was assumed for the Thy-1 lipid anchor (from D measured for lipid probes), f_{exo} and f_{anchor} could be calculated for VSV G and for the Thy-1 exodomain. $f_{Thy\text{-}exo}$ was about 0.1 of $f_{Thy\text{-}anchor}$; $f_{G\text{-}ecto}$ and $f_{G\text{-}anchor}$ were about $4 \times f_{Thy\text{-}anchor}$.

All of the forces acting on the exodomains can be characterized as a pericellular viscosity. Molecules contributing to this apparent viscosity include the proteins in the bilayer and the proteins or proteoglycans in the extracellular matrix. The pericellular viscosity for cells (fibroblasts and keratocytes) in primary cultures of fish scales, transformed mouse fibroblasts, and PtK1 cells was estimated as 0.5 to 0.9 P. This was done by comparing D determined by FPR for an inserted phospholipid label with D for the same determined from the random walk of antibody-coated gold beads.[111] Heparinase treatment of keratocytes reduced this to around 0.1 P. Though at first sight this suggests that the major contribution to pericellular viscosity comes directly from the heparin-sensitive matrix proteoglycans, we must remember that these can also influence the behavior of other membrane proteins which, if themelves constrained, can hinder the diffusion of the lipid probe.

The results just summarized suggest that the constraints to D will vary with the amount and nature of pericellular matrix secreted by a cell. Again, this emphasizes the contigency of constraints to lateral diffusion — their dependence upon the molecule for which D is measured and upon the cell in which it is expressed. Recent reviews of the area provide further discussion of this topic.[112,113]

C. FPR IN CELL PHYSIOLOGY

1. Signaling

FPR measurements of the mobility of membrane receptors, other membrane proteins, and even membrane lipids can give information useful in understanding the cell physiology of responses to signals delivered through specific receptors, in understanding membrane changes in differentiation and in understanding cell responses to stresses.

The relevance of FPR to hormone receptor-mediated responses was mentioned at the beginning of this chapter. The fact that membrane proteins were mobile with an average D of magnitude 10^{-10} cm^2 s^{-1} implied that hormone receptors need not be permanently coupled to cytoplasmic transducer enzymes, e.g., adenylate cyclase. The collision-coupling model proposed instead that activated receptor and enzyme met by random walk and collision.[1,2] While the model was devised without a knowledge of G proteins and other proteins of the cytoplasmic face of the cell surface, it still serves to frame discussions about the role of lateral diffusion in signal transmission. Recent versions of the collision-coupling model assume that D measured by FPR is the same on the scale of receptor/transducer interactions, hundreds of nanometers, as it is on the micrometer scale of FPR. In order to accomodate this to the rapid response times measured for receptor-mediated signaling it is assumed that receptors and transducers are confined to limited regions of the cell surface, rather than being free to diffuse over the entire surface.[4,114] These qualitative models are quite incomplete. There has been an attempt to make a quantitative model of the interaction between rhodopsin and transducin.[115] The model began with the assumption that the concentrations of each reactant are optimized. Lower concentrations would reduce frequency of collision and hence reduce signal transmission, while higher concentrations would impede lateral diffusion and, again, reduce signal transmission. In any event not enough information is available to test the model. From our point of view, the key missing data are on the concentration-dependence of D for rhodopsin and transducin. Thus, while existing FPR measurements suggest mechanisms for signal transduction, more measurements of D must be made before quantitative models can be built.

FPR has been used to show transient coupling of receptors to transducer in B-lymphoblasts, cells in which MHC class II molecules may signal — resulting in a translocation of protein kinase C away from the surface membrane. Results mentioned previously[88] show that the inactive receptor is likely to be in transient contact with the transducer, since truncating the cytoplasmic domain of the class II molecules increases D.

At the limit of receptor/transducer confinement, receptors are continually coupled to transducers. There is evidence that this is true for EGF receptors[116] and nerve growth factor receptors.[117] There is also an excellent correlation between 1/R, the immobile fraction, and desensitization to agonist (possibly by internalization of receptors) for EGF and vasopressin receptors[118] and for the N-formyl peptide receptors of neutrophils.[119] There is also a report that binding of purifed platelet-derived growth factor to its receptor on serum-starved or normal cells reduces R.[120]

A model for signal transduction by N-formyl peptide receptors[121] like the models mentioned previously must assume local D similar to that measured by FPR. This has only recently been measured, using fluorescent agonist and antagonist peptides.[55] D for receptors labeled with either label was the same, $\sim 5 \times 10^{-10}$ cm^2 s^{-1}. However, R was significantly lower after agonist binding than after antagonist binding, 40 vs. 60%.

Immobilization of receptors is also seen when FcεR1 receptors, high-affinity IgE receptors, bind IgE and are crosslinked by antigen. Interestingly, truncation of the cytoplasmic domains of the receptor did not change D or R even after ligation by IgE and antigen.[21] The class II MHC molecules, which function both as receptors (see Reference 88) and to present peptide antigen to responding T cells, are also immobilized when they bind immunogenic peptide.[78]

FPR then appears to detect receptor immobilization as an important part of signal transduction. Some populations of receptors, EGF R and NGF R, are immobilized before binding agonist, while others are immobilized after binding agonist. In the first instance the FPR data report a transducing complex, while in the second lower R probably reflects the preliminaries to receptor internalization and consequent desensitization.

FPR of more general probes reports changes in membranes after agonists bind to their receptors, or after other membrane events, but the interpretation of these changes is difficult. For example, D of lipid probes and of general protein labels is reduced to about half of control values when rat testicular cells bind luteinizing hormone.[122] Treatment of adult male schistosomes with serotonin lowers D of a fluorescent fatty acid analog, apparently through a mechanism involving changes in intracellular free Ca^{++} and activation of adenylate cyclase.[123]

2. Membrane Stresses and Insults

Stresses and insults to the cell surface membrane also perturb lateral diffusion of membrane proteins and lipids. Complement attack on erythrocyte membranes reduces D of covalently labeled proteins to about half of control values,[124] from $\sim 4 \times 10^{-11}$ cm^2 s^{-1} to $\sim 2 \times 10^{-11}$ cm^2 s^{-1}. This effect was seen when the complement came from whole serum or when acid-activated C5-9 was used without any additional serum components. D of the C5-9 complex itself was about 1/4 that of the C5-7 complex and D of the C5-8 complex was 3/4 that of C5-7. Thus, lower D correlated with formation of the terminal attack complex in the cell membrane. The reduction in D appears to involve interactions with the cytoskeleton. Removal of spectrin by extraction increased D of erythrocyte proteins about 15-fold in control cells and 25-fold in complement-treated cells. Also, D for the purified C5-7, C5-8, and C5-9 complexes in liposomes was 1.5 to 2.5×10^{-9} cm^2 s^{-1}, consistent with undhindered diffusion of the complexes.

Fusion of viruses or parasites with cell membranes can be detected by FPR measurements of membrane protein and lipid D and R. Sendai virus fusion transiently increases the mobile fraction of erythrocyte membrane proteins and D of a membrane lipid probe.[125] Nonfusigenic virions did not cause these changes.

The protozoan parasite *Schistosoma mansoni* binds, lyses, and ingests erythrocytes. Contact of parasite membranes with erythrocytes immobilizes both membrane lipids and membrane proteins, even if no lysis ensues.[126,127] This immobilization appears to be due to addition of lysophospholipid to the erythrocyte membrane.[62,128]

3. Cell Differentiation

Cell differentiation often results in changes in lateral mobility of membrane proteins and lipids, and many of these changes can be correlated with changes in membrane organization or composition upon differentiation. One nice example for lipids is the differentiation of pluripotential embryonal carcinoma cells to endoderm. The lateral diffusion of the lipid probe diI C16 is reduced about twofold upon cell differentiation, and this reduction correlates well with an increase in plasma membrane cholesterol.[129] On the other hand, D of a fluorescent fatty acid label increases twofold when cultured MDCK epithelial cells reach confluence and polarize functionally and morphologically.[130] The basis for this change is unknown.

Lipid probes also report heterogeneity and developmental changes in gametes and early embryos. The differentiation of spermatocytes and spermatids from round to condensed correlates with reductions in R of a diI probe.[131] This may reflect the shift to the complicated organization of mature sperm, which has been studied in some detail using FPR.[132]

Lipid mobility in the membranes of mollusk,[133] sea urchin,[134] *Xenopus*,[135] and mouse[136] eggs changes at fertilization. These changes also report regionalization of gamete membranes, evidence for membrane domains.

FPR detects changes in lateral mobility of membrane proteins when cells differentiate. The mobility (D) of proteins in the basal surface of MDCK cells is reduced, compared to the mobility of the same markers in the apical surface of unpolarized cells, when MDCK cells reach confluence.[137] R of the Na^+,K^+-ATPase is reported not to change upon cell differentiation;[66] but R of the same enzyme is reduced from 65 to 40% when chicken photoreceptor cells differentiate.[74] In another system investigated by FPR R of both lipid-anchored and peptide-anchored forms of the adhesion molecule N-CAM was greatly reduced, from ~90 to ~20% when cultured neuroblastoma cells differentiated.[175] Note that in no instance is all of a marker protein immobilized. This raises problems for the maintainence of the polarized phenotype which are beyond the scope of this review.

Some studies of cell aging show changes in D and R of lipid probes and of proteins labeled with succinyl concanavalin A.[138] Membrane lipid composition changes as cells age, changing membrane viscosity and ratios of phospholipid species to one another. However, control experiments on young cells, deliberate changes in fluidity or lipid composition, did not mimic results obtained with aged cells.

Other effects of aging, differentiation, and transformation on D and R have been noted by a number of authors.[42,139-144]

D. FPR AND MEMBRANE DOMAINS

If we begin with the assumption that the lipid bilayer of cell membranes is a fluid continuum, the assumption of the prevailing model of membranes,[145] then we expect that the chemical and physical properties of samples of the bilayer lipids will be the same in all regions of the membrane. Many different experiments suggest that this is not the case. Rather, membranes appear to be patchy; their lipids and proteins concentrate selectively in domains. We understand little about the mechanisms that create this patchiness, especially those that create and maintain lipid domains. The problems of[146] defining domains have recently been reviewed and this review is a good introduction to the general topic as are other reviews in the same volume[147] and elsewhere.[148,149]

The nature of membrane domains has been investigated by many physical and biochemical techniques (see the summary in Reference 150). Given the scope of this review, we can only deal with the use of FPR to detect and define membrane domains. Domains, defined as local heterogeneities in membranes, may occur on a size scale from a few nanometers, in the case of a cluster of glycolipids,[151] to many micrometers in the case of the apical surface of a polarized epithelial cell. FPR is suited to detection of domains on the scale of a fraction of a micrometer, to a few micrometers, the scale set by the area of the focused laser beam. This scale also allows use of FPR to probe mobility of membrane molecules within larger membranes domains, already visible at the light microscope level, e.g., regions of sperm, or on surfaces of epithelial cells. Here we will only deal with the uses of FPR to detect membrane heterogeneities that are not apparent by conventional microscopy.

1. Lipid Domains

If cell membrane bilayers are fluid lipid *continua*, then FPR measurements of lipid probes will show similar behavior for all probes and for all experimental geometries. Deviations from this result can be taken as evidence for heterogeneity of membrane environments for membrane domains. Factors other than membrane domains, e.g., association of some lipid probes, but not others with immobile membrane proteins, may contribute to the results. Therefore, it is well to keep in mind that membrane domains of a particular sort are defined by the conditions of the FPR experiment used to detect heterogeneity. Independent confirmation of the existence of domains defined by FPR should be made, whenever possible (see Reference 152 for confirmation by calorimetry and Reference 153 for confirmation of FPR results on domains by another optical technique).

Though FPR recovery curves are usually fit to one D and one R, it is possible to detect two different D and R in high quality data. If a single species of lipid probe gives recoveries best fit to two components, rather than one, we may argue that the probe resides in two different environments in the membrane. This approach was used to show that soybean protoplast membranes appear to contain lipid domains, one population of which is only probed by relatively high concentrations of NBD lipids and diI.[154] The two D fit to recovery curves of a number of probes differed tenfold. The smaller D, in the range 10^{-10} cm^2 s^{-1}, could reflect binding of the lipid probe to protein. However, similar experiments on 3T3 cells detected only one D even when using high concentrations of lipid probe. These sorts of results have been replicated in experiments on maize and rose protoplasts.[155,156]

A second approach to defining membrane domains is to compare the lateral diffusion of homologous lipid probes, differing only in acyl chain length or headgroup. This approach has been most successful using the diI series of probes (Figure 6), which are available with acyl chains from 10 to 22 carbons in length.[157] The phase partition preferences of the series of diIs determined in well characterized liposomes showed that C10 and C12 diIs prefered fluid phases, C18 prefered gel phases, and C14 and C16 had no phase preference. Armed with this information it was possible to use the series of diIs to show that membrane domains were present in sea urchin and mouse eggs, and that the composition/organization of these domains changed after fertilization.[134,136]

A third approach to detecting lipid domains is to compare the behavior of a single species of lipid probe at several different regions of what appears to be a continuous membrane. This approach has given indications of lipid domains in sperm[132,146] and in mollusk[133] and *Xenopus*[135] eggs.

A fourth approach to lipid domains is to compare the behavior of two different lipid probes in the same membrane. If the two probes behave differently, it is postulated that they are in different domains. This argument has been used to characterize domains in plant protoplasts,[158] and multicellular parasite membranes.[123,159]

An elegant variation on this approach was developed in which endogenous cell lipids were biosynthetically labeled by incoroporation of anthryl fatty acids, and their lateral diffusion in CHO cells was compared with that of exogenous probes.[160] The mobile fraction, R, of endogenous lipids was significantly lower than R of exogenous probes, indicating that some fraction of the endogenously synthesized lipids could incorporate into domains which were inaccessible to exogenous probes.

While all of the approaches described thus far are indirect, they have advantages over other methods for detecting membrane heterogeneity in that they can be used on small numbers of intact cells. However, they are only applicable to membrane lipids. We expect that D and R of membrane proteins will be heterogeneous, simply because of the many interactions between membrane proteins and their neighbors within and to either side of the bilayer.

2. Protein Domains

The approach to membrane proteins in domains, also applicable to membrane lipids, derives from the geometry of the spot FPR experiment itself, rather than from considerations of membrane architecture. Equation 1 described the relationship between half-time for recovery of fluorescence, D, and the area of the bleached spot. The half-time for recovery certainly will depend upon the area of the spot, but this is inserted as a constant in Equation 1. Otherwise, we expect that for most experimental geometries on a uniformly mixed bilayer D and R will be independent of spot area. The only major exception to this is in the case where the area of the spot approaches the area of the membrane being probed.[39,150] However, if the membrane being probed consists of domains which are stable for seconds to minutes, then R will depend upon the area of the bleached spot. R will be high for spots smaller than the average area of a domain, but will be much lower when using spots that can completely cover one or more domains (Figure 6). If all the area of a domain is bleached, then it cannot contribute to recovery of fluorescence.

Dependence of R on spot size has been shown for membrane lipids of mammalian cells[161] and of parasites[162] and has been used to argue for the presence of micrometer-scale domains in these membranes. In human fibroblasts R for the lipid analog NBD-PC varied with spot size, suggesting that these lipids were in domains, but R for another analog, diI C16, showed no dependence on spot size. Further details emerged from the analysis of D. For very small spots there was a distribution of D around two means, while for spots $>\sim 0.6$ μm only a single distribution was found. This could be interpreted in terms of lipid-rich and lipid-poor domains, with D of the lipid probe restricted by the high relative concentration of proteins in the lipid-poor domains.

R for the proteins of mammalian cells also shows a dependence on spot area. This was first demonstrated using a rabbit anti-human cell antibody on normal[161] and transformed (M. Edidin, unpublished) fibroblasts. A later study compared the distribution of peptide-anchored and GPI-anchored class I MHC molecules.[92] The two forms of the MHC molecule are >85% homologous, differing mainly in their

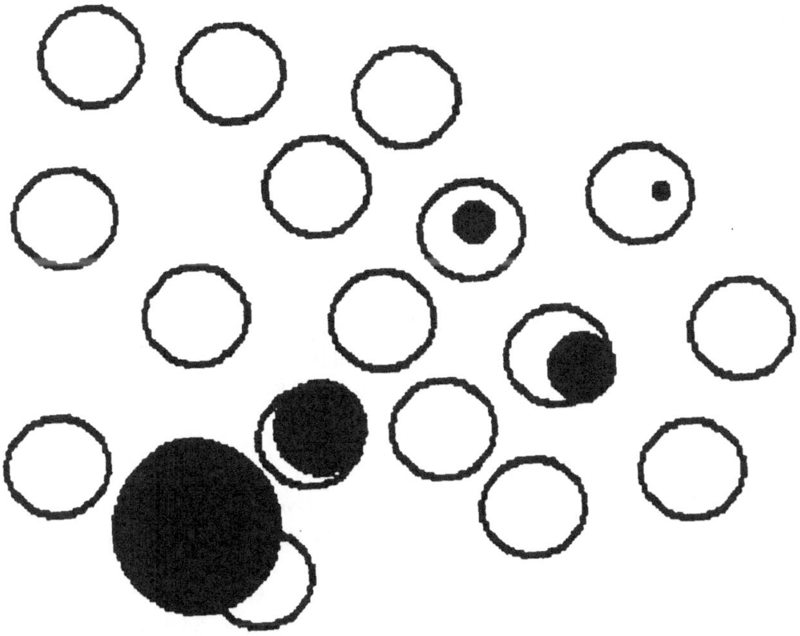

Figure 6. The bleaching spot and membrane domains. The open circles represent membrane domains. Molecules contained in these domains have a low probability of crossing their boundaries in the time required for an FPR measurement. It can be seen that small bleaching spots — filled circles — may bleach part of the area of a domain and so recovery of fluorescence will be nearly complete. On the other hand, large spots, such as those at the lower left of the figure, will bleach large fractions of individual domains and recovery will be correspondingly lower.[161]

anchoring sequences. While the conventionally anchored class I molecules were confined to membrane domains, the lipid-anchored molecules were able to cross the boundaries of domains. R for the latter was independent of spot size. This strongly suggested that the barriers to lateral movement of membrane proteins were in the cell cytoplasm or in the bilayer. Further analysis of this system by other methods indicated that the barriers are dynamic, with the barrier-free path of molecules increasing about sixfold for a 10°C increase in T.[153]

D of the markers analyzed in these experiments also varied unexpectedly with spot size. It was tenfold higher when measured with laser spots 1 μm or more in radius than when measured with smaller spots. This variation appears to reflect an underlying heterogeneity of D for the population of probed molecules. Molecules with large D will contribute to a very rapid fluorescence recovery that cannot be resolved when using a small laser spot. On the other hand, when using a large spot, molecules with low D will never arrive in the bleached area in the time of an experiment. We have been able to show this effect in 3D FPR measuring the D in a mixture of free fluorescein, mol wt ~350, and fluorescein IgM, mol wt ~900,000 Da, with D in pure solutions of each.[163]

Recently, the heterogeneity of D in FPR experiments has been considered theoretically under two different heads. An analysis of "long tail kinetics"[164] predicts the results on D and spot area actually obtained. Another approach analyzed FPR recovery curves in terms of fractal time tails[165] characterized in terms of a percolation exponent, α, in the expression for diffusion of a single molecule

$$\langle r^2 \rangle = \Gamma(\Delta t)^{\alpha} \tag{5}$$

where $\alpha < 1$.

VII. BEYOND FPR

The analysis of fractal time tails though it is applied to FPR arises from another approach to lateral mobility of membrane proteins, particle tracking. As we emphasised at the beginning of this review, FPR measures the properties of populations of labeled molecules. The behavior of thousands of individual molecules contributes to the recovery curve; FPR takes an average of this behavior. In recent years it has become possible to label single molecules or clusters of molecules with fluorescent LDL,[34,35,166,167] fluorescent plastic beads,[168] or gold particles[169-171] and to analyze the tracks swept out by the markers with time. The particles show a number of different sorts of motion, both Brownian and directed,[172] and the track of a single particle may be a mixture of these motions. In addition particles often appear to be percolating through a sea of obstacles, rather than random walking without hinderance. All of these motions contribute to FPR recovery curves. This realization has affected recent thinking about FPR measurements as summarized briefly in the preceeding section. The theory for analyzing particle tracks is just developing.[173,174]

As it develops, and as more versatile particle labels develop, we will learn a good deal more about the dynamics of cell membrane lipids and proteins. I expect that the experiments with this newer method will supplement, rather than supplant, FPR as a probe of cell surfaces.

REFERENCES

1. **Cuatrecasas, P.,** Membrane receptors, *Annu. Rev. Biochem.*, 43, 169, 1974.
2. **Tolkovsky, A. M. and Levitski, A.,** Mode of coupling between the β-adrenergic receptor and adenylate cyclase in turkey erythrocytes, *Biochemistry*, 17, 3795, 1978.
3. **Gupte, S., Wu, E.-S., Hoechli, L., Hoechli, M., Jacobson, K., Sowers, A. E., and Hackenbrock, C. R.,** Relationship between lateral diffusion, collision frequency and electron transfer of mitochondrial inner membrane oxidation-reduction components, *Proc. Natl. Acad. Sci. U.S.A.*, 81, 2606, 1984.
4. **Peters, R.,** Lateral mobility of proteins and lipids in the red cell membrane and the activation of adenylate cyclase by beta-adrenergic receptors, *FEBS Lett.*, 234, 1, 1988.
5. **Frye, L. D. and Edidin, M.,** The rapid intermixing of cell surface antigens after formation of mouse-human heterokaryons, *J. Cell Sci.*, 7, 319, 1970.
6. **Edidin, M. and Wei, T.,** Lateral diffusion of H-2 antigens on mouse fibroblasts, *J. Cell Biol.*, 95, 458, 1982.
7. **Liebman, P. A. and Entine, G.,** Lateral diffusion of pigment in photoreceptor disk membranes, *Science*, 185, 457, 1974.
8. **Poo, M.-M. and Cone, R. A.,** Lateral diffusion of rhodopsin in the photoreceptor membrane, *Nature (London)*, 265, 602, 1974.
9. **Peters, R., Peters, J., Tewes, K. H., and Bahr, W.,** A microfluorimetric study of translational diffusion in erythrocyte membranes, *Biochim. Biophys. Acta*, 367, 282, 1974.
10. **Jovin, T. M. and Vaz, W. L. C.,** Rotational and translational diffusion in membranes measured by fluorescence and phosphorescence methods, *Methods Enzymol.*, 172, 471, 1988.
11. **Matko, J., Szöllősi, J., Trón, L., and Damjanovich, S.,** Luminescence spectroscopic approaches in studying cell surface dynamics, *Q. Rev. Biophys.*, 21, 479, 1988.
12. **Wolf, D. E.,** Designing, building and using a fluorescence recovery after photobleaching instrument, *Methods Cell Biol.*, 30, 271, 1989.
13. **Zagyanksy, Y. and Edidin, M.,** Lateral diffusion of concanavalin A receptors in the plasma membrane of mouse fibroblasts, *Biochim. Biophys. Acta*, 433, 209, 1976.
14. **Yguerabide, J., Schmidt, J. A., and Yguerabide, E. E.,** Lateral mobility in membranes as detected by fluorescence recovery after photobleaching, *Biophys. J.*, 40, 69, 1982.
15. **Woda, B. A., Yguerabide, J., and Feldman, J.,** Mobility and density of AgB "Ia" and Fc receptors on the surface of lymphocytes from old and young rats, *J. Immunol.*, 123, 2161, 1979.
16. **Axelrod, D., Koppel, D. E., Schelssinger, J., Elson, E., and Webb, W. W.,** Mobility measurements by analysis of fluorescence photobleaching recovery experiments, *Biophys. J.*, 16, 1055, 1976.
17. **Barisas, B. G. and Leuther, M. D.,** Fluorescence photobleaching recovery measurement of protein absolute diffusion constants, *Biophys. Chem.*, 10, 221, 1979.
18. **Van Zoelen, E. J., Tertoolen, L. G. J., and DeLaat, S. W.,** Simple computer method for evaluation of lateral diffusion coefficients from fluorescent photobleaching recovery kinetics, *Biophys. J.*, 42, 103, 1983.
19. **Johansson, N., Sundqvist, T., and Magnusson, K. E.,** Regulation of the lateral diffusion of WGA-labeled glycoconjugates in human leukocytes. Comparison between adult granulocytes and differentiating promyelocytic HL60 cells, *Cell Biophys.*, 10, 233, 1987.

20. **Dubinsky, J. M., Loftus, D. J., Fischbach, G. D., and Elson, E. L.,** Formation of acetylcholine receptor clusters in chick myotubes: migration or new insertion?, *J. Cell Biol.*, 109, 1733, 1989.
21. **Mao, S.-Y., Varin-Blank, N., Edidin, M., and Metzger, H.,** Immobilization and internalization of mutated IgE receptors in transfected cells, *J. Immunol.*, 146, 958, 1991.
22. **Smith, B. A. and McConnell, H. M.,** Determination of molecular motion in membranes using periodic pattern photobleaching, *Proc. Natl. Acad. Sci. U.S.A.*, 75, 2759, 1978.
23. **Davoust, J., Devaux, P. F., and Leger, L.,** Fringe pattern photobleaching, a new method for the measurement of transport coefficients of biological molecules, *EMBO J.*, 1, 1233, 1982.
24. **Koppel, D. E.,** Redistribution after photobleaching. A new multipoint analysis of membrane translational dynamics, *Biophys. J.*, 28, 281, 1979.
25. **Koppel, D. E., Sheetz, M. P., and Schindler, M.,** Lateral diffusion in biological membranes. A normal-mode analysis of diffusion on a spherical surface, *Biophys. J.*, 30, 187, 1980.
26. **Koppel, D. E.,** Normal-mode analysis of lateral diffusion on a bounded membrane surface, *Biophys J.*, 47, 337, 1985.
27. **Kapitza, H. G., G. McGregor, G., and Jacobson, K. A.,** Direct measurement of lateral transport in membranes by using time-resolved spatial photometry, *Proc. Natl. Acad. Sci. U.S.A.*, 82, 4122, 1985.
28. **Duband, J. L., Nuckolls, G. H., Ishihara, A., Hasegawa, T., Yamada, K. M., Thiery, J. P., and Jacobson, K.,** Fibronectin receptor exhibits high lateral mobility in embryonic muscle cells but is immobile in focal contacts and fibrillar streaks in stationary cells, *J. Cell Biol.*, 107, 1385, 1988.
29. **Chattopadhyay, A.,** Chemistry and biology of N-(7-nitrobenz-2-oxa-1,3-diazol-4-yl)-labeled lipids: fluorescent probes of biological and model membranes, *Chem. Phys. Lipids*, 53, 1, 1990.
30. **Sims, P. J., Waggoner, A. S., Wang, C. H., and Hoffman, J. F.,** Studies on the mechanism by which cyanine dyes measure membrane potential in red blood cells and phosphatidylcholine vesicles, *Biochemistry*, 13, 3315, 1974.
31. **Axelrod, D.,** Cabrocyanine dye orientation in red cell membranes studied by microscopic fluorescence polarization, *Biophys. J.*, 26, 557, 1979.
32. **Chattopadhyay, A. and London, I.,** Parallax method for direct measurement of membrane penetration depth utilizing fluorescence quenching by spin-labeled phospholipids, *Biochemistry*, 26, 39, 1987.
33. **Edidin, M., Zagyansky, Y., and Lardner, T. J.,** Measurement of membrane protein lateral diffusion in single cells, *Science*, 191, 466, 1976.
34. **Barak, L. S. and Webb, W. W.,** Fluorescent low density lipoprotein for observation of dynamics of individual receptor complexes on cultured human fibroblast, *J. Cell Biol.*, 90, 595, 1981.
35. **Barak, L. S. and Webb, W. W.,** Diffusion of low density lipoprotein-receptor complex on human fibroblasts, *J. Cell Biol.*, 95, 846, 1982.
36. **Fowler, V. and Branton, D.,** Lateral mobility of human erythrocyte integral membrane proteins, *Nature (London)*, 268, 23, 1977.
37. **Golan, D. E. and Veatch, W.,** Lateral mobility of band 3 in the human erythrocyte membrane studies by fluorescence photobleaching recovery: evidence for control by cytoskeleton interactions, *Proc. Natl. Acad. Sci. U.S.A.*, 77, 2537, 1980.
38. **Schindler, M., Koppel, D. E., and Sheetz, M.,** Modulation of membrane protein lateral mobility by polyphosphates and polyamines, *Proc. Natl. Acad. Sci. U.S.A.*, 77, 1457, 1980.
39. **Wey, C.-L., Edidin, M. A., and Cone, R. A.,** Lateral diffusion of rhodopsin in photoreceptor cells by fluorescence photobleaching and recovery (FPR), *Biophys. J.*, 33, 225, 1981.
40. **Philips, E. and Cone, R. A.,** Do diffusion rates in the cytoplasm or the membrane limit the response speed of the vertebrate rod?, *Biophys. J.*, 49, 277a, 1986.
41. **Golan, D.,** Red blood cell membrane protein and lipid diffusion, in *Red Blood Cell Membranes*, Agre, P. and Parker, J. C., Eds., Marcel Dekker, New York, 1989, 367.
42. **Corbett, J. D. and Golan, D. E.,** Band 3 and glycophorin are progressively aggregated in density-fractionated sickle and normal red blood cells, *J. Clin. Invest.*, 91, 208, 1993.
43. **Schlessinger, J., Koppel, D. E., Axelrod, D., Jacobson, K., Webb, W. W., and Elson, E. L.,** Lateral transport on cell membranes: mobility of concanavalin A receptors on myoblasts, *Proc. Natl. Acad. Sci. U.S.A.*, 73, 2409, 1976.
44. **Edidin, M.,** Fluorescent labeling of cell surfaces, *Methods Cell Biol.*, 29, 87, 1989.
45. **Chasis, J., Mohandas, N., and Shohet, S. B.,** Erythrocyte membrane rigidity induced by glycophorin A-ligand interaction. Evidence for a ligand-induced association between glycophorin A and skeletal proteins, *J. Clin. Invest.*, 75, 1919, 1985.

46. **Chasis, J. A., Reid, M. E., Jensen, R. H., Mohandas, N., and Martinicic, K.,** Signal transduction by glycophorin A: role of extracellular and cytoplasmic domains in a modulatable process, *J. Cell Biol.*, 107, 1351, 1988.
47. **Bierer, B., Hermann, S. H., Brown, C. S., Burakoff, S. J., and Golan, D.,** Lateral mobility of class I histocompatibility antigens in B-lymphoblastoid cell membranes: modulation by cross-linking and effect of cell density, *J. Cell Biol.*, 105, 1147, 1985.
48. **Zhang, F., Crise, B., Su, B., Hou, Y., Rose, J. K., Bothwell, A., and Jacobson, K.,** Lateral diffusion of membrane-spanning and glycosylphosphatidylinositol-lonked proteins: toward establishing rules governing the lateral mobility of membrane proteins, *J. Cell Biol.*, 115, 75, 1991.
49. **Chakrabarti, A., Matko, J., Rahman, N. A., Barisas, B. G., and Edidin, M.,** Self-association of class I major histocompatiility complex molecules in liposome and cell surface membranes, *Biochemistry*, 31, 7182, 1992.
50. **Capps, G. G., Robinson, B. E., Lewis, K. D., and Zuniga, M. C.,** *In vivo* dimeric association of class I MHC heavy chains: possible relationship to class I MHC molecule dissociation from antigenic peptide and β_2-microglobulin, *J. Immunol.*, 151, 159, 1993.
51. **Saffman, P. G. and Delbruck, M.,** *Proc. Natl. Acad. Sci. U.S.A.*, 72, 3111, 1975.
52. **Hillman, G. M. and Schlessinger, J.,** Lateral diffusion of epidermal growth factor complexed to its surface receptors does not account for the thermal sensitivity of patch formation and endocytosis, *Biochemistry*, 21, 1667, 1982.
53. **Angelides, K.,** Fluorescent analogs of toxins, *Methods Cell Biol.*, 29, 29, 1989.
54. **Angelides, K. J., Elmer, L. W., Loftus, D., and Elson, E.,** Distribution and lateral mobility of voltage-dependent sodium channels in neurons, *J. Cell Biol.*, 106, 1911, 1988.
55. **Johansson, B., Wymann, M. P., Holmgren-Peterson, K., and Magnusson, K.-E.,** N-formyl peptide receptors in human neutrophils display distinct membrane distribution and lateral mobility when labeled with agonist and antagonist, *J. Cell Biol.*, 121, 1281, 1993.
56. **Shechter, Y., Schlessinger, J., Jacobs, S., Chang, K.-J., and Cuatrecasas, P.,** Fluorescent labeling of hormone receptors in viable cells: preparation and properties of highly fluorescent derivatives of epidermal growth factor and insulin, *Proc. Natl. Acad. Sci. U.S.A.*, 75(5), 2135, 1978.
57. **Salas, P. J. I., Vega-Salas, D. E., Hochman, J., Rodriguez-Boulan, E., and Edidin, M.,** Selective anchoring in the specific plasma membrane domain: a role in epithelial cell polarity, *J. Cell Biol.*, 107, 2363, 1988.
58. **Hochman, J. H., Schindler, M., Lee, J. G., and Ferguson-Miller, S.,** Lateral mobility of cytochrome c on intact mitochondrial membranes as determined by fluorescence redistribution after photobleaching, *Proc. Natl. Acad. Sci. U.S.A.*, 79, 6866, 1982.
59. **Gupte, S.,** Multidimensional diffusion modes and collision frequencies of cytochrome c with its redox partners, *J. Biol. Chem.*, 263, 5241, 1988.
60. **Maxfield, F.,** Fluorescent analogs of peptides and hormones, *Methods Cell Biol.*, 29, 13, 1989.
61. **Axelrod, D.,** Cell surface heating during photobleaching recovery experiments, *Biophys. J.*, 18, 129, 1976.
62. **Bloom, J. A. and Webb, W. W.,** Photodamage to intact erythrocyte membranes at high laser intensities: methods of assay and suppression, *J. Histochem. Cytochem.*, 32, 608, 1984.
63. **Sheetz, M. P. and Koppel, D. E.,** Membrane damage caused by irradiation of fluorescent concanavalin A, *Proc. Natl. Acad. Sci. U.S.A.*, 76, 3314, 1979.
64. **Jacobson, K., Hou, Y., and Wojieszyn, J.,** Evidence for lack of damage during photobleaching measurements of the lateral mobility of cell surface components, *Exp. Cell Res.*, 116, 178, 1978.
65. **Wolf, D. E., Edidin, M., and Dragsten, P. R.,** Effect of bleaching light on measurements of lateral diffusion in cell membranes by the fluorescence photobleaching recovery method, *Proc. Natl. Acad. Sci. U.S.A.*, 77, 204, 1980.
66. **Leslie, R. J., Saxton, W. M., Mitchison, T. J., Neighbors, B., Salmon, E. D., and McIntosh, J. R.,** Assembly properties of fluorescein-labeled tubulin *in vitro* before and after fluorescence bleaching, *J. Cell Biol.*, 99, 2146, 1984.
67. **Simon, J. R., Gough, A., Urbanik, E., Wang, F., Lanni, F., Ware, B. R., and Taylor, D. L.,** Analysis of rhodamine and fluorescein-labeled F-actin diffusion *in vitro* by fluorescence photobleaching recovery, *Biophys. J.*, 54, 801, 1988.
68. **McIntosh, J. R. and Koonce, M. P.,** Mitosis, *Science*, 246, 622, 1989.
69. **Criado, M., Vaz, W. L. C., Barrantes, F. J., and Jovin, T. M.,** Translational diffusion of acetylcholine receptor (monomeric and dimeric forms) of *Torpedo marmorata* reconstituted into phospholipid bilayers studied by fluorescence recovery after photobleaching, *Biochemistry*, 21, 5750, 1982.

70. **Peters, R. and Cherry, R. J.,** Lateral and rotational diffusion of bacteriorhodopsin in lipid bilayers: experimental test of the Saffman-Delbruck equations, *Proc. Natl. Acad. Sci. U.S.A.*, 79, 4317, 1982.
71. **Vaz, W. L. C., Clegg, R. M., and Hallmann, D.,** Translational diffusion of lipids in liquid crystalline phase phosphatidylcholine multibilayers. A comparison of experiment with theory, *Biochemistry*, 24, 781, 1985.
72. **Vaz, W. L. C. and Criado, M. A.,** comparison of the translational diffusion of a monomer and an oligomer of the acetylcholine receptor protein reconstituted into soybean lipid bilayers, *Biochim. Biophys. Acta*, 819, 18, 1985.
73. **Gennis, R. B.,** *Biomembranes*, Springer-Verlag, New York, 1989, chap. 5.
74. **Madreperla, S. A., Edidin, M., and Adler, R.,** Na+,K+-adenosine triphosphatase polarity in retinal photoreceptors: a role for cytoskeletal attachments, *J. Cell Biol.*, 109, 1483, 1989.
75. **Weis, R. E., Roberts, W. M., Stuhmer, W., and Almers, W.,** Mobility of voltage-dependent ion channels and lectin receptors in the sarcolemma of frog skeletal muscle, *J. Gen. Physiol.*, 87, 955, 1986.
76. **Damjanovich, S., Tron, L., Szollosi, J., Zidovetski, R., Vaz, W. L. C., and Jovin, T. M.,** Distribution and mobility of murine histocompatibility H-2^k antigen in the cytoplasmic membrane, *Proc. Natl. Acad. Sci. U.S.A.*, 80, 5985, 1983.
77. **Wier, M. L. and Edidin, M.,** Constraint of the translational diffusion of a membrane glycoprotein by its external domains, *Science*, 242, 412, 1988.
78. **Mecheri, S., Edidin, M., Dannecker, G., and Hoffman, M. K.,** Immunogenic Ia-binding peptides immobilize the Ia molecule and facilitate its aggregation on the B-cell membrane. Control by the Mls-1 gene, *J. Immunol.*, 144, 1361, 1990.
79. **Henis, Y. I. and Gutman, O.,** Lateral diffusion and patch formation of H-$2K^k$ antigens on mouse spleen lymphocytes, *Biochim. Biophys. Acta*, 762, 281, 1983.
80. **Henis, Y.,** Mobility modulation by local concanavalin A binding, *J. Biol. Chem.*, 259, 1515, 1984.
81. **Menon, A. K., Holowka, D., Webb, W. W., and Baird, B.,** Clustering, mobility and triggering activity of small oligomers of immunoglobulin E on rat basophilic leukemia cells, *J. Cell Biol.*, 102, 534, 1985.
82. **Edidin, M.,** Aspects of plasma membrane fluidity, in *Membrane Research*, Fox, C. F., Ed., Academic Press, New York, 1972, 15.
83. **Sheetz, M., Schindler, M., and Koppel, D.,** Lateral mobility of integral membrane proteins is increased on spherocytic erythrocytes, *Nature (London)*, 285, 510, 1980.
84. **Golan, D.,** Red blood cell membrane protein and lipid diffusion, in *Red Blood Cell Membranes*, Agre, P. and Parker, J. C., Eds., Marcel Dekker, New York, 1989, 367.
85. **Tsuji, A. and Ohnishi, S.,** Restriction of the lateral motion of band 3 in the erythrocyte membrane by the cytoskeletal network: dependence on spectrin association state, *Biochemistry*, 25, 6133, 1986.
86. **Shields, M., La Celle, P., Waugh, R. E., Scholz, M., Peters, R., and Passow, H.,** *Biochim. Biophys. Acta*, 905, 181, 1987.
87. **Gupta, S. K. and Woda, B. A.,** Ligand-induced association of surface immunoglobulin with the detergent insoluble cytoskeleton may involve alpha-actinin, *J. Immunol.*, 140, 176, 1988.
88. **Wade, W. F., Freed, J. H., and Edidin, M.,** Translational diffusion of class II major histocompatibility complex molecules is constrained by their cytoplasmic domains, *J. Cell Biol.*, 109, 3325, 1989.
89. **Tank, D., Wu, E.-S., and Webb, W. W.,** Enhanced molecular diffusibility in muscle membrane blebs: release of lateral constraints, *J. Cell Biol.*, 92, 207, 1982.
90. **Scullion, B. F., Hou, Y., Puddington, L., Rose, J. K., and Jacobson, K.,** Effects of mutations in three domains of the vesicular stomatitis virus viral glycoprotein on its lateral diffusion in the plasma membrane, *J. Cell Biol.*, 105, 69, 1987.
91. **Edidin, M. and Zuniga, M.,** Lateral diffusion of wild-type and mutant L^d antigens in L cells, *J. Cell Biol.*, 99, 2333, 1984.
92. **Edidin, M. and Stroynowski, I.,** Differences between the lateral organization of conventional and inositol phospholipid-anchored membrane proteins. A further definition of micrometer scale membrane domains, *J. Cell Biol.*, 112, 1143, 1991.
93. **Livneh, E., Benveniste, M., Prywes, R., Felder, S., Kam, Z., and Schlessinger, J.,** Large deletions in the cytoplasmic kinase domain of the EGF-receptor do not affect its lateral mobility, *J. Cell Biol.*, 103, 327, 1986.
94. **Saxton, M.,** Lateral diffusion in an archipelago: effects of impermeable patches on lateral diffusion in a cell membrane, *Biophys. J.*, 39, 165, 1982.

95. **Saxton, M.,** Lateral diffusion in an archipelago. Distance dependence of the diffusion coefficient, *Biophys. J.*, 56, 615, 1989.
96. **Saxton, M. J.,** Lateral diffusion and aggregation. A Monte Carlo study, *Biophys. J.*, 61, 119, 1992.
97. **Thompson, T. E., Sankaram, M. B., and Biltonen, R.,** Biological membrane domains: functional significance, *Comments Mol. Cell Biophys.*, 8, 1, 1992.
98. **Melo, E. C., Lourtie, I. M., Sankaram, M. B., Thompson, T. E., and Vaz, W. L.,** Effects of domain connection and disconnection on the yields of in-plane bimolecular reactions in membranes, *Biophys. J.*, 63, 1506, 199, 1992.
99. **Happel, S. and Brenner, H.,** *Low Reynolds Number Hydrodynamics*, Prentice-Hall, Englewood Cliffs, NJ, 1965, chap. 9.
100. **Ford, T. F.,** Viscosity-concentration and fluidity-concentration relationships for suspensions of spherical particles in Newtonian fluids, *J. Phys. Chem.*, 64, 1168, 1960.
101. **Edidin, M.,** Rotational and translational diffusion in membranes, *Annu. Rev. Biophys. Bioeng.*, 3, 179, 1974.
102. **Pink, D.,** Protein lateral movement in lipid bilayers: simulation studies of its dependence upon protein concentration, *Biochim. Biophys. Acta*, 818, 200, 1985.
103. **Minton, A.,** Lateral diffusion of membrane proteins in protein-rich membranes. A simple hard particle model for concentration dependence of the two-dimensional diffusion coefficient, *Biophys. J.*, 55, 805, 1989.
104. **Abney, J. R. and Scallettar, B. A.,** Molecular crowding and protein organization in biological membranes, in *Thermodynamics of Membrane Receptors and Channels*, Jackson, M. B., Ed., CRC Press, Boca Raton, FL, 1993.
105. **Tank, D. W., Wu, E.-S., Meers, P. R., and Webb, W. W.,** Lateral diffusion of gramicidin C in phospholipid bilayers, *Biophys. J.*, 40, 129, 1982.
106. **Benveniste, M., Livneh, E., Schlessinger, J., and Kam, Z.,** Overexpression of epidermal growth factor receptor in NIH- 3T3-transfected cells slows its lateral diffusion and rate of endocytosis, *J. Cell Biol.*, 106, 1903, 1988.
107. **Thomas, J. L., Feder, T. J., and Webb, W. W.,** Effects of protein concentration on IgE receptor mobility in rat basophilic leukemia cell plasma membranes, *Biophys. J.*, 61, 1402, 1992.
108. **Stanley, P.,** Glycosylation mutants of animal cells, *Annu. Rev. Genet.*, 18, 525, 1984.
109. **Barbour, S. and Edidin, M.,** Cell-specific constraints to the lateral diffusion of a membrane glycoprotein, *J. Cell. Physiol.*, 150, 526, 1992.
110. **Fahnestock, M. L., Tamir, I., Narhi, L., and Bjorkman, P. J.,** Thermal stability comparison of purified empty and peptide-filled forms of a class I MHC molecule, *Science*, 258, 1658, 1992.
111. **Lee, G. M., Zhang, F., Ishihara, A., McNeil, C. L., and Jacobson, K. A.,** Unconfined lateral diffusion and an estimate of pericellular matrix viscosity revealed by measuring the mobility of gold-tagged lipids, *J. Cell Biol.*, 120, 25, 1993.
112. **Zhang, F., Lee, G. M., and Jacobson, K.,** Protein lateral mobility as a reflection of membrane microstructure, *BioEssays*, in press.
113. **Sheetz, M. P.,** Glycoprotein mobility and dynamic domains in fluid plasma membranes, *Annu. Rev. Biophys. Biomol. Struct.*, 22, 417, 1993.
114. **Gross, W. and Lohse, M. J.,** Mechanism of activation of A_2 adenosine receptors. II. A restricted collision-coupling model of receptor-effector interaction, *Mol. Pharmacol.*, 39, 524 1991.
115. **Saxton, M. J. and Owicki, J. C.,** Concentration effects on reactions in membrane: rhodopsin and transducin, *Biochim. Biophys. Acta*, 979, 27, 1989.
116. **Rees, A. R., Gregoriou, M., Johnson, P., and Garland, P. B.,** High afffinity epidermal growth factor receptors on the surface of A431 cells have restricted lateral diffusion, *EMBO J.*, 3, 1843, 1984.
117. **Venkatakrishnan, G., McKinnon, C. A., Pilapil, C. G., Wolf, D. E., and Ross, A. H.,** Nerve growth factor receptors are preaggregated and immobile on responsive cells, *Biochemistry*, 30, 2748, 1991.
118. **Jans, D. A.,** The mobile receptor hypothesis revisited: a mechanistic role for hormone receptor lateral mobility in signal transduction, *Biochim. Biophys. Acta*, 1113, 271, 1992.
119. **Jesaitis, A. J., Tolley, J. O., Bokoch, G. M., and Allen, R. A.,** Regulation of chemoattractant receptor interaction with transducing proteins by organizational control in the plasma membrane of human neutrophils, *J. Cell Biol.*, 109, 2783, 1989.

120. **Ljungquist-Hoddelius, P., Lirvall, M., Wasteson, A., and Magnusson, K.-E.,** Lateral diffusion of pdgfβ-receptors in human fibroblasts, *Biosci. Rep.*, 11, 43, 1991.
121. **Jesaitis, A. J. and Allen, R. A.,** Activation of the neutrophil respiratory burst by chemoattractants: regulation of the N-formyl peptide receptor in the plasma membrane, *J. Bioenerg. Biomembr.*, 20, 679, 1988.
122. **Zlatanov, I. V., Foley, M., Birmingham, J., and Garland, P. B.,** Effect of some polypeptide hormones on the lateral mobility in plasma membranes of rat testes, L-cells and liver. Influence of development and aging, *Int. J. Biochem.*, 21, 1269, 1989.
123. **Sands, W. A. and Kusel, J. R.,** Changes in the lateral diffusion of fluorescent lipid analogues in the surface membrane of adult male *Schistosoma mansoni*, *Mol. Biochem. Parasitol.*, 53, 233, 1992.
124. **Liu, Z.-Y., Sanders, M. E., and Hu, V. W.,** Effect of complement on the lateral mobility of erythrocyte membrane proteins. Evidence for terminal complex interaction with cytoskeletal components, *J. Immunol.*, 142, 2370, 1989.
125. **Aroeti, B., Gutman, O., and Henis, Y. I.,** Transient alterations in the lateral mobility of erythrocyte membrane components during Sendai virus-mediated fusion, *J. Biol. Chem.*, 267, 13272, 1992.
126. **Golan, D. E., Brown, C. S., Cianci, C. M. L., Furlong, S. T., and Caulfield, J. P.,** Schistosomula of *Schistosoma mansoni* use lysophosphatidylcholine to lyse adherent red blood cells and immobilize red cell membrane components, *J. Cell Biol.*, 103, 819, 1986.
127. **Thatte, H., Kasschau, M. R., Furlong, S. T., Byam-Smith, M. P., Williams, D. F., and Golan, D. E.,** *Schistosoma mansoni*: membranes from adult worms reversibly perturb shape, volume and membrane organization of intact human red blood cells, *Exp. Parasitol.*, 76, 13, 1993.
128. **Golan, D. E., Furlong, S. T., Brown, C. S., and Caulfield, J. P.,** Monopalmitoylphosphatidylcholine incorporation into human erythrocyte ghost membrane causes protein and lipid immobilization and cholesterol depletion, *Biochemistry*, 27, 2661, 1988.
129. **Searls, D. B. and Edidin, M.,** Lipid compositon and lateral diffusion in plasma membranes of teratocarcinoma-derived cell lines, *Cell*, 24, 511, 1981.
130. **Jesaitis, A. J. and Yguerabide, J.,** The lateral mobility of the (Na^+,K^+)-dependent ATPase in Madin-Darby canine kidney cells, *J. Cell Biol.*, 102, 1256, 1986.
131. **Wolf, D. E., Scott, B. K., and Millette, C. F.,** The development of regionalized lipid diffusibility in the germ cell plasma membrane during spermatogenesis in the mouse, *J. Cell Biol.*, 103, 1745, 1986.
132. **Wolf, D. E.,** Corraling the drunken beggar, *BioEssays*, 6, 116, 1986.
133. **Speksnijder, J. E., Dohmen, M. R., Tertoolen, L. G. J., and De Laat, S. W.,** Regional differences in the lateral mobility of plasma membrane lipids in a molluscan embryo, *Dev. Biol.*, 110, 207, 1985.
134. **Wolf, D. E., Kinsey, W., Lennarz, W., and Edidin, M.,** Changes in the organization of the sea urchin egg plasma membrane upon fertilization: indications from lateral diffusion rates of lipid-soluble fluorescent dyes, *Dev. Biol.*, 81, 133, 1981.
135. **Dictus, W. J. A. G., van Zoelen, E. J. J., Tetteroo, P. A. T., Tertoolen, L. G. J., DeLaat, S. W., and Bluemink, J. G.,** Lateral mobility of plasma membrane lipids in *Xenopus* eggs: regional differences related to animal/vegetal polarity become extreme upon fertilization, *Dev. Biol.*, 101, 201, 1984.
136. **Wolf, D. E., Edidin, M. A., and Handyside, A. H.,** Changes in the organization of the mouse egg plasma membrane upon fertilization and first cleavage: indications from the lateral diffusion rates of fluorescent lipid analogs, *Dev. Biol.*, 85, 195, 1981.
137. **Salas, P. J., Vega-Salas, D. E., Hochman, J., Rodriguez-Boulan, E., and Edidin, M.,** Selective anchoring in the specific plasma membrane domain: a role in epithelial cell polarity, *J. Cell Biol.*, 107, 2363, 1988.
138. **Yechiel, E., Barenholz, Y., and Henis, Y. I.,** Lateral mobility and organization of phospholipids and proteins in rat myocyte membranes, *J. Biol. Chem.*, 260, 9132, 1985.
139. **Magnusson, K.-E., Gustafsson, M., Holmgren, K., and Johansson, B.,** Small intestinal differentiation in human colon carcinoma HT29 cells has distinct effects on the lateral diffusion of lipids (ganglioside GM_1) and proteins (HLA class 1, HLA class 2, and neoplastic epithelial antigens) in the apical cell membrane, *J. Cell. Physiol.*, 143, 381, 1990.
140. **Ohta, M. and Kitani, K.,** Age-dependent decrease in the hepatic uptake of taurocholic acid resembles that for ouabain. A possible role for surface membrane protein mobility, *Biochem. Pharmacol.*, 39, 1223, 1990.

141. **Kawasaki, Y., Zundel, J.-L., Miyakawa, T., and Sakaguchi, K.,** Reduced lateral mobility of a fluorescent lipid analog in cell membranes of rat fibroblasts transformed by simian virus 40, *FEBS Lett.*, 246, 65, 1989.
142. **Tournier, J. F., Lopez, A., Gas, N., and Tocanne, J. F.,** The lateral motion of lipid molecules in the apical plasma membrane of endothelial cells is reversibly affected by the presence of cell junctions, *Exp. Cell Res.*, 181, 375, 1989.
143. **Cohen, E., Ophir, I., Henis, Y. I., Bacher, A., and Ben-Shaul, Y.,** Effect of temperature on the assembly of tight junctions and on the mobility of lipids in membranes of HT29 cells, *J. Cell Sci.*, 97, 119, 1990.
144. **Boullier, J. A., Peacock, J. S., Roess, D. A., and Barisas, B. G.,** Protein and lipid lateral diffusion in normal and Rous sarcoma virus transformed chick embryo fibroblasts, *Biochim. Biophys. Acta*, 1107, 193, 1992.
145. **Singer, S. J. and Nicolson, G.,** The fluid mosaic model of the structure of cell membranes, *Science*, 175, 720, 1972.
146. **Wolf, D. E.,** Lipid domains: the parable of the blind men and the elephant, *Comments Mol. Cell. Biophys.*, 8, 83, 1992.
147. **Jacobson, K. and Vaz, W. L. C.,** Eds., Domains in biological membranes, *Comments Mol. Cell. Biophys.*, 8, 1992.
148. **Tocanne, J.-F., Dupou-Cezanne, L., Lopez, A., and Tournier, J.-F.,** Lipid lateral diffusion and membrane organization, *FEBS Lett.*, 257, 10, 1989.
149. **Edidin, M.,** Patches, posts and fences: proteins and plasma membrane domains, *Trends Cell Biol.*, 2, 376, 1992.
150. **Tocanne, J. F.,** Detection of lipid domains in biological membranes, *Comments Mol. Cell. Biophys.*, 8, 53, 1992.
151. **Thompson, T. E. and Tillack, T. W.,** Organization of glycosphingolipids in bilayers and plasma membranes of mammalian cells, *Annu. Rev. Biophys. Bioeng.*, 14, 361, 1985.
152. **Wolf, D. E., Maynard, V. M., McKinnon, C. A., and Melchior, D. L.,** Lipid domains in the ram sperm plasma membrane demonstrated by differential scanning calorimetry, *Proc. Natl. Acad. Sci. U.S.A.*, 87, 6893, 1990.
153. **Edidin, M., Kuo, S. C., and Sheetz, M. P.,** Lateral movements of membrane glycoproteins restricted by dynamic cytoplasmic barriers, *Science*, 254, 1379, 1991.
154. **Metcalf, T. N., Wang, J. L., and Schindler, M.,** Lateral diffusion of phospholipids in the plasma membrane of soybean protoplasts: evidence for membrane lipid domains, *Proc. Natl. Acad. Sci. U.S.A.*, 83, 95, 1986.
155. **Dugas, C. M., Li, Q., Khan, I. A., and Nothnagel, E. A.,** Lateral diffusion in the plasma membrane of maize protoplasts with implications for cell culture, *Planta*, 179, 387, 1989.
156. **Walko, R. M. and Nothnagel, E. A.,** Lateral diffusion of proteins and lipids in the plasma membrane of rose protoplast, *Protoplasma*, 152, 46, 1989.
157. **Klausner, R. D. and Wolf, D. E.,** Selectivity of fluorescent lipid analogues for lipid domains, *Biochemistry*, 19, 6199, 1980.
158. **Furtula, V., Khan, I. A., and Nothnagel, E. A.,** Selective osmotic effect on diffusion of plasma membrane lipis of maize protoplasts, *Proc. Natl. Acad. Sci. U.S.A.*, 87, 6532, 1990.
159. **Proudfoot, L., Kusel, J. R., Smith, H. V., Harnett, W., Worms, M. J., and Kennedy, M. W.,** The surface lipid of parasitic nematodes: organization and modifications during transition to the mammalian host environment, *Acta Tropica*, 47, 323, 1990.
160. **Dupou, L., Lopez, A., and Tocanne, J. F.,** Comparative study of the lateral motion of extrinsic probes and anthracene-labelled constituitive phospholipids in the plasma membrane of Chinese hamster ovary cells, *Eur. J. Biochem.*, 171, 669, 1988.
161. **Yechiel, E. and Edidin, M.,** Micrometer scale domains in fibroblast plasma membranes, *J. Cell Biol.*, 105, 755, 1987.
162. **Caulfield, J. P., Chiang, C.-P., Yacono, P. W., Smith, L. A., and Golan, D. E.,** Low density lipoproteins bound to *Schistosoma mansoni* do not alter the rapid lateral diffusion or shedding of lipids in the outer surface membrane, *J. Cell Sci.*, 99, 167, 1991.
163. **Edidin, M.,** Lateral mobility of membrane proteins — a journey from heterokaryons to laser tweezers, in *The Legacy of Cell Fusion*, Gordon, S., Ed., Oxford University Press, in press.
164. **Nagle, J.,** Long tail kinetics in biophysics?, *Biophys. J.*, 63, 366, 1992.

165. **Brust-Mascher, I., Feder, T. J., Slattery, J. B., Baird, B., and Webb, W. W.**, Constrained diffusion or immobile fraction on the cell surface: a new interpretation, *Mol. Biol. Cell*, 3, 306a, 1992.
166. **Ghosh, R. N. and Webb, W. W.**, Evidence for intra-membrane constraints to cell surface LDL receptor motion, *Biophys. J.*, 57, 286a.
167. **Anderson, C. M., Gerogiou, G. N., Morrison, I. E. G., Stevenson, G. V., and Cherry, R. J.**, Tracking of cell surface receptors by fluorescence digital imaging microscopy usign a CCD camera. Low density lipoprotein and influenza virus mobility at 4°C, *J. Cell Sci.*, 101, 415, 1992.
168. **Fein, M., Chuang, F., Sassaroli, M., DaCosta, R., Väänänen, H., Eisinger, J., and Unkeless, J.**, Mobility of membrane components determined by recording trajectories of tethered fluorescent microbeads, *Biophys. J.*, 64, A75, 1993.
169. **Sheetz, M. P., Turney, S., Qian, H., and Elson, E. L.**, Nanometre-level analysis demonstrates that lipid flow does not drive membrane glycoprotein movements, *Nature (London)*, 340, 284, 1989.
170. **De Brabander, M., Nuydens, R., Ishihara, A., Holifield, B., Jacobson, K., and Geertz, H.**, Lateral diffusion and retrograde movements of individual cell surface components on single motile cells observed with nanovid microscopy, *J. Cell Biol.*, 112, 111, 1991.
171. **Lee, G., Ishihara, A., and Jacobson, K. A.**, Direct observation of Brownian motion of lipids on a membrane, *Proc. Natl. Acad. Sci. U.S.A.*, 88, 6274, 1991.
172. **Haft, D. and Edidin, M.**, Modes of particle transport, *Nature (London)*, 340, 262, 1989.
173. **Saxton, M. J.**, Lateral diffusion in an archipelago: single particle diffusion, *Biophys. J.*, 64, 1766, 1993.
174. **Cherry, R. J.**, How to detect nonrandom motion of proteins in membranes, *Biophys. J.*, 64, 1651, 1993.
175. **Pollerberg, G. E., Schachner, M., and Davoust, J.**, Differentiation state-dependent surface mobilities of two forms of the neural cell adhesion molecule, *Nature (London)*, 324, 463, 1986.
176. **Fire, E., Zwart, D. E., Roth, M. G., and Henis, Y. I.**, Evidence from lateral mobility studies for dynamic interactions of a mutant influenza hemagglutinin with coated pits, *J. Cell Biol.*, 115, 1585, 1991.

Chapter 4

Fluidity/Viscosity of Biological Membranes

János Szöllősi

CONTENTS

0-8493-4931-1/94/$0.00+$.50

I. INTRODUCTION

The structure of the cytoplasmic membrane has been the focus of interest in cell biology. Membranes are responsible for a wide variety of biological, biochemical, and biophysical processes including respiration, vision, photosynthesis, cell-cell recognition, and nerve impulse transmission. Not surprisingly, then, considerable efforts have been spent over the years in trying to characterize molecular structures of the individual membrane components and their intermolecular interactions in an attempt to relate the structure of membranes to their function. One of the most successful models of membrane structures was introduced by Singer and Nicolson in 1972.[1,2] Their fluid mosaic model regarded the membranes as two-dimensional solutions of oriented lipids and globular proteins, the lipids being the solvents and the proteins. According to this model the lipids and proteins are free to diffuse and, as a consequence, are randomly distributed in the cell membrane. In addition, the lipid phase can be characterized simply with the term "fluidity".

In fact, membrane organization is certainly much more complex. The current concept of biological membranes is a dynamic molecular assembly characterized by the coexistence of structures with highly restricted mobility and components having great rotational freedom. The major constituents of membranes, such as proteins, lipids, and polysaccharides are in dynamic interaction with each other.[3] In functionally polarized cells (spermatozoa, eggs, epithelia, etc.), the membrane lipids and proteins are organized into domains whose protein and lipid composition differ from the bulk of the membrane.[4,5] Recent evidence suggests, that this type of microcompartmentalization can be observed on other cell types as well, and the size of these domains is on a micrometer scale.[6,7] The heterogeneity of the membrane questions the applicability of the term "fluidity" as a bulk, uniform property of the lipid phase.

Nevertheless, in this chapter every attempt is made to define the "fluidity" of biological membranes. The static and dynamic properties of the membrane components will be related to the order parameter and the microviscosity. Various types of measuring techniques are available for the investigation of model and biological membranes such as spectroscopic techniques, differential scanning calorimetry, electron microcopy, permeability measurements, and membrane deformability measurements. With the exception of the spectroscopic techniques, the above-mentioned physical measurements give essentially macroscopic information. Therefore this review will be limited to spectroscopic approaches, providing valuable information about structure and dynamics of membranes. Among the spectroscopic techniques nuclear magnetic resonance (NMR) and electron paramagnetic resonance (EPR) are mentioned briefly, while fluorescence is discussed in detail. The proposed models for the interpretation of the results of fluorescence measurements are also summarized. A critical comparison is made between various types of time-resolved and steady-state fluorescence measurements. The spectral and other physical characteristics of generally applied fluorescent probes are also described. This review summarizes our knowledge about the location of these probes in the lipid bilayers and describes how these probes can be used for characterization of membrane dynamics and structure. Among the numerous experimental approaches special attention is paid to instruments where live cells can be analyzed, since this review focuses on the biological membrane. The latest developments of flow cytometry and image microscopy are presented in relation to membrane fluidity measurement through specific examples. After discussing the conceptual and experimental details of membrane fluidity measurements the scope of this review is limited to

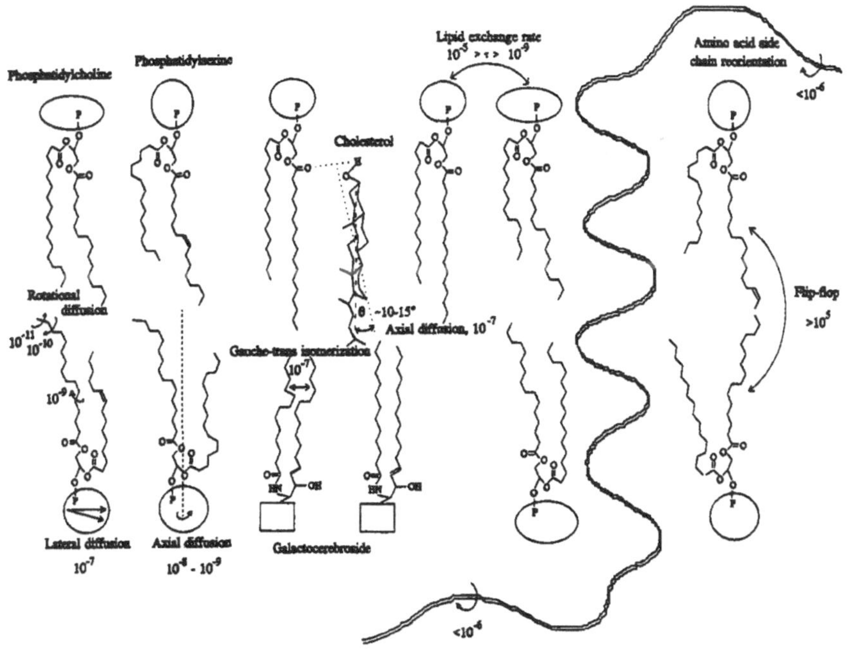

Figure 1. Microscopic, dynamic model of a biological membrane. Correlation times of various dynamic motions are given in seconds. (From Smith, R. L. and Oldfield, E., *Science*, 225, 280, 1984. With permission.)

selected aspects of cellular functions and membrane fluidity. Without aiming at completeness the last part of the review abstracts the important points of a significant number of papers dealing with a possible correlation between the "fluid-state" of biological membranes and membrane functions such as enzyme activity, transport processes, hormone action, and immune response and the physiological state of cells such as cell cycle, differentiation, malignant transformation, and metastasis.

II. ESTIMATION OF MEMBRANE FLUIDITY

A. GENERAL CONCEPTS

The term "membrane fluidity" is usually attributed mostly or solely to the dynamic features of lipid constituents, and it is often called "lipid fluidity".[8,9] Membrane fluidity is governed by a complex pattern of the components' mobilities. These include three different types of motions of individual lipid molecules: lateral, rotational, and transversal diffusion.[10] These types of motions and their corresponding correlation times are shown in Figure 1. The chemical structure of a lipid molecule highly influences its kinetic behavior:[11] the presence and nature of the polar head, the length, and the nature of hydrocarbon tails, whether they are saturated or polyunsaturated fatty acyl chains. The immediate environment of the lipid molecule introduces additional parameters to this motion: the lipid's polar head may interact with the aqueous medium and ions, or glycolipids may form a network of hydrogen bonds at the membrane interface. Membrane proteins also interact with the surrounding lipids.

Lateral diffusion of lipids in biological membranes refers to the two-dimensional translocation of the molecules in the plane of the membrane. One of the most widely used techniques for the determination of lateral diffusion coefficients is the "Fluorescence Photobleaching and Recovery" (FPR) method. This technique is thoroughly discussed in Chapter 3.

The term fluidity is closely related to the rotational motions in the membrane. *Rotational diffusion* of lipid molecules is restricted to the plane of biological membranes and its correlation time is in the 1- to 100-ns time interval.[12] Several spectroscopic techniques can be used to monitor this type of motion, such as EPR, NMR, and fluorescence. Rotational diffusion of molecules with a size of a typical protein in biological membranes, where the viscosity is two or three orders of magnitude higher than that of the water, is a relatively slow process covering the microsecond-millisecond time domain. Thus, detection

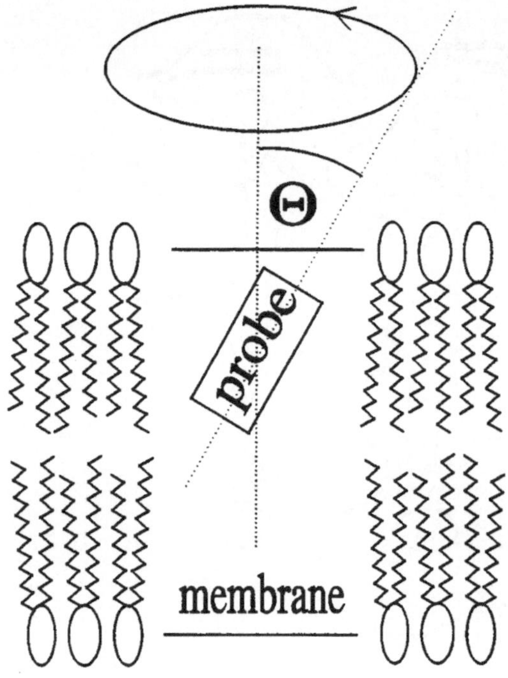

Figure 2. Determination of S order parameter. Orientation of the long molecular axis of a probe molecule at an angle of θ with respect to the normal of the bilayer.

of these "slow" motions requires an appropriate "spectroscopic clock", since the correlation time of this time domain usually differs from the time supplied by fluorescence and nuclear magnetic relaxation processes. Phosphorescence anisotropy decay measurements (discussed thoroughly in Chapter 5, Section II.B) proved to be very effective in these studies.

The third type of molecular motion is the *out of plane rotation*, redistribution of lipid molecules between of the two leaflets of the bilayer. The rate of this "flip-flop" motion is very low in lipid bilayers, on the order of a day or more. In biological membranes, this process is mediated by specific enzymes called "flippases".[13] These enzymes help maintain the asymmetric distribution of lipids in the two leaflets of the membrane at the expense of ATP hydrolysis.

The membrane fluidity is generally considered to consist of two major components. These components may change independently and to date there is no reliable means of calculating one from the other. The first component is the "order parameter" (S), which is a measure of angular range of rotational motion. This parameter is also called the structural, static, or range component of membrane fluidity. More tightly packed chains result in a more ordered or less fluid bilayer. The second component is known as the "microviscosity", dynamic component of membrane fluidity, or rate. This component measures the rate of rotational motion, and is a more accurate reflection of membrane microviscosity.

1. Order Parameter

The order parameter (S) was first defined by Saupe using NMR measurements on liquid crystals.[14] S is a measure of the degree of molecular orientational order, which is closely related to the organization of the hydrocarbon chains in the lipid bilayers. Measured by NMR or time-resolved fluorescence anisotropy, S is described as the first and most important parameter in the characterization of the orientational distribution function. It gives the probability of finding the orientation of the long molecular axis of the probe molecule at an angle of θ (Figure 2) with respect to the normal of the bilayer.[15-20] The order parameter can be expressed as

$$S = \left\langle \frac{[3\cos\theta - 1]}{2} \right\rangle \qquad (1)$$

where the angular brackets represent the ensemble average. The relationship between the range of motion (closely related to the S) and acyl chain motion can be readily understood by considering a simple, fully saturated phospholipid. Below the phase transition temperature, characteristic for the phospholipid, acyl chains are in the extended all-*trans* conformation and are closely packed — the range of motion is small. Above the phase transition, the chains contain a number of *gauche* configurations; the packing of the chains is therefore looser and the range of angular motion open for the acyl chain is increased.[21] In a perfect smectic liquid crystal the value of the order parameter approaches one.[22] In biological membranes where the lipids are not perfectly oriented, this parameter can be obtained by averaging this function over the whole membrane. Values obtained in biological membranes are in the range of 0.3 to 0.6, indicating a relatively low order in organization of the lipid acyl chains in the bilayers.

When the value of the order parameter is close to one in a smectic liquid crystal of a single type of molecule, each molecule in a two-dimensional membrane model possesses only a few degrees of freedom. The presence of different types of lipid molecules in a cell membrane decreases the order in the bilayer. The volume ratio of the polar head to the hydrocarbon chains, the nature of the polar head, saturation of the fatty acyl chains, and the number of *trans* and *gauche* positions of the C-C bonds in the chain all influence the angle of rotation, θ, and thus the order parameter.[11,22,23] The proteins embedded in the membrane cause dramatic changes in the structure of the lipid bilayer[24,25] not only by their different size and structure but also by their specific interactions with the surrounding lipids[26] and cytoskeleton.[27] All these defects in the lipid bilayer modify the order parameter as well as the degree of freedom for the diffusion of each molecule. These effects are specific to the structure of the lipid and its environment but also dependent on the time of residence in that environment.

2. Microviscosity

Microviscosity should be regarded as an operational term which was introduced for the application of classical hydrodynamic expressions to microscopic fluid regions, like the lipid bilayer of biological membranes. This term was first described by Shinitzky and co-workers[8,9,28,29] to characterize the fluidity parameters of the phospholipid matrix. They used the Perrin theory of fluorescence polarization to translate the experimentally observed rotational relaxation times to an average viscosity of the given lipid region, to the "microviscosity".

Viscosity derived from the classical hydrodynamic expression is valid for macroscopic, isotropic fluids:

$$\eta = \frac{F}{A}\left[\frac{dV}{dz}\right]^{-1} \tag{2}$$

where η is the coefficient of viscosity or briefly viscosity, F is the longitudinal force required to maintain a dV difference in flow velocity between two parallel fluid planes (with area A) over a distance dz. The general experimental strategy for obtaining quantitative estimates on membrane viscosity is based on the original relation derived by Einstein which connects rotational correlation time, φ, to the viscosity of the embedding medium:

$$\eta = \frac{\varphi kT}{8\pi r^3} \tag{3}$$

where k is the Boltzman constant, T is the absolute temperature, and r is the radius of the rotating object. The fluidity is defined as a reciprocal of viscosity.

This equation was derived for spherical objects of considerably larger size than that of the particles of the medium, i.e., solvent molecules. Furthermore, it holds true only when the solvent is isotropic. These criteria are fulfilled in the case of rotation of a spherical macromolecule of protein size in an aqueous solution. In biological membranes, however, the solvent is far from being isotropic. Two other limitations in applying Einstein's equation may arise due to the nonspherical shape and the size (relative to the size of the solvent molecules) of the probe molecules. Both factors influence the rotational frictional properties.[30] Therefore, we can deal only with apparent viscosity of membranes. The diffusion constants are usually extracted from measurements of lipid in lipid (self diffusion) and protein in lipid diffusion processes. In both cases one should be cautious in assigning viscosities to membranes, since the molecular interactions affecting the motion of the probe may be local and thus the data do not necessarily reflect such a bulk property as the viscosity is generally treated. Viscosity estimates slightly depend on the measuring technique as well, probably due to the different extent of membrane perturbation by the applied probes. Despite these reservations assignment of apparent viscosities to membranes may give some ideas about the general environment of membrane molecules and especially indicating its perturbation. The apparent diffusion constants themselves are informative of the (local) motion of the probe species in the membrane, independently of the interpretation of diffusion rate in terms of membrane viscosity.

The microviscosity, η_m, formulated by Shinitzky and co-workers,[8,9,29,30] is based on the application of the Perrin equation since most of their studies were performed applying fluorescence techniques. The Perrin equation describes the relationship between fluorescence depolarization and the hydrodynamic properties of the embedding medium as follows:

$$\frac{r_0}{r} = 1 + \frac{3\tau}{\varphi} = 1 + \frac{\tau kT}{V_h \eta} \tag{4}$$

where r is the actual anisotropy of the fluorescence emitted by the probe, r_0 is the limiting initial anisotropy corresponding to the motionless state of the fluorophore, τ is the fluorescence lifetime, φ is the harmonic mean of rotational relaxation times while V_h and η are the solvated molecular volume of the rotating species and the viscosity of the embedding medium, respectively. This original form of Perrin equation holds only for rotation of spheres at isotropic conditions.

Conditions of isotropic rotations, however, are not fulfilled in biological membranes due to their more or less anisotropic structures.[30-32] Thus, the application of the Perrin equation to rotational diffusion of membrane probes faces limitations discussed above. Taking into account that lipid layers in membranes are far from being isotropic, the term microviscosity, η_m, was derived as a geometrical mean of the apparent viscosities along the different symmetry axes:

$$\eta_m = \prod_{i=1}^{n} \eta_i^{1/n} \tag{5}$$

Another analogy to macroscopic fluids was assumed by these authors in order to derive the "flow activation energy", ΔE from the empirical temperature dependence:

$$\eta_m = Ae^{\frac{\Delta E}{RT}} \tag{6}$$

where η_m is the averaged microviscosity, R is the gas constant, and T is the temperature. ΔE is regarded as an energetic parameter characteristic of the degree of order in the given lipid region.[28]

In a simplified case of Equation 5 the microviscosity can be presented as a tensorial parameter:

$$\bar{\eta} = \frac{1}{3}\eta_{\parallel} + \frac{2}{3}\eta_{\perp} \tag{7}$$

where $\eta_{\parallel}$ and $\eta_{\perp}$ are the principal viscosity vectors across and along the plane of the lipid bilayer. Partial resolution can be obtained with a disc-like fluorophore (e.g., perylene), by determination of the apparent viscosities for its in-plane and out-of plane rotations.[33-35] Alternatively, and more directly, measurements of lateral diffusion coefficients of fluorescent[36] or EPR probes[37] can be correlated predominantly with $\eta_{\perp}$.

B. GENERAL EXPERIMENTAL STRATEGY

The lipid fluidity can be studied by analyzing the movement of each atom or segment of the lipid chains with the use of proton and C NMR.[38-41] The resulting microscopic view of lipid fluidity is a complex combination of position, orientation, order, and motional freedom of various hydrocarbon moieties.[38-41] In other methods, such as EPR[42-44] and fluorescence depolarization,[8,9,12,15,19,29,45] a foreign probe is introduced which, by its localization and characteristic motion, averages out most of the microscopic details and therefore reports on the submacroscopic level of lipid fluidity.[8,45]

1. NMR

The use of NMR for the study of membranes was introduced by Chapman and Salsbury.[46] Since then, several hundred reports of observations of ^1H, ^2H, ^{13}C, ^{19}F, and ^{31}P spectra on phospholipid vesicles and several natural membranes have been published. The majority of these papers have addressed various aspects of the problems of order, orientation, mobility, and lateral diffusion in phospholipid bilayers and multilayers. Others have dealt with the conformations of the polar head groups of phospholipids, transmembrane exchange (flip-flop) between two layers, interactions of small molecules with model and natural membranes, function of ion channels in reconstituted membranes, and the nature of protein-lipid interactions.

A semiquantitative characterization of the gradient of mobility was possible from ^{13}C measurements,[47,48] and detailed models of fatty acid side-chain motions were proposed, although, strictly speaking, relaxation data do not permit a clear distinction between the effects of packing on the amplitudes and on the rates of motion at each point in the chain, i.e., between order and mobility. Such a distinction can, to some extent, be made in the studies of quadrupole splittings in ^2H spectra. Theoretical background appropriate for considerations of the ^2H NMR spectra of lipid membranes is discussed in detail in the literature.[10,38,41,49-51] The fact that quadrupolar interactions of deuterons are much greater than dipolar and chemical shift anisotropic interactions renders the analysis of deuterium NMR parameters particularly simple. The presence of molecular motions averages the deuterium quadrupole powder-spectrum to a lineshape which is characteristic of the rate and the mechanism of molecular motion. Thus, the structure and dynamics of a given carbon segment can be deduced from the NMR parameters of deuterons attached directly to the specific carbon atom.[41,49-51] For the ^2H nucleus (with spin $I = 1$ and in C-D bonds an asymmetry parameter $\eta = 0$) the allowed transitions correspond to $+1 \leftrightarrow 0$ and $0 \leftrightarrow -1$, and the residual quadrupole splitting of the pattern, $\Delta\nu_Q$, of the NMR absorption lines is related to the C-D bond orientational order parameter S_{CD} by the following relation:

$$\Delta\nu_Q = \frac{3}{4}\left(\frac{e^2qQ}{h}\right)S_{CD} \qquad (8)$$

with the segmental order parameter, S_{CD}, given by

$$S_{CD} = \left\langle \frac{\left[3\cos^2\theta(t) - 1\right]}{2} \right\rangle \qquad (9)$$

where $e^2qQ/h = 170$ kHz is the deuterium quadrupole coupling constant of the methylene deuterons and $\theta(t)$ is the angle between the C-D bond and the bilayer normal at time t. The angular brackets represent an ensemble average. In the liquid-crystalline state the C-D bond is perpendicular to the normal of the bilayer and the expected S_{CD} is 0.5, corresponding $\Delta\nu_Q = 60$ kHz, which is much broader than 30 kHz measured in pure phospholipid dispersions. The reduced S_{CD} is attributed to the presence of *trans-gauche* isomerization which gives rise to the fluid nature of the acyl chains, characteristic of the liquid crystalline phase.[49,50] Thus, S_{CD} can be taken as a measure of the lipid order and allows the determination of the average orientation of the C-D bond with respect to the lipid bilayer's normal. In brief, the size of the quadrupole splitting increases as the order parameter of the system increases. From the observed quadrupole splitting it is possible to define an average order parameter S_{CD} for each methylene group of a fatty acid side chain in a bilayer. A plot of S_{CD} as a function of chain length clearly shows a progressive decrease in order from the surface to the interior of a bilayer.

Dynamic properties of the lipid bilayers can be deduced from the rotational relaxation times T_1 (spin lattice relaxation times) and T_{2e} (transverse relaxation time). In biological membranes the presence of proteins does not significantly change the T_1, but markedly decreases the T_{2e} even at small concentrations of proteins. Adapting the two site exchange model,[52] the correlation time, τ_C, can be estimated from T_{2e} using the order parameter determined from the residual splittings and assuming that the lipid molecules are equally likely associated with protein as well as with lipid environment according to equation:

$$\frac{1}{T_{2e}} = \left(4\pi^2\Delta\nu_Q^{2/5}\right)\chi_p\tau_c \qquad (10)$$

where $\chi_p = 0.5$ is the fractional time that a lipid is in the protein environment. The correlation time may reflect the time scale of motion on the protein surface and/or the time scale of the exchange of lipid between the protein surface and the free lipid. Using another model, the reduced T_{2e} can be correlated to the retardation in local chain dynamics.[53]

In summary, among the various physical techniques in use for examining membranes, NMR is also powerful, since it is the least perturbing of the techniques and furthermore in ^{13}C- and ^{2}H-NMR the motional characteristics of each C-H bond along the chain can be examined and compared with that of other locations. The technique, however, has low sensitivity; synthesis of specifically deuterated phospholipids is required and data collection time is long, so it is not widely applied in mammalian cell membrane studies. The other drawback is that, although the complete description of an acyl chain in terms of the order parameter is possible by ^{2}H-NMR, the rate of motion, calculated from spin-lattice and transverse relaxation times is less reliable.

2. EPR

First, spin-labeled steroids were applied to give estimates of viscosity from rotational diffusion parameters of these probes in membranes using EPR technique.[54] Later this approach was extended to spin-labeled fatty acids and phospholipids as well.[42] While the conventional EPR technique is ideally suited to monitor the nanosecond rotational motions of fluid lipid hydrocarbon chains or small protein segments, the technique is also insensitive to the slower motions (often in the microsecond range) that are typical for gel-phase lipids and intact membrane proteins.[43] In contrast, saturation transfer EPR (ST-EPR) has its optimum sensitivity in the microsecond region.[44] This technique is a powerful tool in the study of membrane dynamics, particularly the rotational motions of integral membrane proteins.

Similarly to the NMR method, the conventional EPR technique provides many ways to calculate parameters to characterize the static ("packing") and kinetic behavior of lipid molecules in model and biological membranes.[42-44,55-58] One possibility to calculate the order parameter, S, is to use the standard formula relating S to both inner and outer extrema:[56]

$$S = \frac{T'_{\parallel} - \left(T'_{\perp} + C\right)}{T'_{\parallel} + \left(2T'_{\perp} + C\right)} \times 1.66 \qquad \textbf{(11)}$$

where the constant $C = 1.4 - 0.053\ (T'_{\parallel} - T'_{\perp})$ is an empirical correction for the difference between the true and apparent values of $2T'_{\perp}$, and the factor 1.66 is a solvent polarity correction factor.[57] $2T'_{\parallel}$ and $2T'_{\perp}$ are the measured inner and outer extrema in the EPR spectrum.

The kinetic behavior of lipid molecules can be characterized with the effective correlation time τ_r which can be calculated from the formula:

$$\tau_r = \left(6.5 \times 10^{-10}\right) W_0 \left[\left(\frac{h_0}{h_{-1}}\right)^{1/2} - 1\right] \qquad \textbf{(12)}$$

where W_0 is the peak-to-peak line width of the midfield line, and h_0 and h_{-1} are the peak-to-peak line heights in the mid- and high-field regions of the spectrum, respectively.

A new development in EPR technique is the spectral-spatial EPR imaging for diffusion of spin probes in systems with substantial spin concentration. Spectral-spatial ESR imaging provides a way to resolve the concentration-dependent spectral variation in a two-dimensional fashion: along one axis (the spatial axis) it gives the spin concentration profile, whereas along the other axis it gives the EPR spectrum at that position. This technique has been successfully applied for simultaneous measurement of both the macroscopic and microscopic diffusion coefficients, D_{macro} and D_{micro}, respectively, of a spin-labeled phospholipid in a model membrane utilizing a substantial spin concentration. Initially the spin probe is concentrated in a small region of the sample; its distribution will tend, over time, to become homogeneous via translational diffusion. By measuring several spectral images at different times, the spread of the concentration profile can be analyzed as a function of time to obtain the macroscopic diffusion coefficient from the diffusion equation. In addition, the macroscopic diffusion will produce a continuous variation of spin concentration along the spatial axis. Thus, at a later time one can obtain an entire set of concentration-dependent EPR spectra from the spectral-spatial image of just a single sample. The EPR line broadening that results from spin relocation induced by spin-spin interactions can be analyzed to obtain the microscopic diffusion coefficient. The finding that D_{micro} (which involves wagging of the chain causing spin-exchange collision) is larger than D_{macro} (which involves motions of the whole-labeled lipid over long distances) may reflect important details in the microscopic molecular dynamics.[59] The observed differences between microscopic and macroscopic diffusion constants were interpreted by Vaz and Almeida using the free volume model for diffusion in lipid bilayer.[60,61]

The sensitivity of the EPR is much higher than that of the NMR, since it permits the detection of one labeled lipid molecule per protein molecule in biological membranes. In addition, using EPR, unambiguous differentiation between boundary and bulk lipids can be achieved, which would not be possible with NMR. Moreover, the time scale of the EPR matches better with the lipid-lipid and lipid-protein collision rates than that of NMR. Using EPR, however, requires the introduction of an extrinsic probe into the system. This probe should be specifically synthesized and the perturbing effect of the EPR probe on the system under investigation should be analyzed with care. In addition, in membranes of live cells the interference with metabolites possessing unpaired electrons restricts the employment of EPR technique. The high cost of EPR instrumentation also limits widespread application of the technique.

Magnetic resonance studies (NMR and EPR) on membranes discover links between biophysical properties of the complex structures and function. In some cases, unique information can be obtained by direct studies on biological membranes despite their complexity. Many magnetic resonance studies on membranes have, however, been made on aqueous dispersions of polar lipids or on purified membrane proteins incorporated into model bilayer membranes. These limitations are largely dictated by the compositional complexity of biological membranes and the need of simplification in order to control the experimental system.

3. Fluorescence

The measurement of fluorescence depolarization provides another means of obtaining information about the motional preferences of membrane-incorporated fluorophores, with respect to both the rate of the rotational motion (dynamics) and the range or extent of the restricted motion (structure). The original application of the technique aimed to measure the microviscosity of the interior of detergent micelles;[35] perylene was used as a fluorescent probe, and Weber's theory[62] was applied to show that the rate of the in-plane and out-of-plane motions of the planar fluorophore differed by a factor of 10. This technique has undergone considerable development and has been applied to a wide variety of artificial and natural membranes.[8,9,12,29,45,63,64] New probes have been synthesized and characterized, and now it is possible, within certain limits, to investigate almost any membrane system after choosing the appropriate probe for a particular application.

The fluorescence depolarization technique has a series of important advantages over NMR and EPR. It can be applied to complex model systems as well as isolated biological membranes, the polarized signal is highly sensitive and reproducible, and the data are promptly interpretable. Similarly to other biophysical methods, the resolution of this method is also limited since the fluidity regions in biological membranes are highly heterogeneous. In the common steady-state approach the results are of an overall average nature and even then are not explicit. Time-resolved fluorescence depolarization is more convenient for determining the dynamic and structural parameters. The information on the rate of motion is obtained from the initial slope of the decay of fluorescence anisotropy, and the information on the range of the motion is obtained from the nonzero asymptotic value which the fluorescence anisotropy reaches at infinite time. Since the information is obtained on the nanosecond time scale, during the fluorescence lifetime almost no lateral diffusion occurs, and the observed signal is thus the ensemble average over labels in various fixed environments.[63,64]

The fluorescence depolarization method also has some disadvantages. The fluorescence probe introduced into the membranous system might perturb the environment. The localization and the orientation of the fluorophores are not always well defined, and the heterogeneity of the probe environments is difficult to analyze. In addition, the required size of the fluorophore precludes a high-resolution statement about the local dynamics and order.[63,64]

The details of the time-resolved fluorescence anisotropy decay and steady-state fluorescence polarization measurements as well as the applicable fluorescent probes and techniques will be discussed in the subsequent sections of this chapter. For detailed coverage of the principles of fluorescence spectroscopy, readers are referred to the monograph by Lakowicz.[65]

III. FLUORESCENCE ANISOTROPY AND EVALUATION OF MEMBRANE VISCOSITY

A. TIME-RESOLVED FLUORESCENCE ANISOTROPY MEASUREMENTS

1. Pulse Fluorimetry

The physical principles of the method are briefly the following. A short pulse of linearly polarized light falls on an isotropic suspension of membranes containing fluorescent molecules, either extrinsic probes or intrinsically fluorescent molecules. Only fluorophores in the membranes with transition absorption dipole moments approximately parallel to the direction of polarization will be able to absorb. In this way, a highly ordered, anisotropic distribution of excited fluorophores is selected at $t = 0$ out of an originally isotropic (random) distribution of unexcited fluorophores (photoselection). The excited fluorophore fluoresce and the fluorescence from this oriented population is highly polarized at $t = 0$, the intensity I is recorded as a function of time t for two polarization planes, one parallel and the other perpendicular to the plane of polarization of excitation light. The fluorescence anisotropy r_t is obtained as

$$r_t = \frac{I_{\parallel}(t) - I_{\perp}(t)}{I_{\parallel}(t) + 2I_{\perp}(t)} \tag{13}$$

and

$$I_{tot}(t) = I_{\parallel}(t) + 2I_{\perp}(t) \tag{14}$$

where $I_{tot}(t)$ shows the total intensity in time. The anisotropic angular distribution of excited fluorophores is not the equilibrium distribution. It will relax in time by rotational diffusion to the less-ordered equilibrium distribution which, in the case of membranes, is not expected to be isotropic. As a consequence, r_t, which had a high initial value r_0, will decay with the rate determined by the viscosity and the size of the fluorophore to a nonzero asymptotic value r_∞, reached at infinite time, which is a measure of the degree of orientational constraint in the equilibrium state. The limiting initial anisotropy r_0, the virtually constant value of r_t in the absence of rotation, is expressed by

$$r_0 = \frac{3\cos^2\alpha - 1}{5} \tag{15}$$

where α is the angle between the absorption and emission dipole moments of the fluorescent molecule. The simplest way to extract information about the dynamic behavior of the fluorophore is to use the Perrin equation (Equation 4). Application of the Perrin equations assumes that the environment of the fluorescent dipoles is isotropic and the final distribution will also be isotropic, and r_t decreases to zero. A real decay curve of the fluorescence anisotropy of probes embedded into membranes does not reach zero but a finite level r_∞ as shown in Figure 3. Using a single exponential approximation the decay curve can be described with the equation:

$$r_t = (r_0 - r_\infty)e^{-t/\varphi} + r_\infty \tag{16}$$

where ϕ is the rotational correlation time, as mentioned earlier.

For labeling membranes, one of the very frequently applied fluorescent probes is the 1,6-diphenyl-1,3,5-hexatriene (DPH), which is a rod-like molecule. Its distribution and motion in the membranes are far from being isotropic. To characterize the restricted motion of fluorophores, the wobbling-in-cone model has been widely accepted.[66-68]

a. *Wobbling-In-Cone Model*

In the wobbling-in-cone model fluorophores can move in a given cone area which is expressed by a cone angle (θ_c) and the rate of the motion is expressed with the wobbling diffusion rate D_w as demonstrated in Figure 4. The cone angle can be derived from the experimentally determined r_∞ as follows:

$$\frac{r_\infty}{r_0} = \frac{[\cos\theta_c(1 + \cos\theta_c)]^2}{4} \tag{17}$$

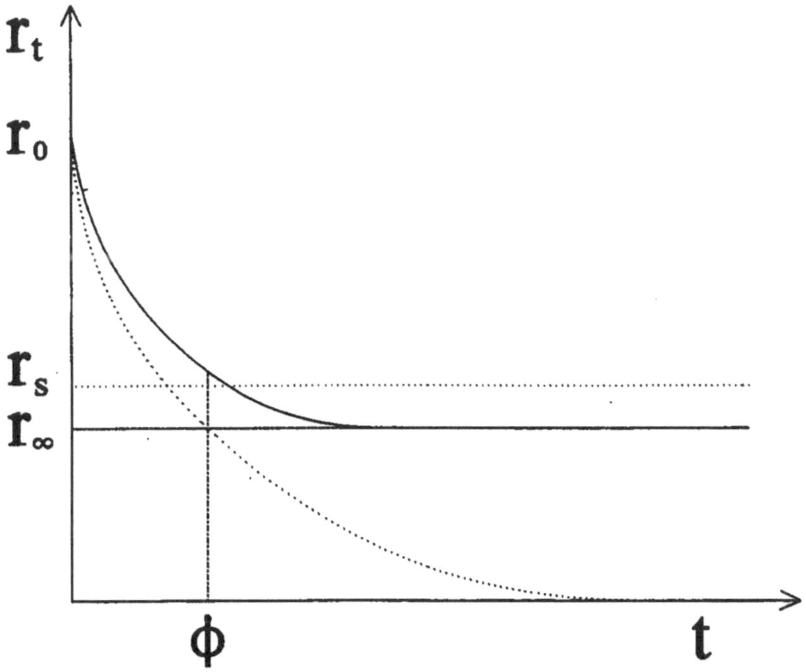

Figure 3. Time dependence of the fluorescence anisotropy (r_t). In case of isotropic environment the r_t decays to zero (dotted line) according to Perrin's law. r_∞ is the finite level of fluorescence anisotropy, φ is the rotational correlation time, r_0 is the limiting initial anisotropy, and r_s is the steady-state fluorescence anisotropy.

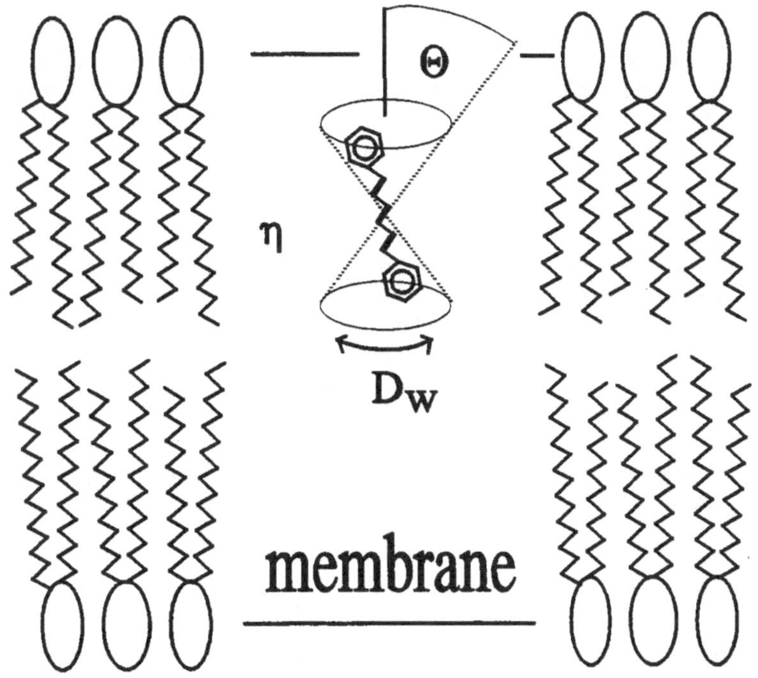

Figure 4. Wobbling-in-cone model of DPH motion in lipid bilayer. The fluorophores can move in a given cone area determined by the cone angle (θ_c) with a wobbling diffusion rate, D_w, in an environment of η viscosity.

The experimentally obtained rotational correlation time can then be translated to the wobbling diffusion constant using the following equation,[69] which is equivalent to the original numerical results of Kinosita et al.[66]

$$D_w \varphi \frac{r_0 - r_\infty}{r_0} = \frac{-x^2(1+x)^2\left\{\ln\left[\frac{1+x}{2}\right] + \frac{(1-x)}{2}\right\}}{[2(1-x)]} + \frac{(1-x)(6+8x-x^2-12x^3-7x^4)}{24} \quad \textbf{(18)}$$

where $x = \cos \theta_c$.

The value r_∞ is related to an order parameter, S. In this respect it is similar to the order parameter of deuterium NMR (Equation 9), which refers to the restricted motion of individual methylene segments along a phospholipid acyl chain. However, the NMR order parameter is averaged over a time span of about 10^{-4} s, whereas the motion of the fluorescent probe is averaged over the nanosecond time scale. Lipari and Szabó[69] have derived a model independent from the above for the expression of r_∞:

$$r_\infty = \frac{2}{5} P_2(\cos\theta_a) P_2(\cos\theta_e) S^2 \quad \textbf{(19)}$$

$$S = \langle P_2(\cos\theta) \rangle = \left\langle \frac{(3\cos^2\theta - 1)}{2} \right\rangle \quad \textbf{(20)}$$

$$S^2 = \frac{r_\infty}{r_0} \quad \textbf{(21)}$$

where θ_a and θ_e are the angles between the absorption and emission moments and the axis of rotation, θ is the angle between the molecular symmetry axis and the membrane normal, and the P_2 is the second Legendre polynomial ($P_2 = (3x^2 - 1)/2$). In the case of DPH, the absorption and emission moments are practically parallel to the molecular symmetry axis, so all three angles can be characterized with one angle as θ_c if we want to use the wobbling-in-cone model. The evaluation of the static equilibrium distribution of a fluorescent probe in a bilayer in terms of order parameter S makes it possible to compare fluorescence depolarization data for such probes with data obtained by EPR and to some extent with data from NMR experiments. Heyn did a limited comparison along these lines for DPH in lipid bilayers prepared from 1,2-dipalmitoyl-glycero-3-phosphocholine (DPPC).[70] For comparison, the order parameters were calculated for the C_{10} and C_{12} methylene carbon of DPPC from NMR data and the average order parameter along the chain was determined for DPPC.[70,71] It was shown that S calculated from r_∞ fit the NMR data reasonably well. Because of the size of the fluorophore and the mutual ordering effect of probe and lipid fatty acyl chains, the order parameter measured by a probe such as DPH can only indirectly reflect the order parameter of the fatty acyl chains in the lipid bilayer. It can only be compared with the average order parameter over a length of fatty acyl chain at least equivalent to the length of the DPH molecule. This type of comparison should be performed with some caution, since there is great variation in order parameter along the methylene groups from C_1 to C_{12} and the exact location of the DPH in the bilayer is not known.

From the wobbling diffusion constant, D_w, the "viscosity in the cone" can be estimated using a modified form of Equation 3. The equation expressed below is valid for an ellipsoid instead of a sphere:

$$D_w = \frac{kT}{6\eta_c V_e f} \quad \textbf{(22)}$$

where η_c denotes the "viscosity in the cone", V_e and f denote the effective volume and shape factor of the probe. The value of V_e can be determined in liquid paraffin assuming that the hydrocarbon chains of the lipids have similar molecular nature in membrane and in liquid paraffin. The viscosity range calculated by Kinosita[68] proved to be an order of magnitude smaller (0.3 to 0.8 P) than microviscosity obtained from steady-state anisotropy measurements. This discrepancy originates probably from the inherent property of the "microviscosity" that it does not discriminate between the dynamic friction and the static rotational constraint. Therefore, characterization of anisotropic medium such as biomembranes requires at least two parameters, e.g., viscosity in cone and the cone angle.

b. WOBHOP Model

When the fluorescence anisotropy decay data cannot be fitted with a single decay function, multiexponential decay functions are employed, which have the general form of

$$r(t) = r_\infty + (r_0 - r_\infty)\Sigma A_i e^{t/\varphi_i} \tag{23}$$

where ϕ_i represents the rotational correlation times of the fluorophore and A_i is the corresponding preexponential factor. It is not always straightforward to fill this general equation with detailed molecular information and to provide parameters on the orientational packing and dynamics of the lipids. Assigning molecular information to the mathematical parameters of equations describing model systems should be performed with special caution, always considering the characteristics of a given experimental system. For example, when fluorophores are embedded in lipid where both micelles and two-dimensional hexagonal lattice tubes (H_{II} phase) are present, both the wobbling and lateral diffusion may influence the rate of fluorescence anisotropy decay.[72,73] This effect is more enhanced when the radius of the hexagonally packed tubes is similar to the physical dimension of the lipid. In this case, the magnitude of the rotational diffusion (hopping) constant D_H of the excited fluorophores around the highly curved lipid tubes should be comparable with that of the wobbling diffusion constant D_W. The axes of rotation for D_H and D_W are along the long symmetry axis of the lipid tubes and perpendicular to the local symmetry axis of the lipid, respectively.[72,73] Using these parameters, the so-called WOBHOP model has been introduced to describe fluorescence anisotropy decay curve of fluorophores incorporated in coexisting lamellar and hexagonal lipid phases:

$$r(t) = \frac{r_0\left(1-\langle P_2\rangle^2\right)e^{-t/\varphi_W} + 3r_0\langle P_2\rangle^2 e^{-t/\varphi_H}}{4} + \frac{3r_0\left(1-\langle P_2\rangle^2\right)e^{-t/(\varphi_W+\varphi_H)} + r_0\langle P_2\rangle^2}{4} \tag{24}$$

where $\langle P_2\rangle$ is the second rank order parameter (see Equation 20), and φ_W and φ_H represent the correlation times of the fluorophore due to wobbling and hopping motions and are $(1 - \langle P_2\rangle^2)/(6D_W)$ and $1/(4D_H)$, respectively. With the help of this theoretical approach Chen et al.[73] analyzed the time-resolved fluorescence anisotropy of DPH conjugated phosphatidylcholine (DPH-PC) in unsaturated phosphatidyl-ethanolamine/diacylglycerol binary lipid mixtures forming liquid crystalline lamellar and inverted hexagonal phases.[73] At least two correlation times were required to describe the anisotropy decay curve of DPH-PC in the hexagonal phase. The motional dynamics of DPH-PC were described by a steady-state orientational factor $\langle P_2\rangle$ and the wobbling and hopping diffusion constants. The hopping diffusion constant, D_H, is closely related to the lateral diffusion constant, D_L, since $D_H = D_L/R^2$ (where R is the radius of the highly curved cylindrical tubes of hexagonal phase). They could estimate D_L from independent experiments by measuring the excimer formation rate of pyrene-conjugated phosphatidylcholine. Upon comparing the values of D_H and D_L, the value of R was determined and agreed with that derived from X-ray diffraction.[72,73] These data suggested that the attempt to fill formulas describing time-resolved anisotropy data with real molecular parameters was successful.

2. Differential Polarized Phase Fluorometry

When a fluorophore is excited with a beam of sinusoidally modulated light, the emitted light will be modulated at the same frequency but shifted in phase angle (ϕ_p) and decreased in amplitude by a demodulation factor (m). These two parameters can be used to calculate the phase and modulation lifetimes, τ_p and τ_m, respectively. For a single exponential decay, the true lifetime, $\tau = \tau_p = \tau_m$ and

$$\tau_p = \omega^{-1}\tan\phi_p \tag{25}$$

$$\tau_m = \omega^{-1}\left[\left(1/m^2\right) - 1\right] \tag{26}$$

where ω is the circular frequency of the exciting light. When coupled with phase-sensitive detection, phase fluorometry is particularly useful for resolving lifetimes and spectra of components of a binary mixture since the detector can be set out of phase with one or the other components and thus be transparent to its presence.

Differential phase fluorometry provides alternative method to investigate the time-dependent decay of fluorescence anisotropy.[66] In this case the sinusoidally modulated light, used to excite samples, is polarized, and the phase difference ($\Delta = \phi_\perp - \phi_\parallel$) between the perpendicular and parallel components of emitted light is measured. This phase difference depends upon the modulation frequency, the fundamental (limiting) anisotropy, r_0, the rate of fluorophore rotation, and the freedom (isotropy) of these rotations. In case of a hindered isotropic rotator, such as DPH embedded in a membrane, tan Δ measurements can be used to calculate the extent of hindrance, as described by r_∞. Predictions of tan Δ for a hindered rotator are readily made using the sine and cosine transforms of the impulse response function and the differential tangent is described by

$$\tan\Delta = \frac{3\omega\tau\left(r_0 - r_\infty\right)\left(\dfrac{\tau}{\varphi}\right)}{m_0\left(1+\omega^2\tau^2\right) + S\left(\dfrac{\tau}{\varphi}\right) + m_\infty\left(\dfrac{\tau}{\varphi}\right)^2} \tag{27}$$

where $m_\infty = (1 + 2r_\infty)(1 - r_\infty)$ and $S = 2 + r_0 - r_\infty(4r_0 - 1)$. The maximum value of tan Δ is given by

$$\tan\Delta_{max} = \frac{3\omega\tau\left(r_0 - r_\infty\right)}{S + 2\left[m_0 m_\infty\left(1+\omega^2\tau^2\right)\right]^{1/2}} \tag{28}$$

In lipid bilayers the maximum observed tan Δ occurs near the transition temperature. Usually, the valuable information from these equations are the rotational correlation time of the probe (φ) and the degree to which these rotations are hindered (r_∞). Since the maximum of tan Δ can be a result of changes in either φ or r_∞, the tan Δ values alone cannot be used to determine either φ or r_∞. This uncertainty can be eliminated by the separate measurement of the steady-state anisotropy. This calculation is described in detail in Reference 65.

Although the differential phase measurements yield less easily interpreted results than time-resolved fluorescence anisotropy measurements, they can detect anisotropic rotations on the subnanosecond time scale. In addition, data collection takes less time than time-resolved anisotropy measurements.

B. STEADY-STATE MEASUREMENTS

Application of the steady-state fluorescence polarization measurements to assess lipid structure and dynamics in membranes is widespread because of the simplicity of the measurements and the availability of the relatively inexpensive instruments in many laboratories. In steady-state measurements the emission of the fluorophore is measured under continuous illumination. The steady-state fluorescence anisotropy (r_s) is determined by the emission intensities through an analyzer-oriented parallel ($I_\parallel$) and perpendicular ($I_\perp$) to the direction of polarization of the excitation light:

$$r_s = \frac{I_\parallel - I_\perp}{I_\parallel + 2I_\perp} \tag{29}$$

Comparing the fluorescence anisotropy of fluorophores in membrane system with that observed in a macroscopic-isotropic system, a reference mineral oil was used to estimate the apparent microviscosity of the membrane interior applying the classical hydrodynamic expression of Perrin (Equation 4),[8,9,28,29,45,74-84] although the implicit assumption that the freedom of the depolarizing rotations of a fluorophore in membranes and in mineral oils is similar appeared not to be valid.[66-68] Time-resolved fluorescence anisotropy (r_t) decay measurements revealed the complex and restricted motion of the fluorophore in lipid bilayers.[66-73] With the help of information gathered with anisotropy decay measurements, a more realistic interpretation of the steady-state fluorescence polarization data was given, in which both a dynamic and a static component of the measured anisotropy appeared.[70,84,85]

1. The Relationship between r_s and r_t

Having measured and fitted the anisotropy decay curve, it is possible to calculate from it the expected steady-state anisotropy r_s by a time integration

$$r_s = \frac{\int_0^\infty r_t I_t \, dt}{\int_0^\infty I_t \, dt} \tag{30}$$

where I_t is the time-dependent total fluorescence intensity. In the case of a single exponential decay of the fluorescence, r_s can be expressed as a sum of two terms:

$$r_s = r_f + r_\infty = \frac{r_0 - r_\infty}{1 + \tau/\varphi} + r_0 S^2 \tag{31}$$

The limiting fluorescence anisotropy, r_∞, reflects the range or amplitude of random librational motion, and r_f depends primarily on the rate of rotation and reflects essentially the frequency of the librational motions, although it depends on r_∞ as well. In the case of the rodlike fluorescence probe, DPH, where at least one of the absorption and emission dipoles lies along the long axis of the molecule, the relationship between r_∞ and the order parameter (S), $r_\infty = r_0 S^2$ is valid. For other probes where the geometry of the absorption and emission dipoles is more complicated, instead of r_0 another factor should be introduced.[69] From the above equation it is apparent that in steady-state fluorescence experiments the information on the dynamics and on the order is no longer separable and that in general neither r_∞ nor φ can be determined from a measurement of r_s. For extreme cases depending on the relationship between the rotational correlation time and the fluorescence lifetime the theoretical limits of the r_s can be determined.[71] When $\tau/\varphi \gg 1$, i.e., when the rotational correlation time is short in comparison with the fluorescence lifetime, diffusion and depolarization can occur, r_∞ is reached, and $r_s = r_\infty$. However, when $\tau/\varphi \ll 1$, the rotational correlation time is much longer than the fluorescence lifetime, the probe rotation is effectively frozen on the scale of the fluorescence lifetime, no depolarization occurs, and $r_s = r_0$. In all other cases, r_s lies between r_∞ and r_0. If from other measurements it is known that $\tau/\varphi \gg 1$, then a measurement of r_s is sufficient to determine r_∞ and the order parameter S. Taking the kinetic contribution to r_s by assuming $\tau/\varphi = 8$ for DPH in lipid phase, Jähnig transformed Equation 31 into the empirical equation:[15]

$$r_\infty = \frac{9r_s}{8} - \frac{1}{20} \tag{32}$$

This equation described the behavior of the DPH in dipalmitoyl phosphatidylcholine vesicles with a convincingly good approximation.[15] Comparing experimental data for the steady-state fluorescence anisotropy and the limiting finite anisotropy measured in a variety of artificial and biological membranes, van Blitterswijk et al.[84] obtained an empirical curve for r_∞ as a function of r_s. In the region $0.13 < r_s < 0.28$ the curve is linear, corresponding to the formula

$$r_\infty = \frac{4}{3} r_s - 0.10 \tag{33}$$

The majority of data for biological membranes and liposomes containing phophatidylcholine fit the curve very well. The same group improved the empirical equation further, taking into consideration some theoretical predictions that the order parameter influences the rotational correlation times.[84,85] They introduced a new parameter, m, which expresses the difference between the rotational diffusion of the probe in the membrane and that in the isotropic reference oil.[85]

$$r_\infty = \frac{r_0 r_s^2}{r_0 r_s + \frac{(r_0 - r_s)^2}{m}} \tag{34}$$

When $m = 0$ ($r_\infty = 0$), the original Perrin equation is valid; when $m = \infty$ ($r_\infty = r_s$), the rotational diffusion does not contribute at all to the fluorescence anisotropy decay. For DPH, used in saturated phospholipids without cholesterol, in unsaturated phospholipids with cholesterol, and in biological membranes essentially the same m (about 1.8) values were found. In single-component unsaturated phospholipids, m for DPH was 0.6, whereas in saturated phospholipids with cholesterol m was 7.[85] This method to describe correlations between time-resolved and steady-state fluorescence anisotropy measurements is applicable not only to DPH, but to other fluorophores as well and is based upon an extended Perrin equation and may be helpful in discussing the dynamics of fluorescent probes undergoing restricted rotation.[85]

2. Angle-Resolved Fluorescence Depolarization

A special version of the steady-state fluorescence anisotropy measurements is the angle-resolved fluorescence depolarization (AFD). AFD experiments on oriented lipid systems provide an alternative to time-resolved experiments on vesicle systems even under steady-state (continuous illumination) condition.[86-94] In these experiments fluorophore molecules lying at different orientation relative to the bilayer normal can be selected by varying the angle of incidence (θ) of the excitation beam. At the same time, the polarized emission from a selected population of excited molecules can be observed by changing the angle between the directions of incidence and observation (ϕ). The theory behind the experiments has been discussed in detail previously.[86,87,90] The angle-resolved polarization ratios are given by

$$R_e = \frac{I_\perp}{I_\parallel} = \left[1 - R_1 + (R_3 + R_2)\sin^2\theta\right]$$

$$\times\left[1 + R_1 + (R_2 - R_1)\sin^2\phi + (R_3 - R_1)\sin^2\theta + R_5\sin^2\theta\sin^2\phi + R_4\sin 2\theta\sin 2\phi\right]^{-1} \tag{35}$$

where $I_\perp$ and $I_\parallel$ are the intensities of the fluorescence emission polarized, respectively, perpendicular to and parallel to the plane of incidence.

The parameters R_1, ..., R_5 contain all the information about the orientational order and dynamics of the fluorophores in the membrane system.[88] The polarization ratio R_e is measured for various combinations of θ and ϕ and affords the determination of the parameters R_1, ..., R_5 from which two order parameters (the second rank order parameter of the absorption and the emission transition moments of the fluorophore, respectively) and three correlation functions (describing the dynamic behavior of the fluorophore) can be calculated.[88] A thorough explanation of parameters R_1, ..., R_5, order parameters, and the correlation functions can be found in References 86 through 88.

The AFD method was applied mostly by Levine and co-workers to characterize the behavior of DPH and TMA-DPH probes incorporated in planar multibilayers.[88,89] They have found that the depolarization ratios of DPH in multibilayers of unsaturated phosphatidylcholine exhibited a weak angular dependence, in marked contrast to those observed when DPH was incorporated in multibilayers of saturated phosphatidylcholine.[88] It was suggested that DPH in bilayers of unsaturated lipids undergoes slower reorientational motions around its long molecular axis than in bilayers of saturated phosphatidylcholines. Furthermore, it was found that the reorientational motion of TMA-DPH molecules is slower than that of DPH molecules in the same bilayer systems.[89] Their results indicated that TMA-DPH is a more useful probe than DPH in systems containing saturated and unsaturated phosphatidylcholines.

Applying ESR experiments on similar systems resulted in detailed information about both molecular order and rotational dynamics.[18] The ESR results were compared with data obtained independently from AFD experiments using TMA-DPH as a fluorescent probe.[18,91] It was found that the orientational order and the rotational dynamics obtained with both techniques agree well. This agreement also suggested that the probe molecules do not perturb the local bilayer structure significantly and indeed they reflect the intrinsic behavior of the lipid molecules. This comparison also indicated that changes in membrane molecular order and reorientational dynamics have to be considered separately and are not necessarily correlated.[18,91]

The same group used this technique for the investigation of molecular order and reorientational dynamics of TMA-DPH embedded in lamellar L_α and hexagonal phases ($H_\parallel$) of lipid/water mixtures.[93] The orientational motion rates of the fluorophores in hexagonal phase were comparable with those in the lamellar phase. In addition, in hexagonal phase the lateral diffusion of TMA-DPH was determined and found to be considerably slower than the reorientational motion.[93]

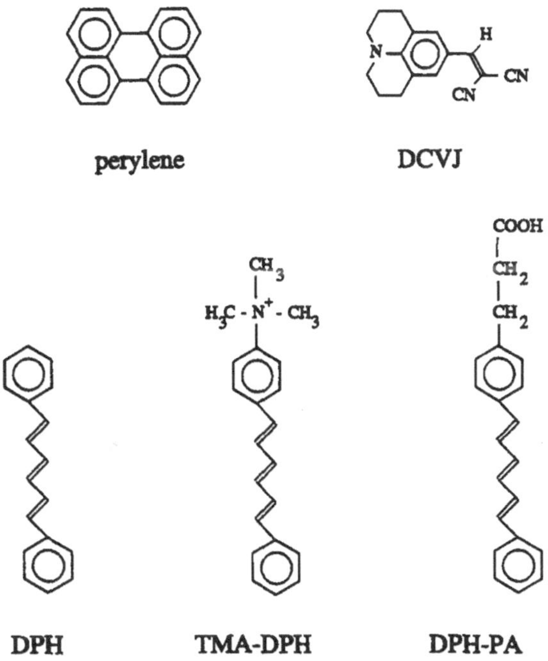

Figure 5. The chemical structures of the most frequently applied fluidity probes. I. Abbreviations used are DCVJ, 9-(dicyanovinyl)julodinine; DPH, 1,6-diphenyl-1,3,5-hexatriene; TMA-DPH, 1-(4-trimethylammoniumphenyl)-6-phenyl-1,3,5-hexatriene; DPH-PA, 3-(*p*-(phenyl)-1,3,5-hexatrienyl)phenyl-propionic acid.

C. FLUORESCENCE PROBES

A great number of fluorescent molecules have been applied as probes for studying membranes.[12,29,31,36,63] For fluidity measurements by fluorescence anisotropy the probes employed should possess well-defined spectral and physical properties, such as high absorption coefficient, high quantum yield, long-enough fluorescence lifetime (comparable to the correlation time of the motion under study), appropriate separation between absorption and emission bands, photostability, and solubility in lipid phase.[35] These fluidity probes can be divided into two main groups. The first group comprises fluorophores which incorporate spontaneously into the well-defined lipid region of membranes. The fluorescence anisotropy characteristics of such probes directly relate to the structural and dynamic properties of the labeled regions. The second major group consists of fluorescent molecules which are covalently attached to fatty acids or phospholipids at known loci. Application of fluorescent probes of the latter group proved to be very useful in studying the heterogeneity in the dynamic behavior of biological membranes. The chemical structures of the most frequently applied fluidity probes, whose characteristics are discussed in detail later, are shown in Figures 5 and 6.

1. Characteristics of Fluidity Probes

a. Perylene

One of the very first probes applied for measuring microviscosity of membranes is perylene, which is a flat aromatic molecule with a rigid disc-like structure of 8 Å in diameter (Figure 5).[33-35,95,96] The disc-like fluorophores embedded in lipid bilayers have two principal modes of rotation: about the axis perpendicular to the plane of the molecule (in-plane) and about an axis in the plane of the ring system (out-of-plane).[35,95,96] The rotation of perylene is anisotropic; the rate of in-plane rotation is larger than the rate of out-of-plane rotation because the out-of-plane rotation requires a larger volume change than the in-plane rotation.[35] The limiting fluorescence anisotropy (r_0) of perylene depends upon the excitation wavelength since the angle between the absorption and emission transition moments is varied from 22 to 90° by appropriately choosing the excitation within the 270 to 410-nm range, so perylene is somewhat unique in that one can observe the in-plane and out-of-plane rotation separately by changing the excitation wavelength.[35,95,96] The probe has a high molar absorption coefficient and fluorescence quantum yield close to one. Perylene is a very stable fluorophore with a fluorescence lifetime of 7.2 ns, which is not too sensitive to temperature or fluidity changes. Because of its disc-like shape perylene is less favorable for monitoring small changes in fluidity than DPH with a hydrodynamically sensitive rod-like shape.[29]

b. DPH and DPH Analogues

The most frequently applied fluidity probe is 1,6-diphenyl-1,3,5-hexatriene (DPH) which possesses many of the features of an ideal probe (Table 1).[33] DPH is a polyene hydrocarbon with *all-trans* configuration

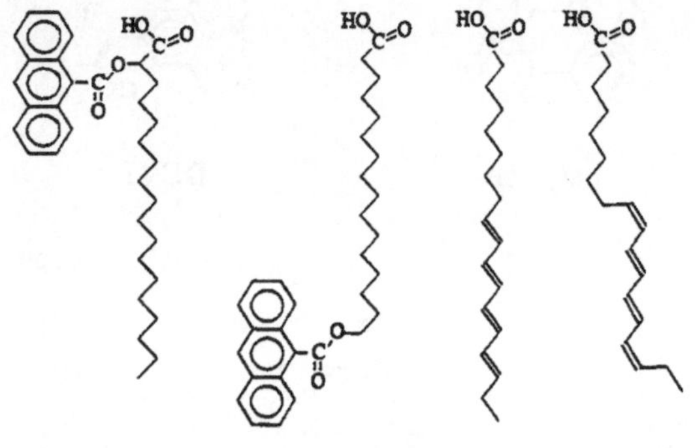

2-AS 16-AP cis-PA trans-PA

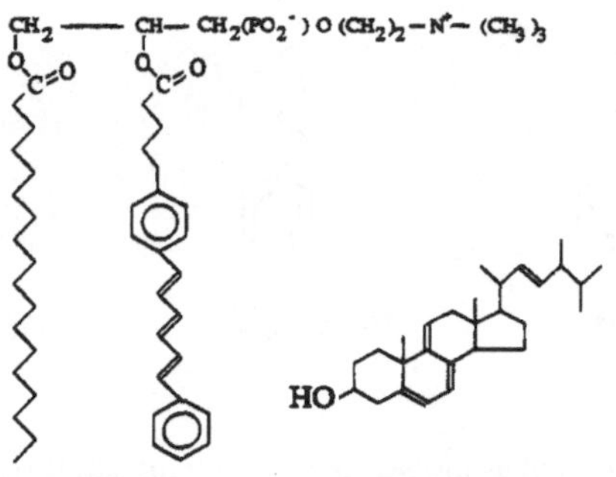

DPH-pPC dehydroergosterol

Figure 6. The chemical structures of the most frequently applied fluidity probes. II. Abbreviations used are 2-AS, 2-(9-anthroyloxy) stearic acid; 16-AP, 16-(9-anthroyloxy)-palmitate; *cis*-PA, 9,11,13,15,-*cis,trans,trans,cis*-octadecatetraenoic acid; *trans*-PA 9,11,13,15,-*all-trans*-octadecatetraenoic acid; DPH-pPC, 1-palmitoyl-2-[2-[4-(6-phenyl-*trans*-1,3,5-hexatrienyl)phenyl]ethyl]carbonyl]-3-*sn*-phosphatidyl-choline; dehydroergosterol, $\Delta^{5,7,9(11)22}$-ergostatetraen-3-β-ol.

Table 1. **Spectral and Physical Properties of the Most Frequently Applied Fluidity Probes**

Features	DPH, DPH analogs	Parinaric acids	Anthroyloxy fatty acids
Molar absorption coefficient (ε) [M^{-1} cm^{-1}]	91,000	74,000[a] 81,000[b]	6500–8300
Quantum efficiency (Q)	≃1	0.2,[c] ≃1[d]	0.3–0.7
Fluorescence lifetime (τ) (ns)	11	5,[c] 50[d]	7–11
Initial anisotropy (r_0)	0.362	0.37–0.39	0.285
Stokes shift (nm)	100	90	110
Solubility	Lipid soluble	Lipid solublel	Lipid soluble
Stability	Photoisomerization	Perishable (O_2 sensitive)	Stable
Ref.	29, 33, 63, 97, 99–103	110–121, 126, 137	32, 124–126, 137, 140

[a]*cis*-Parinaric acid.
[b]*trans*-Parinaric acid.
[c]Dissolved in liquid-crystalline lipid phase.
[d]Dissolved in gel lipid phase.

and an elongated rod-like shape (Figure 5). DPH belongs to the first group of the probes, i.e., spontaneously incorporates into lipid bilayers. Its partition coefficient between water and lipid phase is high, 1.3×10^6, as it was determined by Huang and Haugland.[97] In addition, DPH partitions equally well into solid or fluid lipid domains.[29] DPH dispersed in aqueous media is practically void of fluorescence.[29,33] When DPH dispersion is mixed with a lipid-containing system, the incorporation of DPH into the hydrophobic regions can be monitored with a sharp increase in fluorescence intensity.[29] DPH, dissolved in organic solvents and lipids, has a high extinction coefficient and quantum yield. Its absorption and emission spectra are well separated and the problems associated with the effect of energy transfer on fluorescence anisotropy can be minimized. Absorption and emission dipoles of DPH are almost parallel; therefore, its limiting initial anisotropy (0.362)[29,33] is close to the theoretical maximum.[29,33,63,64,68,70] Fluorescence decay time of DPH is around 11 ns, which provides an appropriate time window for studying dynamics of lipids in membranes.

Illuminated continuously, DPH may undergo *cis-trans* isomerization causing a decrease in fluorescence intensity. This photoisomerization is reversible to a large extent and strongly depends on the viscosity of the medium. In media with high viscosity, the *cis-trans* isomerization is inhibited, and thus the fluorescence quantum yield of DPH is close to 1.[29,63] Photoisomerization of DPH may cause a dramatic increase in its fluorescence anisotropy for the following reasons. In cells DPH is distributed over all lipidic components of the cell.[98] Photoisomerization of DPH into the nonfluorescent state increases with decreasing viscosity, so fluorescence will significantly decrease in the more fluid parts of the cell and only the molecules in the most viscous parts will remain intact, and hence the observed anisotropy will increase. This interpretation was supported by the observation that at low temperature the radiation effect is practically eliminated due to the increased viscosity of all parts of cells.[98]

When labeling intact cells with DPH, in the early phase of DPH uptake most DPH molecules are located in the cytoplasmic membrane but later on they translocate into the intracellular membranes and neutral lipid stores. Charged derivatives of DPH were synthesized to overcome this problem. 1-(4-Trimethylammoniumphenyl)-6-phenyl-1,3,5-hexatriene (TMA-DPH) and 3-[*p*-(phenyl)-1,3,5-hexatrienyl]phenyl-propionic acid (DPH-PA) (Figure 5) are positively and negatively charged, respectively, at neutral Ph.[99-103] The terminal charge of these molecules ensures that the probes are anchored at the lipid-water interphase and the orientation of the fluorophore portion of the molecules is perpendicular to the plane of the membrane. Most of the fluorescence properties of these cationic and anionic DPH derivatives are very similar to that of the parental DPH molecule. TMA-DPH and DPH-PA have the advantage over DPH that their position in the lipid bilayers can be better characterized as it will be discussed below. In contrast to DPH, the fluorescence anisotropy of TMA-DPH or DPH-PA does not increase with illumination time due to photoisomerization, since they do not enter different lipidic compartments of cells. It should be mentioned that, though TMA-DPH is rapidly and specifically incorporated into the plasma membrane and anchors by ionic interactions between the cationic moiety and the negatively charged phospholipid heads, it is further internalized more or less rapidly, depending on the cell type.[104,105] For example, it has been shown that TMA-DPH is internalized by mouse bone marrow macrophages via fluid-phase pinocytosis and the probe remains associated with the plasma membrane throughout its internalization-recycling flow and it is finally concentrated in lysosomes.[104] The internalization of TMA-DPH does not cause a drastic decrease in fluorescence anisotropy since the charged TMA-DPH molecule does not enter neutral lipid compartments of low viscosity inside the cell, unlike the uncharged DPH.

Another way to anchor DPH to the surface of plasma membrane was the synthesis of the sugar derivative of DPH, the glyco-DPH (diphenylhexatrienylpropanoylhydrazylstachyose), whose large hydrophilic tetrasaccharide headgroup should make it even less prone to transmembrane translocation than TMA-DPH.[106]

Although the solubility of DPH, TMA-DPH, and DPH-PA is very low in aqueous phase, these molecules are able to exchange freely through solution and redistribute into unlabeled lipid bilayers. To overcome this problem, a phospholipid analogue incorporating the DPH fluorophore was synthesized.[107] The new compound was unable to translocate freely through solution when incorporated as a probe in a subset of phospholipid vesicles and was exploited to study the fusion of phospholipid vesicles to form larger structures.[107] Further characteristics of DPH-containing phospholipid analogues are discussed in the subsequent subsection.

c. Lipid Analogues

Using the advantageous fluorescence properties of DPH, a phosphatidylcholine analogue (1-palmitoyl-2-[[2-[4-(6-phenyl-*trans*-1,3,5-hexatrienyl)phenyl]ethyl]carbonyl]-3-*sn*-phosphatidylcholine, DPHpPC) was synthesized, which incorporates the DPH moiety into a phospholipid at the 2-position, replacing a normal hydrocarbon chain (Figure 6).[108] The fluorophore portion of this molecules is oriented perpendicular to the bilayer plane similarly to TMA-DPH. DPHpPC shows a three-fold preference to partition into fluid phase relative to solid phase of mixed-lipid bilayers.[108] This probe should not significantly perturb bilayer packing since it is a lipid-like molecule. This is only partially true, because, while DPHpPC has a negligible perturbing effect on the cooperative phase behavior of membranes above phase transition temperature, it appears to perturb its local environment below phase transition, and may segregate into probe-enriched domains that melt at somewhat lower temperatures that the bulk lipid.[108] Analysis of rotational motion of DPHpPC showed a smaller cone angle, related to the range of acyl chain motion as compared to DPH reflecting the difference in probe location.[101,107-109]

Parinaric acid is a polyene fatty acid isolated from the fruit of the plant *Parinarium glabberiumum*. This nonaromatic fluorophore is most similar to natural fatty acids. Its two isomeric forms and their esters are widely used in membrane research: *cis*-parinaric acid (9,11,13,15,-*cis,trans,trans,cis*-octadecatetraenoic acid, *cis*-PA) and *trans*-parinaric acid (9,11,13,15,-*all-trans*-octadecatetraenoic acid, *trans*-PA) (Figure 6, Table 1).[109-120] *trans*-PA partitions preferentially into the solid phase of mixed phase lipid aggregates, while the methyl ester form of *cis*-PA is distributed equally between the gel and liquid-crystalline phases in mixed phase membranes.[110-112,116] Since it is present in both phases of a membrane throughout the phase-transition temperature range, the temperature dependence of its fluorescence anisotropy can be used to detect the onset and completion of phase transition in phospholipid dispersions.[111] The quantum yield and the fluorescence lifetime of parinaric acids, especially that of *trans*-PA, increases two to four times when fluid to gel transition occurs in lipid bilayers.[110-112] This unique feature of *trans*-PA can be used to detect the onset of phase-transition by measuring its fluorescence intensity.[115] Parinaric acids have no effect on the transition temperature if less than 1 probe molecule per 30 phospholipids is present. The emission spectrum and the lifetime are unaffected by the viscosity of the solvent environment. In fluid liquid-crystalline phase the fluorescence lifetime of parinaric acids is around 5 ns; in solid (gel) phase it can increase up to 50 ns.[121] The limiting initial anisotropy value ($r_0 = 0.37$ to 0.39) of parinaric acid is close to the theoretical limit (0.4). Fluorescence anisotropy decay curve of parinaric acid shows a relatively fast decay to a finite value of r_∞ in the fluid phase.[121] The major drawback of parinaric acids is that they are quite perishable and air sensitive due their extended conjugated polyene structure.

The anthroyl fatty acid isomers and homologs are among the most frequently applied fluorescent fatty acid analogues. The first member of this family, 12-(9-anthroyloxy) stearic acid, was introduced by Waggoner and Stryer.[122] In the case of this probe the fluorophore is introduced at a defined transverse position in the lipid bilayers.[122] Since then, a whole family of *n*-(9-anthroyloxy) fatty acids (*n* = 2, 3, 6, 7, 9, 11, 12, and 16) have been synthesized and characterized (Figure 6, Table 1).[32,123-139] These probes are expected to intercalate between phospholipid chains and to locate their anthracene fluorophores at a graded series of depths in the bilayer.[124,126] The molar absorption coefficients of these analogues are moderately high, ranging from 6500 to 8300 M^{-1} cm^{-1}, and significant Stokes shift can be observed between the excitation and emission maxima.[32,124-126] The position of the emission maximum has little dependence on solvent viscosity, but viscosity changes the degree of vibrational structure seen in the emission spectrum. Absorption and emission dipoles of anthracene are not parallel; both are in the plane of the anthracene ring at an angle of 30°; therefore, the limiting initial anisotropy ($r_0 = 0.285$)[126,137,140] is not as close to the theoretical maximum ($r_0 = 0.4$)[29,33,63,64,68] as in the case of DPH.[29,33] The fluorescence lifetime of *n*-(9-anthroyloxy) fatty acids is around 11 ns and its value slightly increases as the fluorophore is moved deeper into the bilayer, indicating that a gradient of polarity exists from the surface to the center of the bilayer.[124,126] Similarly, the fluorescence quantum yield of anthroyloxy fatty acids also increases as the fluorophores are deeper in the bilayer as would be expected for the polarity gradient.[124] The solvent viscosity has little effect on the fluorescence lifetime of the probe.[32,124-126] When both gel and liquid-crystalline phases are present, these probes appear to favor the fluid, liquid-crystalline phase.[141] Anthroyloxy fatty acids slightly perturb the membrane structure because their bulky anthracene ring protrudes as a pendent group from the acyl chain.[142] This perturbation is less when the anthracene ring is linked to the ω-carbon atom as in the case of 16-(9-anthroyloxy)-palmitate.[138,143]

Frequency domain fluorescence spectroscopy has been applied for studying the effect of Ph change on the fluorescence characteristics of 2-(9-anthroyloxy) stearic acid in phospholipid vesicles.[109] The fluorescence lifetime distribution curve was shifted significantly to shorter lifetimes when the Ph of the solution changed from 7.0 to 3.0. The apparent pK_a of this probe was 6.3, implying that the stearic acid probes exist predominantly as a single ionized population in membranes at physiological pH.[109]

d. Other Probes

Dehydroergosterol, $\Delta^{5,7,9(11)22}$-ergostatetraen-3-β-ol, the fluorescent analogue of cholesterol, is similar in structure to cholesterol, a primary constituent of most mammalian plasma membranes (Figure 6). The probe is very suitable for monitoring structural and rotational dynamic properties of cholesterol in model membranes[144-148] and in biological membranes.[120,144,149,150] In the range of 0 to 5 mol % fluorescent sterol, the fluorescence intensity per mole of dehydroergosterol, increases with increasing concentration, while the quantum yield, steady-state anisotropy, limiting finite anisotropy, lifetime, and rotational rate remain relatively constant in phosphatidylcholine vesicles.[146] In vesicles containing 5 to 33 mol % dehydroergosterol, the fluorescent sterol interacts to self-quench, thereby decreasing the fluorescence intensity, quantum yield, steady-state anisotropy, limiting finite anisotropy, and increasing the rotational rate.[146] The fluorescence lifetime of dehydroergosterol remains unchanged. At higher concentrations, above 33 mol % sterol, the dehydroergosterol is laterally segregated and only a single type of sterol phase exists in phospholipid vesicles.[146] In the mixture of phospholipids with gel and fluid phases, the fluorescent sterol below 0.5 mol % interacts preferentially with the fluid phase lipid.[148]

Two other fluorophores, pyrene and 9-(dicyanovinyl)julodinine (DCVJ), can be applied to estimate membrane viscosity through measuring fluorescence parameters other than fluorescence anisotropy.

In case of pyrene and pyrene analogues, the excimer formation is utilized. The diffusion-controlled formation of a binary complex between two pyrene molecules in ground and excited state, respectively, leads to spectral changes that can be exploited to determine the lateral diffusion constant of the pyrene in a lipid bilayer.[151-153] The monomer and the excimer emits in different spectral regions and the ratio of fluorescence intensities monitored at appropriate wavelengths is the measure of the lateral diffusion rate. Under conditions of constant temperature and equal concentrations of pyrene the excimer to monomer ratio decreases when the membrane viscosity increases.[151-153]

The other probe, DCVJ, is a well-characterized charge-transfer malonitril derivative (Figure 6).[154] The fluorophore behaves like a molecular rotor, whose fluorescence decay is governed by torsional relaxation and its fluorescence quantum yield is viscosity dependent. The relationship between the quantum yield of DCVJ and the solvent viscosity has been established and applied for estimating effective microviscosity in dipalmitoylphosphatidylcholine vesicles.[154]

1. Location of Fluidity Probes

Before any probe can be useful in reporting the structural and dynamic properties of membrane systems, the localization of the probe in the system should be determined. Even if the probe is "dissolved" in an artificial membrane made from one component only, problems are encountered. The probe may or may not be dispersed homogeneously, depending on the probe and membrane characteristics. In the case of more than one component, lateral segregation and/or asymmetry in the distribution of fluidity probes among the inner and outer leaflets of lipid bilayer may arise. When protein molecules are embedded into artificial membranes, new possible locations emerge for the probes: the annular lipid region surrounding the protein molecules and the hydrophobic regions of the proteins. Methods available to determine the location of fluidity probes include taking microscopic pictures, analyzing spectral characteristics, fluorescence lifetime measurements, fluorescence resonance energy transfer methods, angle-resolved fluorimetry, and different types of quenching. Applications of these techniques to determine the location of various fluidity probes in membrane systems are discussed below.

a. Perylene

The solubility of perylene and its fluorescence properties suggested that the probe is located in the hydrophobic portion of membranes.[35] The extent of association of hydrophobic proteins with these probes has not been established.[29,35] Investigation of wavelength-dependent fluorescence anisotropy decay of perylene in vesicles of dimyristoylphosphatidylcholine revealed that both in-plane and out-of-plane rotations are partially hindered on the nanosecond time scale, none of them being completely free.[95] It has been suggested that the preferred orientation of the X-Y plane of perylene is parallel to the long axes of the lipid acyl chains, since the order parameter, S^2, did not exceed 0.25.[96]

b. DPH and DPH Analogues

The spectral characteristics of DPH in lipid systems below and above the phase transitions were very similar to those in benzene and high paraffin solvents. In addition, the fluorescence intensity of DPH was practically unquenched in the presence of charged quenchers such as pyridinium or iodide salts. On the other hand, retinal, partitioning into the lipid hydrophobic region, quenches DPH fluorescence.[29] These findings strongly indicate that DPH is localized almost exclusively among the hydrocarbon cores of the lipids.

Application of DPH for monitoring plasma membrane viscosity of intact cells raised criticism with respect to the interpretation of fluorescence anisotropy data. Visualization of DPH labeled intact cells by fluorescence microscopy indicated that all lipid material of cells, including cytoplasmic lipid droplets, was labeled with DPH.[155] The fluorescence signal from inside the cells is a large fraction of the total cell fluorescence. These results indicate that fluorescence anisotropy data obtained from intact cells, using DPH as probe, give information on the total lipid material of the cells rather than exclusive information on the structure and dynamics of plasma membranes.[155] This notion was also supported by the finding that fluorescence anisotropy values of DPH embedded into isolated plasma membranes always report about higher viscosity than measured in intact cells.[100,156-158] Changes in fluorescence anisotropy during the course of DPH uptake in intact cells were also compatible with a model assuming that the plasma membrane and the cytoplasm constitute two compartments, with some DPH translocation from one compartment to the other. In short-term incubation, fluorescence parameters of DPH characterize the plasma membrane. In long-term incubation, a large fraction of fluorescent signal is due to DPH embedded in intracellular lipids being in a more fluid state than membrane lipids.[159]

The intracellular distribution of TMA-DPH is much more favorable than that of DPH as demonstrated by fluorescence micrography and fluorescence intensity measurements.[100,160] TMA-DPH is incorporated very rapidly, within 1 min, into the plasma membrane and remained restrictively localized in it for at least 30 min.[160] After 30 min this probe is progressively internalized through fluid-phase pinocytosis depending on the cell type.[104,105] These data suggest that TMA-DPH reliably monitors plasma membrane fluidity using not too long incubation times (<30 min).[104,105] Since TMA-DPH usually exhibits higher anisotropy values than DPH, TMA-DPH and DPH are probably located in different lipid regions.[158] Using a fluorescence quenching method, the fluorescence of probes embedded in either plasma membranes or cytoplasm can be distinguished.[158,161] 2,4,6-Trinitrobenzenesulfonic acid (TNBS) used in nonpenetrating conditions binds to amino groups on the cell surface and quenches fluorescence of DPH and TMA-DPH in the plasma membrane. Under these conditions, fluorescent molecules embedded in cytoplasmic elements contribute to the emitted fluorescence only. Plasma membrane fluidity can be determined using a procedure considering fluorescence anisotropy values and fluorescence intensities of untreated and TNBS-treated samples.[158,161]

By comparing the fluorescence anisotropy and mean lifetime of DPH in membranes of intact human erythrocytes and ghost membranes, the structural and dynamic behavior of the outer hemi-leaflet of the membrane was estimated.[162] Since nonradiative energy transfer occurs at much higher probability from DPH embedded in the inner leaflet of the membrane than from DPH in the outer leaflet of the membrane to heme acceptors at the endofacial surface of intact cells, the fluorescence observed in intact cell suspension was weighted in favor of outer leaflet fluorophores. Lower fluorescence anisotropy values were measured in the case of intact erythrocytes than in case of ghosts, free of heme, indicating that the outer leaflet lipids have greater motional freedom.[162] Three other fluidity probes, 12-anthroyloxystearate, 2-anthroyloxystearate, and pyrenedecanoic acid, applied in this system, gave similar results.[162] The asymmetry in the fluidity of the inner and outer leaflets of the platelet membrane was just the opposite.[103] The inner leaflet of the bilayer was less rigid as monitored with DPH, cationic TMA-DPH, and anionic DPH-PA. Since the negatively charged phospholipids of almost all mammalian plasma membranes are localized preferentially and almost exclusively in the inner leaflet, TMA-DPH should bind preferentially to the cytoplasmic surface of the membrane and DPH-PA should be repelled from this region by electric expulsion (Figure 7). Incubating platelets with DPH-PA, the fluorescence intensity reached equilibrium in 2 min, and the fluorescence anisotropy remained constant afterward. Labeling with TMA-DPH resulted in gradual increase in fluorescence intensity due to the penetration of the TMA-DPH into the cytoplasmic site of the platelet membrane, whereas the fluorescence anisotropy gradually decreased, indicating that the fluidity of the inner leaflet is higher than that of the outer one.[103] Similar results were obtained when LM fibroblasts were labeled with TMA-DPH and DPH-PA.[163] Contradicting results were obtained when TNBS quenching technique was applied to differentiate between fluidities of inner and outer leaflets of LM fibroblast plasma membranes. These results suggested that the order sensed by TMA-DPH and DPH-PA is intrinsically different (Figure 7).[163]

In addition to the distribution of the probe within the cells, the transverse location of DPH in the lipid bilayer should also be considered (Figure 7). A comparison of order parameters derived from fluorescence anisotropy and ^{2}H-NMR studies suggests that DPH lies close to the C_{10}-C_{12} region of the acyl chains.[70] This conclusion was reached based on the assumption that DPH, which is 13 Å long in its extended all-*trans* form, senses some average order parameter of the hydrocarbon segments in its environment. This is only an approximation because of the size of the fluorophore relative to the thickness of the bilayer.[71] Furthermore, the variation in order parameter along the methylene groups from C_1 to C_{12}

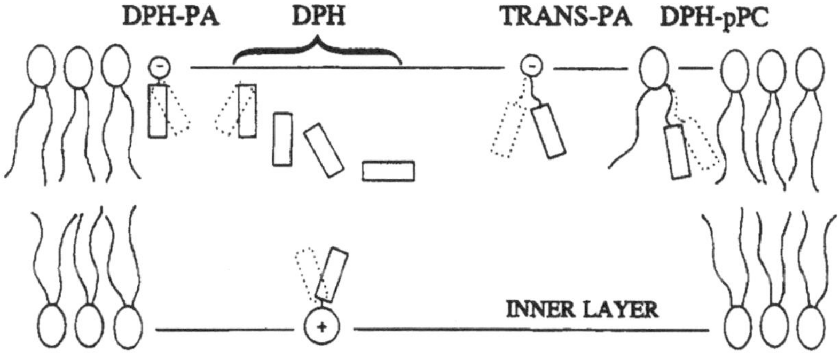

Figure 7. Schematic diagram of proposed locations of the fluorescent probes. The broken lines represent possible types of motion. The abbreviations used are the same as in legends to Figures 5 and 6.

is such that the reasonable fit of the order parameter of DPH with that of C_{10} and C_{12} can even be considered as accidental. DPH probably occupies some average position relative to the fatty acyl chains due to the vertical diffusion of the DPH molecule.[71] Because of its positive charge, TMA-DPH is anchored at the lipid/water interface with the DPH moiety intercalated between the upper portions of the fatty acyl chains, probing essentially the fatty acyl chain regions as far down as C_8-C_{10}.

The depth of a particular probe in the bilayer can be estimated with special quencher molecules. The iodide ion, a very efficient quencher of excited state species, is highly soluble in aqueous environment and much less soluble in the apolar environment of the bilayer. The efficiency of quenching of lipid-soluble probes such as DPH, TMA-DPH, phosphatidylcholine analogue of DPH (DPH-PC), and cholesterol analogue of DPH (DPH-CH) with iodide ion was used to assess the depth of the particular probe in the bilayer.[164] TMA-DPH was most highly quenched, indicating that the position of TMA-DPH is closer to the surface as expected from its polar nature. DPH-CH was quenched the least; therefore, the probe must be situated in the center of the bilayer. Since the DPH molecules were quenched only slightly better than the DPH-PC probes, DPH molecules probably occupy the region near the center of the bilayer.[164] This is consistent with the interpretation that, unlike TMA-DPH, DPH is not "anchored" to the surface of the bilayer (Figure 7).

The fluorescence resonance energy transfer method was also applied to determine the transverse location of the DPH in model membrane systems. The separation between the planes containing the fluorescence acceptor moiety of the fluoresceinthiocarbamyl adduct of a natural phosphatidylethanolamine located at the surface of the bilayer and the donor DPH buried in the hydrophobic interior of the membrane was measured. The estimates obtained for the apparent separation of donor and acceptor planes suggest that the location of DPH is close to the center of the bilayer.[165] Fluorescence resonance energy transfer was measured between peptides containing tryptophan as donor, and DPH molecules as acceptor, in phosphatidylserine vesicles.[166] The positively charged peptide binds strongly to the negatively charged vesicles and these electrostatic interactions allow the aromatic ring of tryptophan to insert into a more hydrophobic environment. The efficiency of fluorescence energy transfer increased with the increasing surface density of the acceptor DPH, indicating that the closest distance of the tryptophan and the DPH probe is around 1 nm.[166] Similarly, significant fluorescence resonance energy transfer from tryptophan residues of intrinsic membrane proteins to DPH was demonstrated in rat erythrocyte ghosts.[166]

The orientational order of DPH was studied from the decay of fluorescence anisotropy in membrane suspensions. The decay curve was analyzed applying a new model for the second- and fourth-order parameters on nonoriented systems. The results indicated that a fraction of DPH was perpendicular to the preferred direction of the acyl chains in the lipid bilayer.[167] Similar results were obtained when angle-resolved fluorescence depolarization experiments were carried out on DPH and TMA-DPH molecules embedded in macroscopically oriented multilayers of saturated or unsaturated phosphatidylcholines.[90,92] Consideration of the fourth-order parameter indicated the DPH molecules have heterogeneous distribution in bilayers of unsaturated lipids, with a significant fraction of the molecules lying with their long axes parallel to the bilayer planes.[90,92] This tendency was much less profound in the case of TMA-DPH, indicating that TMA-DPH is a more useful probe than DPH, especially in systems containing unsaturated lipids.

Heterogeneity in the lipid organization in lipid bilayers and cell membranes can be probed with DPH and DPH analogues by analyzing their fluorescence decay determined by multifrequency phase fluorimetry.[168-176] When the fluorescence decay of DPH in phospholipid vesicles was analyzed by assuming that the decay is a discrete sum of two exponential components, the fit of the decay curve gave a relatively large error.[168] The original assumption that the two exponential terms obtained from the usual lifetime analysis correspond to well-defined molecular surroundings of the DPH molecule may be an oversimplification. DPH lifetime is strongly dependent upon the dielectric constant of the medium.[169] Since the dielectric constant is a function of the position in the membrane, a wide range of lifetime values of DPH is possible. Better results were obtained when the fluorescence decay of DPH in phospholipid vesicles was analyzed by assuming one or more continuous distributions of lifetime. Out of the three symmetric distribution functions (uniform, Gaussian, and Lorentzian) the Lorentzian distribution function of fluorescence lifetimes gave the best description of the decay curve.[168] In most cases investigated, it was necessary to include a second distribution centered at shorter lifetime values. The origin of the short lifetime component is still debated. It has been suggested that this component (1) belongs to photochemical derivatives of the DPH;[170] (2) is rendered to a fraction of probe molecules localized at the surface in a very polar environment or in the boundary between environment of different physicochemical features where the permeability of water molecules is increased;[171] and (3) is an artifact resulting from the assumption of a symmetric distribution.[168] At temperatures above the phase transition the width of the major components of lifetime distributions of DPH is relatively narrow because, in liquid-crystalline phase, the probe molecule can freely diffuse perpendicular to the membrane and rapidly averages different surroundings during the excited lifetime.[168] Similar results were obtained when the heterogeneity of the erythrocyte membrane was studied using the fluorescence lifetime distribution of DPH.[172] For TMA-DPH the width of the lifetime distribution is broader than for DPH because TMA-DPH is anchored to the membrane surface with its cationic residues and the DPH moiety resides in a region with a large polarity gradient near the surface.[173] The phosphatidylcholine analogue of DPH (DPH-PC) has a homogeneous lifetime distribution probably due to a more fixed location along the membrane normal or to a more fluid environment in the middle of the bilayer.[168,174] Increasing concentration of cholesterol decreased the width of the DPH-PC lifetime distribution in vesicles of palmitoyloleoyl phosphatidylcholine, indicating that cholesterol not only rigidifies the membrane but may also act as a membrane homogenizer.[175] Using DPH-PC-labeled hepatocyte plasma membranes and microsomes, it was possible to observe two separate lifetime medians for lifetime distributions, which correspond to the fractions of DPH-PC residing in the bulk and the boundary lipid regions.[176]

c. Lipid Analogues, Fluidity Gradients

cis- and *trans*-parinaric acids, which are polyunsaturated conjugated fatty acids, are aligned parallel to the phospholipid acyl chains with their fluorophores close to the bilayer center and their carboxyl group is located at the interface between the polar and apolar regions of the bilayers (Figure 7).[110,111] Since the fluorescence anisotropy decay curve of parinaric acid can be characterized with a relatively fast decay to a finite value of r_∞ in the fluid phase, the parinoyl chromophore can be regarded as a rigid stick attached to a flexible string that itself is tethered to the bilayer surface by the hydrophilicity of the carboxyl group. The chromophore is situated in a mobile environment relatively deep in the bilayer where *trans-gauche* isomerizations of methylene carbons can occur.[121] Comparing the order parameter for C_9 obtained from ^{2}H-NMR with that calculated from r_∞ for fluorescent parinaric acids showed very good agreement which is not surprising, since these polyene probes can be considered both rigid and fully extended and centered near C_9 of the phospholipid acyl chain.[121] Fluorescence resonance energy transfer measurements between the single tryptophan residue of the M13 virus coat protein and parinaric acid embedded into dimyristoylphosphatidycholine vesicles revealed the perturbing effect of the protein.[114] The *cis* and *trans* isomers of parinaric acid were equally efficient as quenchers at 30°C when the bilayer was in fluid state, but at 10°C quenching by the *cis* isomer was much more efficient than by the *trans* form. While *cis*-parinaric acid distributes equally between the gel and fluid-crystalline lipid phases, *trans*-parinaric acid partitions preferentially into the gel phase, thus avoiding the disordered annular lipid region surrounding the protein donor.[114] Lifetime heterogeneity studies of *cis*- and *trans*-parinaric acids embedded in placental microvillus membrane vesicles detected both long (>25 ns) and short (<6 ns) components,

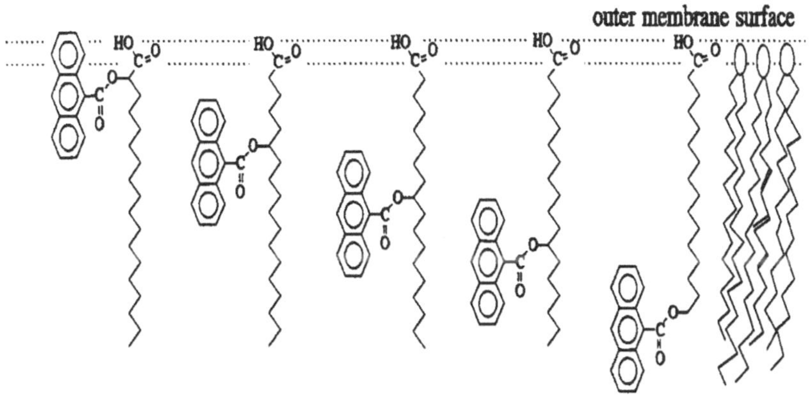

Figure 8. Possible relative transverse positions of *n*-(9-anthroyloxy) fatty acids in a lipid bilayer. The abbreviations used are the same as in the legend to Figure 6.

corresponding to the probes in solid and fluid lipid phases.[177] These data demonstrated that the analysis in the heterogeneity of fluorescence lifetimes of parinaric acid provided information about the lipid domain structure.[177]

Several lines of evidence support that *n*-(9-anthroyloxy) fatty acids ($n = 2, 3, 6, 7, 9, 11, 12$, and 16) intercalate with phospholipid chains, locate their fluorophores at a graded series of depths in the bilayer, and monitor the local environment (Figure 8).[134] Aqueous quenching of anthroyloxy fluorescence with Cu^{2+} in isolated beef mitochondria was in the order $2 > 6 > 9 > 12 > 16$.[178] In the case of dimethylanaline, a quencher that partitions into the lipid phase near the bilayer center, the order of quenching was reversed.[178] In model bilayers, using spin-labeled fatty acids as paramagnetic quenchers, the decreasing order of quenching was always 2, 6, 9, 12, and 16, with respect to alkyl chain position.[179] The fluorescence of *n*-(9-anthroyloxy) fatty acid probes associated with vesicles was not quenched with acrylamide, indicating that the 9-anthroyloxy moiety of the long chain probes is buried within the hydrocarbon region and thus it is well shielded from the aqueous phase.[180] Fluorescence quenching in dipalmitoylphosphatidylcholine bilayers data revealed that only the 2-(9-anthroyloxy) stearate had portions of the anthroyloxy moiety in both the aqueous and the lipid phases, whereas other probes ($n > 2$) were entirely in the lipid phase.[124] The fluorescent membrane potential probe, 3,3′-dipropylthiodicarbocyanine [diS-C_3-(5)] quenched the fluorescence of *n*-(9-anthroyloxy) fatty acids in renal brush-border membrane vesicles.[181] The quenching of *n*-(9-anthroyloxy) fatty acids increased with n ($16 > 12 \gg 6 > 2$) at higher diS-C_3-(5) concentration where the probe is mostly dimerized. diS-C_3-(5) dimers lie perpendicular to the acyl chains of the phospholipids near the center of the bilayer, allowing efficient collision quenching between the dimers and the 16- or 12-(9-anthroyloxy) fatty acids.[181]

In intact, living HeLa cells labeled with anthroyloxy fatty acid probes, it has been demonstrated, using the aqueous phase quencher Cu^{2+}, that the anthroyloxy moieties of the probes are located at a graded series of depths in the outer leaflet of the plasma membrane.[138] The observed quenching behavior indicated that the relative location of anthroyloxy fatty acid probes in the HeLa plasma membrane is the same as in model bilayer system, i.e., $2 \geq 3 > 6 \geq 7 > 9 > 12 > 16$. Therefore, it was concluded that these probes are suitable for measuring dynamic parameters at different depths in the plasma membranes of living cells as well as model membrane systems.[138]

The fluorescence resonance energy transfer method has been also applied to investigate the position of fluorophores in the transverse plane of the lipid bilayer.[182,183] Using *N*-stearyltryptophan as donor located on the outer surface of the lipid bilayer and *n*-(9-anthroyloxy) fatty acids ($n = 2, 6, 9$, and 12) as acceptors, significant transfer efficiencies were measured between the one donor multiple acceptor systems.[182] Although these data indicated that the donor and acceptor probes are within the range of efficient energy transfer process (7 nm), the method did not discriminate between the positions of the fluorophores in the transverse plane of the bilayer.[182] In another experimental system, where erythrocytes were labeled with 12-(9-anthroyloxy) stearic acid, the fluorophore served as donor and the hemoglobin molecule, bound to the inner surface of the red blood cell membrane, served as acceptor.[183] From the measured transfer efficiency values, after considering the possible orientations of hemoglobin, it was concluded that the minimal distance between acceptor and donor is around 4 to 6 nm. This range is in good agreement with the assumptions concerning the location of 12-(9-anthroyloxy) stearic acid in the outer lipid leaflet and the location of hemoglobin on the interior surface of the membrane.[183]

A set of *n*-(9-anthroyloxy) fatty acids (n = 2, 6, 9, 12, and 16) have been used to examine fluidity gradient in synthetic phospholipids using steady-state fluorimetry.[32,126] In bilayers composed of dipalmitoylphosphatidylcholine, the fluorescence anisotropy decreases (the microviscosity decreases) as the fluorophore moves deeper into the bilayer, indicating greater rotational motion rate of the fluorophore in the hydrocarbon core of the bilayer. In the crystalline state at 20°C there is a linear decrease in anisotropy from position 2 to 12. From 12 to 16 there is a steeper decrease in anisotropy, because, while the independent motion of the anthracene ring relative to the acyl chain is sterically hindered at positions 2 through 12, such motion is possible when the fluorophore ring is linked via the terminal carbon atom as in 16-(9-anthroyloxy) stearic acid. In the liquid-crystalline phase at 50°C, the decline in anisotropy is linear between positions 2 to 16, the gradient being steeper than that between positions 1 and 12 at 20°C. The anisotropy value for 16-(9-anthroyloxy) stearic acid may be underestimated relative to other probes because of the greater intrinsic mobility of its fluorophore. At 50°C, the values of anisotropy are so low for 16-(9-anthroyloxy) stearic acid that the difference in this region would be difficult to detect.[32,126] When these probes were embedded in vesicles of distearoyl- and dioleoylphosphatidylcholine, the anisotropy profile of the fluorophore showed a maximum near carbon 6.[125] These data suggested that in the case of the fluorescent probe technique the stiffening effect of the *cis* double bond is manifested above the actual position in the acyl chain. The steady-state fluorescence anisotropy technique cannot distinguish between a decrease in microviscosity caused by increased motion within lipid molecules and a decrease in microviscosity due to the increased molecular separation in the unsaturated phospholipid bilayer.[125]

In addition to the steady-state measurements, the time-resolved fluorescence anisotropy of *n*-(9-anthroyloxy) fatty acids (n = 2, 7, 9, 12, and 16) in vesicles of dipalmitoylphosphatidylcholine has been monitored at 21, 37, and 47°C.[127] The two modes of rotation, "in-plane" and "out-of-plane", of the anthroyl ring have been detected by varying the excitation wavelength.[127] The out-of-plane rotation, monitored at 316-nm excitation, was unhindered as demonstrated by an anisotropy decay profile asymptotic to zero. Thus, using this excitation wavelength, the evaluation of membrane fluidity parameter at different depths of the bilayer is possible, even in steady-state measurements. The out-of-plane rotation, which would mainly occur around the ester linkage, should be affected only by the frictional forces of the phospholipid chains. At 386-nm excitation, when mainly the in-plane rotation contributes to the anisotropy decay, the motional hindrance is evidenced by the existence of an r_∞ value in the anisotropy decay. The r_∞ value is characteristic of the in-plane rotation and reflects the hindrance exerted by the phospholipid organization on the reorientational motion of the fatty acid chain. In the pretransition temperature range (around 37°C), the results demonstrated the existence of structural lipid changes mainly localized in the hydrophobic core of the bilayer. The main transition leads to a complete disappearance of the hindrance on the in-plane rotation.[127] Time-resolved studies have shown that cholesterol increases both the correlation time and the value of r_∞.[184] This effect is most apparent in the C_7-C_9 region and is minimal at the surface (C_2) and near the center (C_{12}-C_{16}) of the bilayer, so that profiles of the parameters against bilayer depth pass through a maximum.[184] A similar fluidity profile has also been observed in unsaturated phospholipid bilayers.[132]

While the anthroyloxy fatty acid probes have been widely applied for studying model membranes or membrane preparations, very few studies have been published using these probes to examine the plasma membranes in intact, living cells. Continuous, very shallow gradients (i.e., decreasing anisotropy order of 2, 3, 6, 9, 12, and 16) were observed in erythrocyte ghosts[125] and in intact erythrocytes.[185] Microviscosity gradient of decreasing order of 6, 2, 9, and 12 was observed for mouse thymocytes.[125] It was suggested that the significant amount of unsaturated phospholipid in the plasma membrane of thymocytes led to a maximum in the anisotropy profile.[125] Using flow cytometry, the shape of the fluidity gradient of the outer hemi-leaflet of the plasma membrane of living HeLa cells has also been studied with a more complete set of *n*-(9-anthroyloxy) fatty acid probes (n = 2, 3, 6, 7, 9, 10, 11, 12, and 16).[136] Anisotropy values were constant from 5 to 30 min after addition of the various probes, and the magnitudes were in the order $7 > 6 > 9 = 10 > 2 = 3 > 11 > 12 > 16$, indicative of the shape of the fluidity gradient.[136] A similarly thorough study has been performed by the same group on human white blood cells taking advantage of the flow cytometric approach in resolving the populations of cells in heterogeneous mixture by dual-angle light scatter.[137] The observed anisotropy maxima were $n = 6$ for neutrophils, $n = 7$ for lymphocytes, and $n = 9$ for monocytes.[137]

d. Other Probes

The presence of two lifetime components for dehydroergosterol in small unilamellar vesicles of 1-palmitoyl-2-oleoylphophatidylcholine and the results of acrylamide quenching indicated that a small portion of dehydroergosterol is in a pool with polarity considerably different from that for the majority of sterols.[147] The two lifetime components of dehydroergosterol do not represent sterol in the outer and inner face of the phospholipid vesicles. Data were consistent with the interpretation that dehydroergosterol exists in both outer and inner monolayers of the vesicle membrane in two pools, with a short lifetime and a long lifetime, respectively. As shown by the experimental data for dehydroergosterol exchange between 1-palmitoyl-2-oleylphosphatidylcholine small unilamellar vesicles, the fast spontaneous exchange of sterol between membranes appears to occur from this small pool, sensitive to medium polarity. At any time only a small portion of the dehydroergosterol molecules is located at the surface so that the molecules are sensitive to the polarity of the medium and may undergo spontaneous desorption into the aqueous phase.[147]

D. MEASURING TECHNIQUES

While discussing the principles of membrane fluidity, evaluation of fluorescence anisotropy and characteristics of data obtained with fluorescent probes in model membrane systems proved to be very useful. From now on data obtained from measurements performed mostly on live cells will be reviewed. Therefore, the measuring techniques of fluorescence anisotropy will be discussed in terms of their applicability for monitoring fluorescence anisotropy in live cell membranes. This type of measurement can be performed on cell suspensions in fluorimeters where results are averaged over the whole cell population or data can be obtained on single cell level using a flow cytometer or an image microscope.

1. Cell Populations

Fluorescence anisotropy measurements of fluorescently labeled cell suspensions can be performed in conventional fluorimeters equipped with polarizers. Two types of geometrical arrangements are available for monitoring fluorescence anisotropy, the L-format and the T-format instruments. In both cases, the exciting light is vertically polarized by means of a Glan-Thomson prism or a polaroid film before reaching the cuvette containing the cell suspension. The depolarization of the emitted light can be determined as the ratio of the fluorescence intensities passing through the emission polarizer (analyzer), positioned vertical (I_{VV}) and horizontal (I_{VH}) with respect to the excitation polarizer. The first subscript labels the position of the excitation polarizer; the second subscript that of the emission polarizer. In L-format instruments, where a single photomultiplier is employed, the two components of the fluorescence intensity (I_{VV} and I_{VH}) cannot be measured simultaneously, rather just sequentially, by rotating the analyzer with 90°. This fact may cause some problems in the accuracy of the measurements, since the intensity of the light source may change between readings and the lack of 100% reproducibility of aligning the analyzer can also be a source of error. In addition to the polarizer and analyzer, other optical elements essential for the spectrofluorimeters such as lenses, filters, and gratings might have some depolarizing effect which should be taken into account as a correction factor (G). The value of G can be determined by measuring the vertical and horizontal emission intensities during excitation with horizontally polarized light (i.e., I_{HV} and I_{HH}). In this experimental setup both orientations of the analyzer are perpendicular to the plane of the excitation light and thus see equal intensities. The G factor can be incorporated in the equation describing fluorescence anisotropy:

$$G = \frac{I_{HV}}{I_{HH}} \tag{36}$$

$$r = \frac{I_{VV} - GI_{VH}}{I_{VV} + 2GI_{VH}} \tag{37}$$

In T-format instruments two photomultipliers are applied, on opposite sides of the sample cuvette, so that the two components of the emission intensity can be monitored simultaneously. In this arrangement it is much easier to follow time- and temperature-dependent changes in fluorescence anisotropy than in L-format instruments, where time is required to change the orientation of the analyzer. In T-format instruments, the photomultipliers should be matched and their gains individually adjusted so that excitation of the sample cuvette with horizontally polarized light results in identical intensities. This procedure also corrects for any differential transmission of polarized light by the analyzers. The linearity of the photomultipliers should also be checked. While the T-format instrument measures the anisotropy with the precision of 0.001, the precision of the L-format arrangement is rarely better than 0.005.[63]

Depending on the light source, the T- and L-format instruments can be used to monitor steady-state or time-resolved fluorescence anisotropy. The steady-state fluorescence anisotropy measurement, where the illumination is continuous, is the most frequently applied method for determining fluorescence anisotropy because of its simplicity. Time-resolved fluorescence anisotropy decay can be determined by applying pulse fluorometry where pulsed flash lamps or, in a more advanced technique, picosecond laser systems are used as excitation sources. A valuable review of this technique is provided by Heyn.[64] The other way to perform time-resolved fluorescence anisotropy measurements is the application of differential phase fluorometry. In this case, the continuous excitation light is sinusoidally modulated and therefore the emitted light will also be modulated at the same frequency but shifted in phase angle. A clear exposition of this technique is provided by Lakowicz.[65]

While working on cell suspensions, sedimentation and photobleaching may cause some difficulties. Gentle stirring of the cell suspension is required to prevent the cells from sedimenting and to limit the exposure of cells to the excitation beam. The other source of error comes from the light scattering phenomenon, which is rarely negligible in cell suspensions. The scattered excitation light is highly polarized, and it should be prevented from reaching the photomultiplier by inserting appropriate cut-off filters between the sample cuvette and the analyzer. The remaining scattered light intensity should be corrected for by measuring the intensity of light scattered by cell suspensions that contain no fluorescent probe. The other source of error is the possibility that a fluorophore can be excited by a scattered photon or that the emitted fluorescence itself can be scattered. Generally, scattering of incident or emitted fluorescent light results in decrease of the measured fluorescence anisotropy.[65] Since the intensity of the scattered light depends not only upon the cell concentration but also on the shape of the cells, care must be taken when fluorescence anisotropy values of cell suspensions with the same cell concentration are compared. The best way to compare fluorescence anisotropies of two particular cell suspensions is to measure anisotropy as a function of cell concentration and extrapolate to zero turbidity. The applicability of the correction method has been proven in various cell systems.[100,186,187]

When significant overlap exists between the excitation and emission spectra of the fluorophore, energy transfer process can take place, causing decrease in the measured fluorescence anisotropy values. Fluorescent probes with large Stokes shift should be selected to minimize this effect.

Although fluorescence anisotropy measurements carried out with cell suspensions in spectrofluorimeters provide useful information, the measured data are average values of all cells present in each sample. Using this approach, the heterogeneity of the cell population cannot be studied, and the fluorescence intensities coming from live or dead cells, or debris, cannot be distinguished. In addition, in systems where the fluorescence intensity of the free dye is not negligible, cumbersome procedures are necessary to get rid of the free dye. Even then, the release of the fluorescent probe from the cell membrane to the aqueous phase should be prevented to avoid the unwanted contribution of the fluorescence intensity of the free fluorophore to the measured fluorescence anisotropy values. Measurements on a single cell basis can successfully overcome most of the aforementioned problems.

2. Single Cells

The first quantitative method for the determination of membrane fluidity in live cells on a single cell level was developed by Inbar.[188] A special fluorescent instrument, the Elscint Single Cell Microviscosimeter, was constructed, which is, in essence, a modified fluorescence microscope. In this instrument a Glan-Thomson prism is used to polarize the excitation beam horizontally. On the emission side a Wollaston prism splits the light into two perpendicular polarized beams which are detected with two photomultipliers simultaneously. Since in the microscope the excitation beam is focused at a small spot, special care was taken to avoid the unnecessary photobleaching.[188] After calibrating the instrument with DPH-labeled liposomes of known anisotropy, membrane fluidities of normal lymphocytes and malignant lymphoma cells were compared.[188] From this instrument two types of experimental approaches have been developed for monitoring fluorescence anisotropy on a cell-by-cell basis: (1) flow cytometry, where thousands of cells can be analyzed within a second, and (2) image cytometry, where intracellular distribution of the parameter studied can be determined.

a. Flow Cytometry

Automated high-speed analysis and separation of cells on the basis of spectroscopic parameters can be performed with the use of flow cytometers. The general features of the instruments for the analysis of cells suspended in a fluid and distinguishable on spectroscopic grounds are described in Chapter 2. In this instrument cells passing through a nozzle are illuminated with a laser beam (which is inherently vertically polarized) and the anisotropy of the fluorescence light emitted from a single cell is determined by recording the vertical and horizontal components of the emission through polarizers with two photomultipliers simultaneously. In the first application of this technique it has been demonstrated that upon differentiation of Friend erythroleukemia cells the cell membrane becomes less fluid.[189] Cells at progressing degree of differentiation could then be sorted out according the their DPH fluorescence anisotropy values.[189]

A rigorous theory has been applied to measurements of fluorescence anisotropy with flow cytometers.[190] For measurements on single cells, the signals from the two photomultipliers detecting, respectively, $I_{\parallel}$ and $I_{\perp}$ have to be normalized by measuring a signal corresponding to totally depolarized emission similarly to that discussed in steady-state fluorometry. Three simple devices are available for equalizing $I_{\parallel}$ and $I_{\perp}$.[133] First, the direction of polarization of the laser light may be turned 90° by means of a 1/2 λ retardation plate. Since most retardation plates function for one wavelength only, separate plates have to be used for different excitation wavelengths. This also poses a problem with UV laser excitation, where two wavelengths are simultaneously used (351.1 and 363.6 nm). Second, it is possible to insert a third polarizer filter in front of the beam splitter oriented at 45° with respect to both analyzers. This requires very careful orientation of the three polarization filters with respect to each other. Third, application of a pseudodepolarizer in front of the beam splitter has distinct advantages, because it can be used regardless of the excitation wavelength, and it does not require the extremely accurate alignment of the analyzers.[133]

The number of membrane fluidity studies utilizing flow cytometry is slowly increasing. An Epics V flow cytometer was used to observe decreases in anisotropy of brush-border membrane vesicles of rabbit enterocytes, as a function of time after DMSO addition.[191] The fluorescence anisotropy of mouse F9 cells stained with DPH and TMA-DPH increased following differentiation as observed with a FACS II flow cytometer.[133] However, there was no change in fluorescence anisotropy upon differentiation when F9 cells were stained with *n*-(9-anthroyloxy) stearic acid (*n* = 2, 6, 9, and 12).[133] Using EPICS V, Fox and Delohery[192] demonstrated fluorescence anisotropy changes, i.e., membrane fluidity changes, in DPH-stained CHO and HeLa cells by varying the temperature of the sheath fluid and the sample container. Cells labeled with DPH after 24-h incubation in medium lacking serum showed a significantly higher fluorescence anisotropy than cells in medium containing serum.[192] This is consistent with the finding that cells can take up alkyldiacylglycerols and triacylglycerols from the serum and sequester them in lipid vesicles. During DPH staining these vesicles also become labeled, and they are more fluid than the plasma membrane, leading to a lower fluorescence anisotropy value.[192] Collins and Grogan[135] compared spectrofluorimetric and flow cytometric kinetic measurements of fluorescence anisotropy of HeLa cells stained with DPH, TMA-DPH, and 12-(9-anthroyloxy) stearic acid. After 1 min the measured anisotropy values of DPH and 12-(9-anthroyloxy) stearic acid labeled cells rapidly decreased, while that of TMA-DPH-labeled cells changed only slightly, suggesting that DPH and 12-(9-anthroyloxy) stearic acid, unlike TMA-DPH, do not remain anchored in the HeLa cell plasma membrane, but translocate to more fluid environments inside the cells.[135] There was no significant difference between the results obtained with the flow cytometer and those obtained with the fluorimeter.[135] Using an Epics 753 flow cytometer, the shape of the fluidity gradient of the outer hemi-leaflet of the plasma membrane of living HeLa cells[136] and human white blood cells[137] was determined using a series of n-(9-anthroyloxy) fatty acid probes where n = 2, 3, 6, 7, 9, 10, 11, 12, and 16. Neutrophils, lymphocytes, and monocytes have been identified using the capability of the flow cytometer to differentiate between them on the basis of forward angle and right angle-scattered light intensity.[137]

Flow cytometry, under appropriate conditions, can be as reliable as the much more widely used spectrofluorimetry method for fluorescence anisotropy measurements. Two obvious advantages of flow cytometry over spectrofluorimetry for membrane studies are the ability to detect subpopulations (due to different cell cycle stages, damaged cells, etc.) within a "homogeneous" cell culture and the ability to select and analyze simultaneously specific populations of cells in heterogeneous mixtures such as blood, bone marrow, liver, etc., as resolved by forward and right angle light scatter. In addition, the effect of photobleaching is much less in flow cytometry than in spectrofluorimetry, since the cells reside in the laser beam only for microseconds. Due to the small illumination volume (picoliter range) and the dilution by the sheath fluid, the contribution of the fluorescence intensity of the free fluorophore in aqueous phase to the fluorescence intensity of cells is negligible.

b. Image Microscopy

Time-resolved, spectrofluorimetric steady-state, and flow cytometric steady-state fluorescence anisotropy measurements have been applied almost exclusively to cells in suspension. There are, however, several classes of biological systems that are not amenable for study in suspension; examples of these are adherent cells, whose properties may change after detaching them from the surface, and intact tissues having cellular heterogeneity. In the original microscopic version of fluorescence anisotropy measurements on single cell basis,[188] all the information about the distribution of fluorescence anisotropy across a cell was lost, because only one signal pair was detected from one illuminated cell, so that a spatially averaged measurement was performed. Due to the latest developments in computer and image analysis techniques it has become possible to map fluorescence anisotropy of nonoriented fluorophores incorporated into cultured cells using an epifluorescence microscope equipped with excitation and emission polarizers and an image analysis system.[193] To correct for the polarization dependence of detector sensitivity, two methods were used. In the first method, unpolarized light was passed through an interference filter, a dichroic mirror, and emission polarizers and detected by an intensified camera. The differential detector sensitivity (G) was determined from the ratio of corrected intensities calculated for perpendicular and parallel orientations of the emission polarizer. In the second method, the fluorescence anisotropy of a fluorophore in aqueous solutions having near zero anisotropy was measured. Here the detected light intensity should be nearly independent of the orientation of the emission polarizer. The background signal determined at the same experimental condition should be subtracted from the measured fluorescence intensity. The microscopic components were chosen with special care to minimize polarization artifacts and background autofluorescence. Contribution of autofluorescence to the fluorescence signal can also be decreased by appropriately choosing the excitation wavelength.[194] It has been demonstrated that objective depolarization effects are very small in these studies since similar fluorescence anisotropy values were measured in cells with objectives having widely different numerical apertures.[193] The mapping capability of the instrument was checked with capillary tubes filled with fluorophores in solutions of differing viscosities. It has been demonstrated that the fluorescence anisotropies measured in different parts of the capillary tube can be resolved spatially within a microscope field and these values were equivalent to those measured in a fluorimeter.[193]

The applicability of the techniques has been successfully demonstrated in studying the membrane fluidity of isolated perfused proximal tubule from rabbit kidney.[194] Since the anisotropy images of the TMA-DPH labeled tubule, obtained using a single plane of excitation, showed dramatic dependence on tubule orientation, orientation-independent anisotropy image was generated by changing the polarization plane of exciting illumination.[194] Based on fluorescence image, steady-state, and fluorescence lifetime measurements, it was concluded that in kidney proximal tubules the apical membrane fluidity was significantly lower than basolateral membrane fluidity.[194] Correlation between fluidity and water permeability was also studied by comparing the fluorescence anisotropy images of vasopressin-sensitive cortical collecting tubule of rabbit kidney with those of water-impermeable cortical thick ascending limbs.[195]

Using another specially designed video-microscope system, the surface membrane fluidity of sand dollar eggs after exocytosis was monitored by measuring the fluorescence anisotropy of isolated cortices stained with hexadecanoylaminofluorescein (HDAF).[196] HDAF binds to membranes with fluorophore in the hydrophilic membrane surface and the fatty tail embedded in the hydrophobic region. Because of the flexibility of the part connecting the fluorophore to the hydrophobic tail, the fluorescence anisotropy is considered to reflect the microviscosity of the polar head groups of the membrane lipids more than the hydrophobic region of acyl chains. The cytoplasmic surface of the plasma membrane of the sand dollar egg is lined with a layer of cortical vesicles which fuse within seconds of fertilization by a transient rise of intracellular Ca^{2+}. The fluorescence anisotropy of isolated cortices increased during exocytosis induced by Ca^{2+}, suggesting that the fluidity of the plasma membrane decreases after exocytosis. The data obtained also indicated that the fluidity of the cytoplasmic leaflets of cortical organelles was higher than that of the plasma membrane. For interpretation, it was suggested that only the outer leaflet of the plasma membrane becomes less fluid after exocytosis.[196]

IV. MEMBRANE FLUIDITY AND CELLULAR FUNCTIONS

A. FACTORS INFLUENCING CELL MEMBRANE FLUIDITY

Membrane fluidity itself, though its regulatory mechanisms are far from being understood, is thought to play an important role in various cellular functions. Many physical and chemical factors regulate the fluidity properties of cytoplasmic membranes, e.g., temperature, pressure, membrane potential, fatty acid composition, protein incorporation, or Ca^{2+} concentration.

The effect of temperature and pressure on the viscosity of lipid bilayers has been studied mostly on model membranes and the reader is referred to the literature.[9,21,96,197,198]

Depolarization of electric potential across model and biological membranes caused in some cases a decrease in membrane microviscosity,[9,199-202] although there is a paper reporting that the membrane microviscosity was not influenced by changing the plasma membrane potential.[203]

The lipid composition of biological membranes is always heterogeneous and includes phospholipids and cholesterol as well as several other species such as glycolipids, free fatty acids, neutral lipids, etc. The permutation of the various head groups, chain lengths, and degrees of chain unsaturation also contributes to the overall "bulk" fluidity monitored in biological membranes. The incorporation of protein molecules into the lipid bilayers increases the heterogeneity of the membrane structure by creating domains and annular lipid layers. For reviews see References 9, 20, 21, and 204 to 208.

Calcium influences the structure of membranes containing acidic phospholipids by nonspecifically crosslinking the negative charges. Increasing calcium concentration may induce structural rearrangements[9,113,209,210] and a decrease in membrane fluidity.[9]

Data have been presented on the correlation between the aforementioned factors and the "fluid state" of biological membranes as it is summarized in References 9, 20, and 21.

Using some typical examples, this review will mainly concentrate on the question how the physical state (bulk dynamics) of the membrane influences membrane functions, such as intramembrane enzyme activity, transport processes, hormone action, immunological processes, and cell physiology such as cell proliferation, differentiation and maturation, malignant transformation, and metastasis.

B. MEMBRANE FUNCTIONS

1. Enzyme Activity

Many enzymes embedded in membranes of cells are sensitive to the chemical and physical properties of the lipid components of the membrane. The functions of these integral membrane-bound enzymes can be regulated to some extent by changes within the lipid portions of biological membranes. The functions of membrane-bound proteins can be manipulated by modifications of their intimate environment. The effect of membrane fluidity on the enzymatic activity has been studied on various membrane-bound enzyme systems; the observations are summarized in Table 2.

a. ATPases

One of the most thoroughly studied membrane-bound enzymes is the Na-K-ATPase, which consists of two subunits and is sensitive to the physical properties of the surrounding lipids. Alterations in membrane fluidity by several different maneuvers have been shown to affect the activity of Na-K-ATPase. An increase in the ratio of polyunsaturated fatty acid chains increases both membrane fluidity and Na-K-ATPase activity.[211-213] Incorporation of cholesterol decreases Na-K-ATPase activity.[214-216] Alteration in pressure and temperature can influence Na-K-ATPase activity.[217,218] Fluorescence anisotropy of hydrophobic probes increased with increasing pressure reversibly.[217] Plots of the activity of enzymes prepared form dog kidney and eel electroplax vs. fluorescence anisotropy showed an inverse relationship, indicating that enzyme activity was directly related to the fluidity of the membrane as measured by lipid probes. The investigators concluded that increased pressure results in decreased membrane fluidity, which hindered conformational transitions associated with rate-limiting steps in the enzyme reaction.[217] Na-K-ATPases from deep-sea fish were less inhibited by pressure than those of shallow-living species.[218] The data were consistent with a model of increased membrane fluidity resulting in reduced pressure-sensitivity of enzymes from deep-sea species.[218]

Table 2. **Effects of Lipid Fluidity on the Activity of Membrane Enzymes**

Enzymes	Modulation of membrane fluidity	Changes in activity by increasing fluidity	Ref.
Na-K-ATPase	Polyunsaturated fatty acid	Increase (↑)	211–213
	Cholesterol	Increase (↑)	214–216
	Pressure	Increase (↑)	217
	Alcohol diet	No correlation	219
	Benzyl alcohol	Increase (↑)	222
	Adriamycin	Increase (↑) then decrease (↓)	223
	Essential hypertension	No correlation	224
	Diabetes	Decrease (↓)	225
	Obesity	Decrease (↓)	227
Ca-ATPase	Melittin	Slight increase (↑)	228, 229
	Essential fatty acid deficiency	Increase (↑)	230
Ca-Mg-ATPase	1-Octanol	Increase (↑) then decrease (↓)	231
	2-Dodecanol	Increase (↑)	231
Adenylate cyclase	Benzyl alcohol	Increase (↑)	233
	Ethanol	Increase (↑) or decrease (↓)	235, 236
	Ketamine	Decrease (↓)	235, 236
	Gangliosides	Increase (↑)	237
	Olfactory membrane	Insensitive	239
Acetylcholine esterase	Acidic phospholipids	Increase (↑)	240
	Cetiedyl	No correlation	241
	Riboflavin deficiency	Decrease (↓)	242
	SAD-128	Decrease (↓)	243
Alkaline phosphatase	Cholesterol	Decrease (↓)	244
	Benzyl alcohol	No correlation	244
	Dietary fish oil	No correlation	245
Sphingomyelinase	Lipid composition	No correlation	247
Phospholipase A	Lipid peroxidation	Decrease (↓)	249
	Hydrogenation	Increase (↑)	251
Methyltransferase	Transcortin	Increase (↑)	256

The effects of various drugs and physiological conditions on the activity of Na-K-ATPase have been partially or fully attributed to changes in membrane fluidity. When the activities of Na-K-ATPases prepared from brain microsomes of rats maintained on an alcohol diet and of control rats were compared, no difference was found.[219] However, in the presence of low concentration of ethanol, the activity of the enzyme was stimulated to a lesser extent than the enzyme from control rats. Molecular tolerance to ethanol was observed in the membranes from rats on alcohol diet by resistance to the disordering effects of added ethanol.[219] Owing to apparent adaptation to chronic ethanol exposure the membrane fluidity is reduced, resulting in greater K^+ affinity, more favorable transition to the phosphatase form of the enzyme, a smaller entropy difference between native and K^+-activated enzyme, and reduced sensitivity to ethanol.[220,221] Benzyl alcohol acts similarly to ethanol; it stimulates the Na-K-ATPase of the apical surface of transporting epithelia by fluidizing the membrane structures.[222] Adriamycin also increased the lipid fluidity of dog brain synaptosomal plasma membrane as monitored with DPH, and stimulated the Na-K-ATPase activity at lower concentrations. At higher adriamycin concentration a progressive inhibition of enzyme activity was observed. The investigators proposed that adriamycin achieved these effects through asymmetric perturbation of the membrane lipid structure and that the change in membrane fluidity may be an early event in adriamycin-induced neurotoxicity.[223] The effect of membrane fluidity on Na-K-ATPase activity is not always as straightforward as mentioned above. When low lipid fluidity was brought about by some kind of chronic state such as essential hypertension,[224] diabetes,[225] gangliosidosis,[226] and obesity,[227] the activity of Na-K-ATPase was either not changed significantly[224] or it was impaired.[225-227] In these cases, the chronic illnesses probably produced structural changes in the enzyme and the membrane, in addition to lowering the lipid fluidity, which eventually decreased Na-K-ATPase activity.

The Ca-ATPases and Ca-Mg-ATPases are less sensitive to membrane fluidity than Na-K-ATPases. When the effect of melittin, a basic membrane-binding peptide, on the Ca-ATPase activity was studied, at a concentration completely inhibiting the enzyme activity, only a slight increase in the lipid hydrocarbon chain order parameter was observed.[228,229] The investigators suggested that the increase in lipid viscosity is too small to account for the large decrease in Ca-ATPase activity, so the melittin inhibited the ATPase primarily through its capacity to aggregate the enzyme.[228,229] Dietary essential fatty acid deficiency decreased the membrane fluidity, as estimated by the reciprocal of the order parameter S in rat liver microsomal membrane as compared to that of control rats.[230] The Ca-ATPase activity was also significantly lower in essential fatty acid-deficient rats than in controls. The investigators concluded that fatty acid composition may be a major factor contributing to membrane fluidity and may play a role in regulating intrinsic membrane protein activities.[230] Ca-Mg-ATPase activity in rat brain synaptosomal membranes was affected by various alkanols. n-Alkanols exhibited an increased inhibition of the enzyme activity with an increase in the carbon chain length up to 1-octanol. 1-Octanol and 1-decanol caused a biphasic effect on the enzyme activity depending on the alkanol concentration, whereas 1-dodecanol caused stimulation. Since all alkanols studied caused an increased fluidity of membranes, the investigators concluded that changes in the membrane fluidity did not seem to be a prerequisite of the Ca-Mg-ATPase inhibition.[231] Lee et al. also came to the conclusion after thoroughly testing the effect of membrane fluidity and structure on the Ca-Mg-ATPase of skeletal muscle sarcoplasmic reticulum that there was no evidence for the fluidity of the surrounding lipid having any significant effect on the enzyme activity.[232]

b. *Adenylate Cyclase*

The membrane-bound adenylate cyclase activity can be significantly influenced by changes in lipid fluidity. Fluidizing the membrane with benzyl alcohol produced a linear decrease in the forskolin-stimulated activity of adenyl cyclase in plasma membranes from bovine adrenal cortex.[233] By contrast, NaF stabilized adenylate cyclase activity against the perturbations induced by benzyl alcohol.[233] Parathyroid hormone-sensitive adenylate cyclase system of canine kidney was stimulated with benzyl alcohol progressively; up to fourfold increase in the basal enzyme activity was achieved. Parallel to this observation, fluorescent polarization studies using DPH showed a decrease in fluorescence anisotropy in the presence of benzyl alcohol. The results suggested that increasing membrane fluidity facilitates the interaction of the components of the adenylate cyclase system.[234] The effect of ethanol on prostaglandin E1- and isoproterenol-sensitive adenylate cyclase activity could not be interpreted simply with changes in lipid fluidity.[235,236] While adenylate cyclases from L6 cells and smooth muscle cells were stimulated with ethanol, enzymes from human platelets were unaffected. In addition, ketamine decreased fluorescence anisotropy similarly to ethanol, but unlike ethanol caused a decrease in adenylate cyclase activity. It was concluded that in these cases ethanol interacts directly with the membrane proteins and that its effect is mediated not only via perturbation in membrane fluidity.[235,236] Incorporation of GT1b gangliosides in bovine thyroid plasma membranes induced a substantial decrease in cholera toxin-stimulated adenylate cyclase activity and to a lesser degree a decrease in NaF-stimulated activity. The inhibitory effects were paralleled by change in fluorescence anisotropy: GT1b modification of the plasma membranes provoked a slight increase in the TMA-DPH anisotropy and a substantial enhancement in DPH anisotropy. These results suggested that gangliosides might interfere with the coupling between the α-subunit of the stimulatory GTP-binding regulatory protein and the catalyst of the adenylate cyclase system by affecting the membrane fluidity.[237] A proper transmembrane Ca^{2+} gradient was essential for the optimal fluidity of the lipid bilayer, favoring a transition to a suitable conformation of adenylate cyclase with higher enzyme activity.[238] At physiological temperatures, the activity of the olfactory adenylate cyclase was fairly insensitive to changes in temperature and membrane fluidity. These are desirable properties for a transduction system which must function in an exposed environment.[239]

c. *Acetylcholinesterase*

The activity of brain synaptosome-associated acetylcholinesterase was stimulated with acidic phospholipids. It has been demonstrated that acidic phospholipids do not act on acetylcholinesterase directly, but rather indirectly, affecting the membrane fluidity.[240] Cetiedil, a vaso-erythroactive drug, lowered the acetylcholinesterase activity in erythrocyte membrane but had no effect on microviscosity. Since cetiedil did not inhibit the enzyme activity directly, it has been suggested that as an amphophilic drug it perturbs the membrane properties in the annular lipid region surrounding the enzyme without affecting the overall physical state of the erythrocyte membrane.[241] Riboflavin deficiency resulted in higher levels of peroxidation products and lower membrane fluidity in rat red blood cells. This decreased fluidity was accompanied by an increase in the activity of the acetylcholinesterase.[242] SAD-128, a reversible inhibitor and allosteric modifier of the acetylcholinesterase, is an effective protector of the enzyme against inhibition by soman. SAD-128 increases the fluidity of human erythrocyte membranes as monitored by EPR technique.[243] The investigators concluded that the protective effects of SAD-128 against inhibition of acetylcholinesterase are related to the structural changes of the active site and can be additionally moderated by the microviscosity changes of the membrane.[243]

d. Other Enzymes

The activity of alkaline phosphatase proved to be very sensitive to the lipid composition of the membrane.[177,244,245] Increasing cholesterol/phospholipid molar ratios decreased the enzyme activity and increased the membrane fluidity. However, when the increased membrane fluidity caused by the elevated cholesterol/phospholipid ratio was decreased to a normal level by addition of benzyl alcohol, no alteration in alkaline phosphatase activity was observed.[244] In addition, dietary fish oil supplementation increased the alkaline phosphatase activity in intestinal brush-border membranes without affecting membrane fluidity.[245] These data suggest that the effects of changes in membrane lipid composition on alkaline phosphatase activity appear to result from changes in the local lipid environment of the enzyme rather than from changes in the bulk physical properties of the membrane.

Similarly to the behavior of alkaline phosphatase the activity of sphingomyelinase[246] and lecithin:cholesterol transferase[247,248] is not influenced significantly by the bulk membrane fluidity, but rather by the composition of the lipid bilayer. While elevating the membrane viscosity with lipid peroxidation[249] or high intracellular calcium concentration[250] significantly enhanced the phospholipase A activity, hydrogenation[251] of fatty acid double bonds markedly inhibited the enzyme activity. Increasing membrane fluidity was associated with increasing activity of carnitine palmitoyltransferase[252-254] and a large decrease in the cooperativity of UDP-glucuronyl transferase.[255] Decrease in the activities of methyltransferase[256] and 3-hydroxy-3-methylglutaryl-CoA reductase[257] was observed when the membrane viscosity was increased with transcortin treatment and with modulation of cholesterol content and phospholipid composition, respectively.

In conclusion, the results reported in the papers discussed above suggest that membrane lipids influence the function of membrane-bound enzymes either through alterations in the bulk physical structure of the membrane or through direct lipid protein interaction. The general assumption that increasing membrane fluidity increases the enzyme activity holds in certain cases (i.e., Na-K-ATPase, alkaline phosphatase, etc.) but is not a universal concept. The effect of membrane fluidity on membrane-bound enzymes must be considered for each specific enzyme, taking into account the origin of the enzyme as well. In addition, the way the changes in membrane fluidity are brought about can significantly influence the interpretation of the effect of fluidity on protein function in membranes.

2. Transport Processes

Some of the enzyme reactions (Na-K-ATPase, Ca-ATPase, Ca-Mg-ATPase, etc.) discussed in the previous section are tightly connected to ion transport processes taking place across cellular membranes. In addition to the above-mentioned ion pumps, other transport processes, such as sugar transport, amino acid uptake, phosphate (P_i) transport, and water permeability, have been thoroughly investigated in conjunction with membrane fluidity. The effects of membrane fluidity on transport processes are summarized in Table 3.

a. Glucose Transport

One of the most thoroughly studied glucose transport systems is the Na-dependent glucose carrier found in intestinal and renal cortical apical membranes. Increasing the membrane viscosity nutritionally (high cholesterol/phospholipid ratio) decreased the maximum velocity of Na-dependent D-glucose transport.[153,258] Addition of 2-[(2-methoxyethoxy)ethyl]-*cis*-8-(2-octylcyclopropyl)octanoate preferentially fluidized the cytofacial (inner) leaflet of human proximal small intestinal brush-border membrane vesicles and enhanced the Na-gradient-dependent D-glucose uptake. Interestingly, a decrease in L-glutamic acid uptake and no change in L-leucine transport were observed in the same concentration range of the fluidizing drug.[259,260] Chlorpromazine also increased the D-glucose uptake in rat intestinal brush-border vesicles. Chlorpromazine increased the fluorescence anisotropy of DPH located in the hydrophobic portion of the membrane, but decreased that of 1-anilino-8-naphthalene sulfonate located at the surface of the membrane, suggesting that the increase of membrane surface fluidity might affect the sugar transport.[261] Gentamicin inhibited Na-dependent D-glucose transport through binding to the superficial region of

Table 3. **Effects of Lipid Fluidity on Membrane Transport Processes**

Transported ligand	Modulation of membrane fluidity	Changes in rate of transport by increasing fluidity	Ref.
Glucose (Na dependent)	Cholesterol/phospholipid	Increase (↑)	153, 258
	Cytofacial fluidizer	Increase (↑)	259, 260
	Chlorpromazine	Decrease (↓)	261
	Gentamicin	Decrease (↓)	262
	Diabetes mellitus (streptozocin)	No correlation	263
Glucose (Na independent)	Cholesterol depletion	Decrease (↓)	267
	Cholesterol addition	Increase (↑)	258
	Benzyl alcohol	Decrease (↓)	268
Serotonin	Cholesterol	Increase (↑)	269
	cis-Vaccenic acid	Increase (↑)	269
	Hypoxia	Increase (↑)	270
	Hyperoxia	Increase (↑)	271
	Vitamin E	Increase (↑)	272
Tyrosine	Protein free diet	Decrease (↓)	273
Glutamine	Protein free diet	No correlation	273
Alanine	Benzyl alcohol	No correlation	274
	Chronic ethanol consumption	No correlation	275
P_i (phosphate)	Benzyl alcohol	Increase (↑)	276
	1,25-Dihydroxyvitamin D	Increase (↑)	278
	Phosphate deprivation	Increase (↑)	279, 280
	Gentamicin	Decrease (↓)	281
Water	Chloroform	Increase (↑)	283
	Benzyl alcohol	Increase (↑)	284
	Cortical collecting tubules	No correlation	194, 195 282
LDL	*cis*-Unsaturated fatty acids	Increase (↑)	285

brush-border membranes, thus decreasing membrane fluidity.[262] However, no obvious relationship could be established between alteration in membrane fluidity and increased Na-dependent uptake of D-glucose in acute diabetes mellitus induced by streptozotocin.[263] It has been shown that both Na^+ and glucose bind to different sites on the carrier. Na^+ binding induces a rapid conformational change in the carrier which increases its affinity for D-glucose.[264,265] Changes in membrane fluidity, especially on the membrane surface, could therefore alter the equilibrium of Na^+ binding, the conformational change, the affinity of D-glucose binding, or the actual translocation of glucose through the membrane.

Similar effects of fluidity have been observed on Na-independent glucose carriers, which mediate facilitated glucose transport of erythrocytes, hepatocytes, fibroblasts, adipocytes, and basolateral membranes of epithelial cells.[266] Yuli et al.,[267] using human erythrocytes and 3T3 mouse fibroblasts, demonstrated that membrane fluidization accomplished by cholesterol depletion resulted in reduced 2-deoxyglucose uptake via a decrease in the apparent V_{max}. Membrane rigidification by addition of cholesterol or cholesterol ester had the opposite effect. Increased cholesterol content greatly inhibited the activity of the Na-independent glucose transport in guinea pig intestinal brush border.[258] The inhibiting effect of cholesterol was interpreted as the increase in microviscosity imparted by cholesterol in the bulk lipid phase environment where these proteins operate.[258] An opposite observation was described in the paper by van Valen and Keck, where the fluidizing effect of the local anesthetic drug benzyl alcohol provoked inhibition of the glucose uptake.[268]

b. Serotonin and Amino Acid Uptake

Serotonin (5-hydroxytryptamine, 5-HT), a biologically active aromatic monoamine, is taken up into pulmonary endothelial cells by a carrier-mediated plasma membrane process. Cholesterol, which decreased the fluidity in the subpolar head group and central and midacyl side-chain regions of the plasma membrane, decreased serotonin uptake, whereas *cis*-vaccenic acid, which increased the fluidity in the central and the midacyl side-chain regions of the plasma membrane, increased serotonin transport.[269] The investigators concluded that the physical state of the central and midacyl chains within the pulmonary artery endothelial cell plasma membrane influence transmembrane transport of serotonin.[269] Similarly, close relationship has been observed between membrane fluidity and serotonin transport in other systems as well.[270-272] In hypoxic cells, higher membrane fluidity was accompanied by elevated 5-HT uptake.[270] The increases in fluidity and HT-transport were reversible after returning to control conditions. Under hyperoxic conditions, the membrane fluidity was lower than that of control pulmonary artery endothelial cells, and the specific 5-HT transport was significantly inhibited.[271] Vitamin E, a dietary antioxidant, which incorporates into the lipid bilayer of biological membranes, significantly increased the fluorescence anisotropy of DPH and TMA-DPH and caused a reduction in 5-HT uptake.[272] The authors interpreted the observation that, in addition to its role as an antioxidant, vitamin E alters the physical state of the membrane and modulates 5-HT transport.[272] The protein-free diet increased the membrane fluidity of synaptosomes isolated from rat cerebrum and reduced the uptake of tyrosine, while the glutamate transport was unaffected.[273] Fluidization of membrane by benzyl alcohol[274] or chronic ethanol consumption[275] had no effect on Na-dependent alanine uptake.

c. P_i Transport

Similarly to glucose uptake, P_i transport also proved to be very sensitive to membrane fluidity in many systems. Apical membranes of renal epithelial cells are more rigid than other plasma membranes, partly due to the abundance of sphingomyelin among their constituents of phospholipids. Tight junctions play a key role in maintaining differences between the apical and basolateral domains of the plasma membrane with respect to their lipid composition and fluidity. Opening tight junctions or adding benzyl alcohol increased membrane fluidity, which resulted in an increase in P_i uptake, i.e., elevated P_i transport.[276] Interestingly, in the same system, the fluidizing effects caused a decrease in the α-methyl-D-glucopyranoside uptake.[276] Greater sodium gradient-dependent P_i transport has been observed in brush-border membrane of superficial cortex than in that of juxtamedullary cortex. The observed heterogeneity in P_i uptake closely correlated with the dynamic behavior of the membranes, i.e., the superficial cortical membrane had lower fluidity as monitored by fluorescence anisotropy of DPH.[277] 1,25-Dihydroxyvitamin D stimulated the sodium-dependent P_i transport in renal outer cortical brush-border membranes by lowering membrane fluidity.[278] Dietary phosphate deprivation decreased the cholesterol content and the rigidity of renal proximal brush-border membrane and increased the Na-dependent P_i transport, suggesting that changes in brush-border membrane fluidity are critical for adaptation to P_i restriction.[279,280] While in all the above-mentioned P_i transport systems the increased fluidity paralleled the enhancement of P_i uptake, gentamicin, although decreasing the membrane fluidity of the brush-border membrane, caused a significant decrease in the V_{max} of sodium-gradient dependent P_i transport, probably due to other adverse effects of the drug.[281]

d. Water Permeability

No obvious correlation between plasma membrane fluidity and the regulation of kidney tubule water permeability has been observed.[194,195,282] In cortical collecting tubules the fluorescence anisotropy of the apical membrane was similar to that of the basolateral membrane. Vasopressin at a dose that increased water permeability tenfold did not effect the fluorescence anisotropy in apical membrane. Fluorescence anisotropy values measured in the cortical thick ascending limb were similar to that determined in the cortical collecting tubule. These results suggested that regulation of bulk fluidity in the apical membrane of the cortical collecting tubule of rabbit kidney is not a component of the hydrosmotic action of vasopressin.[195] Addition of chloroform[283] or benzyl alcohol[284] fluidized the membrane of granular cells of toad bladder and that of Madin-Darby canine kidney cells, and an increase in water permeability was observed in both cases.

Dietary *cis*-unsaturated fatty acids increased the membrane fluidity of hepatocytes and augmented low-density lipoprotein (LDL) binding, uptake, and degradation in comparison to unmodified cells. Conversely, stearate or palmitate enrichment resulted in a decreased plasma membrane fluidity and the LDL binding, uptake, and degradation varied with plasma membrane fluidity in a highly correlated manner.[285]

In conclusion, at least two possible ways exist to interpret the effect of the dynamic structural state of membrane lipids on membrane transporters. First, the overall lipid structure (fluidity) may influence the carrier conformation resulting in reduced or enhanced accessibility of binding sites. Second, various carriers may be embedded in membrane compartments of different microenvironment. Alterations may occur in only one microcompartment and thereby selectively affect certain transporters.

Table 4. **Modulation of Membrane Fluidity by Hormone Action**

Hormone	Membrane system	Membrane fluidity: Short-term effect	Membrane fluidity: Long-term effect	Ref.
Hyperthyroidism	Skeletal muscle		Increase (↑)	75
	Hepatic mitochondria		Increase (↑)	286
	Rat erythrocytes		Increase (↑)	287
	Rat brain		Increase (↑)	288
Thyrotropin	Rat thyroid line (FRTL-5)		Increase (↑)	289
Aldosterone	Liver microsomes		Decrease (↓)	290
Ethinyl estradiol	Liver plasma membrane		Decrease (↓)	291, 292
Adrenalin	Human fibroblasts		Increase (↑)	293
Cortisol	Dog brain synaptosomes	Increase (↑)		294
Estradiol	Myometrial plasma m.	Increase (↑)		295
Prostaglandin E_2	Myometrial plasma m.	Increase (↑)		296
	Luteal plasma membrane	Increase (↑)		297
	Brain synaptosomes	Increase (↑)		298
Progesterone	Brain synaptosomes	Decrease (↓)		298
Medroxy-progesterone	Mammary cancer cells	Decrease (↓)		299
	Granulocytes	Decrease (↓)		300
Insulin	Erythrocytes	Increase (↑)		301
	Rat liver plasma m.	Increase (↑)		302–304
Glucagon	Rat liver plasma m.	Decrease (↓)		302
Atropine	Rat brain fragments	Increase (↑)		305
Carbachol	Rat brain fragments	Increase (↑)		305
Lutropin	Testis plasma membrane	Decrease (↓)		306
	L-cells from rat	Decrease (↓)		306
Neuropeptide Y	Aortic smooth muscle	Decrease (↓)		307

3. Hormone Action

The interaction of hormones with their membrane receptor is of particular interest, since not only can the binding and/or the subsequent metabolic processes lead to changes in membrane fluidity but also the binding itself and hormone action can be modulated by altered membrane fluidity. Binding of a hormone can modulate the lipid fluidity directly within a short time (less than an hour) or can change it in the long run (hours, days) by altering metabolic processes, which will change the fatty acid composition of cell membranes and therefore the membrane fluidity as well. In addition, changes in membrane fluidity may influence the hormone action by up or down regulating the receptor sites of the cell membrane.

a. Effects of Hormones on Membrane Fluidity

The effects of hormones on lipid fluidity have been studied in various membrane systems and are summarized in Table 4.

Hyperthyroidism causes a decrease in cholesterol and an increase in the phospholipid content of the membrane, which in turn increase the membrane fluidity.[75,286,287] Induced hypothyroidism resulted in decreased lipid fluidity of rat brain membrane.[288] The reduced fluidity seemed to be related more to alterations in the ratios between phosphatidylcholine, sphingomyelin, and phosphatidylethanolamine than to those of cholesterol/phospholipid, and protein/phospholipid ratios or fatty acid unsaturation index.[288] Continued exposure of cells of a rat thyroid line (FRTL-5) to medium lacking thyrotropin caused a decrease in membrane fluidity, which reached its minimum in approximately 8 d.[289] The change in lipid fluidity was due to increased cholesterol phospholipid ratio and increased ratio of saturated to unsaturated fatty acids of the membrane phospholipids. Both membrane fluidity and lipid composition can be restored by thyrotropin to their original levels, i.e., levels measured under continuous exposure to the hormone, after 48-h treatment.[289] Aldosterone or 11-deoxycorticosterone inhibited the specific activity of linoleyl-CoA desaturase 24 h after injection and the alterations in fatty acid composition resulted in a decrease of liver microsomal membrane fluidity.[290] A synthetic estrogen, ethinyl-estradiol, decreased the lipid fluidity in liver plasma membrane due to induced cholestasis.[291,292] Administration of isoproterenol and adrenalin increased the fluidity of human adult fibroblast membrane owing to the change in the lipid metabolism, i.e., decreasing the total amount of phosphatidylcholine.[293]

Direct effect of a hormone has been observed when dog brain synaptosomal plasma membranes were treated with cortisol.[294] Cortisol binding caused an increase in bilayer fluidity.[294] Binding sex steroids (estradiol, estrone) to purified myometrial plasma membranes also increased the membrane fluidity.[295] While prostaglandin E_2 treatment resulted in an increase in membrane fluidity in myometrial plasma membranes,[296] luteal plasma membranes[297] and brain synaptosomal plasma membranes,[298] progesterone decreased the fluidity in brain synaptosomal plasma membranes.[298] Medroxyprogesterone acetate and various other steroid hormones caused a significant decrease in the lateral mobility of a lipid probe in CAMA-1 mammary cancer cells. The effect of hormones decreased in the following order medroxyprogesterone > 5a-DHT = 17-β-estradiol > dexamethasone, and roughly paralleled their lipid solubility as estimated by partition coefficients in octanol-water systems. The investigators assumed that any involvement of classical steroid hormone receptors should be excluded and the nongenomic steroid effects on the plasma membranes were responsible for the observed phenomenon.[299] Similarly, extreme high doses of corticosteroids were shown to reduce granulocyte membrane fluidity, assessed by electron paramagnetic resonance. The investigators proposed that the unique effects of the extremely high-dose corticosteroid are not mediated via the glucocorticoid receptor, but rather result from physicochemical effects of the drug on the membranes of effector cells.[300]

A more specific effect has been observed when human erythrocytes were treated with insulin.[301] Binding of insulin to its high affinity, low-capacity receptors resulted in a significant reduction in microviscosity.[301] Addition of insulin to bilayer lipid membranes modified by rat liver plasma membrane fragments decreased the viscoelastic parameters of membranes, for example, the modulus of elasticity in the direction perpendicular to the membrane plane, $E_\perp$.[302-304] The effect of insulin antagonist glucagon on modified bilayer membrane resulted in opposite changes in the relative value of $E_\perp$ compared to the effect of insulin.[302] Upon addition of carbachol and atropine at low concentrations to rat brain membrane fragments, the fluorescence polarization of the lipid probe decreased significantly, indicating that binding of these ligands to the muscarinic receptors increases the fluidity of its lipid environment.[305] Addition of lutropin to testis plasma membrane reduced the pyrene and DPH mobilities.[306] Similarly, incubating intact L-cells from rats significantly decreased the mobility of lipids and proteins in the plasma membrane.[306] Neuropeptide Y also decreased the lipid fluidity in isolated aortic smooth muscle cell membranes according to the increased fluorescence anisotropy of DPH.[307]

The modulation of lipid fluidity in the vicinity of an occupied receptor may come from the conformational change of the occupied receptor, which can change the packing density of the neighboring lipid layer. The resulting fluidity changes fade away from the receptor creating a fluidity gradient laterally. When hormone binding is accompanied by receptor aggregation, clustering of receptors may also affect the lipid fluidity.

b. Effects of Membrane Fluidity on Hormone Action

Dietary modulation of lipid composition of membranes highly correlated with membrane fluidity and insulin binding (Table 5).[308-310] The increased lipid fluidity of erythrocyte ghost membranes prepared from women fed a low-fat diet with high polyunsaturated/saturated fatty acid ratio (P/S) paralleled with the increased insulin binding.[309] Similarly, cells from animals fed the high P/S diet bound significantly more insulin than cells from animals fed the low P/S diet. From Scatchard analysis of binding in animals, increased insulin binding was associated with an increased number of available high-affinity, low-capacity insulin binding sites.[310] In obese adults, significant inverse relationship was observed between insulin binding and cell microviscosity monitored by DPH anisotropy.[311] Local anesthetics (lidocaine, procaine) known to fluidize plasma membrane reversibly enhanced the binding capacity for insulin in hepatoma cells[312] and cultured chick embryo fibroblasts.[313] Y79 retinoblastoma cells enriched with docosahexaenoic acid possessed higher membrane fluidity and showed increased insulin binding as compared to control cells.[314] Following NO_2 exposure of porcine pulmonary artery endothelial cells, membrane fluidity, monitored by the steady-state anisotropy of TMA-DPH, decreased extensively. At the same time, specific binding of ^{125}I-insulin to cells exposed to NO_2 was reduced significantly compared to binding in control cells. Scatchard analysis of the binding data indicated that NO_2 exposure caused a fivefold reduction in insulin receptor binding sites in endothelial cells.[315] Increased membrane fluidity, induced by sex steroids, decreased the binding of the uterus stimulatory prostaglandins E_2 and $F_{2\alpha}$ by

Table 5. **Effects of Membrane Fluidity on Hormone Binding**

Hormone	Target membrane	Modulation of membrane fluidity	Effect of increased membrane fluidity on receptor accessibility	Ref.
LH/hCG[e]	Human lymphocytes	Cholesteryl hemisuccinate	Decrease (↑)	309
Insulin	Erythrocytes	Low fat diet with high P/S[a]	Increase (↑)	309
	Adipocytes	Diet with high P/S	Increase (↑)	310
	Mononuclear leukocytes	Obesity	Increase (↑)	311
	Hepatoma cells	Lidocaine, procaine	Increase (↑)	312
	Chick embryo fibroblasts	Lidocaine, procaine	Increase (↑)	313
	Y79 retinoblasts	Docohexanoic acids	Increase (↑)	314
	Porcine artery endothelial cells	NO_2 exposure	Increase (↑)	315
PG E_2[b]	Granulocytes	Sex steroids	Decrease (↓)	295
PG F_2[c]	Granulocytes	Sex steroids	Decrease (↓)	295
VIP[d]	Rat prostatic epithelial cells	Cholesteryl hemisuccinate	Decrease (↓)	316
	Human lymphocytes	Phenythoin	Decrease (↓)	317
LH/hCG[e]	Rat ovary cells	Pseudopregnancy	Decrease (↓)	318

[a]Polyunsaturated/saturated fatty acid ratio.
[b]Prostaglandin E_2.
[c]Prostaglandin $F_{2\alpha}$.
[d]Vasoactive intestinal peptide.
[e]Luteinizing hormone/human chorionic gonadotropin.

membrane receptors. The investigators postulated the sex steroid-induced reduction of uterine contractions as a "nongenomic" mechanism.[295] The receptor for neuropeptide vasoactive intestinal peptide (VIP) behaves in response to membrane fluidity in a way just opposite to the insulin receptor. Experiments on VIP binding indicated that increasing cholesterol concentration, accompanied with increasing microviscosity, led to a higher VIP binding capacity without changes in the affinity of VIP receptors. At the same time, the increased cholesterol level significantly inhibited the stimulatory effect of VIP on cyclic AMP accumulation. These data suggested that, although the membrane cholesterol incorporation enhanced the VIP binding, the decreased membrane fluidity diminished the coupling efficiency between adenylate cyclase and VIP-receptor complex.[316] Conversely, plasma membrane fluidity of lymphocytes from phenytoin-treated patients decreased as compared to controls and changes in membrane fluidity significantly decreased the VIP binding.[317] Changes in membrane fluidity influence the LH/hCG receptor exposure similarly to VIP receptors. In ovarian membranes, prepared from immature rats during hormonally induced pseudopregnancy, significant positive correlation was found between lipid rigidity and the number of LH/hCG receptors. It has been suggested that decreased fluidity of the membrane lipid during pseudopregnancy may maximally expose ovarian LH/hCG receptors maintained in a cryptic form.[318]

From the above-mentioned examples it is clear that the direction of the induced change in hormone binding depends on whether the lipid modulator fluidizes or rigidifies the membrane and on the "type" of the hormone receptor. At least three types of hormone receptors can be identified: (1) receptors which are up regulated by the increase in membrane fluidity and down regulated by the decrease in fluidity. Muller and Krueger have categorized these receptors displaying positive correlation with lipid fluidity as "syndromic proteins".[319] (2) Receptors which are down regulated when the fluidity increases and up regulated when the fluidity decreases have been termed as "antidromic proteins". (3) A third group of proteins appears to be indifferent to changes in membrane lipid fluidity. An example of a syndromic membrane receptor would be the insulin receptor, while neuropeptide receptors are antidromic.

To interpret the underlying mechanisms of this phenomenon Shinitzky proposed a very interesting model in which the lateral and vertical receptor displacements are highly regulated by alterations in the lipid fluidity.[9,320] Depending on the structure of integral and transmembrane proteins, changes in membrane fluidity increase the exposition of the protein toward the outer side of the membrane or toward the inner surface.[9] As a result, cryptic binding sites will be exposed or earlier active sites become unavailable for their appropriate ligand.

4. Immune Response

In the immune network receptor-ligand interactions play important roles in determining the relationship between each lymphoid cell and its partners. These reactions regulate not only cell activation, but also communication between the cells. The interactions between lymphoid cell receptors and various ligand are accompanied and followed by a range of membrane events such as endocytosis, phagocytosis, exocytosis, secretion of various substances, activation of membrane-bound enzymes, cytotoxic lysis, etc. Many of these events are modulated, or paralleled by biophysical changes in the membrane. Many reports have been published about monitoring membrane structure and dynamics during immune reactions.

a. Effects of Lipid Composition on Immune Functions

Modulation of membrane fluidity can alter the expression of various proliferation and differentiation antigens in the plasma membranes of lymphoid cells (Table 6). Immunogenicity of mouse tumor cell lines has been significantly enhanced by increasing the membrane rigidity with cholesteryl hemisuccinate

Table 6. **Effects of Membrane Fluidity on Surface Expression of Membrane Proteins**

Membrane proteins	Membrane system	Modulation of membrane fluidity	Effect of increased membrane fluidity on the expression of proteins	Ref.
Antigenic proteins	Lymphocytes	Cholesteryl hemisuccinate	Decrease (↓)	321
	Mammary adenocarcinoma	Cholesteryl hemisuccinate	Decrease (↓)	321
	Human renal carcinoma	Cholesteryl hemisuccinate	Decrease (↓)	324, 325
		25-Hydroxy-cholesterol	No change	324, 325
Malaria-related antigens	Human red blood cells	Cholesterol	Decrease (↓)	323
H-2K^k	GRSL cells	Hydrogenation	Decrease (↓)	326
n-fMLP[a]	Polymorphonuclear leukocytes "adult"	Aliphatic alcohols	Increase (↑)	327, 328
	Polymorphonuclear leukocytes "neonatal"	Aliphatic alcohols	No change	327
H-2 antigens	Mouse spleen cells	Cholesteryl hemisuccinate	Increase (↑)	329
		Lecithin	Increase (↑)	329
Thy-1.2	Mouse spleen cells	Cholesteryl hemisuccinate	Decrease (↓)	329
		Lecithin	No change	329
sIgG	Rabbit spleen cells	Phosphatidylcholine	Decrease (↓)	330
		Cholesterol	Decrease (↓)	330
CD4	RDM-4 lymphoma cells in AKR mice	Beef tallow diet	Decrease (↓)	336
		Fish oil diet	Decrease (↓)	336
	CD4$^+$ lymphocytes	Phenytoin	Decrease (↓)	317
		AL271 liposomes	Decrease (↓)	337
		Dextran sulfate	Decrease (↓)	338
	Lymphocytes from primate strains	Cholesterol	Decrease (↓)	339

[a] *n*-Formylmethionyl-leucyl-phenylalanine.

treatment.[321] Rigidification of the cell membrane lipid bilayer induced passive shedding of erythrocyte antigens[322] and lead to marked increase in the cell surface antigen reactivity of malaria-infected human red blood cells.[323] Rigidification of human renal carcinoma cells by treatment with cholesteryl hemisuccinate or 25-hydroxycholesterol enhanced the expression of tumor-associated antigens, although the increased immunogenicity of carcinoma cells seemed not to be related only to membrane rigidification.[324,325] Decreasing the membrane fluidity by catalytic hydrogenation of fatty acyl chains in plasma membranes increased the expression of a 15-kDa antigen on the surface of viable GRSL leukemia cells, while the expression of H-2^k antigens was much less affected by this treatment.[326] Changes in membrane properties influenced the chemoattractant-mediated cellular responsiveness of polymorphonuclear leukocytes. Fluidization of the plasma membrane by treatment with short chain aliphatic alcohols enhanced the binding of *n*-formylmethionyl-leucyl-phenylalanine to polymorphonuclear leukocytes[327,328] and also affected the binding-induced subsequent responsiveness including migration, lysosomal enzyme release, and superoxide anion production.[327] Treatment of plasma membranes of neonatal polymorphonuclear leukocytes with aliphatic alcohols did not affect the binding the *n*-formylmethionyl-leucyl-phenylalanine nor the post-binding events.[327] The membrane fluidity of neonatal polymorphonuclear leukocytes was much lower than that of adult polymorphonuclear leukocytes and it was not influenced by alcohol. The authors concluded that abnormal membrane fluidity of neonatal polymorphonuclear leukocytes caused the impaired functional dynamics of the chemoattractant receptors, which appears to underlie the defective modulation of cell functions by the membrane fluidizers in these cells.[327] Muller et al.[329] have shown that an increase in lipid fluidity resulted in an increased expression of major histocompatibility complex determinants on spleen cells, and a decreased expression of Thy-1.2 antigens. Conversely, a decrease in lipid fluidity led to a reduced expression of H-2 determinants without effecting the expression of Thy-1.2 antigens.[329] Membrane fluidity also seems to play a role in regulation of cell surface immunoglobulin expression in rabbit spleen cells: fluidization by phosphatidylcholine liposomes decreased the number of cells bearing surface IgG, and this effect could be reversed by rigidification of the membrane with cholesterol.[330] These changes in antigenic expression may have influence on the effectiveness of cell-cell interactions during immune responses.

Oxysterols exhibit potent immunosuppressive effects via influencing the order and dynamics of lipids in plasma membrane. Lymphocytes treated with 7-β,25-dihydroxycholesterol showed a decrease of the relative protein kinase C (PKC) activity in the particulate fractions compared to control cells, and a reduction in the phosphorylation endogenous PKC substrates.[331] Thus, oxysterols interfere with two membrane-related phenomena, i.e., the modification of membrane PKC activity and the inhibition of the phosphorylation of the substrates of PKC located in the membrane. The alteration of cell membrane by incorporation of oxysterol could impair the signal transduction and may explain the immunosuppressive activity of oxysterol.[331] The effect of membrane fluidity on the mitogen responsiveness of human peripheral blood lymphocytes is not as straightforward. Mitogen responses were enhanced by a special lipid mixture designated as active lipid (AL721), which is a reputedly potent membrane fluidizing agent, but unaffected by cholesterol. It was concluded that AL721 does not exert its effect through enhanced triggering of cells by altered membrane fluidity but rather that lymphocytes require an exogenous source of phospholipids/triglycerides for optimal growth *in vitro*.[332,333] The type of dietary fat dramatically affects the function and lipid composition of lymphoid cells.[334,335] Autoimmune disease development in lupus-prone hybrid female mice was significantly slowed in mice fed a diet containing ω-3-fatty acids; at 10 months of age no or only mild glomerulonephritis was found in treated animals, while 94% of mice fed a saturated fat diet had succumbed to glomerulonephritis. It was suggested that ω-3-fatty acids of fish oil reduce immune-complex-induced glomerulonephritis through production of prostaglandin metabolites with attenuated activity and/or through altering cell membrane fluidity, which modulates the responsiveness of immune cells.[334] Dietary fat concentration and degree of saturation influenced the antibody response to a T-dependent antigen via changes in the number of cells producing antibody and the level of antibody produced, while the total number of splenic B-cells remained unchanged.[335] Lymphocytes from mice fed a diet containing a high level of polyunsaturated fatty acids exhibited an increase in membrane fluidity and lower number of antibody producing cells.[335]

The human immunodeficiency virus (HIV-1), responsible for AIDS, infects lymphoid cells through binding to the CD4 antigen expressed on the cell surface. Therefore, it is of importance how the expression of CD4 antigens can be modulated by altered membrane fluidity. When $CD4^+$ RDM-4 lymphoma cells were grown as ascites in adult AKR mice fed diets supplemented either with saturated fat (beef tallow) or polyunsaturated fatty acids (fish oil) the membrane fluidity of cells from the fish oil-fed donors was much higher than that of the cells from the beef tallow-fed animals. The higher membrane fluidity correlated with the diminished expression of CD4 cell surface marker. The percentage of CD4 positive lymphoma cells exhibiting sites associated with "high" fluorescence intensity was the only one affected by diet, whereas the percentage of cells exhibiting "low" fluorescence CD4 sites remained unaffected.[336] Fluidizing the membranes of $CD4^+$ lymphoid cells with phenytoin,[317] AL721 liposomes,[337] or dextran sulfate[338] abrogates HIV-1 adsorption. Conversely, membrane rigidification *in vitro* induces HIV-1 infectibility of otherwise insusceptible lymphocytes isolated from two different primate strains.[339] These data clearly demonstrate that membrane fluidity can modulate the CD4 receptor function and that the plasma membrane properties have a great impact on HIV-1 adsorption and infection.

b. Effects of Ligand Binding on Membrane Fluidity

Similarly to hormones binding of various ligands (antibodies, mitogens, lymphokines, etc.) to their receptors on immune cells can modulate the membrane fluidity directly, within a short time (less than an hour) or indirectly during a longer time scale (hours, days) through altered metabolic processes resulting in changes in the fatty acid composition of cell membranes, and therefore in the membrane fluidity as well (Table 7).

It has been demonstrated the membrane fluidity of sheep erythrocytes was increased by specific IgM immunoglobulins and decreased by non-specific proteins.[79] The following model has been suggested to account for the observed phenomena. The membrane becomes more rigid when nonspecific proteins incorporate in the lipid regions. IgM immunoglobulins bound to the surface receptors have a quite different effect. The uniformly distributed antigen receptors may form a more dense islet upon binding of specific antibodies; thus, larger lipid mosaics free of proteins will be formed, showing higher fluidity. This model has a limited validity because it could not be used in the interpretation of the effect of IgE binding on the membrane fluidity of rat mast cells.[78] Binding of IgE to its receptors on rat mast cells caused a distinct decrease in membrane fluidity. Addition of antigen specifically recognized by IgE further decreased the membrane fluidity within a minute. After 10 min this decrease was followed by an increase in membrane fluidity, reaching values below that of control cells.[78] Similarly, binding specific immunoglobulins to human lymphocytes caused a decrease in membrane fluidity parallel with the

Table 7. **Modulation of Membrane Fluidity by Ligand Binding**

Ligand	Membrane system	Membrane fluidity short-term effect	Ref.
IgM	Sheep erythrocytes	Increase (↑)	79
IgE	Rat mast cells	Decrease (↓)	78
IgG	Human lymphocytes	Decrease (↓)	77
α-IgM	B lymphocytes	Increase (↑)	340
α-H-$2K^k$	Mouse myeloma X5563	Increase (↑)	341
	Mouse lymphoma RDM4	Increase (↑)	341
α-sIg	Burkitt lymphoma EB-3	Decrease (↓)	342
PHA[a]	Equine, bovine, canine lymphocytes	Slight increase (↑)	343
Con A[b]	Rat thymocytes	Increase (↑)	344
	Calf T-lymphocytes (intact)	Increase (↑)	346
	Calf T-lymphocytes (purified plasma membrane)	Increase (↑)	346
IL-2[c]	Mouse T lymphocytes (LC7 and CTLL-2)	Increase (↑)	347
IL-3[d]	Murine FDCD-Mix cell line[e]	Decrease (↓)	348
	Murine lymphocyte clone 17-I[f]	No effect	348
β-Thromboglobulin	Polymorphonuclear leukocytes	Decrease (↓)	349

[a]Phytohemagglutinin.
[b]Concanavalin A.
[c]Interleukin-2.
[d]Interleukin-3.
[e]IL-3-dependent line.
[f]IL-3-independent line.

antibody concentration.[77] This phenomenon, however, was observed only when antibody was added to DPH-treated lymphocytes. Upon changing the sequence of treatments, i.e., adding first the antibody and then the DPH, there was practically no observable change in membrane viscosity. In addition, capping of antibody-receptor complexes was observed when DPH treatment followed the antibody binding, but no capping developed when cells were treated first with DPH. Thus, DPH treatment produces certain changes in cellular membranes which prevent capping. The authors suggested that binding of antibody to its receptors results in a stiffening of membrane structure in the vicinity of the complex. This effect can be monitored by the DPH molecules near the complexes. When capping has developed, only a few DPH molecules are close enough to the complexes to indicate the change, while the remaining majority show an unchanged membrane fluidity.[77]

Binding of anti-IgM antibody stimulates B cells and also causes increase in membrane fluidity that occurs more promptly than the increase in intracellular calcium concentration. The increase in membrane fluidity was inhibited by 20-min pretreatment of phorbol myristate acetate (PMA), which activates PKC. These results indicated that membrane fluidity changes are closely related to the increased intracellular calcium concentration after cross-linkage of membrane IgM on the B cell surface.[340] Using stopped-flow fluorometry with three different fluorescent probes, it has been shown that membrane fluidities of tumor cells increased first after binding of anti-H-$2K^k$ monoclonal antibody. After binding of antibody, an increase in the intracellular calcium concentration was also observed.[341] The interaction of poly- and monoclonal antibodies against the L-chain of human Ig expressed on Burkitt lymphoma EB-3 cell surface resulted in increased fluorescence polarization of anthrylvinyl-labeled sphingomyelin, incorporated into the plasma membrane.[342] Detectable decrease in membrane fluidity was observed even when binding was less than 100 antibody molecules per cell.[342]

Plant lectins such as concanavalin A (Con A) and phytohemagglutinin (PHA) stimulate T lymphocytes to proliferation and differentiation. Within minutes after bindings of mitogens to the cell surface receptors, multiple changes are observed in the plasma membrane. Membrane fluidity of equine, bovine, and canine lymphocytes stimulated with PHA increased only slightly, which was not significant enough to accelerate the diffusion velocity of receptors in the plasma membranes.[343] Stimulation of rat thymocytes with Con A resulted in a prompt decrease in the order parameter of ESR probe incorporated to the plasma membrane. Changes in the order parameter were closely associated with the clustering of the lipid probe, induced by Con A treatment. The time course of changes in order parameter during a 6-h incubation with Con A showed that after a sharp decrease the S parameter gradually increases toward the control value within the first 30 min. After 30 min the S gradually decreases again.[344] The activity of Δ 9-desaturase enzyme, which converts stearic acid to oleic acid, in Con A or Con A + interleukin-2-(IL-2) stimulated mouse thymocytes was activated only from the fourth hour of stimulation. Therefore, the increased membrane fluidity in the first 30 min of activation is probably not due to enrichment of oleic acids in the plasma membranes.[345] Goppelt-Strübe and Resch[346] warned of the possible pitfalls in measuring fluorescence polarization in mitogen-stimulated T-lymphocytes. Con A stimulation caused an increase in the fluorescence polarization of DPH. This, however, was not due to the changes in membrane fluidity, but due to the interaction between the mitogen and the DPH, because in purified plasma membranes a decrease rather that an increase in fluorescence polarization was observed.[346] In the early stage of T-lymphocyte activation after IL-2 binding, the membrane mobilities of cells increased first, then the intracellular calcium concentration increased due to release from intracellular stores later on due to the calcium influx from the extracellular medium.[347] The other lymphokine, IL-3, showed a decrease in membrane fluidity of IL-3-dependent hemopoietic cell lines, for which it serves as growth factor, and no change at all in the membrane fluidity of IL-3 independent cell line.[348] The observed very rapid change in membrane fluidity could not be the direct physical effect of IL-3 binding since the number of receptor sites for IL-3 in the cells used was very small (on the order of 500 to 1000 per cell). The authors suggested that the mobilization of stored lipid, e.g., from the Golgi apparatus, into the membrane might be involved in the rapid rigidification of plasma membrane.[348] Binding of β-thromboglobulin, one specific platelet release product, to polymorphonuclear leukocytes caused a decrease of the membrane fluidity and the chemiluminescence response of the cells.[349] These results suggested that β-thromboglobulin could play a role in the interaction between platelets and polymorphonuclear cells.[349]

c. Membrane Fluidity of Cells in Immune Reactions

Changes in the membrane fluidity of RAW264 macrophages during endocytosis were followed using time-resolved fluorescence anisotropy measurements. With this approach both rotational correlation time (dynamics) and order parameter (structure) could be determined simultaneously. Binding and endocytosis of immune complexes increased both the rotational correlation time and the order parameter in macrophages as compared to that in control cells, indicating the rigidification of the plasma membrane.[350] Receptor-independent phagocytosis of latex beads, however, decreased the rotational correlation time and the order parameter, suggesting a significant increase in membrane fluidity. Based upon the results of independent experiments, it has been suggested that the decrease in membrane fluidity, caused by immune complexes, is closely correlated with the H_2O_2 production.[350] Phagocytosis of oil emulsion and polystyrene beads by rabbit polymorphonuclear leukocytes also induced an increase in membrane fluidity. The presence of colchicine during phagocytosis abolished the increase in membrane fluidity, suggesting that the phagocytosis is accompanied by a microtubule-dependent reorganization of membrane lipids.[351] While the rate of endocytosis of membrane immunoglobulin by B cells was not affected by chemical effectors increasing membrane fluidity, rigidification of plasma membrane resulted in the slowing down and eventually the blocking of the endocytosis.[352] The authors suggested that a step depending upon membrane fluidity may be involved in the endocytosis and this step may become rate limiting when membranes become less fluid.[352]

Natural killer (NK) cell-mediated cytotoxicity results from membrane interactions between NK effector and target cells. The changes in the membrane fluidity of either effector or target cells result in significant alteration of various stages of NK-mediated cytotoxicity.[353-355] In NK cell-mediated cytotoxicity the interaction of NK effector and target cells involves the intimate contact between the plasma membrane of both cells and the physical state of these membranes influences the cytotoxic reaction. Fluidization of NK effector cell membrane with phosphatidylcholine resulted in a significant inhibition of cytotoxic activity due to a decrease in the frequency of NK-target conjugates.[353] Rigidification of the NK effector cell membranes by the treatment with cholesteryl hemisuccinate also inhibited the cytotoxicity. However, this inhibition was post binding, because binding was increased and lysis was abrogated.[353] Later on, it has been demonstrated that inhibition the NK cell-mediated cytotoxicity by rigidification of the NK effector cell resulted in defective trigger for activation of the NK cytotoxic factor release mechanism.[355] Fluidization of K562 target cell membranes sightly but insignificantly increased their lysis by NK cells without affecting the binding step. On the other hand, rigidification of target cell membranes decreased the sensitivity of these target cells to lysis. This inhibition proved to be post binding; the altered membrane structure interfered with the lysis by NK cytotoxic factor.[353] The fact that the membrane structure of the target cell is important in this process was also supported by the observation that transferring the membranes from an NK-sensitive target cell to an NK-resistant cell renders this cell sensitive to NK cytotoxicity.[354] The inhibition of NK cell-mediated cytotoxicity by rigidification of the target cell membrane was not due to loss of binding sites of NK cytotoxic factor on the target cells because target cells with rigid membranes absorbed more NK cytotoxic factor than control cells, but it was due to the inhibition in processing-bound NK cytotoxic factor to lysis.[355] Some oxysterols (oxidation products of cholesterol) inhibited the NK cell-mediated cytotoxicity without affecting the viability of effectors cells. Oxidation of cholesterol at C-5, C-6, and C-7 positions generated very effective inhibitors whereas oxysterols that are oxidized at the C-20 and C-25 positions had little or no inhibitory effect. The authors concluded that oxidation in the cholesterol nucleus which is situated closer to the phospholipid headgroups at the lipid bilayer-aqueous interface results in a more profound effect on the plasma membrane physical structure.[356]

The mechanism whereby cytotoxic T lymphocytes specifically lyse target cells involves several discrete stages such as target cell recognition, cell contact, a calcium-dependent lytic stage, and dissociation of the killer-target cells. When mouse cytotoxic T cells bound to their specific target cells, cytotoxic cells increased their membrane fluidity and then calcium was released from intracellular stores.[357] In the calcium-dependent stage perforin is released, which inserts into the target cells forming pores on it and finally lysing the target cells. One single lymphocyte may lyse multiple target cells in a sequential manner without self-destruction. The mechanism underlying the resistance of cytotoxic T lymphocytes to perforin-mediated lysis has been studied by Ojcius et al.[358] They have compared the lipid head group packing and the fluidity of membrane hydrocarbon region of cytotoxic T lymphocytes and several target cells using merocyanine 540 and TMA-DPH. The investigators concluded that the membrane lipid structure could not account for the unusually high resistance of cytotoxic T lymphocytes, since these cells have looser lipid head group packing than some target cell lines and lower membrane fluidity than some target cells.[358] In another experimental system the relationship between membrane fluidity and susceptibility of villous trophoblast to lysis by natural cytotoxic cells has been studied.[359] It has been observed that cells treated with cholesteryl hemisuccinate (CHS) were more susceptible for natural lymphocyte cytotoxicity than untreated controls. Rigidification of trophoblasts by CHS also increased binding of lymphocytes to them. Preincubation with progesterone prevented the increase in membrane fluidity caused by CHS treatment and also restored the impaired resistance of CHS-treated trophoblast cells to lysis.[359]

The release of histamine from mast cells, involved in inflammatory responses, is also sensitive to membrane fluidity. Propranolol liberated histamine from isolated mast cells by increasing the membrane fluidity significantly.[360] In the presence of antigen, interferon-α/β inhibited the histamine release by reducing the fluidity of the cell membrane.[361] The development of delayed hypersensitivity of lymphocytes to staphylococci also depended on the membrane fluidity and antioxidation resistance of membrane lipids.[362] The increase in microviscosity inhibited the development of delayed hypersensitivity leading to the aggravation of the infectious process. The increase of membrane fluidity and the accumulation of antioxidants facilitated the development of delayed hypersensitivity leading to a milder course of the infectious process.[362]

Taking the advantage of the flow cytometric approach in resolving the populations of cells in heterogeneous mixture by dual-angle light scatter, the shape of the fluidity gradient of the outer hemileaflet of the plasma membrane of human white blood cells was studied by the use of *n*-(9-anthroyloxy) fatty acid probes where $n = 2, 3, 6, 7, 9, 10, 11, 12$, and 16.[137] The order of the anisotropy values describing the shapes of the fluidity gradient were, for neutrophils, $6 > 7 > 9 > 2 = 3 = 11 = 12 > 16$, for lymphocytes, $7 > 6 > 9 > 11 > 2 = 3 > 11 = 12 > 16$, and, for monocytes, $9 > 7 > 6 > 11 > 2 = 3 > 12 > 16$.[137] Probes with an *n*-value less than or equal to the maxima ($n = 6$, neutrophils; $n = 7$, lymphocytes; $n = 9$, monocytes) rapidly (1, 2 min) reached equilibrium uptake, whereas probes with *n*-values grater than the maxima took progressively longer times to equilibrate as *n* increased. This behavior is consistent with the existence of an energy barrier just below the approximate region sensed by the probes as fluidity minima (anisotropy maxima), i.e., below C6 for neutrophils, C7 for lymphocytes, and C9 for monocytes.[137] It was suggested that the observed differences in fluidity profile for neutrophils, lymphocytes, and monocytes was closely correlated with the diverse functions of these cells demanding different membrane compositions.[137]

Altered membrane fluidity of lymphocytes from patients of rheumatoid arthritis[363] and from patients after thermal injury[364] has been observed. In rheumatoid arthritis lymphocytes showed a significant decrease in membrane fluidity as compared to controls. The authors suggested that the changes in lymphocyte membrane fluidity could be involved in the pathogenetic mechanism of rheumatoid arthritis.[363] After thermal injury, however, a marked increase in membrane fluidity of lymphocytes was observed as compared to normal (control) subjects. Simultaneously, elevated cortisol levels were found in patients after thermal injury. Since the membrane fluidity of lymphocytes increased with short- or long-term exposure to cortisol *in vitro*, the investigators concluded that the increase in membrane fluidity is a direct effect of cortisol, which may explain, in part, the cellular dysfunction and immunologic suppression that is observed after thermal injury.[364]

The relationship between the microenvironment and the physical properties of mouse T lymphocytes has also been examined.[365] T cells from spleen exhibited significantly lower membrane fluidity and significantly higher resting level of intracellular calcium concentration than T cells of lymph node origin. It has been suggested that the physical differences between T lymphocytes from distinct lymphoid compartments may be intrinsic, more "fluid" cells preferentially homing to the spleen as opposed to lymph nodes or, alternatively, the difference may be imposed on cells by the environment. The observed differences may be responsible for the finding that T cells from murine spleen and lymph nodes are differentially responsive to antigen priming.[365]

Activation of platelet with ADP caused a decrease in membrane fluidity which paralleled with the shape change induced.[366] The effect of ADP on membrane fluidity was completely abolished by preincubation of cells with cytochalasin B, a cytoskeleton depolymerizing agent, which also inhibited the platelet shape change. α-Thrombin caused a two-phase changes in membrane fluidity of platelets depending on the dose applied. At lower doses a decrease, while at higher doses an increase, in membrane fluidity was observed. The former effect was sensitive to cytochalasin B and was correlated to shape changes. The latter effect resulted from granule fusion with the plasma membrane during exocytosis, since a similar effect was not observed with ADP. The investigators concluded that the membrane fluidity changes were associated with platelet shape changes.[366] It has been pointed out, however, that these data should be handled with caution.[186] Feijge et al.[186] have shown that membrane fluidity changes of intact platelets are not a consequence of modified lipid composition but are, at least partially, caused by changed optical properties of the platelet suspension. Measurements of membrane fluidity of platelets by fluorescence polarization is seriously hindered by the high turbidity of the platelet suspension and also by changes in the turbidity and platelet morphology during the activation process.[186] Using DPH, TMA-DPH and DPH-PA as membrane fluidity probes, Kitagawa et al.[103] demonstrated that the cytoplasmic side of the platelet membrane is more fluid than the outer leaflet of the plasma membrane. The investigators concluded that the asymmetry of fluidity in the plasma membrane seemed to be suitable for the roles of each hemileaflet. The outer leaflet serves as a barrier to the extracellular environment, whereas the inner leaflet provides sites for various enzymatic reaction and for fusion between plasma membrane and intracellular granule membranes.[103]

C. CELL PHYSIOLOGY

1. Cell Cycle

Numerous events occur during a cell cycle, several of which are localized at the level of cell surface membrane. The membrane activities necessary to accomplish a cell division cycle comprise among others transport processes, enzymic reactions, hormone-receptor interactions, etc., and are inherently sensitive to changes in membrane fluidity. In addition, the cell-cell contact, known to regulate cell proliferation (contact inhibition), might induce significant changes in the dynamic structure of the plasma membrane. These observations stimulated several investigators to study membrane fluidity changes during cell cycle and to investigate how various agents influencing membrane fluidity can modulate cell proliferation. Although several papers have been published in this area, no generally applicable correlation has been established between membrane fluidity and cell cycle phases (Table 8).

In one of the earliest reports dealing with this subject synchronized murine neuroblastoma cells were studied *in situ* on a glass substratum.[367] Low membrane fluidity was observed during mitosis, which increased significantly in G1 phase and remained constant during *S* phase. At the end of *S* phase, in G2 phase, membrane fluidity decreased again.[367] A different observation has been published about K562, an erythroid precursor cell line.[368] Using flow microfluorimetry, single cell measurements were performed on synchronized cells. Membrane fluidity decreased after cell division and was minimal during S phase. Concomitant with the transition from *S* to G2+M phase an increase in membrane fluidity was observed.[368]

When membrane fluidity of human and rat fibroblast grown on glass coverslips was studied, a different fluidity profile was observed in relation to cell cycle progression.[369] Membranes of growing cells were more fluid than those of stationary cells. Collard et al.[370] obtained similar results using three different cell systems. Mouse lymphocytes were stimulated with Con A, contact-inhibited 3T3 fibroblasts were stimulated with serum, and proliferation of liver cells were induced by partial hepatectomy. In all three systems, although the stimulation was achieved in different ways, cycling cells exhibited a higher membrane fluidity than noncycling cells.[370] Similar findings have been described for hepatoma cells,[371] avian fibroblasts,[372] and Chinese hamster ovary cells.[373] A more accurate approach has been applied by Pajor et al.[374] to establish correlation between fluidity changes and cell cycle phases. Cell cycle progression was defined by monitoring RNA, protein, and DNA synthesis as well as the expression of interleukin-2 and transferrin receptors. The increase in membrane fluidity correlated with RNA and protein synthesis as well as with the cell surface density of growth factor receptors. After the initial changes, the membrane fluidity remained unaltered throughout the cell cycle although DNA synthesis occurred during this time period. The data indicated that the structural change in the cytoplasmic membrane is an early event of cell

Table 8. **The Relationship between Membrane Fluidity and Cell Cycle**

Cell type	Membrane fluidity during cell cycle (*S* phase)	Ref.
Murine neuroblastoma cells	Elevated (↑)	367
K562 erythroid precursor cells	Decreased (↓)	368
Rat fibroblasts	Elevated (↑)	369
Murine lymphocytes	Elevated (↑)	370
3T3 fibroblasts	Elevated (↑)	370
Rat brain cells	Elevated (↑)	370
Hepatoma cells	Elevated (↑)	371
Avian fibroblasts	Elevated (↑)	372
Chinese hamster ovary cells	Elevated (↑)	373
Human lymphocytes	Elevated (↑)	374
Murine L 1210 cells (synchronized)	No change	375

activation and occurs in the course of G0-G1 transition as part of the growth cycle.[374] No change in membrane fluidity has been observed during the cell cycle progression of synchronized murine L 1210 cells,[375] since only a few percent of continuously proliferating leukemic cells resided in the G0 phase where lower membrane fluidity could have been monitored than in other phases in the cell cycle. We should consider that the synchronization was induced by a high dose of colcemid, although not causing cell death, may evoke drastic changes in membrane structure disturbing the correlation between membrane fluidity and cell cycle.

The data discussed above indicated that the profile of membrane fluidity changes during cell cycle depends upon the cell type, experimental conditions (i.e., treatment of cells, synchronization, methods for monitoring membrane fluidity). No unequivocal conclusion can be drawn on cell cycle-dependent fluidity changes in cytoplasmic membranes. It should be noted, however, that the majority of observations support the model that cycling cells possess higher membrane fluidity than noncycling ones, in order to facilitate metabolic transport across the membrane and enzyme reactions taking place in the membrane, promoting RNA, protein, and DNA synthesis during cell cycle.

2. Differentiation and Maturation

During cellular differentiation cell proliferation ceases to exist and maturation processes take place with a pace progressively slowing down. In a completely differentiated cell, most cellular processes reach basal activity. Cell differentiation and maturation entail a series of membrane events, which are modulated by the dynamic structure of the plasma membrane. In addition, the metabolic processes accompanying cellular differentiation usually change the composition of cellular membranes, new lipid and protein components arise, and the abundance of other components, previously present in the system, will decrease or they completely disappear, all together resulting in an altered dynamic structure of the plasma membrane (Table 9).

Several experimental systems have been studied to investigate the correlation between differentiation and maturation processes and membrane fluidity. When the dynamic properties of plasma membrane in epidermal cells were monitored by means of electron spin resonance, close correlation was observed between the keratinization process and membrane fluidity.[376] Three subsets of guinea pig epidermal keratinocytes were isolated. Morphological observation and cytofluorometric DNA analysis showed that the high-, intermediate-, and low-density fractions were rich in basal, spinous, and granular cells, respectively.[377] Keratinocytes proliferate in the basal cell layer of the epidermis and mature into spinous and granular cells. In different fractions of Percoll gradient-separated keratinocytes the order parameter (S) increased from the lower to the upper layers while the Na-K-ATPase activity gradually decreased in the same direction. Accordingly, the differentiation of epidermal cells in the keratinization process was associated with a decrease in plasma membrane fluidity and a decline of Na-K-ATPase activity.[376] Treating keratinocytes with topical hexadecane resulted in hyperproliferation accompanied by higher membrane fluidity and Na-K-ATPase activity than in control cells. Membrane composition changed in agreement with altered membrane fluidity, elevated levels of palmitic acid, stearic acid, and ω-3

Table 9. **Changes in Membrane Fluidity during Differentiation and Maturation**

Cell type	Membrane fluidity	Ref.
Guinea pig keratinocytes	Decreased (↓)	376, 378, 379
Rat enterocytes	Decreased (↓)	380
Rat intestinal brush border	Decreased (↓)	382–386
Rabbit ileum brush border	Decreased (↓)	387, 388
Friend erythroleukemia cells	Decreased (↓)	189
HL-60 human promyelocytic cells	Decreased (↓)	389, 390
F9 teratocarcinoma cells	Decreased (↓)	133
B16 melanoma cells	Decreased (↓)	392
Rabbit alveolar macrophages	Decreased (↓)	393
C1300 mouse neuroblastoma cells	Increased (↑)	394
Hypertrophic "ossifying" chondrocytes	Increased (↑)	395
Human liver cells	Increased (↑)	396
Chick embryo neurons	No change	397

unsaturated fatty acid, and a significant decrease in the molar ratio of free cholesterol to phospholipid was observed.[378] Similar findings were reported when the fluidity changes during keratinization were monitored by fluorescence polarization of DPH. These results also indicated that membrane fluidity of guinea pig keratinocytes decreased during their maturation.[379] Interestingly, the cytoplasmic viscosity, monitored by fluorescence polarization of intracellular fluorescein, decreased in keratinocytes during the process of differentiation.[377]

The other experimental system which has been thoroughly studied was the maturation of mammalian enterocyte membrane. Brasitus and Dudeja[380] investigated the structural and dynamic properties of brush-border membranes prepared from rat small intestinal villus and crypt cells by monitoring fluorescence polarization of three lipid-soluble fluorescent probes. Crypt cell membranes exhibited higher membrane fluidity than those of villus cells as measured with each probe. Lower cholesterol/phospholipid and saturated/unsaturated fatty acyl chain ratios were observed in membranes from crypt cells. These results indicated that fluidity of brush-border membranes decreased during differentiation.[380] Similarly, rigidification of the membrane has been observed when the physical properties of rabbit enterocytes maturing and migrating up the crypt-villus axis have been compared.[381] The membrane fluidity of villus tip, mid-villus, and crypt enterocytes decreased in this order. At the same time, maximal rates of glucose transport were greatest in microvillus membrane vesicles prepared from mature cells.[381] Postnatal maturation profiles of functional and biochemical properties of rat small intestinal brush border membrane have been studied in various age groups: newborns, sucklings, weaned, juveniles, and adults.[382,383] Parallel to the increase in cholesterol-to-phospholipid molar ratio from birth to the weaned age, a significant decrease in membrane fluidity was observed in adult vs. newborn rats and the lipid permeability also changed throughout the maturation. The results suggested that during the first 3 weeks of life the jejunal microvillus membrane undergoes a maturation process that is well suited to the change in diet.[382,383] The closure of the intestinal barrier to immunoglobulin is a normal maturation process in the rat and can be promoted by thyroid hormone[384] and glucocorticoids.[385,386] The induced maturation of enterocyte membrane coincided with the decrease in membrane fluidity in the above cases. These treatments induced structural and functional maturation of enterocyte membrane and enhanced the intestinal mucosal barrier by decreasing the penetration of antibodies.[384-386] Schwarz et al.[387] examined the ontogenesis of bile acid transport in brush-border membranes from rabbit ileum and demonstrated a temporal association between the decrease of membrane fluidity and maturation of taurocholate uptake during aging. The same group studied the physicochemical properties of the small intestinal basolateral cell surface during postnatal development.[388] Comparing the membrane composition and fluidity of jejunal and ileal enterocytes from suckling and weanling-mature rabbits showed that enterocyte membranes of matured rabbits were more rigid and that ileum exhibited lower membrane fluidity compared to jejunum in both age groups. These data demonstrated developmental as well as regional differences in small intestinal basolateral membrane fluidity.[388]

Similarly to the aforementioned examples, significant decrease in membrane fluidity has been found during differentiation of Friend erythroleukemic cells,[189] promyelocytic leukemic cells,[389,390] and F9 teratocarcinoma cells,[133] which was associated, at least in part, with the increase in cholesterol/phospholipid and saturated/nonsaturated fatty acid ratios. Comparative studies on three Friend erythroleukemia clones demonstrated that the membrane fluidity of DMSO-sensitive clones [FLC and F(+) clones] was lower than that of a DMSO-resistant clone (R(3) at normal serum concentration (15%). Growing the DMSO-resistant clone R(3) in low serum concentration decreased the membrane fluidity and the cell surface expression of insulin receptors increased to the level seen with differentiable clones.[391] B16 melanoma cells induced to differentiate by a plant glycoside, gindensenoside Rh2, exhibited lower membrane fluidity than nontreated cells. Parallel to the fluidity changes, Rh2 inhibited cell growth, induced melanogenesis, and increased cell-to-cell and cell-to-substrate adhesiveness.[392] Ricardo et al.[393] studied the membrane fluidity changes during postnatal development of rabbit alveolar macrophages by using TMA-DPH. Gradual but significant decrease of membrane fluidity and increase of cholesterol to phospholipid ratio in alveolar macrophages was observed along with the functional maturation of cells.[393]

In contrast to the examples discussed above, fluorescence polarization studies with DPH on neuroblastoma cells[394] revealed a progressive increase in membrane fluidity coinciding with morphological and cytochemical differentiation. Moreover, treatment of the cells with dipalmitoyl phosphatidylcholine vesicles reversibly inhibited morphological differentiation, suggesting that an optimal (high) fluidity is a prerequisite of differentiation.[394] Similarly, increased membrane fluidity was observed in hypertrophic ("ossifying") chondrocytes induced toward cartilage mineralization as compared to that of resting chondrocytes.[395] Kaputilnik et al.[396] compared the fluidity and lipid composition of human hepatic microsomal membranes isolated from fetal and adult livers. The more fluid adult microsome had lower cholesterol-to-phospholipid and higher phosphatidylcholine-to-sphingomyelin molar ratios than those of the more rigid fetal microsome.[396] No change was detected in membrane fluidity or in phospholipid pattern of plasma membranes isolated from chick embryo neurons during maturation in culture.[397]

Differentiation and maturation is a multifaceted process but the changes in physical and biochemical structure of plasma membrane seem to play a central role in this context. No general trends are discernible in differentiation and maturation regarding the direction of alteration in membrane fluidity. It can be hypothesized that differentiating cells are aimed to reach membrane fluidity optimal for the special membrane functions of completely matured cells.

3. Transformation and Malignancy

Differences in the physical and biochemical characteristics of the plasma membrane of normal, and transformed cells have received much interest. As discussed earlier, changes in the membrane fluidity may alter enzyme activity, transport processes, and expression of cell surface receptors and antigens, which, in turn, may cause malignant transformation where transformed cells with modified cell surface escape immune surveillance and regulating mechanisms of normal cell growth. The relationship between membrane fluidity and malignant transformation has been studied in solid tumors and in leukemic cells as well (Table 10).

A solid tumor which has been extensively studied with respect to the fluidity of the plasma and mitochondria membrane is the hepatoma. Plasma membrane of both slowly and rapidly proliferating hepatomas consistently shows an increased cholesterol-to-phospholipid molar ratio as compared to normal liver plasma membranes. Parallel to the changes in lipid composition, lipid fluidity, monitored by DPH fluorescence polarization, was significantly lower in these membranes than that measured in membranes from normal liver cells.[156] Similarly, rigidification of plasma membranes of Morris hepatoma cells have been observed. The investigators suggested that uncontrolled cholesterogenesis caused an overproduction of the sterol, which could result in an enrichment of tumor cell membranes with cholesterol; thus, the hepatomas can be characterized by a decreased membrane fluidity.[398]

Membrane fluidity of plasma membranes of brain tumors including gliomas and meningiomas was investigated and compared with those of white and gray matter.[399] The fatty acid composition of brain

Table 10. **The Relationship between Membrane Fluidity and Malignant Transformation**

Cell type	Changes in membrane fluidity	Ref.
Solid Tumors		
Hepatoma cells	Decrease (↓)	300, 156
Morris hepatoma cells	Decrease (↓)	398
Brain tumors	Increase (↑)	399
Rous sarcoma virus transformed chicken	Increase (↑)	
fibroblasts	Decrease (↓)	400
4013T3 fibroblasts transformed with SV40	Decrease (↓)	402
3T3 fibroblasts transformed with SV101	Decrease (↓)	403
Rat fibroblasts transformed with SV40	Decrease (↓)	404
Lymphomas		
Mouse lymphoma cells	Increase (↑)	28, 188, 405, 406
Chronic lymphatic leukemia cells	Increase (↑)	407, 408
Chronic granulocytic leukemia cells	Increase (↑)	408
Hairy cell leukemia	Increase (↑)	409
Lymphocytes transformed with Friend virus	Increase (↑)	410
Lymphocytes transformed with Moloney virus	Increase (↑)	410
Acute lymphoblastoid cells	Increase (↑)	413

tumors was characterized by an increase in linoleic and arachidonic acids when compared to the control brain. The order parameter calculated from ESR spectra changed in the following order: meningioma < gray matter < glioma < white matter, indicating that brain tumors have higher membrane fluidity than the corresponding normal tissues.[399]

Altered membrane fluidity has been observed in Rous sarcoma virus-transformed chicken embryo fibroblasts.[400,401] Pessin et al.[400] reported that fluorescence polarization of DPH showed an inverse correlation with the amount of triacylglycerols and alkyldiacylglycerols present in cells. In isolated plasma membranes containing negligible amounts of triacylglycerols and alkyldiacylglycerols, fluorescence polarization of DPH was lower than that measured in intact cells. Membrane fluidity in plasma membranes isolated from transformed cell was lower than that in membranes from normal growing cells.[400] On the contrary, Porat et al.[401] found increased membrane fluidity in Rous sarcoma virus-transformed cells as compared to the control cells. This contradiction might come from the fact that, while in the first report the membrane fluidity was monitored with DPH reporting about a bulk membrane fluidity, multifrequency differential phase fluorimetry of a special membrane protein, the adenosine deaminase enzyme complex, was used to determine the dynamic properties of plasma membrane from the rotational mobility of the protein in the latter report. It is conceivable that the rotation of the enzyme complex is rather influenced by the microenvironment than the bulk membrane fluidity. Transformation of mouse 3T3 fibroblasts with simian virus 40[402] and 101[403] significantly decreased the membrane fluidity as compared to normal cells, in coherence with increased cholesterol level. Similarly, transformation of rat fibroblast with simian virus resulted in decreased membrane fluidity.[404]

Differences in plasma membrane fluidity between normal and leukemic cells have also been reported. Using fluorescence polarization of DPH, it was demonstrated that the fluidity of mouse lymphoma cell membranes was greater than that of normal lymphocytes.[28,188,405,406] These differences were conserved when comparison was made between liposomes prepared from lipids extracted from normal and transformed cells. In one case, the relative amount of cholesterol in lymphoma cells was less than that of the normal lymphocytes, a factor which might account for alteration in membrane fluidity.[28] In another case, the concentration of free fatty acids and glycerides was particularly high, increasing the fluidity of the cells.[406] The membrane of human leukemic cells was also more fluid than that of normal lymphocytes, and the differences in fluidity could be detected both in intact cells and in isolated plasma membranes.[407] *In vitro* incubation of leukemic lymphocytes in normal serum resulted in a decrease in the fluidity of cellular membranes, whereas incubation of normal lymphocytes in leukemic serum resulted in increased lipid fluidity. The investigators concluded that the fluidity difference between normal and leukemic lymphocytes is controlled by components in the blood serum.[407] Saturation indices (molar ratio of saturated fatty acids over unsaturated fatty acids) of white blood cells in specimens from patients with chronic granulocytic leukemia,[408] patients with chronic lymphocytic leukemia,[408] and patients with hairy-cell leukemia[409] were significantly and consistently lower than that of the control specimens. These findings provided indirect evidence for the increase in membrane fluidity of cells from patients with chronic leukemias. Similarly, fluorescence polarization measurements showed a significant increase in lipid fluidity in lymphocytes transformed with Friend virus or Moloney virus as compared to that in nontransformed lymphocytes.[410] Localization of the lymphoid tumors cells is also an important parameter determining membrane fluidity. Mouse thymic-derived ascitic leukemia (GRSL) cells grown in ascites showed much higher lipid fluidity than GRSL cells which were grown in the spleen.[157] These cells also possessed an interesting feature that extracellular membranous vesicles exfoliated from GRSL cells showed much lower membrane fluidity than normal thymocytes or GRSL cells.[157] Similarly, extracellular membranes derived from B-lymphoblastoid cell lines (Raji, WI HF2 729) and a T-lymphoblastoid cell line (Jurkat) were more rigid as assessed by fluorescence polarization of DPH and were highly enriched in cholesterol when compared with the corresponding plasma membrane. The investigators suggested that extracellular membranes are shed from specialized regions of the lymphocyte plasma membrane.[411] This shedding process can provide a way for lymphoma cells to regulate their membrane fluidity. From the presented data it can be concluded that the membranes of leukemic cells are more fluid than that of lymphocytes from controls. This conclusion, however, should be handled with caution, because in these studies membrane fluidity of leukemic cells was always compared to that of resting lymphocytes instead of using a real control, e.g., nonleukemic blast cells. Blastic transformation of human T-lymphocytes caused a significant decrease in the order parameter of DPH and corresponding changes in lipid composition.[412] The membrane fluidity of mitogen-induced blasts was very similar to that of acute malignant lymphoblastoid cells.[413] These examples show that sometimes it is difficult to decide whether the observed fluidity changes are the consequences of maturity of malignancy.

In summary, the malignant transformation is a very complex phenomenon where a vast number of factors (genetic, environmental, etc.) play an important role. The membrane fluidity is only one of these parameters, and it is not surprising that the relationship between malignancy and membrane fluidity varies depending on type, physiological state, and culture conditions of the tumor. Among solid tumors, while the hepatomas have lower, the brain tumors and the virus transformed fibroblasts have higher membrane fluidity than their corresponding normal (control) cells. In lymphomas, the lipid fluidity of cell membranes was usually higher than that of normal lymphocytes. In all cases, changes in membrane fluidity correlated well with the alteration in lipid composition. The papers discussed in this section show that plasma membrane composition and fluidity are altered in tumor cells although no common pattern of alteration has been observed.

4. Metastasis

Plasma membrane plays an important role in the social behavior of cells in living organisms. Such behavior regulates the interactions among cells and at least three elements contribute to their relationship: cellular adhesion, aggregation, and movements. The invasive growth of neoplastic cells is a deviation from normal social behavior of cells. The importance of this type of deviation is especially obvious in the case of metastatic tumor cells. In metastatic tumor progression, the intercellular relationships are completely out of control; metastatic cells escape the regulatory mechanisms coordinating normal cell behavior. Several investigations have been performed on the structure and dynamics of plasma membranes of metastatic cells in order to obtain information about physical and biochemical characteristics of the membrane responsible for the impaired intercellular communication (Table 11).

Comparison of the order parameters obtained from ESR spectra of membranes of metastatic brain tumors, meningioma, glioma, and white and gray matter showed that metastatic brain tumor cells had the lowest order parameter, i.e., the highest lipid fluidity.[399] Similarly, ESR measurements have been used to characterize metastatic subclones of AH66F and Yoshida sarcoma (YS) rat cells.[414] In both YS and AH66F strains, the cell membrane fluidity of metastatic cancer cells was higher than that of nonmetastatic, control cells. The results, obtained in this study, suggested that the change of the cell membrane fluidity of cancer cells is closely related to metastases.[414] Ward et al.[415] used a micropipette aspiration technique to investigate the deformation properties of a panel of nontransformed and transformed rat fibroblasts derived from the same normal cell line. Their results suggested that direct correlation exists between the

Table 11. **The Relationship between Membrane Fluidity and Metastasis**

Cell type	Membrane fluidity of metastatic cells compared to control cells	Ref.
Brain tumors	Increased (↑)	399
AH66F rat cells	Increased (↑)	414
Yoshida rat cells	Increased (↑)	414
	No correlation	427
Rat fibroblasts	Increased (↑)	415
Murine lymphoma ESb	Increased (↑)	416
LM fibroblasts	Increased (↑)	120, 150, 418, 419
Moloney sarcoma-virus-transformed MS2 cells	Increased (↑)	420
mFS6 murine sarcoma	Increased (↑)	420
Lewis lung tumor	Decreased (↓)	421
HM tumor cells	Decreased (↓)	422
C1-30 and LC-AH hepatoma cells	Increased (↑)	423
MO4 mouse fibrosarcoma cells	Decreased (↓)	424
Mouse lymphoma cells	Decreased (↓)	425
Rat adenocarcinoma cells	Decreased (↓)	425
B16 melanoma cells	No correlation	426
L5178 lymphoma cells	No correlation	426

increase in deformability and the progression of the transformed phenotype from a nontumorigenic cell line into a tumorigenic, metastatic cell line.[415] Biochemical analysis of plasma membrane fragments from two variants of a murine lymphoma, Eb and ESb, with different metastatic capacity, showed altered lipid composition and lipid-to-protein ratio, indicating a more fluid state of the plasma membranes derived from the highly metastatic tumor line ESb.[416] The investigation of the lipid composition of plasma membranes isolated from primary tumors (L-929, LM, A-9, and C3H) and nine metastatic cell lines revealed that cultured primary tumor and metastatic cells differed twofold in sterol/phospholipid molar ratios.[150] Application of a fluorescent sterol, dehydroergosterol, in conjunction with covalently linked trinitrophenol groups as selective quenchers, demonstrated an asymmetric transbilayer distribution of sterol in the plasma membranes. The inner monolayer of the plasma membranes from both cultured primary and metastatic tumor cells was enriched in sterol as compared with the outer monolayer. As a consequence, the inner monolayer was distinctly more rigid. Treatment with phenobarbital or prilocaine, drugs that selectively fluidize the outer and inner monolayers of LM plasma membranes, respectively, did not change the transbilayer distribution of dehydroergosterol.[150] Highly metastatic tumor cells had significantly lower sterol carrier protein content and plasma membrane sterol/phospholipid molar ratios as compared with tumor cell lines.[417] Plasma membrane properties of cultured local LM fibroblast tumors and lung metastases from athymic (nude) mice were also studied by multifrequency phase and modulation fluorimetry with DPH.[120,418] In order to minimize the effect of complicating factors such as site of growth, presence of host inflammatory cells, and variation in nutrition, the excised local tumors and lung metastases were cultured in chemically defined, serum-free medium. Plasma membranes of metastatic cells were more fluid (lower limiting anisotropy) than those of local tumor cells. The investigator concluded that the higher fluidity of metastatic as compared to local tumor plasma membrane was not due to difference in site of growth, host cell contamination, and nutrition.[120,418] The effect of *in vitro* phospholipid manipulations of these LM cell lines on tumor growth and metastasis was examined in nude mice.[419] Serum and choline-fed cells metastasized more frequently in mice than monomethylethanolamine- and ethanolamine-fed cells. In addition, metastases from cells cultured with serum or choline, but not from monomethylethanolamine or ethanolamine, were extensive and highly invasive. The plasma, microsomal, and mitochondrial membranes from lung metastases differed markedly from those of local tumors with respect to membrane enzymes, phospholipid content, cholesterol content, phosphatidylethanolamine content, and unsaturated fatty acid content, indicating that the metastatic cell plasma membranes were more fluid than those of the local tumor cells.[419] Taraboletti et al.[420] also investigated the effect of membrane fluidity on the metastatic capacity of murine tumor-cell lines. Correlation was observed in cell lines with different metastatic potential, and was confirmed when their lung-colonizing ability was altered by modulating the membrane lipid composition. Cholesteryl hemisuccinate was used as membrane rigidifier and egg-yolk lecithin as fluidizer. In both cases, the investigators observed the expected alteration in membrane fluidity, with a parallel alteration of the lung-colonizing potential, the fluidized cells being more metastatic and the rigid cells less metastatic than the untreated ones.[420]

While in all of the above-discussed papers the highly metastatic cells possessed higher membrane fluidity as compared to the corresponding original cell lines, there are several studies also reporting about negative correlations between metastasis and membrane fluidity. Using ESR studies, Aszalos et al.[421] have shown that the motional freedom of lipid and protein molecules of the cellular membrane of Lewis lung tumor cells is higher in the less-metastatic cell populations. This finding argues for altered membrane dynamics in cell population with low metastatic potential.[421] Similarly, increased membrane fluidity was observed when the metastatic capability of HM tumor cells was highly reduced with HUdR.[422] As a consequence of the membrane alterations evoked by HUdR treatment, a reduced adhesiveness and spreading on extracellular matrix components was detected. In addition, HM cells treated with HUdR showed diminished capacity to invade fibroblast monolayers *in vitro*.[422] Treating rat ascites hepatoma cells with methylthiodeoxyadenosine (MTA) and with five MTA analogs decreased the membrane fluidity and inhibited the invasion of the tumor cells without affecting their proliferation.[423] Three other MTA analogs neither affected the invasiveness of the tumor cells nor altered membrane fluidity. The investigators concluded that the decrease in invasiveness of tumor cells by MTA and its analogs is due to alteration in the phospholipid composition and fluidity of the tumor cell membranes.[423] Racemic 1-*O*-octadecyl-2-*O*-methylglycero-3-phosphocholine had an antimetastatic effect in MO4 mouse fibrosarcoma cells and its antimetastatic activity is accompanied with an overall increase of membrane fluidity.[424] Haeffner and co-workers[425] studied the membrane properties of mouse lymphoma and rat adenocarcinoma cell variants with low and high metastatic potential and found that less-malignant cell lines exhibited greater membrane fluidity than highly malignant cell lines.

When the metastatic potential of B16 melanoma and L5178 lymphoma variants was studied in relation to their membrane fluidity, no clear correlation was found between these two parameters.[426] The membranes of tumor cells with higher metastatic potential had a lower cholesterol/phospholipid ratio but greater unsaturated phospholipid content. Since the rigidifying effect of cholesterol was compensated with the elevated unsaturated phospholipid content, steady-state polarization measurements showed no marked differences in bulk fluidity of metastatic cell variants. The investigators suggested that instead of the bulk fluidity the mosaicism might be relevant to the metastatic potential of tumors.[426] According to this model the plasma membrane is a mosaic of domains possessing different degrees of microviscosity. Modifying the lipid composition of the plasma membrane of YS cells significantly altered their metastatic characteristics.[427] Exogenous modification of cells with unsaturated fatty acids resulted in an increase of membrane fluidity and the modified cells produced fewer metastatic foci in the lung as compared to that of unmodified cells. On the other hand, although exogenous eicosapentaenoic acid and docosahexaenoic acid treatments induced a slight increase in membrane fluidity of the cells, the metastatic potential of YS cells in lung was significantly increased. These data suggested that the metastatic capacity of cells did not correlate with the membrane fluidity but rather with the specific lipid composition of the plasma membrane of cells.[427]

As discussed above, we could see positive, negative, or no correlation between membrane fluidity and metastatic potential of tumor cells. These contradicting results may come from fact that data are not always comparable because of the completely different experimental conditions. Very few attempts have been made to maintain similar growth conditions for metastatic and original tumors.[120,418] In these cases the local tumors and lung metastases were cultured in chemically defined, serum-free medium but this artificial condition may cause another problem. Culturing tumor cells *in vitro* may result in clonal selection, after a few passages the cultured cells might have completely different characteristics from the original tumor. At this time it is premature to relate differences in membrane properties of metastatic vs. local tumor cells to the known biological properties of metastatic cells. This difficulty is largely due to paucity of data on purified plasma membranes and other membranes from metastases obtained either *in vitro* or *in vivo*. Application of the latest developments in experimental techniques for generating new data and their detailed and thorough analysis may result in better understanding of the correlation between membrane structure and dynamics and the metastasis.

REFERENCES

1. **Singer, S. J. and Nicolson, G.,** The fluid mosaic nature of the structure of cell membranes, *Science*, 175, 720, 1972.
2. **Singer, S. J.,** The structure and function of membranes — a personal memoir, *J. Membr. Biol.*, 129, 3, 1992.
3. **Alberts, B., Bray, D., Lewis, J., Roberts, K., and Watson, J. D.,** *Molecular Biology of the Cell*, Garland, London, 1983, 255.
4. **Edidin, M.,** Tissue architecture and lateral diffusion, *Comments Mol. Cell. Biophys.*, 2, 285, 1984.
5. **Wolf, D. E.,** Overcoming random diffusion in polarized cells — corralling the drunken beggar, *BioEssays*, 6, 116, 1986.
6. **Yechiel, E. and Edidin, M.,** Micrometer-scale domains in fibroblast plasma membranes, *J. Cell Biol.*, 105, 755, 1987.
7. **Edidin, M. and Stroynowski, I.,** Differences between the lateral organization of conventional and inositol phospholipid-anchored membrane proteins. A further definition of micrometer scale membrane domains, *J. Cell Biol.*, 112, 1143, 1991.
8. **Shinitzky, M. and Henkart, P.,** Fluidity in model membranes: current concepts and trends, *Int. Rev. Cytol.*, 60, 121, 1979.
9. **Shinitzky, M.,** Membrane fluidity and cellular functions, in *Physiology of Membrane Fluidity*, Shinitzky, M., Ed., CRC Press, Boca Raton, FL, 1984, chap. 1.
10. **Smith, R. L. and Oldfield, E.,** Dynamic structure of membranes by deuterium NMR, *Science*, 225, 280, 1984.
11. **Israelachivili, J. N., Marcelja, S., and Horn, R. G.,** Physical principles of membrane organization, *Q. Rev. Biophys.*, 13, 121, 1980.

12. **Matkó, J., Szöllősi, J., Trón, L., and Damjanovich, S.**, Luminescence spectroscopic approaches in studying cell surface dynamics, *Q. Rev. Biophys.*, 21, 479, 1988.
13. **Devaux, P. F.**, Phospholipid flippases, *FEBS Lett.*, 234, 8, 1988.
14. **Saupe, A.**, Kernresonanzen in kristallinen Flüssigkeiten un in kristallinflüssigen Lösungen. I, *Z. Naturforsch. A.*, 19, 161, 1964.
15. **Jähnig, F.**, Structural order of lipids and proteins in membranes: evaluation of fluorescence anisotropy data, *Proc. Natl. Acad. Sci. U.S.A.*, 76, 6361, 1979.
16. **Johansson, L. B. A. and Lindblom, G.**, Orientation and mobility of molecules in membranes studied by polarized light spectroscopy, *Q. Rev. Biophys.*, 13, 63, 1980.
17. **Ludescher, R. D., Peting, L., Hudson, S., and Hudson, B.**, Time-resolved fluorescence anisotropy for systems with lifetime and dynamic heterogeneity, *Biophys. Chem.*, 28, 59, 1987.
18. **van Ginkel, G., van Langen, H., and Levine, Y. K.**, The membrane fluidity concept revisited by polarized fluorescence spectroscopy on different model membranes containing unsaturated lipids and sterols, *Biochimie*, 71, 23, 1989.
19. **Szabo, A.**, Theory of fluorescence depolarization in macromolecules and membranes, *J. Chem. Phys.*, 81, 150, 1984.
20. **Araiso, T., Saito, H., Shirahama, H., and Koyama, T.**, Analysis for the molecular motion of phospholipid bilayer with picosecond fluorometry, *Biorheology*, 27, 375, 1990.
21. **Stubbs, C. D. and Smith, A. D.**, The modification of mammalian membrane polyunsaturated fatty acid composition in relation to membrane fluidity and function, *Biochim. Biophys. Acta*, 779, 89, 1984.
22. **Alfsen, A.**, Membrane dynamics and molecular traffic and sorting in mammalian cells, *Prog. Biophys. Mol. Biol.*, 54, 145, 1989.
23. **Cullis, P. R. and De Kruijff, B.**, Lipid polymorphism and the functional role of lipids in biological membranes, *Biochim. Biophys. Acta*, 559, 399, 1979.
24. **Kapitza, H. G., Ruppel, D. A., Galla, H. J., and Sackmann, E.**, Lateral diffusion of lipids and glycophorin in solid phosphatidylcholine bilayers. The role of structural defects, *Biophys. J.*, 45, 577, 1984.
25. **Sperott, M. M., Ipsen, J. H., and Mouritsen, O. G.**, Theory of protein-induced lateral phase separation in lipid membranes, *Cell Biophys.*, 14, 79, 1989.
26. **Sackmann, E., Kotulla, R., and Heiszler, F. J.**, On the role of lipid-bilayer elasticity for the lipid-protein interaction and the indirect protein-protein coupling, *Can. J. Biochem. Cell Biol.*, 62, 778, 1984.
27. **Golan, D. E. and Veatch, W.**, Lateral mobility of phospholipid and cholesterol in the human erythrocyte membrane: effects of protein-lipid interactions, *Proc. Natl. Acad. Sci. U.S.A.*, 77, 2537, 1984.
28. **Shinitzky, M. and Inbar, M.**, Microviscosity parameters and protein mobility in biological membranes, *Biochim. Biophys. Acta*, 433, 133, 1976.
29. **Shinitzky, M. and Barenholz, U.**, Fluidity parameters of lipid regions determined by fluorescence polarization, *Biochim. Biophys. Acta*, 515, 367, 1978.
30. **Cantor, C. R. and Schimmel, P. R.**, *Biophysical Chemistry, II*, W. H. Freeman, San Francisco, 1980, 433.
31. **Vanderkoooi, J. and McLaughlin, A.**, Use of fluorescence probes in the study of membrane structures and function, in *Biochemical Fluorescence: Concepts*, Vol. 2, Chen, R. F. and Edelhoch, H., Eds., Marcel Dekker, New York, 1976, 737.
32. **Tilley, L., Thulborn, K. R., and Sawyer, W. H.**, An assessment of the fluidity gradient of the lipid bilayer as determined by a set of n-(9-anthroyloxy) fatty acids (n = 2, 6, 9, 12, 16), *J. Biol. Chem.*, 254, 2592, 1979.
33. **Shinitzky, M. and Barenholz, Y.**, Dynamics of the hydrocarbon layer in liposomes of lecithin and sphingomyelin containing dicetylphosphate, *J. Biol. Chem.*, 249, 2653, 1974.
34. **Cogan, U., Shinitzky, M., Weber, G., and Mishida, T.**, Microviscosity and order in the hydrocarbon region of phospholipid and phospholipid-cholesterol dispersions determined with fluorescent probes, *Biochemistry*, 12, 521, 1973.
35. **Shinitzky, M., Dianoux, A. C., Gitler, C., and Weber, G.**, Microviscosity and order in the hydrocarbon regions of micelles and membranes determined with fluorescence probes. I. Synthetic micelles, *Biochemistry*, 10, 2106, 1971.
36. **Galla, H. J. and Sackmann, E.**, Lateral diffusion in the hydrophobic region of membranes: use of pyrene excimer as optical probes, *Biochim. Biophys. Acta*, 339, 103, 1974.

37. **Devaux, P. F. and McConnell, H. M.,** Lateral diffusion in spin labelled phosphatidylcholine multilayers, *J. Am. Chem. Soc.*, 94, 4475, 1972.
38. **Bloom, M., Evans, E., and Mouritsen, O. G.,** Physical properties of the fluid lipid-bilayer components of cell membranes: a perspective, *Q. Rev. Biophys.*, 24, 293, 1991.
39. **Smith, I. C. P.,** Organization and dynamics of membrane lipids as determined by magnetic resonance spectroscopy, *Can. J. Biochem.*, 57, 1, 1979.
40. **Jardetzky, O.,** NMR in the study of membranes, in *Membranes and Transport*, Vol. 1, Martonosi, A. N., Ed., Plenum Press, New York, 1982, 109.
41. **Oldfield, E.,** NMR of protein-lipid interactions in model and biological membrane systems, in *Membranes and Transport*, Vol. 1, Martonosi, A. N., Ed., Plenum Press, New York, 1982, 115.
42. **McConnell, H. M.,** Molecular motion in biological membrane, in *Spin Labeling: Theory and Applications*, Berliner, L. J. E., Ed., Academic Press, New York, 1976, 525.
43. **Devaux, P. F. and Davoust, J.,** Current views on boundary lipids deduced from electron spin resonance studies, in *Membranes and Transport*, Vol. 1, Martonosi, A. N., Ed., Plenum Press, New York, 1982, 125.
44. **Thomas, D. D.,** Saturation transfer EPR studies of rotational dynamics in membranes, in *Membranes and Transport*, Vol. 1, Martonosi, A. N., Ed., Plenum Press, New York, 1982, 135.
45. **Shinitzky, M. and Yuli, I.,** Lipid fluidity at the submacroscopic level: determination by fluorescence polarization, *Chem. Phys. Lipids*, 30, 261, 1982.
46. **Chapman, D. and Salsbury, N. J.,** Physical studies of phospholipids. V. Protonmagnetic resonance studies of molecular motion in some 2′3′-diacyl-DL-phosphatidylethanolamines, *Trans. Faraday Soc.*, 62, 2607, 1966.
47. **Levine, Y. K., Birdsall, N. J. M., Lee, A. G., and Metcalfe, J. C.,** ^{13}C nuclear magnetic resonance relaxation measurements of synthetic lecithins and the effect of spin-labeled lipids, *Biochemistry*, 11, 1416, 1972.
48. **Wittebort, R. J., Blume, A., Huang, T., DasGupta, S. K., and Griffin, R. G.,** Carbon-13 nuclear magnetic resonance investigations of phase transitions and phase equilibria in pure and mixed phospholipid bilayers, *Biochemistry*, 21, 3487, 1982.
49. **Seelig, J.,** Deuterium magnetic resonance: theory and application to lipid membrane, *Q. Rev. Biophys.*, 10, 353, 1977.
50. **Huang, T., DeSiervo, A. J., and Yang, Q.-X.,** Effect of cholesterol and lanosterol on the structure and dynamics of the cell membrane of *Mycoplasma capriolum*, *Biophys. J.*, 59, 691, 1991.
51. **Pator, R. W., Venable, R. M., and Karlpus, M.,** Model for the structure of the lipid bilayer, *Proc. Natl. Acad. Sci. U.S.A.*, 88, 892, 1991.
52. **Paddy, M. R., Dahlquist, F. W., Davis, J. H., and Bloom, M.,** Dynamics and temperature dependent effects of lipid-protein interactions. Application of deuterium nuclear magnetic resonance and electron paramagnetic resonance spectroscopy to the same reconstitution of cytochrome *c* oxidase, *Biochemistry*, 20, 3152, 1981.
53. **Meier, P., Sachse, J.-H., Brophy, P. J., Marsh, D., and Kothe, G.,** Integral membrane proteins significantly decrease the molecular motion in lipid bilayers: a deuterium NMR relaxation study of membranes containing myelin proteolipid apoprotein, *Proc. Natl. Acad. Sci. U.S.A.*, 84, 3704, 1987.
54. **Hubbel, W. L. and McConnell, H. M.,** Molecular motion in spin labelled phospholipids and membranes, *J. Am. Chem. Soc.*, 93, 314, 1971.
55. **Mátyus, L., Balázs, M., Aszalós, A., Mulhern, S., and Damjanovich, S.,** Cyclosporin A depolarizes cytoplasmic membrane potential and interact with Ca^{2+} ionophores, *Biochim. Biophys. Acta*, 886, 353, 1986.
56. **Ervasti, J. M., Mickelson, J. R., Lewis, S. M., Thomas, D. D., and Louis, C. F.,** An electron paramagnetic resonance study of skeletal muscle membrane fluidity in malignant hyperthermia, *Biochim. Biophys. Acta*, 986, 70, 1989.
57. **Wassal, S. R. and Stillwell, W.,** Interactions of retinoids with phospholipid membranes: electron spin resonance, *Methods Enzymol.*, 189, 383, 1990.
58. **Knowles, P. F. and Marsh, D.,** Magnetic resonance of membranes, *Biochem. J.*, 274, 625, 1991.
59. **Shin, Y.-K., Ewert, U., Budil, D. E., and Freed, J. H.,** Microscopic versus macroscopic diffusion in model membranes by electron spin resonance spectral-spatial imaging, *Biophys. J.*, 59, 950, 1991.
60. **Vaz, W. L. C. and Almeida, P. F.,** Microscopic versus macroscopic diffusion in one-component fluid phase lipid bilayer, *Biophys. J.*, 60, 1553, 1991.

61. **Vaz, W. L. C., Clegg, R. M., and Hallmann, D.,** Translational diffusion of lipids in liquid crystalline phase phosphatidylcholine multilayers. A comparison of experiment with theory, *Biochemistry*, 24, 781, 1985.
62. **Weber, G.,** Theory of fluorescence depolarization by anisotropic Brownian rotations. Discontinuous distribution approach, *J. Chem. Phys.*, 55, 2399, 1971.
63. **Sawyer, W. H.,** Fluorescence spectroscopy in the study of membrane fluidity: model membrane systems, in: *Methods for Studying Membrane Fluidity*, Aloin, R. C. A., Ed., Alan R. Liss, New York, 1988, 161.
64. **Heyn, M. P.,** Order and viscosity of membranes: analysis by time-resolved fluorescence depolarization, *Methods Enzymol.*, 172, 462, 1989.
65. **Lakowicz, J. R.,** *Principles of Fluorescence Spectroscopy*, Plenum Press, New York, 1983.
66. **Kinosita, K., Jr., Kawato, S., and Ikegami, A.,** A theory of fluorescence polarization decay in membranes, *Biophys. J.*, 20, 289, 1977.
67. **Kawato, S., Kinosita, K., Jr., and Ikegami, A.,** Dynamic structure of lipid bilayers studied by nanosecond fluorescence techniques, *Biochemistry*, 16, 2319, 1977.
68. **Kinosita, K., Jr., Kataoka, R., Kimura, Y., Gotoh, O., and Ikegami, A.,** Dynamic structure of biological membranes as probed by 1,6-diphenyl-1,3,5,-hexatriene: a nanosecond fluorescence depolarization study, *Biochemistry*, 20, 4270, 1981.
69. **Lipari, G. and Szabó, A.,** Effect of librational motion on fluorescence depolarization and NMR relaxation in macromolecules and membranes: diffusion in a cone, *Biophys. J.*, 30, 489, 1980.
70. **Heyn, M. P.,** Determination of lipid order parameters and rotational correlation times from fluorescence depolarization experiments, *FEBS Lett.*, 108, 359, 1979.
71. **Engel, L. W. and Prendergast, F. G.,** Values for and significance of order parameters and "cone angles" of fluorophore rotation in lipid bilayers, *Biochemistry*, 20, 7338, 1981.
72. **van der Meer, B. W., Cheng, K. H., and Chen, S.-Y.,** Effects of lateral diffusion on the fluorescence anisotropy in hexagonal lipid phases. I. Theory, *Biophys. J.*, 58, 1517, 1990.
73. **Chen, S.-Y., Cheng, K. H., van der Meer, B. W., and Beechem, J. M.,** Effects of lateral diffusion on the fluorescence anisotropy in hexagonal lipid phases. II. An experimental study, *Biophys. J.*, 20, 1527, 1990.
74. **Perez-Albarsanz, M. A., Lopez-Aparicio, P., Senar, S., and Recio, M. N.,** Effects of lindane on fluidity and lipid composition in rat renal cortex membranes, *Biochim. Biophys. Acta*, 1066, 124, 1991.
75. **Pilarska, M., Wrosek, A., Pikula, S., and Famulski, K. S.,** Thyroid hormones control lipid composition and membrane fluidity of skeletal muscle sarcolemma, *Biochim. Biophys. Acta*, 1068, 167, 1991.
76. **Han, X. and Gross, R. W.,** Alterations in membrane dynamics elicited by amphophilic compounds are augmented in plasmenylcholine bilayers, *Biochim. Biophys. Acta*, 1069, 37, 1991.
77. **Balázs, M., Szöllősi, J., Somogyi, B., and Damjanovich, S.,** Effect of specific antibodies on membrane microviscosity of human lymphocytes, *Acta Biochim. Biophys. Acad. Sci. Hung.*, 14, 213, 1979.
78. **Rédai, I., Balázs, M., and Juszupova, S.,** Changes in membrane fluidity of mast cells by antigen anaphylactic antibody reaction, *Acta Biochim. Biophys. Acad. Sci. Hung.*, 14, 223, 1979.
79. **Rédai, I., Balázs, M., Somogyi, B., and Damjanovich, S.,** Changes in intramembrane viscosity of sheep erythrocytes in the presence of membrane bound proteins, *Acta Biochim. Biophys. Acad. Sci. Hung.*, 15, 177, 1980.
80. **Kiss, C., Balázs, M., and Kéri-Fülöp, I.,** Dexamethasone decreases membrane fluidity of leukemia cells, *Leukemia Res.*, 14, 221, 1990.
81. **Kiss, C. and Balázs, M.,** Changes in leukemia cell membrane fluidity induced by growth-restrictive and permissive culture conditions, *Leukemia Res.*, 14, 293, 1990.
82. **Lakos, Zs., Somogyi, B., Balázs, M., Matkó, J., and Damjanovich, S.,** The effect of transmembrane potential on the dynamic behavior of cell membranes, *Biochim. Biophys. Acta*, 1023, 41, 1990.
83. **Kiss, C., Balázs, M., Kovács, I., and Karmazsin, L.,** Cell membrane fluidity in blast cells of children with acute leukaemia, *Acta Pediatr. Hung.*, 30, 291, 1990.
84. **van Blitterswijk, W. J., van Hoeven, R. P., and van der Meer, B. W.,** Lipid structural order parameter (reciprocal of fluidity) in biomembranes derived from steady-state fluorescence polarization measurements, *Biochim. Biophys. Acta*, 644, 323, 1981.
85. **van der Meer, B. W., van Hoeven, R. P., and van Blitterswijk, W. J.,** Steady-state fluorescence depolarization data in membranes. Resolution into physical parameters by an extended Perrin equation for restricted rotation of fluorophores, *Biochim. Biophys. Acta*, 854, 38, 1986.

86. **van der Meer, B. W., Kooyman, R. P. H., and Levine, Y. K.**, A theory of fluorescence depolarization in macroscopically ordered membrane systems, *Chem. Phys.*, 66, 39, 1982.
87. **Vos, M. H., Kooyman, R. P. H., and Levine, Y. K.**, Angle-resolved fluorescence depolarization experiments on oriented lipid membrane systems, *Biochem. Biophys. Res. Commun.*, 116, 462, 1983.
88. **van de Ven, M. J. M., van Ginkel, G., and Levine, Y. K.**, Angle-resolved fluorescence depolarization of 1,6-diphenyl-1,3,5-hexatriene (DPH) in unsaturated digalactosyl diacylglycerol (DGDG) lipid bilayers, *Biochem. Biophys. Res. Commun.*, 123, 352, 1984.
89. **van de Ven, M. J. M., G. and Levine, Y. K.**, Angle-resolved fluorescence polarization of macroscopically ordered bilayers of unsaturated lipids, *Biochim. Biophys. Acta*, 777, 283, 1984.
90. **Mulders, F., van Langen, H., van Ginkel, G., and Levine, Y. K.**, The static and dynamic behavior of fluorescent probe molecules in lipid bilayers, *Biochim. Biophys. Acta*, 859, 209, 1986.
91. **van Gurp, M., van Ginkel, G., and Levine, Y. K.**, Orientational properties of biological pigments in ordered systems studied with polarized light: photosynthetic pigment-protein complexes in membranes, *J. Theor. Biol.*, 131, 333, 1988.
92. **Deinum, G., van Langen, H., van Ginkel, G., and Levine, Y. K.**, Molecular order and dynamics in planar lipid bilayers: effects of unsaturation and sterols, *Biochemistry*, 27, 852, 1988.
93. **van Langen, H., Schrama, C. A., van Ginkel, G., Ranke, G., and Levine, Y. K.**, Order and dynamics in the lamellar L_α and in the hexagonal H_{II} phase. Dioleoylphosphatidyl-ethanolamine studied with angle-resolved fluorescence depolarization, *Biophys. J.*, 55, 937, 1989.
94. **van Ginkel, G., Korstanje, L. J., van Langen, H., and Levine, Y. K.**, The correlation between molecular orientational order and reorientational dynamics of probe molecules in lipid multibilayers, *Faraday Discuss. Chem. Soc.*, 81, 49, 1986.
95. **Lakowicz, J. R. and Knutson, J. R.**, Hindered depolarizing rotations of perylene in lipid bilayers. Detection by time-resolved fluorescence anisotropy measurements, *Biochemistry*, 19, 905, 1980.
96. **Chong, P. L.-G., van der Meer, B. W., and Thompson, T. E.**, The effects of pressure and cholesterol on rotational motions of perylene in lipid bilayers, *Biochim. Biophys. Acta*, 813, 253, 1985.
97. **Huang, Z. and Haugland, R. P.**, Partition coefficients of fluorescent probes with phospholipid membranes, *Biochem. Biophys. Res. Commun.*, 181, 166, 1991.
98. **Duportail, G. and Weinreb, A.**, Photochemical changes of fluorescent probes in membranes and their effect on the observed fluorescence anisotropy values, *Biochim. Biophys. Acta*, 173, 171, 1983.
99. **Prendergast, F. G., Haugland, R. P., and Callahan, P. J.**, 1-[4-(trimethylamino)phenyl]-6-phenylhexa-1,3,5-triene: synthesis, fluorescence properties, and use as a fluorescence probe of lipid bilayers, *Biochemistry*, 20, 7333, 1981.
100. **Kuhry, J.-G., Duportail, G., Bronner, C., and Laustriat, G.**, Plasma membrane fluidity measurements on whole living cells by fluorescence anisotropy of trimethylammonium-diphenylhexatriene, *Biochim. Biophys. Acta*, 845, 60, 1985.
101. **Stubbs, C. C., Kinosita, K., Munkonge, F., Quinn, P. J., and Ikegami, A.**, The dynamics of lipid motion in sarcoplasmic reticulum membranes determined by steady-state and time-resolved fluorescence measurements on 1,6-diphenyl-1,3,5-hexatriene and related molecules, *Biochem. Biophys. Acta*, 775, 374, 1984.
102. **Allen, M. T., Miola, L., and Whitten, D. G.**, Host-guest interactions: a fluorescence investigation of the solubilization of diphenylpolyene solute molecules in lipid bilayers, *J. Am. Chem. Soc.*, 110, 3198, 1988.
103. **Kitagawa, S., Matsubayashi, M., Kotani, K., Usui, K., and Kametani, F.**, Asymmetry of membrane fluidity in lipid bilayer of blood platelets: fluorescence study with diphenylhexatriene and analogs, *J. Membr. Biol.*, 119, 221, 1991.
104. **Illinger, D., Poindron, P., Fonteneau, P., Modollel, M., and Kuhry, J.-G.**, Internalization of the lipophilic fluorescent probe trimethylamino-diphenylhexatriene follows the endocytosis and recycling of the plasma membrane in cells, *Biochem. Biophys. Acta*, 1030, 73, 1990.
105. **Illinger, D., Poindron, P., Glasser, N., Modollel, M., and Kuhry, J.-G.**, The plasma-membrane internalization and recycling is enhanced in macrophages upon activation with gamma-interferon and lipopolysaccharide; a study using the fluorescent probe trimethylamino-diphenylhexatriene, *Biochem. Biophys. Acta*, 1030, 82, 1990.
106. **Hemetter, A., Rainer, B., Ivessa, E., Kalb, E., Loidl, J., Roscher, A., and Paltauf, F.**, Influence of plasmalogen deficiency on membrane fluidity of human skin fibroblasts: a fluorescence anisotropy study, *Biochem. Biophys. Acta*, 978, 151, 1989.

107. **Morgan, C. G., Thomas, E. M., Moras, T. S., and Yianni, Y. P.**, The use of a phospholipid analogue of diphenyl-1,3,5-hexatriene to study melittin-induced fusion of small unilamellar phospholipid vesicles, *Biochem. Biophys. Acta*, 692, 196, 1982.
108. **Parente, R. A. and Lentz, B. R.**, Advantages and limitations of 1-palmitoyl-2-[[2-[4-(6-phenyl-*trans*-1,3,5-hexatrienyl)phenyl]ethyl]carbonyl]-3-*sn*-phosphatidyl-choline as a fluorescent membrane probe, *Biochemistry*, 24, 6178, 1985.
109. **Squier, T. C., Mahaney, J. E., Yin, J. J., Lai, C. S., and Lakowicz, J. R.**, Resolution of phospholipid conformational heterogeneity in model membranes by spin-label EPR and frequency-domain fluorescence spectroscopy, *Biophys. J.*, 59, 654, 1991.
110. **Sklar, L. A., Hudson, B. S., and Simoni, R. D.**, Conjugated polyene fatty acids as membrane probes. Preliminary characterization, *Proc. Natl. Acad. Sci. U.S.A.*, 72, 1649, 1975.
111. **Sklar, L. A., Hudson, B. S., Peterson, M., and Diamond, J.**, Conjugated polyene fatty acids as fluorescent probes. Spectroscopic characterization. *Biochemistry*, 16, 813, 1977.
112. **Sklar, L. A., Miljanich, G. P., and Dratz, E. A.**, Phospholipid lateral phase separation and the partition of *cis*-parinaric acid and *trans*-parinaric acid among aqueous, solid lipid, fluid lipid phases, *Biochemistry*, 18, 1707, 1979.
113. **Sklar, L. A., Miljanich, G. P., and Dratz, E. A.**, A comparison of the effects of calcium on the structure of bovine retinal rod outer segment membranes, phospholipids, and bovine brain phosphatidylserine, *J. Biol. Chem.*, 254, 9592, 1979.
114. **Kimelman, D., Tecoma, E. S., Wolber, P. K., Hudson, B. S., Wickner, W. T., and Simoni, R. D.**, Protein-lipid interactions. Studies of the M13 coat protein in dimyristoylphosphatidylcholine vesicles using parinaric acid, *Biochemistry*, 18, 5874, 1979.
115. **Cook, S. L., Bouma, S. R., and Huestis, W. H.**, Cell to vesicle transfer of intrinsic membrane proteins: effect of membrane fluidity, *Biochemistry*, 19, 4601, 1980.
116. **Silbert, D. F.**, Partition of parinoyl phospholipids in mixed headgroup systems, *Biochemistry*, 21, 5690, 1982.
117. **Pugh, E. L., Kates, M., and Szabo, A. G.**, Studies on fluorescence polarization of 1-acyl-2-*cis*- or *trans*-parinoyl *sn*-3-glycerophosphorylcholines in model systems and microsomal membranes, *Chem. Phys. Lipids*, 30, 55, 1982.
118. **Welti, R. and Silbert, D. F.**, Partition of parinaroyl phospholipid probes between solid and fluid phospholipid phases, *Biochemistry*, 21, 5685, 1982.
119. **Ben-Yashar, V. and Barenholz, Y.**, The interaction of cholesterol and cholest-4-en-3-one with dipalmitoyl-phosphatidylcholine. Comparison based on the use of three fluorophores, *Biochem. Biophys. Acta*, 985, 271, 1989.
120. **Kier, A. B.**, Membrane properties of metastatic and non-metastatic cells cultured C_3H mice injected with LM fibroblasts, *Biochem. Biophys. Acta*, 1022, 365, 1990.
121. **Wolber, P. K. and Hudson, B. S.**, Fluorescence lifetime and time-resolved polarization anisotropy studies of acyl chain order and dynamics in lipid bilayers, *Biochemistry*, 20, 2800, 1981.
122. **Waggoner, A. S. and Stryer, L.**, Fluorescent probes of biological membranes, *Proc. Natl. Acad. Sci. U.S.A.*, 67, 579, 1970.
123. **Cadenhead, D. A., Kellner, B. M. J., Jacobson, J., and Papahadjopuolos, D.**, Fluorescent probes in model membranes. I. Anthroyl fatty acids in monolayers and liposomes of dipalmitoyl phosphatidylcholine, *Biochemistry*, 16, 5386, 1977.
124. **Thulborn, K. R. and Sawyer, W. H.**, Properties and the locations of a set of fluorescent probes sensitive to the fluidity gradient of the lipid bilayer, *Biochim. Biophys. Acta*, 511, 125, 1978.
125. **Thulborn, K. R., Treloar, F. E., and Sawyer, W. H.**, A microviscosity barrier in the lipid bilayer due to the presence of phospholipids containing unsaturated acyl chains, *Biochem. Biophys. Res. Commun.*, 81, 42, 1978.
126. **Thulborn, K. R., Tilley, L. M., Sawyer, W. H., and Treloar, F. E.**, The use of *n*-(9-anthroyloxy) fatty acids to determine fluidity and polarity gradients in phospholipid bilayers, *Biochim. Biophys. Acta*, 558, 166, 1979.
127. **Vincent, M., de Foresta, B., Gallay, J., and Alfsen, A.**, Nanosecond fluorescence anisotropy decay of *n*-(9-anthroyloxy) fatty acids in dipalmitoylphosphatidyl-choline vesicles with regard to isotropic solvents, *Biochemistry*, 21, 708, 1982.
128. **Vincent, M., de Foresta, B., and Gallay, J.**, Fluorescence anisotropy decays of *n*-(9-anthroyloxy) fatty acids in dipalmitoyl phosphatidylcholine vesicles, *Biochem. Biophys. Res. Commun.*, 107, 914, 1982.

129. **Kutchai, H., Chandler, L. H., and Zavoico, G. B.,** Effects of cholesterol on acyl chain dynamics in multilamellar vesicles of various phosphatidylcholines, *Biochem. Biophys. Acta*, 736, 137, 1983.
130. **Blatt, E. and Sawyer, W. H.,** Rotational motion of *n*-(9-anthroyloxy) fatty acids in phospholipid bilayer vesicles, *Aust. J. Chem.*, 36, 1079, 1983.
131. **Storch, J. and Schachter, D.,** Dietary induction of acyl chain desaturases alters the lipid composition and fluidity of rat hepatocyte plasma membranes, *Biochemistry*, 23, 1165, 1984.
132. **Vincent, M. and Gallay, J.,** Time-resolved fluorescence anisotropy study of effect of a *cis* double bond on structure of lecithin and cholesterol-lecithin bilayers using *n*-(9-anthroyloxy) fatty acids as probes, *Biochemistry*, 23, 6514, 1984.
133. **Schaap, G. H., de Josselin de Jong, J. E., and Jongkind, J. F.,** Fluorescence polarization of six membrane probes in embryonal carcinoma cells after differentiation as measured on a FACS II cell sorter, *Cytometry*, 5, 188, 1984.
134. **Blatt, E. and Sawyer, W. H.,** Depth-dependent fluorescent quenching in micelles and membranes, *Biochim. Biophys. Acta*, 822, 43, 1985.
135. **Collins, J. M. and Grogan, W. M.,** Comparison between flow cytometry and fluorometry for the kinetic measurement of membrane fluidity parameters, *Cytometry*, 10, 44, 1989.
136. **Collins, J. M., Dominey, R. N., and Grogan, W. M.,** The shape of the fluidity gradient in the plasma membrane of living HeLa cells, *J. Lipid. Res.*, 31, 261, 1990.
137. **Collins, J. M., Scott, R. B., and Grogan, W. M.,** Plasma membrane fluidity gradients of human peripheral blood leukocytes, *J. Cell. Physiol.*, 144, 42, 1990.
138. **Collins, J. M. and Grogan, W. M.,** Fluorescence quenching of a series of membrane probes measured in living cells by flow cytometry, *Cytometry*, 12, 247, 1991.
139. **Collins, J. M. and Grogan, W. M.,** Comparison of steady-state anisotropy of the plasma membrane of living cells with different probes, *Biochim. Biophys. Acta*, 1067, 171, 1991.
140. **Schachter, D. and Shinitzky, M.,** Fluorescence polarization studies of rat intestinal microvillus membranes, *J. Clin. Invest.*, 59, 536, 1977.
141. **Huang, N. N., Florine-Casteel, K., Feigenson, G. W., and Sink, C.,** Effect of fluorophore linkage position of *n*-(9-anthroyloxy fatty acids on probe distribution between coexisting gel and fluid phospholipid phases, *Biochim. Biophys. Acta*, 939, 124, 1988.
142. **Kuroda, Y., Matsuzaki, K., Handa, T., and Nakagaki, M.,** Mobility of fluorescent probe molecules in lipid bilayer vesicles as studied by steady-state in time-dependent nuclear Overhauser effect measurements in ^{1}H-nuclear magnetic resonance spectroscopy, *Biochim. Biophys. Acta*, 859, 171, 1986.
143. **Lala, A. K., Dixit, R. R., Koppaka, V., and Patel, S.,** Design, synthesis, and fluorescence studies of fuorenyl fatty acids as new depth-dependent fluorescent probes for membranes: getting over the looping back problem, *Biochemistry*, 27, 8981, 1988.
144. **Hale, J. E. and Schroeder, F.,** Asymmetric transbilayer distribution of sterol across plasma membranes determined by fluorescence quenching of dehydroergosterol, *Eur. J. Biochem.*, 122, 649, 1982.
145. **Schroeder, F.,** Fluorescent sterols: probe molecules of membrane structure and function, *Prog. Lipid Res.*, 23, 97, 1984.
146. **Schroeder, F., Barenholz, Y., Gratton, E., and Thompson, T. E.,** A fluorescence study of dehydroergosterol in phosphatidylcholine bilayer vesicles, *Biochemistry*, 26, 2441, 1987.
147. **Nemecz, G. and Schroeder, F.,** Time-resolved fluorescence investigation of membrane cholesterol heterogeneity and exchange, *Biochemistry*, 27, 7740, 1988.
148. **Schroeder, F. and Nemecz, G.,** Interaction of sphingomyelins and phosphatidylcholines with fluorescent dehydroergosterol, *Biochemistry*, 28, 5992, 1989.
149. **Schroeder, F.,** Use of a fluorescent sterol to probe the transbilayer distribution of sterols in biological membranes, *FEBS Lett.*, 135, 127, 1981.
150. **Kier, A. B., Sweet, W. D., Cowlen, M. S., and Schroeder, F.,** Regulation of transbilayer distribution of fluorescent sterol in tumor cell plasma membranes, *Biochim. Biohys. Acta*, 861, 287, 1986.
151. **Galla, J. J. and Sackmann, E.,** Lateral diffusion in the hydrophobic region of membranes. Use of pyrene excimers as optical probes, *Biochem. Biophys. Acta*, 339, 103, 1974.
152. **Bryszewska, M. and Szosland, K.,** Association between the glycation of erythrocyte membrane proteins and membrane fluidity, *Clin. Biochem.*, 21, 49, 1988.
153. **Brasitus, T. A., Dudeja, P. K., Bolt, M. J. G., Stirin, M. D., and Baum, C.,** Dietary triacylglycerol modulates sodium-dependent D-glucose transport, fluidity and fatty acid composition of rat small intestinal brush-border membrane, *Biochim. Biophys. Acta*, 979, 177, 1989.

154. **Kung, C. E. and Reed, J. K.**, Microviscosity measurements of phospholipid bilayers using fluorescent dyes that undergo torsional relaxation, *Biochemistry*, 25, 6114, 1986.
155. **Collard, J. G. and de Wildt, A.**, Localization of the lipid probe 1,6-diphenyl-1,3,5-hexatriene (DPH) in intact cell by fluorescence microscopy, *Exp. Cell Res.*, 116, 447, 1978.
156. **van Hoeven, R. P., van Blitterswijk, W. J., and Emmelot, P.**, Fluorescence polarization measurements on normal and tumor cells and their corresponding plasma membranes, *Biochim. Biophys. Acta*, 551, 44, 1979.
157. **van Blitterswijk, T., Emmelot, P., Hilkmann, H. A. M., Oomenmeulemans, E. P. M., and Inbar, M.**, Differences in lipid fluidity among isolated plasma membranes of normal and leukemic lymphocytes and membranes of exfoliated from their cell surface, *Biochim. Biophys. Acta*, 467, 309, 1977.
158. **Donner, M., Muller, S., and Stoltz, J. F.**, Fluorescence depolarization method in the study of dynamic properties of blood cells, *Biorheology*, 27, 367, 1990.
159. **Bouchy, M., Donner, M., and André, J. C.**, Evolution of fluorescence polarization of 1,6-diphenyl-1,3,5-hexatriene (DPH) during the labelling of living cells, *Exp. Cell Res.*, 133, 39, 1981.
160. **Kuhry, J. G., Fonteneau, P., Duportail, G., Maechling, C., and Laustriat, G.**, TMA-DPH: a suitable fluorescence polarization probe for specific plasma membrane fluidity studies in intact living cells, *Cell Biophys.*, 5, 129, 1983.
161. **Boudet, G., Lévy-Tolédano, S., Maclouf, J., Rendu, F., and Salesse, R.**, Change in the physical state of platelet plasma membranes upon ionophore A23187 activation. A fluorescence polarization study, *Biochim. Biophys. Acta*, 812, 243, 1985.
162. **Schachter, D., Cogan, U., and Abbot, R. E.**, Asymmetry of lipid dynamics in human erythrocyte membranes studied with permeant fluorophores, *Biochemistry*, 21, 2146, 1982.
163. **Schroeder, W.**, Use of fluorescence spectroscopy in the assessment of biological membrane properties, in *Methods for Studying Membrane Fluidity*, Aloia, R. C. A., Ed., Alan R. Liss, New York, 1988, 193.
164. **Cranney, M., Cundall, R. B., Jones, G. R., Richards, J. T., and Thomas, E. W.**, Fluorescence lifetime and quenching studies on some interesting diphenylhexatriene membrane probes, *Biochim. Biophys. Acta*, 735, 418, 1983.
165. **Davenport, L., Dale, R. E., Bisby, R. H., and Cundall, R. B.**, Transverse location of the fluorescent probe 1,6-diphenyl-1,3,5-hexatriene in model lipid bilayer membrane systems by resonance excitation energy transfer, *Biochemistry*, 24, 4097, 1985.
166. **Doan, T. L., Takasugi, M., Aragon, I., Boudet, G., Montenay-Garestier T., and Helene, C.**, Excitation energy transfer from tryptophan residues of peptides and intrinsic proteins to diphenylhexatriene in phospholipid vesicles and biological membranes, *Biochim. Biophys. Acta*, 735, 259, 1983.
167. **Ameloot, M., Hendricks, H., Herreman, W., Pottel, H., van Cauwelaert, F., and van der Meer, W.**, Effect of orientational order on the decay of the fluorescence anisotropy in membrane suspensions. Experimental verification on unilamellar vesicle and lipid/α-lactalbumin complexes, *Biophys. J.*, 46, 525, 1984.
168. **Fiorini, R., Valentino, M., Wang, S., Glaser, M., and Gratton, E.**, Fluorescence lifetime distributions of 1,6-diphenyl-1,3,5-hexatriene in phospholipid vesicles, *Biochemistry*, 26, 3864, 1987.
169. **Zannoni, C., Arcioni, A., and Cavatorta, P.**, Fluorescence depolarization in liquid crystals and membrane bilayers, *Chem. Phys. Lipids*, 32, 179, 1983.
170. **Parasassi, T., Conti, F., Glaser, M., and Gratton, E.**, Detection of phospholipid phase separation. A multifrequency phase fluorimetry study of 1,6,-diphenyl-1,3,5-hexatriene fluorescence, *J. Biol. Chem.*, 259, 14011, 1984.
171. **Karnovsky, M. J., Kleinfeld, A. M., Hoover, R. L., and Klausner, R. D.**, The concept of lipid domains in membranes, *J. Cell Biol.*, 94, 1, 1982.
172. **Fiorini, R. M., Valentino, M., Gratton, E., and Bertoli, E.**, Erythrocyte membrane heterogeneity studied using 1,6-diphenyl-1,3,5-hexatriene fluorescence lifetime distribution, *Biochem. Biophys. Res. Commun.*, 147, 460, 1987.
173. **Valentino, M., Governa, M., Gratton, E., Fiorini, R., Curatola, G., and Bertoli, E.**, Increased membrane heterogeneity in stimulated human granulocytes, *FEBS Lett.*, 234, 451, 1988.
174. **Barrow, D. A. and Lentz, B. R.**, Membrane structural domains. Resolution limits using diphenylhexatriene fluorescence decay, *Biophys. J.*, 48, 221, 1985.

175. **Kalb, E., Paltauf, F., and Hermetter, A.,** Fluorescence lifetime distributions of diphenylhexatriene-labeled phosphatidylcholine as a tool for the study of phospholipid-cholesterol interactions, *Biophys. J.*, 56, 1245, 1989.
176. **Ho, C., Williams, B. W., and Stubbs, C. D.,** Analysis of cell membrane micro-heterogeneity using the fluorescence lifetime of DPH-type fluorophores, *Biochim. Biophys. Acta*, 1104, 273, 1992.
177. **Illsley, N. P., Lin, H. Y., and Verkman, A. S.,** Lipid domain structure correlated with membrane protein function in placental microvillus vesicles, *Biochemistry*, 26, 446, 1987.
178. **Chatelier, R. C. and Sawyer, W. H.,** The transverse organization of ubiquininones in mitochondrial membranes as determined by fluorescence quenching. Evidence for a two-site model, *Eur. Biophys. J.*, 11, 179, 1985.
179. **Blatt, E., Chatelier, R. C., and Saywer, W. H.,** The transverse location of fluorophores in lipid bilayers and micelles as determined by fluorescence quenching techniques, *Photochem. Photobiol.*, 39, 477, 1984.
180. **Chalpin, D. B. and Kleinfeld, A. M.,** Interaction of fluorescence quenchers with the *n*-(9-anthroyloxy) fatty acid membrane probes, *Biochim. Biophys. Acta*, 731, 465, 1983.
181. **Cabrini, G. and Verkman, A. S.,** Localization of cyanine dye binding to brush-border membranes by quenching of *n*-(9-anthroyloxy) fatty acid probes, *Biochim. Biophys. Acta*, 862, 285, 1986.
182. **Haigh, E. A., Thulborn, K. R., and Sawyer, W., H.,** Comparison of fluorescence energy transfer and quenching methods to establish the position and orientation of components within the transverse plane of the lipid bilayer. Application to the gramicidin A-bilayer interaction, *Biochemistry*, 18, 3525, 1979.
183. **Shaklai, N., Yguerabide, J., and Ranney, H. M.,** Interaction of hemoglobin with red blood cell membrane as shown the a fluorescent chromophore, *Biochemistry*, 16, 5585, 1977.
184. **Vincent, M., de Foresta, B., and Gallay, J.,** Fluorescence anisotropy decays of *n*-(9-anthroyloxy) fatty acids in dipalmitoylphosphatidylcholine vesicles. Localization of effects of cholesterol addition, *Biochem. Biophys. Res. Commun.*, 107, 914, 1982.
185. **Howard, R. J. and Sawyer, W. H.,** Changes in the membrane microviscosity of mouse red blood cells infected with *Plasmodium berghei* using *n*-(9-anthroyloxy) fatty acid fluorescent probes, *J. Parasitol.*, 80, 331, 1980.
186. **Feijge, M. A. H., Heemskerk, J. W. M., and Hornstra, G.,** Membrane fluidity of non-activated and activated human blood platelets, *Biochim. Biophys. Acta.*, 1025, 173, 1990.
187. **Kutchai, H., Huxley, V. H., and Chandler, L. H.,** Determination of fluorescence polarization of membrane probes in intact erythrocytes, *Biophys. J.*, 39, 229, 1982.
188. **Inbar, M.,** Fluidity of membrane lipids: A single cell analysis of mouse normal lymphocytes and malignant lymphoma cells, *FEBS Lett.*, 67, 180. 1976.
189. **Arndt-Jovin, D. J., Ostertag, W., Eisen, H., Klimek, F., and Jovin, T. M.,** Studies of cellular differentiation by automated cell separation. Two model systems: Friend virus-transformed cells and *Hydra attenuata*, *J. Histochem. Cytochem.*, 24, 332, 1976.
190. **Jovin, T. M.,** Fluorescence polarization and energy transfer: theory and application, in *Flow Cytometry and Sorting*, Melamed, M., Mullaney, M., and Mendelsohn M., Eds., John Wiley & Sons, New York, 1979, 137.
191. **Gorvel, J.-P., Mawas, C., Maroux, S., and Mishal, Z.,** Flow cytometry is a new method for the characterization of intestinal plasma membrane, *Biochem. J.*, 221, 453, 1984.
192. **Fox, M. H. and Delohery, T. M.,** Membrane fluidity measured by fluorescence polarization using an EPICS V Cell Sorter, *Cytometry*, 8, 20, 1987.
193. **Dix, J. A. and Verkman, A. S.,** Mapping of fluorescence anisotropy in living cells by ratio imaging. Application to cytoplasmic viscosity, *Biophys. J.*, 57, 231, 1990.
194. **Fushimi, K., Dix, J. A., and Verkman, A. S.,** Cell membrane fluidity in the intact kidney tubule measured by orientation-independent fluorescence anisotropy, *Biophys. J.*, 57, 241, 1990.
195. **Fushimi, K. and Verkman, A. S.,** Relationship between vasopressin-sensitive water transport and plasma membrane fluidity in kidney collecting tubule, *Am. J. Physiol.*, 260, C1, 1991.
196. **Hirano, K.,** Change in membrane fluidity of sand dollar egg cortices caused by Ca^{2+}-induced exocytosis: microscopic analysis with fluorescence anisotropy, *Dev. Growth Differ.*, 33, 451, 1991.
197. **Fulford, A. J. C. and Peel, W. E.,** Lateral pressures in biomembranes estimated from the dynamics of fluorescent probes, *Biochim. Biophys. Acta*, 598, 237, 1980.
198. **Chong, P. L.-G. and Weber, G.,** Pressure dependence of 1,6-diphenyl-1,3,5-hexatriene fluorescence in single-component phosphatidylcholine liposomes, *Biochemistry*, 22, 5544, 1980.

199. **Corda, D., Pasternak, C., and Shinitzky, M.**, Increase in lipid microviscosity upon the creation of membrane potential, *J. Membr. Biol.*, 65, 232, 1982.
200. **Lelkes, P. I.**, Potential dependent rigidity changes in lipid membrane vesicles, *Biochem. Biophys. Res. Commun.*, 90, 656, 1978.
201. **Georgescauld, D. and Duclohier, A.**, Transient fluorescent signals from pyrene labelled pike nerves during action potential: possible implication for membrane fluidity changes, *Biochem. Biophys. Res. Commun.*, 85, 1186, 1978.
202. **Tanabe, H., Kurihara, K., and Kobatake, Y.**, Changes in membrane potential and membrane fluidity in *Tetrahymena pyriformis* in association with chemoreception of hydrophobic stimuli: fluorescence studies, *Biochemistry*, 19, 5339, 1980.
203. **Lakos, Z., Somogyi, B., Balázs, M., Matkó, J., and Damjanovich, S.**, The effect of transmembrane potential on the dynamic behavior of cell membranes, *Biochim. Biophys. Acta*, 1023, 41, 1990.
204. **Quinn, P. J.**, The fluidity of cell membranes and its regulation, *Prog. Biophys. Mol. Biol.*, 38, 1, 1981.
205. **Lenaz, G.**, Lipid fluidity and membrane protein dynamics, *Biosci. Rep.*, 7, 823, 1987.
206. **Devaux, P. F. and Seigneuret, M.**, Specificity of lipid protein interactions as determined by spectroscopic techniques, *Biochim. Biophys. Acta*, 822, 63, 1985.
207. **Devaux, P. F.**, Static and dynamic asymmetry in cell membranes, *Biochemistry*, 30, 1163, 1991.
208. **Mouritsen, O. G. and Jørgensen, K.**, Dynamic lipid-bilayer heterogeneity: a mesoscopic vehicle for membrane function, *BioEssays*, 14, 129, 1992.
209. **Storch, J., Schachter, D., Inoue, M., and Wolkoff, A. W.**, Lipid fluidity of hepatocyte plasma membrane subfractions and their differential regulation by calcium, *Biochim. Biophys. Acta*, 727, 209, 1983.
210. **Sorensen, E. M. B., Acosta, D., and Nealon, D. G.**, Effects of cadmium and calcium on the fluidity of plasma membranes, *Toxicol. Lett.*, 25, 319, 1985.
211. **Solomonson, L. P., Liepkalns, V. A., and Spector, A. A.**, Changes in (Na^+ + K^+)-ATPase activity of Erlich ascites tumor cells produced by alteration of membrane fatty acid composition, *Biochemistry*, 15, 892, 1976.
212. **Szamel, M. and Resch, K.**, Modulation of enzyme activities in isolated lymphocyte plasma membranes by enzymatic modification of phospholipid fatty acids, *J. Biol. Chem.*, 256, 11618, 1981.
213. **Brasitus, T. A., Davidson, N. O., and Schachter, D.**, Variations in dietary triacylglycerol saturation alter the lipid composition and fluidity of rat intestinal plasma membranes, *Biochim. Biophys. Acta*, 812, 460, 1985.
214. **Keefe, E. B., Scharschmidt, B. F., Blankenship, N. M., and Ockner, R. K.**, Studies of relationships among bile flow, liver plasma membrane NaK-ATPase, and membrane microviscosity in the rat, *J. Clin. Invest.*, 64, 1590, 1979.
215. **Simon, F. R., Gonzalez, M., Sutherland, E., Accatino, L., and Davis, R. A.**, Reversal of ethinyl estradiol-induced bile secretory failure with Triton WR-1339, *J. Clin. Invest.*, 65, 851, 1980.
216. **Sinensky, M., Pinkerton, F., Sutherland E., and Simon, F. R.**, Rate limitations of (Na^+ + K^+)-stimulated adenosine triphosphatase by acyl chain ordering, *Proc. Natl. Acad. Sci. U.S.A.*, 76, 4839, 1979.
217. **Chong, P. L., Fortes, P. A., and Jameson D. M.**, Mechanisms of inhibition of (Na,K)-ATPase by hydrostatic pressure studied with fluorescent probes, *J. Biol. Chem.*, 260, 14484, 1985.
218. **Gibbs, A. and Somero, G. N.**, Pressure adaptation of Na^+/K^+-ATPase in gills of marine teleosts, *J. Exp. Biol.*, 143, 475, 1989.
219. **Aloia, R. C., Paxton, J., Daviau, J. S., van Gelb, O., Mlekusch, W., Truppe, W., Meyer, J. A., and Brauer, F. S.**, Effect of chronic alcohol consumption on rat brain microsome lipid composition, membrane fluidity and Na^+-K^+-ATPase activity, *Life Sci.*, 36, 1003, 1985.
220. **Swann, A. C.**, Chronic ethanol and (Na^+-K^+)-adenosine triphosphate: apparent adaptation in cation binding and enzyme conformation, *J. Pharmacol. Exp. Ther.*, 232, 475, 1985.
221. **Swann, A. C.**, Ethanol and (Na^+,K^+)-ATPase: alteration of Na(+)-K+ selectivity, *Alcohol. Clin. Exp. Res.*, 14, 922, 1990.
222. **Sutherland, E., Dixon, B. S., Leffert, H. L., Skally, H., Zaccaro, L., and Simon, F. R.**, Biochemical localization of hepatic surface-membrane Na^+,K^+-ATPase activity depends on membrane lipid fluidity, *Proc. Natl. Acad. Sci. U.S.A.*, 85, 8673, 1988.
223. **Deliconstantinos, G., Kopeikina-Tsiboukidou, L., and Villiotou, V.**, Evaluation of membrane fluidity effects and enzyme activities alterations in adriamycin neurotoxicity, *Biochem. Pharmacol.*, 36, 1153, 1987.

224. **Sakamoto, K.**, Erythrocyte membrane abnormality of essential hypertension with special reference to membrane fluidity and Na(+)-K+ ATPase activity, *Fukuoka. Igaku. Zasshi.*, 82, 562, 1991.
225. **Mazzanti, L., Rabini, R. A., Testa, I., and Bertoli, E.**, Modifications induced by diabetes on the physicochemical and functional properties of erythrocyte plasma membrane, *Eur. J. Clin. Invest.*, 19, 84, 1989.
226. **Wood, P. A., McBride, M. R., Baker, H. J., and Christian, S. T.**, Fluorescence polarization analysis, lipid composition, and Na^+,K^+-ATPase kinetics of synaptosomal membranes in feline GM1 and GM2 gangliosidosis, *J. Neurochem.*, 44, 947, 1985.
227. **Izpisua, J. C., Barber, T., Cabo, J., Hrelia, S., Rossi, C. A., Castelli, G. P., Lercker, G., Biagi, P. L., Bordoni, A., and Lenaz, G.**, Lipid composition, fluidity and enzymatic activities of rat liver plasma and mitochondrial membranes in dietary obese rats, *Int. J. Obes.*, 13, 531, 1989.
228. **Mahaney, J. E. and Thomas, D. D.**, Effects of melittin on molecular dynamics and Ca-ATPase activity in sarcoplasmic reticulum membranes: electron paramagnetic resonance, *Biochemistry*, 30, 7171, 1991.
229. **Voss, J., Birmachu, W., Hussey, D. M., and Thomas, D. D.**, Effects of melittin on molecular dynamics and Ca-ATPase activity in sarcoplasmic reticulum membranes: time-resolved optical anisotropy, *Biochemistry*, 30, 7498, 1991.
230. **Christon, R., Fernandez, Y., Cambon-Gros, C., Periquet, A., Deltour, P., Leger, C. L., and Mitjavila, S.**, The effect of dietary essential fatty acid deficiency on the composition and properties of the liver microsomal membrane of rats, *J. Nutr.*, 118, 1311, 1988.
231. **Edelfors, S. and Ravn-Jonsen, A.**, The effects of alcohols *in vitro* on the nervous cell membrane measured by changes in the (Ca^{2+}/Mg^{2+}) ATPase activity and fluidity of the synaptosomal membrane, *Pharmacol. Toxicol.*, 67, 56, 1990.
232. **Lee, A. G., Michelangeli, F., and East, J. M.**, Tests for the importance of fluidity for the function of membrane proteins, *Biochem. Soc. Trans.*, 17, 952, 1989.
233. **de Foresta, B., Rogard, M., le Maire, M., and Gallay, J.**, Effects of temperature and benzyl alcohol on the structure and adenylate cyclase activity of plasma membranes from bovine adrenal cortex, *Biochim. Biophys. Acta*, 905, 240, 1987.
234. **Martin, K. J., McConkey, C. L., Jr., and Stokes, T. J., Jr.**, Effects of benzyl alcohol on PTH receptor-adenylate cyclase system of canine kidney, *Am. J. Physiol.*, 248, E31, 1985.
235. **Noe, M., Oliva, D., Corsini, A., Soma, M., Fumagalli, R., and Nicosia, S.**, Differential effects of *in vitro* ethanol on prostaglandin E1-sensitive adenylate cyclase from smooth muscle cells and platelets, *J. Cyclic Nucleotide Protein Phosphor. Res.*, 10, 293, 1985.
236. **Rabin, R. A., Bode, D. C., and Molinoff, P. B.**, Relationship between ethanol-induced alterations in fluorescence anisotropy and adenylate cyclase activity, *Biochem. Pharmacol.*, 35, 2331, 1986.
237. **Depauw, H., de Wolf, M., van Dessel, G., Lagrou, A., Hilderson, H. J., and Dierick, W.**, Modification of the adenylate cyclase activity of bovine thyroid plasma membranes by manipulating the ganglioside composition with a nonspecific lipid transfer protein, *Biochim. Biophys. Acta*, 1024, 41, 1990.
238. **Yang, F. Y. and Tu, Y. P.**, A proper transmembrane Ca2+ gradient is essential for the higher enzymatic activity of adenylate cyclase, *Biochem. Biophys. Res. Commun.*, 175, 366, 1991.
239. **Shirley, S. G., Robinson, C. J., and Dodd, G. H.**, The influence of temperature and membrane-fluidity changes on the olfactory adenylate cyclase of the rat, *Biochem. J.*, 245, 613, 1987.
240. **Tsakiris, S.**, Control of the activity of brain synaptosome-associated acetylcholinesterase by acidic phospholipids, *Z. Naturforsch. C*, 40, 97, 1985.
241. **Giannettini, J., Chauvet, M., Dell'Amico, M., and Bourdeaux, M.**, Influence of cetiedyl on erythrocyte membrane microviscosity and acetylcholinesterase activity, *Pharmacol. Res.*, 25, 31, 1992.
242. **Levin, G., Cogan, U., Levy, Y., and Mokady, S.**, Riboflavin deficiency and the function and fluidity of rat erythrocyte membranes, *J. Nutr.*, 120, 857, 1990.
243. **Stalc, A. and Sentjurc, M.**, A contribution to the mechanism of action of SAD-128, *Biochem. Pharmacol.*, 40, 2511, 1990.
244. **Brasitus, T. A., Dahiya, R., Dudeja, P. K., and Bissonnette, B. M.**, Cholesterol modulates alkaline phosphatase activity of rat intestinal microvillus membranes, *J. Biol. Chem.*, 263, 8592, 1988.
245. **Stenson, W. F., Seetharam, B., Talkad, V., Pickett, W., Dudeja, P., and Brasitus, T. A.**, Effects of dietary fish oil supplementation on membrane fluidity and enzyme activity in rat small intestine, *Biochem. J.*, 263, 41, 1989.

246. **Petkova, D. H., Momchilova, A. B., and Koumanov, K. S.**, Phospholipid dependence of the neutral sphingomyelinase in rat liver plasma membranes, *Biochimie*, 68, 1195, 1986.
247. **Ellerbe, P., Murphy, R. B., and Rose, H. G.**, Lecithin:cholesterol acyltransferase regulation. II. Effect of fluidity of egg phosphatidylcholine vesicles, *Chem. Phys. Lipids*, 38, 275, 1985.
248. **Ellerbe, P., Murphy, R. B., and Rose, H. G.**, Lecithin:cholesterol acyltransferase regulation. Effect of fluidity of dimyristoylphosphatidylcholine vesicles, *Biochim. Biophys. Acta*, 817, 282, 1985.
249. **Sevanian, A., Wratten, M. L., McLeod, L. L., and Kim, E.**, Lipid peroxidation and phospholipase A_2 activity in liposomes composed of unsaturated phospholipids: a structural basis for enzyme activation, *Biochim. Biophys. Acta*, 961, 316, 1988.
250. **Sorisky, A., Kucera, G. L., and Rittenhouse, S. E.**, Stimulated cholesterol-enriched platelets display increased cytosolic Ca^{2+} and phospholipase A activity independent of changes in inositol trisphosphates and agonist/receptor binding, *Biochem. J.*, 265, 747, 1990.
251. **Horváth, I., Vigh, L., Pali, T., and Thompson, G. A., Jr.**, Effect of catalytic hydrogenation of *Tetrahymena ciliary* phospholipid fatty acids on ciliary phospholipase A activity, *Biochim. Biophys. Acta*, 1002, 409, 1989.
252. **Brady, L. J., Silverstein, L. J., Hoppel, C. L., and Brady, P. S.**, Hepatic mitochondrial inner membrane properties and carnitine palmitoyltransferase A and B. Effect of diabetes and starvation, *Biochem. J.*, 232, 445, 1985.
253. **Brady, L. J., Hoppel, C. L., and Brady, P. S.**, Hepatic mitochondrial inner-membrane properties, β-oxidation and carnitine palmitoyltransferases A and B. Effects of genetic obesity and starvation, *Biochem. J.*, 233, 427, 1986.
254. **Kolodziej, M. P. and Zammit, V. A.**, Sensitivity of inhibition of rat liver mitochondrial outer-membrane carnitine palmitoyltransferase by malonyl-CoA to chemical- and temperature-induced changes in membrane fluidity, *Biochem. J.*, 272, 421, 1990.
255. **Deliconstantinos, G., Mykoniatis, M., and Papadimitriou, D.**, Carbon tetrachloride modulates the rat hepatic microsomal UDP-glucuronyl transferase activity and membrane fluidity, *Experientia*, 42, 181, 1986.
256. **Levi, F., Benavides, J., Touitou, Y., Quarteronet, D., Canton, T., Uzan, A., Auzeby, A., Gueremy, C., Sulon, J., le Fur, G., and Reinberg, A.**, Circadian rhythm in the membrane of circulating human blood cells: microviscosity and number of benzodiazepine binding sites. A search for regulation by plasma ions, nucleosides, proteins or hormones, *Chronobiol. Int.*, 4, 235, 1987.
257. **Davis, P. J. and Poznansky, M. J.**, Modulation of 3-hydroxy-3-methylglutaryl-CoA reductase by changes in microsomal cholesterol content or phospholipid composition, *Proc. Natl. Acad. Sci. U.S.A.*, 84, 118, 1987.
258. **Minocha, R., Sanyal, S. N., Wali, A., Gumbhir, K., and Majumdar, S.**, Effect of cholesterol and temperature perturbations on membrane hydrolases and transport of calcium and glucose in guinea pig brush border membrane vesicles, *Biochem. Int.*, 16, 1019, 1988.
259. **Dudeja, P. K., Harig, J. M., Wali, R. K., Knaup, S. M., Ramaswamy, K., and Brasitus, T. A.**, Differential modulation of human small intestinal brush-border membrane hemileaflet fluidity affects leucine aminopeptidase activity and transport of D-glucose and L-glutamate, *Arch. Biochem. Biophys.*, 284, 338, 1991.
260. **Dudeja, P. K., Wali, R. K., Harig, J. M., and Brasitus, T. A.**, Characterization and modulation of rat small intestinal brush-border membrane transbilayer fluidity, *Am. J. Physiol.*, 260, G586, 1991.
261. **Iseki, K., Sugawara, M., Saitoh, H., Miyazaki, K., and Arita, T.**, Effect of chlorpromazine on the permeability of β-lactam antibiotics across rat intestinal brush border membrane vesicles, *J. Pharm. Pharmacol.*, 40, 701, 1988.
262. **Moriyama, T., Nakahama, H., Fukuhara, Y., Horio, M., Yanase, M., Orita, Y., Kamada, T., Kanashiro, M., and Miyake, Y.**, Decrease in the fluidity of brush-border membrane vesicles induced by gentamicin. A spin-labeling study, *Biochem. Pharmacol.*, 38, 1169, 1989.
263. **Dudeja, P. K., Wali, R. K., Klitzke, A., and Brasitus T. A.**, Intestinal D-glucose transport and membrane fluidity along crypt-villus axis of streptozocin-induced diabetic rats, *Am. J. Physiol.*, 259, G571, 1990.
264. **Peerce, B. E. and Wright, E. M.**, Sodium induced conformational changes in the glucose transporter of intestinal brush border, *J. Biol. Chem.*, 259, 14015, 1984.
265. **Peerce, B. E. and Wright, E. M.**, Evidence for tyrosyl residues at the Na^+ site on the intestinal Na^+/glucose co-transporter, *J. Biol. Chem.*, 260, 6026, 1985.

266. **Molitoris, B. A.,** Membrane fluidity: measurement and relationship to solute transport, *Sem. Nephrol.*, 7, 61, 1987.
267. **Yuli, I., Wilbrandt, W., and Shinitzky, M.,** Glucose transport through cell membranes of modified lipid fluidity, *Biochemistry*, 20, 4250, 1981.
268. **van Valen, F. and Keck, E.,** Forskolin inhibition of glucose transport in bone cell cultures through a cAMP-independent mechanism, *Bone*, 9, 89, 1988.
269. **Block, E. R. and Edwards, D.,** Effect of plasma membrane fluidity on serotonin transport by endothelial cells, *Am. J. Physiol.*, 253, C672, 1987.
270. **Block E. R., Patel, J. M., and Edwards, D.,** Mechanism of hypoxic injury to pulmonary artery endothelial cell plasma membranes, *Am. J. Physiol.*, 257, C223, 1989.
271. **Sheridan, N. P. and Block, E. R.,** Serotonin transport and fluidity in plasma membrane vesicles: effect of hyperoxia, *Am. J. Physiol.*, 254, C781, 1988.
272. **Patel, J. M., Sekharam, M., and Block, E. R.,** Vitamin E distribution and modulation of the physical state and function of pulmonary endothelial cell membranes, *Exp. Lung. Res.*, 17, 707, 1991.
273. **Felipo, V., Minana, M. D., and Grisolia, S.,** A protein-free diet changes synaptosomal membrane fluidity and tyrosine and glutamate transport, *Neurochem. Res.*, 14, 431, 1989.
274. **Friedlander, G., Le Grimellec, C., and Amiel, C.,** Increase in membrane fluidity modulates sodium-coupled uptakes and cyclic AMP synthesis by renal proximal tubular cells in primary culture, *Biochim. Biophys. Acta*, 1022, 1, 1990.
275. **Smith, D. J. and Ploch, S. A.,** Impaired alanine uptake by basolateral liver plasma membrane vesicles from rats consuming ethanol, *Am. J. Physiol.*, 261, G913, 1991.
276. **Friedlander, G., Shahedi, M., Le Grimellec, C., and Amiel, C.,** Increase in membrane fluidity and opening of tight junctions have similar effects on sodium-coupled uptakes in renal epithelial cells. *J. Biol. Chem.*, 263, 11183, 1988.
277. **Levi, M.,** Heterogeneity of P_i transport by BBM from superficial and juxtamedullary cortex of rat, *Am. J. Physiol.*, 258, F1616, 1990.
278. **Suzuki, M., Kawaguchi, Y., Momose, M., Morita, T., Yokoyama K., and Miyahara, T.,** 1,25-Dihydroxyvitamin D stimulates sodium-dependent phosphate transport by renal outer cortical brush-border membrane vesicles by directly affecting membrane fluidity, *Biochem. Biophys. Res. Commun.*, 150, 1193, 1988.
279. **Levi, M., Baird, B. M., and Wilson, P. V.,** Cholesterol modulates rat renal brush border membrane phosphate transport, *J. Clin. Invest.*, 85, 231, 1990.
280. **Levine, B. S., Knibloe, K. A., Golchini, K., Hashimoto, S., and Kurtz, I.,** Renal adaptation to dietary phosphate deprivation: role of proximal tubule brush-border membrane fluidity, *Am. J. Physiol.*, 260, F613, 1991.
281. **Levi, M. and Cronin, R. E.,** Early selective effects of gentamicin on renal brush-border membrane Na-P_i cotransport and Na-H exchange, *Am. J. Physiol.*, 258, F1379, 1990.
282. **Verkman, A. S. and Masur, S. K.,** Very low osmotic water permeability and membrane fluidity in isolated toad bladder granules, *J. Membr. Biol.*, 104, 241, 1988.
283. **Shi, L. B., Wang, Y. X., and Verkman, A. S.,** Regulation of the formation and water permeability of endosomes from toad bladder granular cells, *J. Gen. Physiol.*, 96, 789, 1990.
284. **Giocondi, M. C. and Le Grimellec, C.,** Water permeation in Madin-Darby canine kidney cells is modulated by membrane fluidity, *Biochim. Biophys. Acta*, 1064, 315, 1991.
285. **Kuo, P., Weinfeld, M., and Loscalzo, J.,** Effect of membrane fatty acyl composition on LDL metabolism in Hep G2 hepatocytes, *Biochemistry*, 29, 6626, 1990.
286. **Paradies, G. and Ruggiero, F. M.,** Enhanced activity of the tricarboxylate carrier and modification of lipids in hepatic mitochondria from hyperthyroid rats, *Arch. Biochem. Biophys.*, 278, 425, 1990.
287. **Marzoev, A. I., Panasenko, O. M., and Andriushchenko, A. P.,** Effect of thyroid hormones on heat-induced changes in the EPR-spectra of erythrocytes, *Probl. Endokrinol. (Mosk).*, 33, 75, 1987.
288. **Tacconi, M. T., Cizza, G., Fumagalli, G., Sarzi Sartori, P., and Salmona, M.,** Effect of hypothyroidism induced in adult rats on brain membrane fluidity and lipid content and composition, *Res. Commun. Chem. Pathol. Pharmacol.*, 71, 85, 1991.
289. **Beguinot, F., Beguinot, L., Tramontano, D., Duilio, C., Formisano, S., Bifulco, M., Ambesi-Impiombato, F. S., and Aloj, S. M.,** Thyrotropin regulation of membrane lipid fluidity in the FRTL-5 thyroid cell line. Its relationship to cell growth and functional activity, *J. Biol. Chem.*, 262, 1575, 1987.

290. **Marra, C. A. and de Alaniz, M. J.**, Mineralocorticoids modify rat liver delta 6 desaturase activity and other parameters of lipid metabolism, *Biochem. Int.*, 22, 483, 1990.
291. **Miccio, M., Orzes, N., Lunazzi, G. C., Gazzin, B., Corsi, R., and Tiribelli, C.**, Reversal of ethinylestradiol-induced cholestasis by epomediol in rat. The role of liver plasma-membrane fluidity, *Biochem. Pharmacol.*, 38, 3559, 1989.
292. **Rosario, J., Sutherland, E., Zaccaro, L., and Simon, F. R.**, Ethinylestradiol administration selectively alters liver sinusoidal membrane lipid fluidity and protein composition, *Biochemistry*, 27, 3939, 1988.
293. **Eggl, P., Wirthensohn, K., and Hirsch, H.**, Effect of hormones on phospholipid metabolism in human cultured fibroblasts, *Biochim. Biophys. Acta*, 862, 399, 1986.
294. **Deliconstantinos, G.**, Cortisol effect on (Na^+ + K^+)-stimulated ATPase activity and on bilayer fluidity of dog brain synaptosomal plasma membranes, *Neurochem. Res.*, 10, 1605, 1985.
295. **Deliconstantinos, G. and Fotiou, S.**, Sex steroid and prostaglandin interactions upon the purified rat myometrial plasma membranes, *Mol. Cell. Endocrinol.*, 45, 149, 1986.
296. **Deliconstantinos, G. and Fotiou, S.**, Effect of prostaglandins E_2 and $F_{2\alpha}$ on membrane fluidity in rat myometrial plasma membranes, *J. Endocrinol.*, 110, 395, 1986.
297. **Greenhalgh, E. A.**, Luteal steroidogenesis and regression in the rat: effects of human chorionic gonadotrophin and phospholipase A2 on cells and plasma membranes, *J. Endocrinol.*, 125, 387, 1990.
298. **Deliconstantinos, G.**, Effects of prostaglandin E_2 and progesterone on rat brain synaptosomal plasma membranes, *Ciba Found. Symp.*, 153, 190, 1990.
299. **Van Bommel, T., Marsen, T., and Bojar, H.**, Effects of high-dose medroxyprogesterone acetate and various other steroid hormones on plasma membrane lipid mobility in CAMA-1 mammary cancer cells, *Anticancer. Res.*, 7, 1217, 1987.
300. **Hammerschmidt, D. E., Knabe, A. C., Silberstein, P. T., Lamche, H. R., and Coppo, P. A.**, Inhibition of granulocyte function by steroids is not limited to corticoids. Studies with sex steroids, *Inflammation*, 12, 277, 1988.
301. **Dutta-Roy, A. K., Ray, T. K., and Sinha, A. K.**, Control of erythrocyte membrane microviscosity by insulin, *Biochim. Biophys. Acta*, 816, 187, 1985.
302. **Hianik, T. and Kavecansky, J.**, Study of physical mechanisms of insulin reception, *Czech. Med.*, 12, 101, 1989.
303. **Hianik, T., Kavecansky, J., Zorad, S., and Macho, L.**, Insulin-induced changes in mechanical characteristics of lipid bilayers modified by liver plasma membrane fragments, *Gen. Physiol. Biophys.*, 7, 191, 1988.
304. **Hianik, T., Zorad, S., Kavecansky, J., and Macho, L.**, Effect of insulin on lipid bilayer viscoelasticity, *Gen. Physiol. Biophys.*, 6, 173, 1987.
305. **Manevich, E. M., Köiv, A., Järv, J., Molotkovsky, J. G., and Bergelson, L. D.**, Binding of specific ligands to muscarinic receptors alters the fluidity of membrane fragments from rat brain, *FEBS Lett.*, 236, 43, 1988.
306. **Zlatanov, I. V., Foley, M., Birmingham, J., and Garland, P. B.**, Effect of some polypeptide hormones on the lateral mobility in plasma membranes of rat testes, L-cells and liver. Influence of development and ageing, *Int. J. Biochem.*, 21, 1269, 1989.
307. **McLean, L. R., Baron, B. M., Buck, S. H., and Krstenansky, J. L.**, Lipid and membrane interactions of neuropeptide Y, *Biochim. Biophys. Acta*, 1024, 1, 1990.
308. **Clandinin, M. T., Cheema, S., Field, C. J., Garg, M. L., Venkatraman, J., and Clandinin, T. R.**, Dietary fat: exogenous determination of membrane structure and cell function, *FASEB J.*, 5, 2761, 1991.
309. **Berlin, E., Bhathena, S. J., Judd, J. T., Nair, P. P., Jones, D. Y., and Taylor, P. R.**, Dietary fat and hormonal effects on erythrocyte membrane fluidity and lipid composition in adult women, *Metabolism*, 38, 790, 1989.
310. **Field, C. J., Ryan, E. A., Thomson, A. B. R., and Clandinin, M. T.**, Diet fat composition alters membrane phospholipid composition, insulin binding, and glucose metabolism in adipocytes from control and diabetic animals, *J. Biol. Chem.*, 265, 11143, 1990.
311. **Neufeld, N. D., Ezrin, C., Corbo, L., Long, D., and Bush, M. A.**, Effects of caloric restriction and exercise on insulin receptors in obesity: association with changes in membrane lipids, *Metabolism*, 35, 580, 1986.
312. **Terris, S., Hofmann, C., and Steiner, D. F.**, Mode of uptake and degradation of ^{125}I-labelled insulin by isolated hepatocytes and H4 hepatoma cells, *Can. J. Biochem.*, 57, 459, 1979.

313. **Raizada, M. K. and Fellow, R. E.**, Alteration of insulin binding and cytoskeletal organization in cultured fibroblasts by tertiary amine local anaesthetics, *J. Supramol. Struct.*, 11, 547, 1979.
314. **Yorek, M., Leeney, E., Dunlap, J., and Ginsberg, B.**, Effect of fatty acid composition on insulin and IGF-I binding in retinoblastoma cells, *Invest. Ophthalmol. Vis. Sci.*, 30, 2087, 1989.
315. **Patel, J. M., Edwards, D. A., Block, E. R., and Raizada, M. K.**, Effect of nitrogen dioxide on surface membrane fluidity and insulin receptor binding of pulmonary endothelial cells, *Biochem. Pharmacol.*, 37, 1497, 1988.
316. **Carmena, M. J., Hueso, C., Guijarro, L. G., and Prieto, J. C.**, Cholesterol modulation of membrane fluidity and VIP receptor/effector system in rat prostatic epithelial cells, *Regul. Pept.*, 33, 287, 1991.
317. **Lehr, H. A., Zimmer, J. P., Hubner, C., Ballmann, M., Hachmann, W., Vogel, W., Baisch, H., Hartter, P., Albani, M., and Kohlschutter, A.**, Decreased binding of HIV-1 and vasoactive intestinal peptide following plasma membrane fluidization of $CD4^+$ cells by phenytoin, *Virology*, 179, 609, 1990.
318. **Kolena, J., Matejcikova, K., Danisova, A., and Virsik, Z.**, Pseudopregnancy-dependent changes in rat ovarian LH/hCG receptors in relation to membrane lipid fluidity, *Reprod. Nutr. Dev.*, 30, 115, 1990.
319. **Muller, C. P. and Krueger, G. R. F.**, Modulation of membrane proteins by vertical phase separation and membrane lipid fluidity. Basis for a new approach to tumor immunotherapy, *Anticancer. Res.*, 6, 1181, 1986.
320. **Shinitzky, M.**, The lipid regulation of receptor functions, *Biomemb. Recep. Mech.*, 7, 135, 1987.
321. **Shinitzky, M., Skornick, Y., and Haran-Ghera, N.**, Effective tumor immunization induced by cells of elevated membrane-lipid microviscosity, *Proc. Natl. Acad. Sci. U.S.A.*, 76, 5313, 1979.
322. **Muller, C. P. and Shinitzky, M.**, Passive shedding of erythrocyte antigens induced by membrane rigidification, *Exp. Cell Res.*, 136, 53, 1981.
323. **Baruch, D. and Cabantchik, Z. I.**, Passive modulation of antigenic expression in the surface of normal and malaria-infected erythrocytes, *Mol. Biochem. Parasitol.*, 36, 127, 1989.
324. **Ludes, B., Staedel, C., Jacqmin, D., Cremel, G., Hubert, P., Bollack, C., and Beck, J. P.**, Increased immunogenecity of human renal carcinoma cells following treatment with cholesterol derivatives, *Eur. Urol.*, 17, 166, 1990.
325. **Ludes, B., Staedel, C., Jacqmin, D., Cremel, G., Hubert, P., Saussine, C., Beck, J. P., and Bollack, C.**, Stimulation of antigenicity of renal adenocarcinoma cells, *J. Urol., (Paris)*, 96, 295, 1990.
326. **Benko, S., Hilkmann, H., Vigh, L., and van Blitterswijk, W. J.**, Catalytic hydrogenation of fatty acyl chains in plasma membranes: effect on membrane lipid fluidity and expression of cell surface antigens, *Biochim. Biophys. Acta*, 896, 129, 1987.
327. **Yasui, K., Masuda, M., Matsuoka, T., Yamazaki, M., Komiyama, A., Akabane, T., Hasui, M., Kobayashi, Y., and Murata, K.**, Abnormal membrane fluidity as a cause of impaired functional dynamics of chemoattractant receptors on neonatal polymorphonuclear leukocytes: lack of modulation of the receptors by a membrane fluidizer, *Pediatr. Res.*, 24, 442, 1988.
328. **Psychoyos, S. and Smith, C. W.**, Enhancement of N-formyl-methionyl-leucyl-phenylalanine (fMLP) binding to isolated human neutrophils, *Agents Actions*, 26, 372, 1989.
329. **Muller, C. P., Stephany, D. A., Shinitzky, M., and Wunderlich, J. R.**, Changes in cell-surface expression of MHC and Thy-1,2 determinants following treatment of lipid modulating agents, *J. Immunol.*, 131, 1356, 1983.
330. **Ostro, M. J., Bessinger, B., Summers, J. F., and Dray, S.**, Liposome modulation of surface immunoglobulins in rabbit spleen cells, *J. Immunol.*, 124, 2956, 1980.
331. **Moog, C., Deloulme, J. C., Baudier, J., Revel, M. O., Bischoff, P., Hietter, H., and Luu, B.**, Membrane-related oxysterol function: preliminary results on the modification of protein kinase C activity and substrate phosphorylation by 7 β,25-dihydroxycholesterol, *Biochimie*, 73, 1321, 1991.
332. **Traill, K. N., Ratheiser, K., Pfeilschifter, R., and Wick, G.**, Lymphocyte membrane lipid composition and mitogen responsiveness in chickens: role of membrane "fluidity", *Eur. J. Immunol.*, 16, 75, 1986.
333. **Traill, K. N., Offner, F., Winter, U., Paltauf, F., and Wick, G.**, Lipid requirements of human T lymphocytes stimulated with mitogen in serum-free medium. Membrane "fluidity" changes are an artefact of lipid (AL721) uptake by monocytes, *Immunobiology*, 176, 450, 1988.
334. **Alexander, N. J., Smythe, N. L., and Jokinen, M. P.**, The type of dietary fat affects the severity of autoimmune disease in NZB/NZW mice, *Am. J. Pathol.*, 127, 106, 1987.
335. **Erickson, K. L., Adams, D. A., and Scibienski, R. J.**, Dietary fatty acid modulation of murine B-cell responsiveness, *J. Nutr.*, 116, 1830, 1986.

336. **Oth, D., Mercier, G., Tremblay, P., Thérien, H. M., Bégin, M. E., Ells, G., and Potworowski, E.,** Modulation of CD4 expression on lymphoma cells transplanted to mice fed (*n*-) polyunsaturated fatty acids, *Biochim. Biophys. Acta*, 1027, 47, 1990.
337. **Sarin, P. S., Gallo, R. C., Scheer, D. I., Crews, F., and Lippa, A. S.,** Effects of a novel compound (AL721) on HTLV-III infectivity *in vitro*, *N. Engl. J. Med.*, 313, 1289, 1985.
338. **Lehr, H. A., Zimmer, J. P., Hubner, C., Reisinger, E. C., Kohlschutter, A., and Schmitz, H.,** Dextran sulphate reduces diphenylhexatriene anisotropy in human peripheral blood lymphocytes: impact on plasma membrane fluidity and HIV-cytopathogenicity, *J. Antimicrob. Chemother.*, 28, 677, 1991.
339. **Zimmer, J. P., Lehr, H. A., Hübner, C., Lindner, S. G., Ramsperger, R., Kohlschütter, A., and Schmitz, H.,** Effects of membrane lipid and fluidity modifications on HIV-1 infectibility of primate lymphoctyes *in vitro*, *Biosci. Rep.*, 10, 263, 1990.
340. **Mizuguchi, J., Utsunomiya, N., Nakanishi, M., and Arata, Y.,** Phorbol myristate acetate inhibits increases in membrane fluidity induced by anti-IgM in B cells, *J. Immunol.*, 140, 2495, 1988.
341. **Utsunomiya, N., Tsuboi, M., and Nakanishi, M.,** Early transmembrane events in tumour cell responses observed by stopped-flow fluorometry, *FEBS Lett.*, 187, 245, 1985.
342. **Tonivetsky, A. G., Manevich, E. M., Zhukova, O. S., Arsen'eva, E. L., Bogacheva, G. T., and Bergelson, L. D.,** A new approach to studies of cell-antibody interactions using fluorescent lipid probes, *FEBS Lett.*, 236, 315, 1988.
343. **Tajima, M., Araiso, T., Koyama, T., Fujinaga, T., Otomo, K., and Koike, T.,** Membrane viscosity of lymphocytes and influence of phytohemagglutinin, *Biorheology*, 26, 45, 1989.
344. **Curtain, C. C., Looney, F. D., and Gordon, L. M.,** Electron spin resonance spectroscopy in the study of lymphoid cell receptors, *Methods Enzymol.*, 150, 418, 1987.
345. **Pajor, L., Kálmán, E., and Farkas, T.,** Microsomal desaturation of stearic acid in relation to lymphocyte activation, *Arch. Biochem. Biophys.*, 245, 348, 1986.
346. **Goppelt-Strübe, M. and Resch, K.,** Pitfalls in measuring fluorescence polarization in mitogen-stimulated T-lymphocytes, *Biol. Chem. Hoppe-Seyler*, 368, 409, 1987.
347. **Utsunomiya, N., Tsuboi, M., and Nakanishi, M.,** Interleukin 2 increases T lymphocyte membrane mobility before the rise in cytosolic calcium concentration, *Biochemistry*, 25, 2582, 1986.
348. **Laskay, G. and Dale, R. E.,** Interleukin-3-specific modification of cell membrane fluidity of haemopoietic cells, *Growth Factors*, 1, 67, 1988.
349. **Allegrezza-Giulietti, A., Serretti, R., Beccerica, E., Ferretti, G., and Cervini, C.,** β-Thromboglobulin and polymorphonuclear leukocytes activation. (Effects on chemiluminescence, release of membrane bound calcium, NADPH-oxidase activity and membrane fluidity), *Biochem. Int.*, 24, 273, 1991.
350. **Petty, H. R., Niebylski, C. D., and Francis, J. W.,** Influence of immune complexes on macrophage membrane fluidity: a nanosecond fluorescence anisotropy study, *Biochemistry*, 26, 6340, 1987.
351. **Berlin, R. D. and Fera, J. P.,** Changes in membrane microviscosity associated with phagocytosis: effects of colchicine, *Proc. Natl. Acad. Sci. U.S.A.*, 74, 1072, 1977.
352. **Elguindi, I., Redziniak, G., Metezeau, P., Meybeck, A., and Goldberg, M. E.,** Modulation of membrane receptor endocytosis by chemical effectors of membrane fluidity, *Biol. Cell.*, 54, 199, 1985.
353. **Roozemond, R. C. and Bonavida, B.,** Effect of altered membrane fluidity on NK cell-mediated cytotoxicity. I. Selective inhibition of the recognition or post recognition events in the cytolytic pathway of NK cells, *J. Immunol.*, 134, 2209, 1985.
354. **Roozemond, R. C. and Bonavida, B.,** Effect of altered membrane structure on NK cell-mediated cytotoxicity. II. Conversion of NK-resistant tumor cells into NK-sensitive targets upon fusion with liposomes containing NK-sensitive membranes, *J. Immunol.*, 136, 3921, 1986.
355. **Roozemond, R. C., Mevissen, M., Urli, D. C., and Bonavida, B.,** Effect of altered membrane structure on NK cell-mediated cytotoxicity. III. Decreased susceptibility to natural killer cytotoxic factor (NKCF) and suppression of NKCF release by membrane rigidification, *J. Immunol.*, 139, 1739, 1987.
356. **Kucuk, O., Stoner-Picking, J., Yachnin, S., Gordon, L. I., Williams, R. M., Lis, L. J., and Westerman, M. P.,** Inhibition of NK cell-mediated cytotoxicity by oxysterols, *Cell. Immunol.*, 139, 541, 1992.
357. **Utsunomiya, N. and Nakanishi, M.,** A serine protease triggers the initial step of transmembrane signalling in cytotoxic T cells, *J. Biol. Chem.*, 261, 16514, 1986.

358. **Ojcius, D. M., Jiang, S., Persechini, P. M., Storch, J., and Young, J. D.-E.,** Resistance to the pore-forming protein of cytotoxic T cells: comparison of target cell membrane rigidity, *Mol. Immunol.*, 27, 839, 1990.
359. **Szekeres-Bartha, J., Németh, A., Varga, P., Csernus, V., Köszegi, T., and Paál, M.,** Membrane fluidity of trophoblast cells and susceptibility to natural cytotoxicity, *Am. J. Reprod. Immunol.*, 19, 92, 1989.
360. **Nocal, R., Ondrias, K., Pecivova, J., and Drabikova, K.,** Histamine liberation and membrane fluidisation of mast cells exposed to the β-adrenoreceptor blocking drug propanolol, *Agents Actions*, 23, 143, 1988.
361. **Swieter, M., Ghali, W. A., Rimmer, C., and Befus, D.,** Interferon-α/β inhibits IgE-dependent histamine release from rat mast cells, *Immunology*, 66, 606, 1989.
362. **Afonina, G. B.,** Changes in the membrane lipids and lymphocyte function in staphylococcal infection and the development of delayed hypersensitivity, *Zh. Mikrobiol. Epidemiol. Immunobiol.*, 6, 76, 1990.
363. **Beccerica, E., Piergiacomi, G., Curatola, G., and Ferretti, G.,** Changes of lymphocyte membrane fluidity in rheumatoid arthritis: a fluorescence polarisation study, *Ann. Rheum. Dis.*, 47, 472, 1988.
364. **Tolentino, M. V., Sarasua, M. M., Hill, O. A., Wentworth, D. B., Franceschi, D., and Fratianne, R. B.,** Peripheral lymphocyte membrane fluidity after thermal injury, *J. Burn. Care Rehabil.*, 12, 498, 1991.
365. **Tomita-Yamaguchi, M., Rubio, C., and Santoro, T. J.,** Regional influences on the physical properties of T cell membranes, *Life Sci.*, 48, 433, 1991.
366. **Kubina, M., Lanza, F., Cazenave, J.-P., Laustriat, G., and Kuhry, J.-G.,** Parallel investigation of exocytosis kinetics and membrane fluidity changes in human platelets with fluorescent probe, *Biochim. Biophys. Acta*, 901, 138, 1987.
367. **de Laat, S. W., van der Saag, P. T., and Shinitzky, M.,** Microviscosity modulation during the cell cycle of neuroblastoma cells, *Proc. Natl. Acad. Sci. U.S.A.*, 74, 4458, 1977.
368. **Muller, C. P.,** Cyclic changes in fluorescence polarization of a membrane probe during the cell cycle of an erythroid precursor cell line, *Exp. Hematol.*, 17, 1, 1989.
369. **Toplak, H., Batchiulis, V., Hermetter, A., Hunziker, T., Honegger, U. E., and Wiesmann, U. N.,** Effects of culture and incubation conditions on membrane fluidity in monolayers of cultured cells measured as fluorescence anisotropy using trimethylammoniumdiphenylhexatriene (TMA-DPH), *Biochim. Biophys. Acta*, 1028, 67, 1990.
370. **Collard, J. G., de Wildt, A., Oomen-Meulemans, E. P. M., Smeekens, J., and Emmelot, P.,** Increase in fluidity of membrane lipids in lymphocytes, fibroblasts, and liver cells stimulated for growth, *FEBS Lett.*, 77, 173, 1977.
371. **Beiderman, B., Whitney, J. O., and Thaler, M. M.,** Regulation of cell surface microviscosity during the cell cycle in hepatoma, *Gastroenterology*, 76, 1275, 1979.
372. **Nicolau, C., Hildenbrand, K., Reimann, A., Johnson, S. M., Vaheri, A., and Friis, R. R.,** Membrane lipid dynamics and density dependent growth control in normal and transformed avian cells, *Exp. Cell Res.*, 113, 63, 1978.
373. **Lai, C. S., Hopwood, L. E., and Swartz, H. M.,** Electron spin resonance studies of changes in membrane fluidity of Chinese hamster ovary cells during the cell cycle, *Biochim. Biophys. Acta*, 602, 117, 1980.
374. **Pajor, L., Kálmán, E., and Kúszegi, T.,** The relationship of membrane fluidity to growth and nuclear cycle in mitogen stimulated lymphocytes, *Acta Biol. Hung.*, 40, 355, 1989.
375. **Obrénovitch, A., Sené, C., Négre, M.-T., and Monsigny, M.,** Fluorescence polarization of 1,6-diphenyl-1,3,5,-hexatriene embedded in mebranes of mouse leukemic L 1210 cells during the cell cycle, *FEBS Lett.*, 88, 187, 1978.
376. **Tanaka, T., Hidaka, T., Ogura, R., and Sugiyama, M.,** Changes of membrane fluidity and Na^+,K^+-ATPase activity during cellular differentiation in the guinea pig epidermis, *Arch. Dermatol. Res.*, 280, 29, 1988.
377. **Hachisuka, H., Nomura, H., Sasai, Y., Shiotsuki, K., and Yokoyama, M. M.,** Change of intracellular fluidity during keratinocyte differentiation measured by fluorescence polarization, *Cell Tissue Res.*, 246, 557, 1986.
378. **Tanaka, T., Ogura, R., Hidaka, T., and Sugiyama, M.,** Changes of electron spin resonance membrane fluidity in hexadecane-induced hyperproliferative epidermis, *J. Invest. Dermatol.*, 93, 682, 1989.

379. **Hachisuka, H., Nomura, H., Mori, O., Nakano, S., Okubo, K., Kusuhara, M., Karashima, M., Tanikawa, E., Higuchi, M., and Sasai, Y.,** Alterations in membrane fluidity during keratinocyte differentiation measured by fluorescence polarization, *Cell Tissue Res.*, 260, 207, 1990.
380. **Brasitus, T. A. and Dudeja, P. K.,** Alterations in the physical state and composition of brush border membrane lipids of rat enterocytes during differentiation, *Arch. Biochem. Biophys.*, 240, 483, 8 1985.
381. **Meddings, J. B., DeSouza, D., Goel, M., and Thiesen, S.,** Glucose transport and microvillus membrane physical properties along the crypt-villus axis of the rabbit, *J. Clin. Invest.*, 85, 1099, 1990.
382. **Hubner, C., Lindner, S. G., Stern, M., Claussen, M., and Kohlschutter, A.,** Membrane fluidity and lipid composition of rat small intestinal brush-border membranes during postnatal maturation, *Biochim. Biophys. Acta*, 939, 145, 1988.
383. **Meddings, J. B. and Theisen, S.,** Development of rat jejunum: lipid permeability, physical properties, and chemical composition, *Am. J. Physiol.*, 256, G931, 1989.
384. **Israel, E. J., Pang, K. Y., Harmatz, P. R., and Walker, W. A.,** Structural and functional maturation of rat gastro-intestinal barrier with thyroxine, *Am. J. Physiol.*, 252, G762, 1987.
385. **Neu, J., Ozaki, C. K., and Angelides, K. J.,** Glucocorticoid-mediated alteration of fluidity of brush border membrane in rat small intestine, *Pediatr. Res.*, 20, 79, 1986.
386. **Pang, K. Y., Newman, A. P., Udall, J. N., and Walker, W. A.,** Development of gastrointestinal mucosal barrier. VII. In utero maturation of microvillus surface by cortisone, *Am. J. Physiol.*, 249, G85, 1985.
387. **Schwarz, S. M., Watkins, J. B., and Ling, S. C.,** Taurocholate transport by brush-border membrane vesicles from the developing rabbit ileum: structure/function relationships, *J. Pediatr. Gastroenterol. Nutr.*, 10, 482, 1990.
388. **Schwarz, S. M., Bostwick, H. E., Danzinger, M. D., Newman, L. J., and Medow, M. S.,** Ontogeny of basolateral membrane lipid composition and fluidity in small intestine, *Am. J. Physiol.*, 257, G138, 1989.
389. **Ip, S. H. C. and Cooper, R. A.,** Decreased fluidity during differentiation of human promyelocytic leukemia cells in culture, *Blood*, 56, 227, 1980.
390. **Tsui, Z. C., Hou, W. H., Yang, L., and Zhu, Z. M.,** Effect of a cell differentiation inducer, ganglioside GM3, on the neutral glycosphingolipid composition and cell membrane fluidity of a human promyelocytic leukemia cell line HL-60, *In Vivo*, 4, 205, 1990.
391. **Simon, I., Brown, T. J., and Ginsberg, B. H.,** Modification of membrane physical properties, biological response and insulin binding in Friend cells by low serum concentration, *Biochim. Biophys. Acta*, 896, 165, 1987.
392. **Ota, T., Fujikawa-Yamamoto, K., Zong, Z. P., Yamazaki, M., Odashima, S., Kitagawa, I., Abe, H., and Arichi, S.,** Plant-glycoside modulation of cell surface related to control of differentiation in cultured B16 melanoma cells, *Cancer Res.*, 47, 3863, 1987.
393. **Ricardo, M. J., Jr., Small, G. W., Myrvik, Q. N., and Kucera, L. S.,** Lipid composition of alveolar macrophage plasma membrane during postnatal development, *J. Immunol.*, 136, 1054, 1986.
394. **de Laat, S. W., van der Saag, P. T., Lelemans, S. A., and Shinitzky, M.,** Microviscosity changes during differentiation of neuroblastoma cells, *Biochim. Biophys. Acta*, 509, 188, 1978.
395. **Pollesello, P., D'Andrea, P., Martina, M., de Bernard, B., and Vittur, F.,** Modification of plasma membrane of differentiating preosseous chondrocytes: evidence for a degradative process in the mechanism of matrix vesicle formation, *Exp. Cell Res.*, 188, 214, 1990.
396. **Kapitulnik, J., Weil, E., Rabinowitz, R., and Krausz, M. M.,** Fetal and adult human liver differ markedly in the fluidity and lipid composition of their microsomal membranes, *Hepatology*, 7, 55, 1987.
397. **Lelong, I., Frechard, E., Cremel, G., Langley, K., Rebel, G., Vincendon, G., and Mersel, M.,** Expression of plasma membrane and cell surface phospholipids and gangliosides of chick embryo neurons grown in primary cultures: developmental studies, *Dev. Neurosci.*, 13, 54, 1991.
398. **Waite, M., Parce, B., Morton, R., Cunningham, C., and Morris, H. P.,** The deacylation and reacylation of phosphoglyceride in microsomes of Morris hepatoma 7777 and host rat liver, *Cancer Res.*, 37, 2092, 1977.
399. **Hattori, T., Andoh, T., Sakai, N., Yamada, H., Kameyama, Y., Ohki, K., and Nozawa, Y.,** Membrane phospholipid composition and membrane fluidity of human brain tumour: a spin label study, *Neurol. Res.*, 9, 38, 1987

400. **Pessin, J. E., Salter, D. W., and Glaser, M.**, Use of a fluorescent probe to compare the plasma membrane properties in normal and transformed cells. Evaluation of the interference by triacylglycerols and alkyldiacylglycerols, *Biochemistry*, 17, 1997, 1978.
401. **Porat, N., Gill, D., and Parola, A. H.**, Adenosine deaminase in cell transformation. Biophysical manifestation of membrane dynamics, *J. Biol. Chem.*, 263, 14608, 1988.
402. **Fuchs, P., Parola, A., Robbins, P. W., and Blout, E. R.**, Fluorescence polarization and viscosities of membrane lipids of 3T3 cells, *Proc. Natl. Acad. Sci. U.S.A.*, 72, 3351, 1975.
403. **Scandella, C. J., Hayward, J. A., and Lee, N.**, Cholesterol levels and plasma membrane fluidity in 3T3 and SV101-3T3 cells, *J. Supramol. Struc.*, 11, 477, 1979.
404. **Kawasaki, Y., Zundel, J. L., Miyakawa, T., and Sakaguchi, K.**, Reduced lateral mobility of a fluorescent lipid analog in cell membranes of rat fibroblasts transformed by simian virus 40, *FEBS Lett.*, 246, 65, 1989.
405. **Yang, Y. Z., Hao, T. L., Qian, M., Dai, W. X., Huangfu, Y. M., and Zhang, Z. H.**, Normal and leukemic lymphocyte membrane fluidity and response to stimulation with Con A and PHA, *J. Tongji. Med. Univ.*, 9, 143, 1989.
406. **Johnson, S. M. and Robinson, R.**, The composition and fluidity of normal and leukaemic or lymphomatous lymphocyte plasma membranes in mouse and man, *Biochim. Biophys. Acta*, 558, 282, 1979.
407. **Inbar, M., Goldman, R., Inbar, L., Bursuker, I., Goldman, B., Akstein, E., Segal, P., Ipp, W., and Ben-Bassat, I.**, Fluidity difference of membrane lipids in human normal and leukemic lymphocytes as controlled by serum components, *Cancer Res.*, 37, 3037, 1977.
408. **Apostolov, K., Barker, W., Catovsky, D., Goldman, J., and Matutes, E.**, Reduction in the stearic to oleic acid ratio in leukaemic cells — a possible chemical marker of malignancy, *Blut*, 50, 349, 1985.
409. **Worman, C. P., Barker, W. R., and Apostolov, K.**, Saturation index of blood cell membrane fatty acids before and after IFN treatment in hairy cell leukemia, *Leukemia*, 1, 379, 1987.
410. **Krueger, G. R., Stolzenburg, T., and Muller, C.**, Cell membrane lipid fluidity and receptor expression in Moloney- and Friend virus-transformed cells, *In Vivo*, 1, 343, 1987.
411. **Armstrong, M. J., Storch, J., and Dainiak, N.**, Structurally distinct plasma membrane regions give rise to extracellular membrane vesicles in normal and transformed lymphocytes, *Biochim. Biophys. Acta*, 946, 106, 1988.
412. **Anel, A., Naval, J., Gonzalez, B., Torres, J. M., Mishal, Z., Uriel, J., and Pineiro, A.**, Fatty acid metabolism in human lymphocytes. I. Time-course changes in fatty acid composition and membrane fluidity during blastic transformation of peripheral blood lymphocytes, *Biochim. Biophys. Acta*, 1044, 323, 1990.
413. **Pajor, L., Kálmán, E., and Kószegi, T.**, Comparison of membrane fluidity and transferrin receptor expression as proliferation markers in lymphoproliferative disorders and in mitogen induced lymphoblasts, *Leuk. Res.*, 12, 517, 1988.
414. **Nakazawa, I. and Iwaizumi, M.**, A role of the cancer cell membrane fluidity in the cancer metastases: an ESR study, *Tohoku J. Exp. Med.*, 157, 193, 1989.
415. **Ward, K. A., Li, W. I., Zimmer, S., and Davis, T.**, Viscoelastic properties of transformed cells: role in tumor cell progression and metastasis formation, *Biorheology*, 28, 301, 1991.
416. **Barz, D., Goppelt, M., Szamel, M., Schirrmacher, V., and Resch, K.**, Characterization of cellular and extracellular plasma membrane vesicles from a non-metastasizing lymphoma (Eb) and its metastasizing variant (ESb), *Biochim. Biophys. Acta*, 814, 77, 1985.
417. **Schroeder, F., Kier, A. B., Olson, C. D., and Dempsey, M. E.**, Correlation of tumor metastasis with sterol carrier protein and plasma membrane sterol levels, *Biochim. Biophys. Res. Commun.*, 124, 283, 1984.
418. **Kier, A. B.**, Plasma membrane properties of cultured local LM cell tumors and metastases from athymic (nude) mice, *Cancer Lett.*, 50, 19, 1990.
419. **Kier, A. B., Parker, M. T., and Schroeder, F.**, Local and metastatic tumor growth and membrane properties of LM fibroblasts in athymic (nude) mice, *Biochim. Biophys. Acta*, 938, 434, 1988.
420. **Taraboletti, G., Perin, L., Bottazzi, B., Mantovani, A., Giavazzi, R., and Salmona, M.**, Membrane fluidity affects tumor-cell motility, invasion and lung-colonizing potential, *Int. J. Cancer*, 44, 707, 1989.
421. **Aszalos, A., Grof, P., Fidy, J., Hajdu, O., Pal, K., Jeney, A., and Lapis, K.**, Comparison of biophysical parameters of primary low and high metastatic Lewis lung tumor cells, *J. Med.*, 17, 179, 1986.

422. **Tímár, J., Pogány, G., Balázs, M., Szöllősi, J., Ladányi, A., Oláh, J., Tímár, F., Lapis, K., and Jeney, A.**, Modulation of membrane phenotype, matrix adhesion and microinvasiveness of metastatic tumour cells by HUdR, *Cell. Biochem. Funct.*, 8, 211, 1990.
423. **Kido, J., Ashida, Y., Shinkai, K., Akedo, H., Isoai, A., Kumagai, H., and Inoue, H.**, Effects of methylthiodeoxyadenosine and its analogs on in vitro invasion of rat ascites hepatoma cells and methylation of their phospholipids, *Jpn. J. Cancer Res.*, 82, 1104, 1991.
424. **Storme, G. A., Berdel, W. E., van Blitterswijk, W. J., Bruyneel, E. A., De Bruyne, G. K., and Mareel, M. M.**, Antiinvasive effect of racemic 1-O-octadecyl-2-O-methylglycero-3-phosphocholine on MO4 mouse fibrosarcoma cells in vitro, *Cancer Res.*, 45, 351, 1985.
425. **Haeffner, E. W., Doenges, K. H., and Buchholz, J.**, Difference in cell surface 5′-nucleotidase activity, protein iodination and membrane properties between mouse lymphoma and rat adenocarcinoma cell variants with low and high-metastatic potential, *Int. J. Biochem.*, 20, 929, 1988.
426. **Sherbet, G. V.**, Membrane fluidity and cancer metastasis, *Exp. Cell Biol.*, 57, 198, 1989.
427. **Sobajima, T., Tamiya-Koizumi, K., Ishihara, H., and Kojima, K.**, Effects of fatty acid modification of ascites tumor cells on pulmonary metastasis in rat, *Jpn. J. Cancer Res.*, 77, 657, 1986.

Chapter 5

Rotational Mobility of Cell Membrane Elements

Sándor Damjanovich

CONTENTS

I. ENERGETIC BACKGROUND AND SIGNIFICANCE OF ROTATIONAL MOBILITIES AT MOLECULAR LEVELS

Rotational and lateral mobility of molecules has its energetic background in thermal energy. The discovery of the Brownian motion (1828) can be considered as the first example, where the limitations of the resolution power of light microscopes was first time overcome. Robert Brown, the ingenious Scottish botanist, by observing molecular motions in liquids bypassed the straightforward resolution power of light microscopes and recognized that the random disordered motion of the pollen powder at the surface of the liquids had its origin in the molecular motions and collisions of individual molecular entities with the pollen particles. This early discovery was cast into a scientific theory by Albert Einstein more than a half century later in his famous paper: "On the movement of small particles suspended in a stationary liquid demanded by the molecular kinetic theory of heat."[1,2]

The contemporary scientific methodology made use (and sometimes abuse) of this property of molecules to obtain information on these complicated phenomena for the interest of physics, chemistry, and in the last approximately 50 years that of biology. The complexity of the random motions in gas, liquid, or even solid environment explains the above-bracketed remark. Careless application of the method or overlooking the limitations and restrictions when one concludes from measured data may produce a number of problems in strict scientific sense. Thus, observing the rules of science are sometimes not trivial and the easy art of drawing simplified conclusions from half-understood mechanisms of molecular mobilities in complex systems demands very thorough knowledge of the available theories. Even accepting the above requisites we have to be humble enough to admit that in many cases the conclusion that can be drawn from mobility measurement under different conditions in a complex system is very limited. Sometimes the only valid statement is that under different conditions a particular element of the system changes its rotational or translational freedom. However, despite these pessimistic sentences, complex biological systems unveil quite a few of their features even through such limited possibilities.

Here, our principal aim is to call the attention of the reader to possibilities offered by this molecular mobility to gain information about intricate biological processes at the molecular level. The easy looking tool to fulfil this desire is to give an idea on the series of information attainable through such investigation. This will be attempted by analyzing briefly a number of outstanding papers of eminent scientists, who contributed much to our understanding in this attractive and very important field of biological research.

0-8493-4931-1/94/$0.00+$.50

Beyond the pure biological and scientific interest of the forthcoming analysis, we are quick to remind ourselves that without lateral and rotational mobility of molecules any type of life is impossible. The elementary steps of metabolism in gas, fluid, and even solid phase is impossible without diffusion, and diffusion definitely means rotational and translational diffusion as well.

The ever-developing scientific methodology provided us with quite a few high-powered methodologies to "visualize" the above-outlined dynamic molecular properties. The absorption techniques in the energy interval of microwaves and also radio frequency, such as electron spin resonance and nuclear magnetic resonance spectroscopies and furthermore the absorption and emission spectroscopies of the ultraviolet, visible, and infrared light, offer a broad range of possibilities to study such parameters. Living cells are very sensitive and apt to react or yield artifacts, if clumsy interferences from the methodical side disturb their "quiet" living conditions. Fortunately, the visible light is the least perturbing agent in the majority of cases, if we intend to obtain delicate information on cellular mechanisms at the molecular level. Even in this seemingly limited area we have to focus our interest basically on those works which aimed at the understanding of molecular mobilities in biological membranes, if we do not want to lose readability of the text (not mentioning the limitations of the competence of the writer).

Emission and absorption techniques will be the major elements of our discussion, although we cannot always avoid at least the reference to some other less-easily applicable techniques and data obtained with them.

II. MEASURING TECHNIQUES AND DATA EVALUATION

Focusing our interest on molecular mobility in living systems and within this area those of the molecules in cellular membranes as revealed by mostly visible light-based methods, one problem immediately arises. Rotational mobility of molecules, having the size of an average protein in, for example, the plasma membrane of different cells is restricted by the relatively high viscosity of the medium. Nevertheless, little doubt is raised concerning the biological significance of rotational freedom of antigen and receptor molecules embedded into the membrane. Receptor aggregations, assemblies, and disassemblies belong to elementary cellular reactions to environmental interactions. The environmental "viscosity" (rather restricted mobility) in the membrane has its origin on one hand in the viscosity of lipid domains and on the other hand in the cytoskeletal connections of the integral membrane proteins. Even if they have none of the latter, complex receptor structures have such multisubunit composition that they can hinder free rotational motions. Rotational mobilities of proteins in the lipid regions, having two to three orders of magnitude higher viscosity than that of pure water or ordinary physiological electrolyte solutions, will take place in the time range of microseconds. In comparison, molecular mobilities have a correlation time in aqueous medium in the nanosecond time range. These time intervals immediately tell us that molecular rotations in solutions can be followed by fluorescence, while in membranes they can be followed by phosphorescence spectroscopy.

The changes in the intermolecular connections will immediately be reported by the altered, increased, or decreased molecular mobilities. An interesting contemporary problem is the intercellular communication and signal transduction from the external world to the cell interior, i.e., the cellular and intercellular information exchange. The former one is frequently termed transmembrane signaling or the transmembrane signal-transducing process. The above brief discussion is enough to recognize the powerful capacity of rotational mobility parameter measurements regarding the mentioned cellular functions.

A. MEASURING MOLECULAR ROTATIONS IN BIOLOGICAL MEMBRANES WITH TRANSIENT ABSORPTION TECHNIQUES

Absorption, fluorescence, and phosphorescence spectroscopic methods have great biological value in order to accumulate molecular information on changes in mobilities inside or at the surface of cells among very different conditions.

The selection of the appropriate spectroscopic tool to assess mobilities depends on their rotational correlation time that is inversely proportional to their rotational diffusion constant:

$$t = \frac{1}{6D_{rot}} \qquad \text{where} \qquad D_{rot} = \frac{kT}{4\pi\eta\rho^3} \tag{1}$$

Both resonance spectroscopic methods (electron spin or nuclear magnetic resonance) as well as optical methods are appropriate to study rotational relaxation processes in biological systems. However, if one works with live cells, optical methods can be less harmful and measurements can be carried out among different yet basically physiological conditions. The measurement of molecular mobility is based upon the principle of photoselection. If we start measuring the time delay between the absorption and emission, assuming a suitable correlation time, the molecular population that was selected by the polarized absorption will decay upon the randomizing effect of the rotational Brownian motion and also by some intrinsic motional modes. These processes result in a multiexponential decay of anisotropy, in time, as follows:

$$r(t) = r_\infty + (r_0 - r_\infty)\sum_{i=1}^{n} a_i e^{-\frac{t}{\varphi_i}} \quad (2)$$

where r_0 is the initial anisotropy at time $t = 0$, r_∞ is the so-called limiting anisotropy, which is the remaining anisotropy of the system even after a relatively longer time period, a_i is the amplitude, and φ_i the rotational correlation time of the i-th molecular entity. Of course, the shape and rotational freedom of the rotating reporter molecule and specific environmental contributions from the membrane (i.e., the embedding medium) perturb the system, hence generally more than one model can describe even the same set of data. (Figure 1)[3-9] The globular sphere model is the simplest approach, but assuming rotational ellipsoids results in better fit for the data in many cases. The several possible (sometimes one to five or more) exponential terms may or may not have biological meaning, but in practice we can rarely handle more than three. A combined mathematical, physical, and biological reasoning will decide which is the most likely, yet acceptable, model. One can find such examples in the review paper of Matko et al.[10]

The not too numerous schools, contributing to this budding branch of biophysics published several basic papers utilizing mostly rotational relaxation measurements through following time-resolved phosphorescence anisotropy decay.[3-13] Techniques based on resonance spectroscopy are summarized by Cantor and Schimmel in their excellently written book.[14]

As it can be seen from the above equations, the rotating molecular bodies are highly dependent on their hydrodynamic volume (size and shape), the viscosity of the surrounding medium, and the environmental heat. Among physiological conditions, or near to that, the membrane viscosities and the average size of membrane-bound proteins will shift the rotational relaxation times into the microsecond time scale as we stated above. Room temperature phosphorescence anisotropy measurements became possible only in the middle 1970s and later by Strambini and Galley and their groups.[15-17] Their technique made possible the measurement of rotating molecules in solutions and also membranes by suitable phosphorescence labels. The phosphorescence labels shifted the monitoring of the rotational relaxations to the right correlation time interval. On the other hand, their invention, the saturation of the fluid measuring medium with argon gas, prevented the oxygen quenching of the phosphorescence molecules (usually eosin or erythrosin, i.e., the brominated and iodinated forms of the fluorescein, respectively). Their invention paid a very high dividend for a better understanding of molecular interactions in solutions and also embedded into membranes. The relatively short time period that was necessary for the measurements did not significantly impair the viability of the system, even that of the living cells.

An early candidate to measure the relatively slow rotational mobility of proteins in biological membranes was the band 3 protein. Cherry and co-workers,[3] overcame the difficulties which limited the measurability of the rotational modes of membrane-bound proteins to a few exceptions having special spectroscopic properties, applying long lifetimes of triplet states of organic molecules. Using eosin, which transfers about 80% of the excitation energy to the triplet states by intersystem crossing and linking it covalently to the band 3 protein which is abundant in erythrocyte membranes (nearly 25% of the total membrane protein), the measurement became feasible.

The method to follow the slow, microsecond rotations of the band 3 was based on the absorption of eosin. The isothyocyanate derivatives of the eosin bound almost exclusively to the band 3. Transient absorption values were calculated from the following equation:

$$r(t) = \frac{A_{pa}(t) - A_{pc}(t)}{A_{pa}(t) + 2A_{pc}(t)} \quad (3)$$

where r(t) is the transient absorption and the $A_{pa}(t)$ and $A_{pc}(t)$ are the absorbance changes at time t for light polarized parallel and perpendicular to the exciting flash, respectively. In order to exclude the unwanted

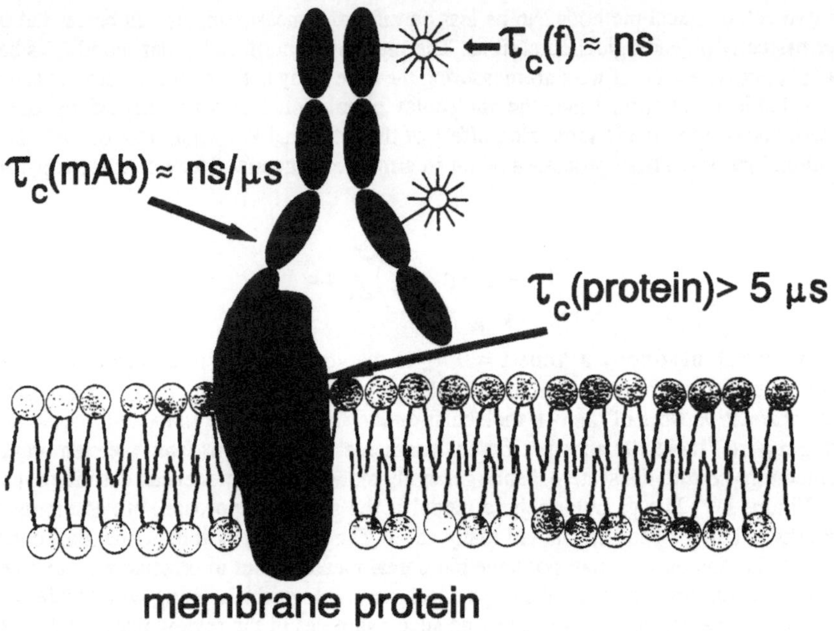

Figure 1. The figure shows a model of an antibody bound to an integral (membrane-spanning) protein. The rotational correlations times (τ_c) are several orders of magnitude different for the dye, the antibody, and the membrane-bound protein. That situation enables us to measure rotational relaxation parameters of membrane-bound antigens by antibodies, labeled with phosphorescent dyes, like eosin or erythrosine, which are halogenated forms of the fluorescein.

interference of the hemoglobin absorbance, ghosts were used instead of erythrocytes. The measurement, the triplet-triplet absorption, was carried out with 650 nm and the ground state depletion was measured with 525-nm wavelengths of the flash lamp.

The analysis of the data showed more than one rotational relaxation component. Besides a time-independent factor, they found a 0.5-ms component. The time-"independent" second timescale mobility could arise from aggregates or populations of band 3, immobilized by cytoskeletal elements. Auxiliary investigations showed that the labeling itself did not cause aggregation and the spectrin did not restrict the rotational mobility of the band 3. Since band 3 was known to be an anion transport protein, the relatively slow motion excluded a direct participation of the rotational mobility in the anion transport that would have demanded one or two orders of magnitude faster rotation.[3] These experiments pioneered the application of transient absorption measurements for determining rotational mobility of an important membrane protein, in particular, but on the other hand clearly demonstrated the resolution power of the rotational measurements even in clarifying molecular transport mechanisms, in general.

B. LUMINESCENCE DEPOLARIZATION METHODS

The basic principles of the phosphorescence anisotropy decay methods are very similar to those of the transient absorption but in this case the emission anisotropy is the very parameter that reports on the rotational mobility of molecules or that of their environment. Phosphorescence anisotropy decay measurement is preferred due to its very favorable life time parameters, having the same correlation time intervals as the molecular motions in most of the biological membranes. Fluorescence anisotropy decay measurements have too fast rotational characteristics for monitoring relaxation of molecules imbedded in such a viscous environment such as cell membranes. Despite this handicap, steady-state fluorescence anisotropy measurements have the capacity to provide data on rotational mobility of specific molecules which merge into the lipid layers of the plasma membrane.[10] Apolar probes such as 1,6-diphenyl-1,3,5-hexatriene (DPH), 1[4-(trimethylamino)phenyl]-6-phenyl-hexa-1,3,5,-triene (TMA-DPH), and perylene are frequently applied to elucidate fluidity changes of plasma membranes as early responses to external stimuli, i.e., cell activation. These probes, and a number of others, can report on the "fluidity" of membranes in different depth along the axis perpendicular to the plane of the membrane. Although, such measurements are easily applicable, they report on bulk, very loosely defined parameters of the cell membranes and are less favored in contemporary research trends. Nevertheless, data obtained so far with such simple methods have been valuable for promoting our current knowledge and understanding of rotational relaxations, viscosity parameters, and intermolecular interactions in cell membranes.[10]

A good example of how to study time-dependent rotational mobility of a particular and functionally very important protein was given by Lo and co-workers.[7] The rotational mobility of the membrane bound-acetylcholine receptor of the *Torpedo electricus* electric organ was investigated by phosphorescence depolarization of the erythrosin, the tetraiodo-fluorescein. This phosphorescence dye was selected, because it has a fivefold higher emission intensity than eosin. In comparison with a resonance spectroscopic method that also has the capacity to monitor time-dependent rotational mobilities, the saturation transfer electron paramagnetic spectroscopy, the sensitivity of the erythrosin-based phosphorescence depolarization is more than 10^4-fold higher.

The α-bungarotoxin (BuTX), the snake venom of *Bungarus multicinctus*, is known to have a specific and irreversible binding to the acetylcholine receptor (AChR). Thus, BuTX was labeled with erythrosin, without impairing its AChR binding capacity. The time-dependent anisotropy of the dye-labeled toxin was measured using receptor-rich membrane preparations at 2°C. The emission anisotropy was calculated by them as

$$r(t) = A_1 e^{-Dt} + A_2 e^{-4Dt} + A_3$$

where

$$A_1 = \frac{6}{5}\sin^2\Theta\cos^2\Theta \qquad (4)$$

$$A_2 = \frac{3}{10}\sin^4\Theta$$

$$A_3 = \frac{1}{10}\left(3\cos^2\Theta - 1\right)^2$$

where D is the rotational diffusion coefficient of the probe carrying protein about an axis perpendicular to the plane of the membrane and Θ is the angle between the axis and the transition moment for absorption. The anisotropy parameter r becomes independent of time either when D^{-1} is large relative to the duration of observation or when $\Theta = 90°$ (whereupon $r = A_3$). If there are two triplet states contributing to the phosphorescence with lifetimes τ_1 and τ_2, then the time-dependent anisotropy is given by[7]

$$r(t) = \frac{a_1 r_1(t) e^{-\frac{t}{\tau_1}} + a_2 r_2(t) e^{-\frac{t}{\tau_2}}}{a_1 e^{-\frac{t}{\tau_1}} + a_2 e^{\frac{t}{\tau_2}}} \qquad (5)$$

where a_1 and a_2 are the phosphorescence intensities of the two components at $t = 0$ and r_1 and r_2 their respective time-dependent anisotropies.

The measurements resulted in at least two components, one with 200- and another with 500-μs lifetimes. The anisotropy was found on the order of 0.1 for 900 μs after the laser flash, and showed further decay; however, beyond this range the signal-to-noise ratio was increased markedly. The authors conclusion was that the AChR at 2°C has a correlation time of several milliseconds or more.

An alkaline extract of the preparation, which was supposed to lack nonmembrane or peripheral proteins had the behavior of the normal membranes at 2°C; however, at 26°C the phosphorescence decay was faster, and more so in the case of the more rapidly decaying component. The removal of a 43 kDa and some other proteins did not change the composition of receptor molecules as judged by electron microscopy, but it permitted rotation of the receptor with rotational correlation times of less than 10 to 20 μs. Other data supported the view that both forms of the receptor retained their functionality; thus, the authors concluded that the synaptic AChR can function whether mobile or not. Furthermore, interactions between neighboring receptor assemblies were not prerequisites of the normal receptor localization.[7]

Similar techniques were applied to study the rotational mobility of H-2K^k antigen of the murine major histocompatibility complex in the plasma membrane of live or fixed HK-22 cells from a murine lymphoma line.[18] A suspension of HK-22 (when the first data were published termed T-41) lymphoma cells labeled with eosin or erythrosin-conjugated antibodies was used to measure the phosphorescence lifetime and anisotropy decay of the labeled cells. Anisotropy decay curves of eosin or erythrosin-labeled antigen-antibody complexes was obtained on the cell surface among antibody saturation conditions. The rotational correlation times were typically between 10 and 20 μs at 25°C with either whole antibody or F_{ab} fragment, demonstrating that this correlation time was not an artifact caused by aggregation due to crosslinking of the whole antibodies. The values of the correlation times were accompanied by an initial anisotropy value (r_0) on the order of 0.05 to 0.11, and the limiting anisotropy (r_∞) was 0.035. A slight increase in the correlation times was observed when the anisotropy values were determined at 4°C. Capping by addition of unlabeled concanavalin A prior to labeling of cells with eosin-conjugated anti-H-$2K^k$ antibody did not change the correlation times at 37 or 4°C, demonstrating that concanavalin A-caused capping did not redistribute (cluster or aggregate) the H-$2K^k$ antigen in the plasma membrane. Addition of unlabeled monoclonal antibodies directed to other epitopes of the H-$2K^k$ as the one that carried the phosphorescence label did not change the observed correlation times. However a dramatic change was observed at 37°C when the eosin-carrying monoclonal antibodies were capped by adding a second antibody, a rabbit anti-mouse IgG. The correlation times measured after this treatment were 200 times longer than those of the uncapped samples, indicating an almost complete immobilization of the complex on this time scale. The necessary control experiments excluded that the observed phosphorescence decay curves were artifacts or had their origin in the autophosphorescence of the cells.[18] These experiments were the first where relative proximity relationships, rotational mobility, and lateral diffusion of the same antigen were investigated together in the plasma membrane of lymphocytes.

Both of the above-described experiments applied suspension of cells or membrane preparations. However, it is clear that any other optical device can be used to gain similar information, let it be a microscopic or a flow system.

C. THEORETICAL ASPECTS OF ANISOTROPY DECAY IN MEMBRANES

In order to understand the real physical meaning of mobility parameters of proteins in cell membranes we have to accept the fact that optical anisotropy decay processes are extremely complicated. Although the aim of a biological analysis, and its practical usefulness, generally do not demand a full and detailed description of all kinds of mobility modes, a physical model of the motional modes has to assumed. Data of luminescence depolarization are highly complex, but the resolution power of such methods deserves a few sentences on the models and their validity.

The general basis of the models is the rotational diffusion of a spherical or ellipsoidal rigid body. However, what we actually observe is the luminescence properties of a fluorescence or phosphorescence dye, fixed more or less rigidly to the surface of such a, for example, protein body. "The rotational diffusion of a membrane anchored protein may consist of two different modes occurring simultaneously: one is restricted wobbling of the whole or part of the protein molecule, and the other is rotation of the whole protein about a fixed axis."[19] The works of Weber,[20,21] Ehrenberg and Riegler,[22] Szabo,[8] Ameloot et al.,[9] Dale,[23] and Piston and Gratton[24] provided a sound theoretical basis for understanding such mobilities in particular cases and also in general. However, one cannot deny that each model and theoretical interpretation has its limitations when they are applied in any specific real system.[8,9,20-24] The more we know about the real shape, size, and structure of a protein, the more precise the model we are able to apply. Generally, this is not the case. However, we have to emphasize again that, without overlooking the basic importance of theoretical foundations, at present even a detection of an altered mobility parameter among well-controlled conditions can have significant biological value. It is enough to refer back to those two examples wherein Cherry analyzed the rotation of the band 3 protein of erythrocyte membranes or Lo et al. that of the acetylcholine receptor of the *T. electricus* electric organ.[3,7]

The restricted mobilities of singlet or triplet excited state probes in the membranes can reflect rotational modes of the neighbor molecules in the membranes. The evaluation of such mobilities is generally based on the assumption that the reporter molecules (or rather their photoselected subpopulation which is oriented more or less parallel to the polarized light that causes excitation) represent a particular type of mobility of membrane-bound molecules. Unfortunately, this is not the case. The membrane-bound molecules are heterogeneous. Lipid molecules are sequestered into lipid domains, which float and sometimes fuse with each other and also accommodate proteins. The lipid moieties have also a flip-flop motion with different correlation times; this can have functional consequences.[25] In the majority of cases the proteins have oligosaccharide moieties which carry electric charges and contribute to the electric forces of the cell surface. The lipid-protein interactions are determined by the hydrophobicity of the polypeptide chains. The membrane-bound peptide chains will be inserted into the lipid regions immediately after or even during their synthesis at the surface of the rough endoplasmic reticulum. Since the neighborhood of the proteins in the membrane is largely determined by their primary structure, one can claim that the "two-dimensional pattern" of the membrane proteins is highly influenced by genetic factors.[26] In the light of the above analysis it is clear that experiments on artificial systems having well-controlled lipid and protein compositions can help us to understand the major mobility patterns of molecules in membrane structures. Detailed knowledge of such features may be indispensable to a proper understanding of the function of living cell membranes in various regulatory or transmembrane signaling processes and of their dysfunction in pathological states, especially those of cancer cells.[23]

Data obtained from fluorescence or phosphorescence depolarization are generally evaluated by mathematical models, based upon the rotational diffusion equations of differently shaped rigid bodies.[24] Monte Carlo simulations (i.e., numerical calculations in computers) supported this assumption fairly well. However, "nonrotational" components of small reporter molecules can also occur.

The so-called jump model of Weber supposes discrete jumps, i.e., sudden transitions between different, but predetermined orientational states.[20] Recently, Piston and Gratton developed a compartmental formalism that can be used to combine the joint effects of rotational motions with discrete jumps between a few preferred orientations.[24] A further complication arises from the mobility of small diffusing probes, like DPH, in perpendicular direction to the plane of the membrane. It is noteworthy that changes in the membrane potential can also influence the mobility of integral membrane elements and probes.[27]

Without being lost in the numerous theoretical details and mathematical formalism of diffusion or discrete jumps, we emphasize again that a major asset of rotational relaxation measurement of membrane molecules is that it enables us to find differences between different physiological and pathological states of cells. In particular cases and types of cells sometimes we have at least some reasonable models as well, which may contribute to our understanding of membrane processes at the molecular level. As a future trend, video-image microscopy is a promising technique to gain information on individual cells in greater details and with the possibility of time kinetics. On the other hand, flow cytometric anisotropy measurements have the capacity to observe the distribution of particular properties over distinct cell populations.

D. POLARIZED FLUORESCENCE PHOTOBLEACHING RECOVERY CAN MEASURE ROTATIONAL DIFFUSION IN SOLUTION AND MEMBRANES

Dale in 1987 and Velez and Axelrod in 1988 studied the possibility of combining the polarized fluorescence photobleaching recovery with rotational diffusion of the unbleached fluorophores.[28,29]

The principle has the inherent beauty of the simplicity. The brief flash of polarized (laser) light can create an anisotropic distribution of the unbleached fluorophores. These entities retaining their fluorescent properties will "relax" by rotational diffusion, in this case an entropy driven process, which finally will lead to an increase in the fluorescence after the bleaching. It is obvious that the fluorescence intensity increase here is not due to any lateral diffusion of the unbleached elements, but the rotational mobility that relaxes after the photoselection provided by the polarized bleaching. The polarized photobleaching recovery can be applied to practically any time scales from microseconds to several seconds.

The method itself resembles very much to that of the ground state depletion method, introduced by Johnson and Garland in 1981.[31,32] The latter makes possible the measuring of triplet relaxation at fluorescent time scales, because the triplet transition "depletes" the available ground state electrons available for fluorescence measurements. In both cases the excitation "pumps" the system through photoselection, and the other "end" of the measuring setup "probes" the output of the excited and relaxing molecules. The method of Johnson and Garland[31,32] does not apply a polarizing filter in the microscope through which the fluorescence is observed. Velez and Axelrod,[29] as well as other authors,[33] employ such filters at the signal receiving end of the experimental system.

An interesting work studied the rotational diffusion of AChRs on cultured rat myotubes utilizing polarized fluorescence recovery after photobleaching.[30] These AChRs are known to exist in two distinct classes, as it was revealed by labeling with rhodamine-α-BuTX, a specific toxin for the receptor. The oligomeric clustered AChR is aggregated in a pattern of highly concentrated speckles and streaks. These clusters occupy an area that exceeds 1 mm in diameter. The other class was not clustered and displayed a diffuse labeling pattern. The polarized fluorescence recovery after photobleaching that was applied according to the principles outlined above revealed a practically immobile population with a characteristic rotational correlation time of several seconds. This population was identified as the clustered pattern of the AChRs. The nonclustered population was far more mobile with a ϕ value between 50 and 100 ms. Addition of tetravalent concanavalin A, which cross-linked the cell surface α-mannopyranosid rings of the oligosaccharide moieties of membrane proteins, immobilized most of the mobile AChRs. On the other hand, treatments with carbachol, which were supposed to disrupt clustering directly, or that with cytochalasin D, which disrupts cytoplasmic actin, failed to disperse clusters already formed in the membrane.

The conclusion of the elegant work showed that polarized FRAP could be applied to study rotational mobility of membrane proteins, using fluorescence probes instead of phosphorescence ones.[30]

III. ROTATIONAL MOBILITY DATA OF DIFFERENT MEMBRANE PROTEINS

A. BAND 3 REVISITED

Since the mid 1970s, more and more data accumulated focusing on rotational relaxations of molecules in solution, but relatively soon in membrane structures as well. Measurements of band 3 protein mobility remained a preferred target, since the high density and easy way for specific labeling of this functionally important anion carrier protein of erythrocyte membranes offered, not only important biological information, but a very good model for testing theoretical concentrations and investigating intermolecular interactions in real membrane structure. Cherry and colleagues carried out a systematic series of such experiments.[34-36]

Using transient dichroism to indicate rotational mobility of eosin-labeled band 3 of erythrocyte ghosts, they selectively removed or severed the connection of membrane proteins with the cytoskeletal elements such as actin, band 4.1, ankyrin (band 2.1), or the α and β spectrin. As was expected, an enhanced decay was observed in those cases when the ghost was stripped of all these connections, and a negligible change occurred in decay curves for only ankyrin-degraded ghosts. Surprisingly, there were a number of different treatments which definitely (and selectively) impaired cytoskeletal elements, and which had no measurable effect on band 3 rotational mobility, i.e., removal of spectrin molecules and actin by low salt treatment, selective proteolytic degradation of ankyrin, or heat treatment of ghosts up to 70°C.

On the other hand, high salt removal of spectrin, actin, ankyrin, and band 4.1 increased the rotational mobility of band 3 by more than 30%. Tryptic digestion can remove the cytoplasmic domain of the band 3 from the ghost membrane and increase its mobility by 26%. Heat treatment of intact cells at 50°C alone increased the rotational mobility in the same extent; however, the combination of tryptic digestion and heat treatment did not further increase the mobility. The heat treatment of intact cells for 24 h at 37°C in combination with tryptic digestion resulted in only a 12% increase of band 3 rotational mobility. Analysis of these data revealed that the cytoskeletal elements must have a flexibility. This flexibility resides principally in spectrin residues of the cytoskeleton. The slow components of the band 3 originate mostly from a slowly rotating aggregate of this molecule. It is worthy to remind the reader that rotational mobilities are especially sensitive to increased volumes, i.e., aggregations. This property can be derived from the rotational diffusional constant of sphere-like molecules, since the rotational diffusion constant has an inverse relationship to the negative third power of the radius of the molecule. Since the removal of ankyrin and band 4.1 increases the rotational mobility of band 3, they are good candidates among the cytoskeletal elements which can aggregate band 3. An interesting relationship was suggested concerning a connection between band 3 and glycophorin, another erythrocyte membrane element. Band 4.1 has a high-affinity binding site for glycophorin and a lower-affinity site for band 3. Both glycophorin and band 3 are either dimeric or even polymeric. Thus, band 4.1-interlinking glycophorin and band 3 may be an important factor in regulating aggregation properties of these molecules in the membrane. In the light of these data the finding that attachment of band 3 to cytoskeleton via ankyrin did not restrict its rotational mobility can be attributed to rapid molecular motions within the cytoskeletal structure.[36]

As we justified above, rotational mobility is sensitive to aggregation of the rotating entities in general, and receptor assemblies and disassemblies are supposed to display this characteristics in cell membranes as well. Thus, one has a strong motivation to detect rotational mobilities of functional elements in cell membranes, because they can provide a simple and very sensitive instrument to monitor the changes in receptor structure and function among physiological conditions.

Clague and Cherry applied a number of different agents aggregating band 3 in erythrocyte membranes and studied the retardation of the rotational mobility of the molecule.[35] Polyvalent cationic species are thought to initiate such aggregation; thus, melittin, polylysine, and Zn^{2+} ions were tested. The abundance of band 3 in erythrocyte membranes increased the detectability of the structural alterations caused by the above agents. The melittin is a basic peptide that bounds to membrane-like surfaces, and possesses an

α-helical structure when engaging membranes. The amphiphilic regions of the melittin adjoin a very hydrophilic C-terminal region. After cleaving these two regions, the effects of the two distinct regions can be studied separately. Applying the transient dichroism technique, absorption anisotropy decays arising from eosin-labeled band 3 in ghost membranes were subjected to increasing concentrations of Zn^{2+}. The decay curves were supposed to reflect a rotational motion of band 3, which supposedly occurred only about an axis perpendicular to the plane of the membrane. Increasing Zn^{2+} concentrations decreased the rate of the decay, i.e., the rotational mobility in the membrane. The same trend was observed over specific concentration ranges of melittin and polylysine.[35] Similarly to their other experiments,[34] the effect of the removal of cytoskeletal elements was tested. Applying low and high salt treatment of the ghosts, and thereby removing particular components of the cytoskeleton, the retardation of the rotational mobility of band 3 in ghosts, which were depleted of spectrin and actin or spectrin, actin, ankyrin, and band 4.1, showed only minor variation as compared to the untreated controls. As we have seen, tryptic digestion significantly enhanced the rotational mobility of band 3.[34] However, when the immobilizing action of the melittin was studied on the same rotational relaxation of band 3 after trypsin treatment of the ghosts, only a mere 20% higher concentration of the melittin was required to achieve the same immobilization. The trypsin treatment was far more effective to cancel the immobilizing effects of either polylysine or Zn^{2+}.

Freeze-fracture electron microscopic studies demonstrated that the mechanism of the mode of action of melittin was an effective clustering of band 3 proteins in the membrane, although it is likely that all polycationic agents have a similar mode of action: melittin is anchored in the membrane, while polylysin and Zn^{2+} follow a less-biased distribution pattern, governed by simpler physicochemical laws. Experiments with fragments of melittin suggested that the 1-19 amino acid residues can confine the 20-26 section to the membrane surface via hydrophobic interactions.[35] The clustering of integral membrane proteins by physiological or artificial agents provides a greater potential for intercellular crosslinking. While polylysine-like agents work mostly at low ionic strength, membrane-anchored agents, such as melittin, which have a hydrophobic helical structure, can remain effective at physiological ionic strength. Thus, the fixed positive charges can neutralize negative surface charges, and the electrostatic repulsive forces between cell surfaces are less effective in preventing intercellular interactions.

Band 3, as a very convenient and easily available model, remained a target even for more recent investigations. Wyatt and Cherry revisited the problem of the influence of cytoskeletal elements in ghosts.[36] As we described above, in their earlier work they found that only particular elements of the cytoskeleton had a significant influence upon the rotational mobility of band 3. However, the experiments, which provided the basis for the conclusions, were carried out with such systems (erythrocyte ghosts), wherein the connection between the measured rotational mobility of band 3 and the cytoskeletal elements was interrupted by removing particular protein or proteins. In this recent work, the authors applied a different approach. The cytoskeletal elements had been removed first, and the rotational relaxation of band 3 was studied after gradual rebinding of them. Rebinding either ankyrin or 4.1 to ghosts had no detectable effect on band 3 mobility. However, rebinding both of these proteins together did reimpose a restriction on band 3 rotation. The effect on band 3 rotational mobility of rebinding ankyrin and 4.1 were similar irrespective of whether or not band 4.2 was removed from the membrane. Again, the explanation for the findings was that ankyrin and band 4.1 together promote the formation of slowly rotating clusters of band 3.[36] Cherry's group preferred the application of the transient dichroism-based method to look for rotational relaxation of band 3.

Matayoshi et al.[37,38] studied this complex and very important model problem of rotational mobility of membrane-bound proteins using also eosin-labeled band 3, but their studies were carried out by following polarized delayed luminescence.

The instrument had a light source, which was built from an excimer laser pumping a dye laser. The excitation pulses had an average energy between 1 and 5 mJ, an approximately 10-ns duration, and a repetition rate of 10 to 20 Hz. Excitation wavelengths of 515 and 385 nm were obtained with the laser dyes Coumarin 307 and Polyphenyl 1. Eosin phosphorescence was isolated with a combination of cutoff and broad band-pass interference filters, while eosin-delayed fluorescence was selected with a combination of broad band-pass filters. Spectral isolation of the latter from the strong immediate fluorescence is not possible; therefore, gated photomultiplier was used.

Data analysis was based on the time-dependent measuring of the two polarized emission components [$I_{vv}(t)$ and $I_{vh}(t)$], which provided the basis to calculate the time-dependent functions of s(t), d(t), and r(t), i.e., the total emission, difference function, and anisotropy, respectively. For a system with one diffusing component, and single lifetime:

$$s(t) = I_{vv}(t) + I_{vh}(t) = \sum_k S_{oj} e^{-\frac{t}{\tau_j}}$$

$$d(t) = I_{vv}(t) - I_{vh}(t) \tag{6}$$

$$r(t) = \frac{d(t)}{s(t)} = \sum_i r_{oi} e^{-\frac{t}{\varphi_i}}$$

here k is the number of chemical species with triplet lifetime τ_j and the index i can have values of 1 to 6.[37] For further details we refer to the References 11 and 37.

The time-resolved anisotropy with both intact erythrocytes and freshly prepared ghosts was studied and found very similar. The correlation times, well resolved for three different components, were around 30 ns, 200 ns, and on the order or longer than 1 ms.

Mild proteolytic removal of the hydrophilic cytoplasmic domain (41 kDa) of band 3 by trypsinization dramatically increased the fractional contribution of the two fastest depolarizing components to the overall rotational mobility. The multicomponent nature of the total intensity decay was underlined with the finding that, while sodium dodecyl sulfate gel analysis of red cell membrane proteins showed only one fluorescent eosin band, the percentual distribution of the correlation times measured as 40, 400, and about 2500 μs had a fractional contribution to the total intensity 0.2, 3, and 97%.

Thus, the gel analysis cannot be expected to resolve the very low amount of fast components. Besides the multicomponent suggestion, the possibility that the eosin reacts with more than one lysine residue cannot be excluded. When using eosin maleimide for covalent labeling of the band 3, the specificity was guaranteed by the low amount of other available proteins. Glycophorin, which has a high-enough density in the cell membrane, has no sulfhydryl (SH) groups, and the labeling of amino groups is minimal. The experimental data, as well as theoretical calculations, based upon a simple cylindrical model of the band 3 dimer, resulted in a rotational correlation time between 23 and 28 μs. The conclusion was that a substantial fraction of the band 3 rotates freely as a dimer, unhindered by (other) protein-protein interactions. Of course, theoretical analysis also supports the necessity to be careful in concluding from the existence of several rotational lifetimes. "The interpretation of the multicomponent r(t) is difficult in principle and in practice. In theory a single diffusional species by itself can give rise to as many as five exponential decay times. Even the simple model of pure cylindrical rotation in a membrane can lead to two exponentials. In general, the apparent presence or absence of exponentials depends on the geometry of the rotation axes relative to the absorption and emission transition moments, since this information is contained within the preexponential amplitudes."[37]

Time-resolved phosphorescence anisotropy decay had been used to study the rotational diffusion of eosin-labeled human erythrocyte band 3 in the presence of two enzymes, which were supposed to be associated with the intracellular moiety of the band 3.[38] Glyceraldehyde-3-phosphate dehydrogenase (GAPD) was added to ghosts in increasing amount prior to measuring rotational parameters of band 3. The infinite time anisotropy r_∞ increased with increasing amounts of the enzyme, and at saturation concentrations the anisotropy decay of band 3 was practically paralyzed. A qualitatively similar effect was caused by addition of aldolase, another enzyme of the glycolytic pathway. Although the existence of membrane-bound glycolytic enzymes in intact erythrocytes was suggested by earlier works, there were contradictory data as well. These experiments provide more direct evidence for the suggestion.[38]

A significant number of experiments concentrated on studying the rotational correlation times of proteins in solutions of different viscosity. Although, these experiments are seemingly unrelated to the behavior of membrane-bound molecules, some of them were prepared from membrane, but on the other hand contributed much to the understanding of the molecular rotations among more complex environmental conditions, such as membranes. The literature of such experiments is far too numerous to discuss individual papers systematically, but a brief survey of a few experiments may be helpful to understand their logic and major concerns.

Clague et al.[39] studied the rotation of spectrin purified from human erythrocytes and labeled with eosin-5-maleimide that covalently binds free SH groups. The rotational diffusion of the spectrin was investigated by observing again transient dichroism following flash excitation of the triplet probe. Measurements were performed at 4°C in solution of varying viscosity and with spectrin rebound to spectrin/actin-depleted erythrocyte membranes. In solution, complex anisotropy decays were observed which could not be satisfactorily fitted by the equations for rod-shaped molecules with dimensions appropriate for spectrin. Upon rebinding the spectrin to the erythrocyte membrane a decay in the anisotropy was still present but it was markedly less sensitive to solution viscosity and the curves were flatter at longer time ranges. In order to overcome the objection that the cytoskeleton is only partially reconstituted when spectrin is rebound, a method was developed by the authors for labeling the spectrin with the same triplet probe *in situ*. Anisotropy values obtained with this system resulted in the same data as those obtained for spectrin labeled in solution and subsequently rebound to erythrocyte membrane. Taken together their results, the conclusion was drawn that the results provide good evidence for the segmental motion of spectrin when incorporated into the erythrocyte cytoskeleton. Elevating the temperature, the initial anisotropy for both rebound and *in situ*-labeled spectrin decreased and above 30°C the measured anisotropies were small. Thus, at physiological temperature range the probe was practically randomized by motions with correlation times less than 10 μs.[39]

Several authors utilized the tryptophan fluorescence and also its phosphorescence in combination with, e.g., quenching. Several proteins were studied by Vanderkooi and co-workers with time-resolved fluorescence and phosphorescence anisotropy experiments, contributing much to our understanding of the significance of long-range electron exchange procedures.[40]

The biological usefulness of rotational relaxation measurements can hardly be questioned, despite all the theoretical difficulties, and as we have seen above they can be utilized in determining important physiological relationships at molecular level.

B. DO F_C RECEPTORS ROTATE IN THE MAST CELL MEMBRANES?

Basophil cells and their special type, the mast cells, are important in developing allergic symptoms, carrying many hormones and hormone-like compounds, e.g., histamine and serotonin, in their basophil granules. On the other hand they are prominent targets to study IgE-type immunoglobulin and receptor interactions, a particular but nevertheless interesting type of immunoglobulin-receptor system. The type I receptor for the $F_{C\varepsilon}$ domains of IgE is an integral plasma membrane complex of four polypeptide chains and it occurs specifically in mast cell and basophil membranes.[13] The basophils and mast cells are very fragile and their disruption releases the mediators of the basophil granules with subsequent severe immunological consequences. The mechanism of the immunoglobulin receptor interactions at the plasma membrane level and that of a triggered cascade process that leads to the degranulation of these cells have been vigorously studied.[13] Although no biochemical or biological effects were detected so far upon the monovalent binding of the immunoglobulin to the above-mentioned $F_{C\varepsilon}$ receptor type I at the cell surface, it was regarded as a priming process of the cascade leading to the mediator release. Plasma membrane receptor clustering was observed many times as the initial step of a cellular stimulus, i.e., the trigger signal for transmembrane signal processing.[10,11,39-42]

This rather general property of the cell surface elements directed the interest of researchers on the relationship between mobility and the state of aggregation of cell surface receptors. Changes in their lateral or rotational diffusion can easily be monitored by time-resolved fluorescent and phosphorescence measurements experimentally, if the membrane components are labeled with suitable fluorescence or triplet probes.[10,11,41-47] The selection of the actual label depends on the correlation time of the expected mobility changes of the membrane-bound molecules. The lifetimes of triplet probes lies in the microsecond to millisecond range that is expected for motional dynamics of macromolecules embedded in the viscous plasma membrane of living cells. The time course of the phosphorescence anisotropy decay, as we have seen above, reflects directly the size and mobility of the triplet-labeled membrane component and hence its state of aggregation.[10,11] Pecht et al.[13] carefully studied the rotational mobility, i.e., the anisotropy decay of erythrosin-labeled monoclonal antibody conjugates, in a high viscosity medium as that of 70% sucrose in water. These anisotropy decay kinetics determined the extent of emission depolarization caused by the relatively fast motions of the triplet probe itself and, furthermore, those resulting from the segmental flexibility of the monoclonal antibody or its F_{ab} fragment. Investigation of three different F_{ab} probes showed positive anisotropy curves that differed significantly from those observed with the same ligands bound to the $F_{C\varepsilon}$ receptor I on the surface of a particular cell type. The difference was caused by an aggregation of the antibodies upon dilution with concentrated sucrose solution. This phenomenon was even more expressed with erythrosin-labeled IgE antibody. Significant, but expected, differences were found between IgG and IgE antibodies and the F_{ab}s, which reflected their different molecular structures. According to the authors, a reasonable agreement with earlier measurements providing data from nanosecond decay of fluorescence anisotropy and underlined the advantage of the phosphorescence measurements. The latter measures the totality of the depolarization process; thus, the longer range global motions are preferentially represented in the mean values of the rotational correlation times (ϕ). Particular molecular components carrying the erythrosin label (Er) may report very sensitively the maximal dimensional changes of the molecule.

The observed patterns of the time-resolved anisotropy decay of cell-bound Er-IgE and F_{ab} fragments showed the characteristics for membrane-associated macromolecules displaying positive initial anisotropies, 30 to 200-μs rotational correlation times, and a non-zero limiting (remnant) anisotropy. The signals were reduced by over 80% when unlabeled monoclonal antibody molecules competed for the specific binding sites, demonstrating that the signals were obtained from specific binding of the labeled molecules to the cell surface receptors. The spontaneous dissociation of the ligands was expected during the several hours of the experiments. Its degree and possible disturbing effects were determined during and after the measurements. Different antibodies and F_{ab} had different dissociation properties. Of course, it was expected that the initial anisotropy, amplitudes, and the limiting anisotropy would follow the dissociation of ligands, but the values of the rotational correlation times had to remain unaffected. Deeper analysis of the decay kinetics showed that, in addition to the inherent dynamics of the ligand-bound receptor, individual differences in the molecular structure of the ligands carrying the triplet probes at different sites contributed due to wobbling and/or segmental motions and different flexibility. This was expressed by a second fast component in the case of the cell surface-bound IgE molecule.

The marked differences observed among the cell bound Er-F_{ab}-s, in terms of both the amplitudes and time dependence of the anisotropy decay, probably reflect nonrandom orientation of these probes relative to the cell membrane, and thus to the uniaxial or more complex axes of the $F_{C\varepsilon}$RI.

Analysis of the whole antibodies revealed such differences in the rotational modes of the molecules that could be attributed to dimeric or oligomeric distribution of the receptors. Such structures have a low abundance before ligand binding to initiate secretion of mediators. In conclusion, the experiments of Pecht et al.[13] described briefly here presented direct and well-aimed evidence that molecular properties of specific ligands, binding particular cell surface receptors, could be followed by spectroscopic means. The applied time-resolved phosphorescence anisotropy measurements could differentiate between distinct receptor-ligand assemblies. The directly observed changes in flexibility and dynamics of cell surface receptor complexes upon ligand binding have a yet-unresolved relationship with the transmembrane signaling events.

C. MOBILITY OF CA^{2+}-ATP-ASE IN MEMBRANES

Ca^{2+}-ATP-ase is one of the most frequently studied membrane-bound enzymes and not without reason. The "pumping" activity of membrane-bound proteins plays key role in transporting ions and molecules against electrochemical gradients, "uphill" at the expense of the metabolic energy of living cells. The mechanisms of these important activities are thought to be closely correlated with the motional characteristics of the molecules, which are instrumental in transporting ions, molecules, drugs, etc.[25]

Birmachu and Thomas[48] investigated the microsecond rotational motions of the Ca^{2+}-ATP-ase in rabbit skeletal sarcoplasmic reticulum by measuring the time-resolved phosphorescence anisotropy decay of erythrosin isothiocyanate (ERITC), covalently and specifically attached to the enzyme. As it was expected the phosphorescence anisotropy decay was best fit by a sum of three exponentials plus a constant term over a wide range of solvent conditions and temperature. The three rotational correlation times were $\phi_1 = 13$ μs; $\phi_2 = 77$ μs; and $\phi_3 = 314$ μs, very similar to those of the F_C receptor measured by Pecht et al.[13] It is likely that average-sized membrane-bound proteins rotate very similarly around their axes. Thus, probably the major factor that regulates rotation is the environmental viscosity. We think that is again an argument for the principle that, in the case of measuring physical parameters in biological systems, the changes in the parameters might have greater practical meaning, than the absolute values themselves. The above data were obtained at 4°C, which helps to keep rotational conditions at steady state during the measurement. The effect of the environmental viscosity was probed by increasing the extramembrane viscosity with glycerol and changing the intramembrane viscosity with diethyl ether. The first treatment was rather ineffective, while the diethyl ether substantially decreased the correlation times, indicating that the decay corresponds to rotation of the protein within the membrane, not the vesicle tumbling. In the case of cells, which are much larger than the vesicles prepared from sarcoplasmic reticulum, such possibility does not arise. The normalized residual anisotropy, r_∞, was insensitive to viscosity and temperature changes, supporting a model for uniaxial rotation of the protein about the axis perpendicular to the membrane. The value of the $r_\infty = 0.20$ indicates that each of the three decay components can be analyzed as a separate rotational species, with a preexponential factor that was found 1.25 times the mole fraction. An empirically accurate measurement of the membrane lipid viscosity was obtained, permitting a theoretical analysis of the correlation times in terms of the sizes of the rotating entities. At 4°C, the dominant correlation time (314 μs) is too large for a Ca^{2+}-ATP-ase monomer, strongly suggesting that the enzyme is mostly in an aggregated, oligomeric form. The similarly large rotational correlation times of other proteins may suggest the possibility that the majority of the membrane-bound proteins are not in a monomeric form.

The Ca^{2+}-ATP-ase has the great advantage of the easy and specific labeling at more than one site even by different luminescence dyes. This property of the enzyme was utilized by Martonosi et al.[49,50] in studying its fluctuation patterns by temperature-normalized fluorescence resonance energy transfer (for details see Chapter 2, Section II.D.1).

IV. CONCLUSION

Of course, the above brief survey is far from being complete, and its only intention was to direct the attention of biologists to this powerful possibility to follow rotational mobilities of individual groups of molecules among physiological conditions. This method can be combined at different levels with lateral mobility measurements, proximity determinations by FRET methods, or other physical means revealing a new aspect of molecules in living systems, i.e., the real time mobility. Functional problems and mechanisms can be discovered and correlations established through this track of science. A recent advancement of fluorescence microscopy made possible even receptor tracking.[51] The complexity of receptor mobilities on cell surfaces promises a new exciting world for the scientist, and useful information for applications in cell biology.

REFERENCES

1. **Einstein, A.,** On the movement of small particles suspended in a stationary liquid demanded by the molecular kinetic theory of heat, *Ann. Phys.*, 17, 549, 1905.
2. **Einstein, A.,** *Investigations on the Theory of the Brownian movement*, Furth, R., Ed., Dover Inc., New York, 1956, 68.
3. **Cherry, R. J., Burkli, A., Busslinger, M., Schneider, G., and Parish, G. R.,** Rotational diffusion of band 3 proteins in the human erythrocyte membrane, *Nature*, 263, 389, 1976.
4. **Cherry, R. J.,** Rotational and lateral diffusion of membrane proteins, *Biochim. Biophys. Acta*, 559, 289, 1979.
5. **Cherry, R. J.,** Transient dichroism and rotational diffusion of macromolecules: applications to membrane proteins, in *Spectroscopy and Dynamics of Molecular Biological Systems*, Bayley, P. M. and Dale, R. E., Eds., Academic Press, London, 1985, 75.
6. **Cherry, R. J.,** Effect of mellitin and membrane potential on the mobility of band 3 proteins in erythrocyte membranes, in *Dynamics of Biochemical Systems*, Damjanovich, S., Keleti, T., and Trón, L., Eds., Elsevier, Amsterdam, 1986, 487.
7. **Lo, M. M. S., Garland, P. B., Lamprecht, J., and Barnard, E. A.,** Rotational mobility of the membrane-bound acetylcholine receptor of Torpedo electric organ measured by phosphorescence depolarisation, *FEBS Lett.*, 111, 407, 1980.
8. **Szabo, A.,** Theory of fluorescence depolarisation in macromolecules and membranes, *J. Chem. Phys.*, 81, 150, 1984.
9. **Ameloot, M., Hendrickx, H., Herreman, W., Pottel, H., Van Cauwelaert, F., and van der Meer, W.,** Effect of orientational order on the decay of fluorescence anisotropy in membrane suspensions, *Biophys. J.*, 46, 525, 1984.
10. **Matkó, J., Szöllősi, J., Trón, L., and Damjanovich, S.,** Luminescence spectroscopic approaches in studying cell surface dynamics, *Q. Rev. Biophys.*, 21, 479, 1988.
11. **Jovin, T. M. and Vaz, W. L. C.,** Rotational and translational diffusion in membranes measured by fluorescence and phosphorescence methods, *Methods Enzymol.*, 172, 471, 1989.

12. **Aoreti, B., Jovin, T. M., and Hennis, Y. I.,** Rotational mobility of Sendai virus glycoprotein in membranes of fused human erythrocytes and in the envelopes of cell bound virions, *Biochemistry*, 29, 9119, 1990.
13. **Pecht, I., Ortega, E., and Jovin, T. M.,** Rotational dynamics of the F_C receptor on mast cells monitored by specific monoclonal antibodies and IgE, *Biochemistry*, 30, 3450, 1991.
14. **Cantor, C. and Schimmel, R.,** *Biophysical Chemistry*, Vol. 2, W. H. Freemann, San Francisco, 1980, 433.
15. **Strambini, G. B. and Galley, W. C.,** Detection of slow rotational motions of proteins by steady state phosphorescence anisotropy, *Nature*, 260, 554, 1976.
16. **Strambini, G. B. and Gonelli, M.,** Tryptophan luminescence from liver alcohol dehydrogenase in its complexes with coenzyme. A comparative study of protein conformation in solution, *Biochemistry*, 29, 196, 1990.
17. **Strambini, G. B. and Gonelli, M.,** Room temperature phosphorescence of trp-314 as a monitor of subunit communications in alcohol dehydrogenase from horse liver, *Biochemistry*, 29, 203, 1990.
18. **Damjanovich, S., Trón, L., Szöllősi, J., Zidovetzki, R., Vaz, W.L.C., Regateiro, F., Arndt-Jovin, D. J., and Jovin, T. M.,** Distribution and mobility of murine histocompatibility H-2K^k antigen in the cytoplasmic membrane, *Proc. Natl. Acad. Sci. U.S.A.*, 80, 5985, 1983.
19. **Komura, S., Ohta, Y., and Kawato, S.,** A theory of optical anisotropy decay in membranes, *J. Phys. Soc. Jpn.*, 59, 2584, 1990.
20. **Weber, G.,** Theory of fluorescence depolarization by anisotropic Brownian rotations. Discontinous distribution approach, *J. Chem. Phys.*, 55, 2399, 1971.
21. **Weber, G.,** Perrin revisited: parametric theory of the motional depolarization of fluorescence, *J. Phys. Chem.*, 93, 6069, 1989.
22. **Ehrenberg, M. and Riegler, R.,** Polarized fluorescence and rotational Brownian motion, *Chem. Phys. Lett.*, 14, 539, 1972.
23. **Dale, R. E.,** Some aspects of excited-state probe emission spectroscopy for structure and dynamics of model and biological membranes, in *Polarized Spectroscopy of Ordered Systems*, Samori, B. and Thulstrup, E. W., Eds., Kluwer Academic, Boston, 1988, 491.
24. **Piston, D. W. and Gratton, E.,** Orientational exchange approach to fluorescence anisotropy decay, *Biophys. J.*, 56, 1083, 1989.
25. **Higgins, C. F. and Gottesman, M. M.,** Is the multidrug transporter a flippase?, *Trends Biochem. Sci.*, 17, 18, 1992.
26. **Hartmann, E., Rapoport, T. A., and Lodish, H. F.,** Predicting the orientation of eukaryotic membrane spanning proteins, *Proc. Natl. Acad. Sci. U.S.A.*, 86, 5786, 1989.
27. **Lakos, Z., Somogyi, B., Balázs, M., Matkó, J., and Damjanovich, S.,** The effect of transmembrane potential on the dynamic behaviour of cell membranes, *Biochim. Biophys. Acta*, 1023, 41, 1990.
28. **Dale, R. E.,** Depolarized fluorescence photobleaching recovery, *Eur. Biophys. J.*, 14, 179, 1987.
29. **Velez, M. and Axelrod, D.,** Polarized fluorescence photobleaching recovery for measuring rotational diffusion in solutions and membranes, *Biophys. J.*, 53, 575, 1988.
30. **Velez, M., Barald, K. F., and Axelrod, D.,** Rotational diffusion of acetylcholine receptors on cultured rat myotubes, *J. Biol. Chem.*, 110, 2049, 1990.
31. **Johnson, P. and Garland, P. B.,** Depolarization of fluorescence depletion, *FEBS Lett.*, 132, 252, 1981.
32. **Johnson, P. and Garland, P. B.,** Fluorescent triplet probes for measuring the rotational diffusion of membrane proteins, *Biochem. J.*, 203, 313, 1982.
33. **Yoshida, T. M. and Barisas, B. G.,** Protein rotational motion in solution measured by polarized fluorescence depletion, *Biophys. J.*, 50, 41, 1986.
34. **Clague, M. J., Harrison, J. P., and Cherry, R. J.,** Cytosceletal restraints of band 3 rotational mobility in human erythrocyte membranes, *Biochim. Biophys. Acta*, 981, 43, 1989.
35. **Clague, M. J. and Cherry, R. J.,** A comparative study of band 3 aggregation in erythrocyte membranes by melittin and other cationic agents, *Biochim. Biophys. Acta*, 980, 45, 1989.
36. **Wyatt, K. and Cherry, R. J.,** Both ankyrin and band 4.1 are required to restrict the rotational mobility of band 3 in the human erythrocyte membrane, *Biochim. Biophys. Acta*, 1103, 327, 1992.
37. **Matayoshi, E. D. and Jovin, T. M.,** Rotational diffusion of band 3 in erythrocyte membranes. I. Comparison of ghosts and intact cells, *Biochemistry*, 30, 3527, 1991.

38. **Matayoshi, E. D., Sawyer, W. H., and Jovin, T. M.,** Rotational diffusion of band 3 in erythrocyte membranes. II. Binding of cytoplasmic enzymes, *Biochemistry*, 30, 3538, 1991.
39. **Clague, M. J., Harrison, J. P., Morrison, I. E., Wyatt, K., and Cherry, R. J.,** Transient dichroism studies of spectrin rotational diffusion in solution and bound to erythrocyte membranes, *Biochemistry*, 29, 3898, 1990.
40. **Vanderkooi, J. M., Englander, S. W., Papp, S., Wright, W. W., and Owen, C. S.,** Long-range electron exchange measured in proteins by quenching of tryptophan phosphorescence, *Proc. Natl. Acad. Sci. U.S.A.*, 87, 5099. 1990.
41. **Zidovetzki, R., Yarden, Y., Schlessinger, J., and Jovin, T. M.,** Rotational diffusion of epidermal growth factor complexed to cell surface receptors reflects rapid microaggregation and endocytosis of occupied receptors, *Proc. Natl. Acad. Sci. U.S.A.*, 78, 6981, 1981.
42. **Zidovetzki, R., Yarden, Y., Schlessinger, J., and Jovin, T. M.,** Microaggregation of hormone-occupied epidermal growth factor receptors on plasma membrane preparations, *EMBO J.*, 5, 247, 1986.
43. **Zidovetzki, R., Bartholdi, M., Arndt-Jovin, D. J., and Jovin, T. M.,** Rotational dynamics of the F_c receptor for immunoglobulin E on histamine releasing rat basophilic leukemia cells, *Biochemistry*, 25, 4397, 1986.
44. **Edidin, M.,** Rotational and translational diffusion in membranes, *Annu. Rev. Biophys. Bioeng.*, 3, 179, 1974.
45. **Edidin, M. and Wei, T.,** Diffusion rates of cell surface antigens of mouse-human heterokaryons. II. Effect of membrane potential on lateral diffusion, *J. Cell Biol.*, 75, 483, 1977.
46. **Edidin, M. and Zuniga, M.,** Lateral diffusion of wild-type and mutant Ld antigens in L cells, *J. Cell Biol.*, 99, 2333, 1984.
47. **Edidin, M. and Weir, M.,** Mobility of membrane proteins and the social life of cells, *Biochem. Soc. Trans.*, 14, 818, 1986.
48. **Birmachu, W. and Thomas, D. D.,** Rotational dynamics of Ca-ATP-ase in sarcoplasmic reticulum studied by time-resolved phosphorescence anisotropy, *Biochemistry*, 29, 3904, 1990.
49. **Jóna, I., Matkó, J., and Martonosi, A.,** Structural dynamics of the Ca^{2+}-ATP-ase of sarcoplasmic reticulum. Temperature profiles of fluorescence polarisation and intramolecular energy transfer, *Biochim. Biophys. Acta*, 1028, 183, 1990.
50. **Martonosi, N. A., Jóna, I., Molnar, E., Seidler, N. W., Buchet, R., and Varga, S.,** Emerging views on the structure and dynamics of the Ca^{2+}-ATP-ase in sarcoplasmic reticulum, *FEBS Lett.*, 268, 365, 1990.
51. **Cherry, R. J.,** Keeping track of cell surface receptors, *Trends Cell Biol.*, 2, 242, 1992.

Chapter 6

Ion Channels and Membrane Potential Changes in Lymphocytes

Sándor Damjanovich

CONTENTS

0-8493-4931-1/94/$0.00+$.50

I. INTRODUCTION

Ion channels are integral membrane proteins spanning membranes. These proteins have virtual holes inside (which actually serve as tunnels), occasionally allowing ions to pass through, thereby circumventing the hydrophobic barrier of the lipid bilayer that separates cell interior from extracellular space. Ion channels of cells from the central nervous system display enormous diversity. Although the ion channels of leukocytes were discovered much later, and they are perhaps less well characterized yet, it is more and more obvious, thanks to studies using patch clamp recording techniques, that a surprising diversity of channel types can also be found in these cells. For instance, channels such as the voltage-gated outward rectifier, inward rectifier, and Ca^{2+}-activated potassium channels have been identified along voltage-gated Na^+ and voltage- and ligand-gated Na^+, Ca^{2+}, and Cl^- channels. Lewis and Cahalan[1] gave a concise summary of the above, while recently Pieri et al.[2] reported on the voltage- and ligand-gated sodium channels, which may belong to a new signal transduction pathway-regulating mechanism. The variety and abundance of these channels are highly dependent on the types of cells and species of their origin. The structure and regulation of these channels will be discussed below, but even before doing that a question immediately arises: what is the purpose of such an abundance of ion channel varieties in the immune system?

As for the central nervous system, the role of receptor and ion channel subtype diversity may contribute to its plasticity, i.e., a kind of dynamic willingness to learn and accommodate cellular functions to the ever-changing environmental conditions.[3] The significance of the multiplicity and diversity of these systems is very likely an increased information handling capacity.

For lymphocytes, an entirely different class of cells, one intuitively feels that a similar role could be assigned to the ion channels. This role could be the handling of the different signal transducing mechanisms and metabolic processes. Thus, we may even conclude that it is not the diversity and variety of the ion channel and coupled receptor systems that is surprising. On the contrary, the diversity of metabolic processes and information transducing biochemical and biophysical pathways must quite naturally demand such a system in the plasma membranes.

II. SUPERFAMILIES OF ION CHANNELS

It is worthy to note that, similarly to those of the central nervous system (CNS), the ion channels of lymphocytes also have the three major kinds of regulatory mechanisms:

1. Ligand
2. Voltage
3. G-protein- (GTP binding protein) regulated ion channels

The classification of ion channels (and also receptors) into such superfamilies also means that these proteins have similar characteristics and have been maintained very conservatively throughout the evolution.

G-protein-coupled receptors, for example, usually have seven membrane-spanning domains. The prototype of ligand-gated ion channels, the nicotinic acetylcholine receptor, has five domains ($\alpha_2\beta\gamma\delta$) and each domain posseses four membrane-spanning regions. Voltage-sensitive ion channel molecules belong to an integral membrane protein superfamily that is built up from four domains, each having six membrane-traversing segments. The purpose and the actual mechanism that made such a structural arrangement feasible is not yet clear. The gating properties of these peptides are frequently connected to amino acid residues carrying positive or negative charges. These are mainly arginine, glutamine, lysine, and histidine residues.

The above-outlined general characteristics of ion channels demonstrate a strong conservatism, and are ubiquitous practically regardless of the origin of cells we study.

The anion or cation specificity of ion channels, although far from being absolute, depends mainly on the size and charge of the individual cations and anions. The size of the ions naturally highly depends on the hydrated shell surrounding the "naked" ion, due to dipole-dipole interactions. The thickness of the hydrate shell does not necessarily depend on the size of the ion. A well-known example is the comparison of the hydrated Na^+ and K^+ ions. The naked Na^+ ion is smaller than the K^+, according to their respective positions occupied in the periodic table of elements. However, due to the greater hydrate shell of Na^+ ions, they have a larger diameter and consequently lower permeability than that of the hydrated K^+ ions. Thus, this relatively small difference in size causes a hundredfold difference in their permeability constants across membranes in favor of the K^+ ions. We also have to keep in mind that the average squared distance covered by a diffusing particle has a linear relationship to diffusion constants and time.

Only a few ion channels of lymphocytes have been characterized in satisfactory details, so far, including their densities in plasma membranes upon expression and also chromosomal localizations.[4,5]

Beyond the enormous theoretical and scientific importance of such investigations there are also practical reasons for studying the ion channel activities of so-called nonexcitable cells. A number of diseases, among them genetic ones, are supposed to be linked possibly causatively to impairments of ion channel activities. Cystic fibrosis is likely to be connected to the impaired regulation of a single chlorine channel.[6] Autoimmune diseases such as systematic lupus erythematosus (SLE), diabetes mellitus, allergic arthritis, and encephalomyelitis are accompanied by altered potassium channel activities as is convincingly demonstrated by Grissmer et al.[4] Of course, whether these alterations in the occurrence rate and functional properties of a particular channel type will only make a good diagnostic marker or some causative relationship will be found between the disease and the observed changes is a question yet to be answered in the majority of cases. The voltage-gated K^+ channels of lymphocytes may even play a part in mitogenesis.[7]

Also, it has been known for a few years that mitogens can cause early changes in cytoplasmic free Ca^{2+} concentration in T cells, and in the membrane potential of T as well as B lymphocytes as it was established by Tsien and co-workers.[8,9] The frequently discussed importance of ion channels in the initiation and regulation of cell activation processes, particularly those of T and B lymphocytes, thereby modulating both cellular and humoral immune responses, is still not completely understood despite intensive efforts in this field. These questions were reviewed extensively by Szöllősi et al.,[10] Matkó et al.,[11] and Grinstein and Dixon[12] relatively recently.

The above-outlined data suggest that dynamic properties of cell membrane elements are also controlled and intricately influenced by ion channel activities and the frequently accompanying transmembrane electric potential changes.[13-22] This statement sounds almost trivial yet very few direct measurements have been made to follow and study dynamic changes in intramembrane- or membrane-bound molecular interactions upon changing ion channel activities and membrane potential. In the case of excitable cells such as muscle or nerve cells, it is generally accepted that a decrease in resting potential facilitates the triggering of responses to stimuli. Lymphocytes have complex transmembrane signaling systems which have a variable number of membrane-bound steps involved. Data will be presented, suggesting that, the higher the number and complexity of membrane-bound steps, the greater are the influence and regulatory role of ion channel activities and membrane potential changes on transmembrane signaling events.[2,19-21,23,24]

III. THE ORIGIN OF MEMBRANE POTENTIAL

Apparently it is a well-known series of physical-chemical interactions across a semipermeable membrane which explains how the transmembrane electric potential is generated. The three major components of the transmembrane potential are known as follows:

1. Donnan potential, which can be calculated from the uneven ion distribution across the plasma membrane.
2. Diffusion potential of particular ions (eminently among them the K^+), which is supposed to be a major contributor to the resting potential (and also the action potentials of excitable cells). This is believed to be a decisive contributing factor since it alone can generate a potential value of over 100 mV across the membrane. In contrast to this the Donnan potential has only a contribution on the order of 10 to 20 mV as can be calculated from the Nernst equation.
3. The ion-pumping activities, which may also modulate membrane potential values. Some of them can be electrogenic such as the Na^+-K^+-ATPase and can drive ions "uphill", i.e., against the electrochemical gradients, exploiting the readily available chemical potential from metabolite catabolism, in the cited case that of ATP dephosphorylation.[25]

The many contradictions originate mostly from the fact that different cell types maintain their membrane potentials relying on the above three components to a different extent. Even the lowest contributor, the ion-pumping mechanism, can be considered as the sole component of membrane potential in specific cases, as in Lettre cells a particular and very malignant type of Ehrlich ascites tumor cells, but also in some other cell types. As we will see in later sections, as Bashford and Pasternak[26] pointed out, the Lettre cells are eminently suited to a study of such questions since they have no K^+ channels at all. Thus, the diffusion potential, mediated by the K^+ channels as a major contributor to the membrane potential, is out of the question in such particular cases.

It is obvious that pumping mechanisms can also provide an extremely sensitive regulatory possibility of the plasma membrane potential through maintaining pseudo-Donnan systems. Contrary to a real Donnan system where electrochemical potential is generated by nonpenetrating protein anions, pseudo-Donnan equilibrium can be generated and maintained by a continuous active ion pump that will create an inequilibrium of ions across the cell membrane. It is easy to accept that any perturbation of the pumping mechanisms will result in an immediate change of equilibrium potential due to the uncompensated ion diffusion.

Living cells are richly endowed with all kinds of such regulatory mechanisms, enabling them to accommodate in a very short time to the ever-changing metabolic and environmental conditions.

Further examples and a little more detailed analysis will be given in this chapter, but before doing that it is mandatory to get acquainted with some theoretical aspects and the measuring techniques of membrane potential, which is the most sensitive indicator of ion channel functions.

It is a remarkable fact that changes in the membrane potential report very sensitively on viability and also metabolic processes of cells in general, thus underlining some practical aspects of the questions to be discussed.

A. ION CONCENTRATION DIFFERENCES ACROSS MEMBRANES

It is generally accepted that biological membranes separate compartments with different ion concentrations. Furthermore, these compartments have electric potential differences. This chapter is intended to give an overview on the potential differences of aqueous compartments, separated by membranes.

In closed systems, among given starting conditions, sooner or later an equilibrium will be formed in ion concentrations and membrane potential at the two sides of a membrane. Of course, a "state", before the equilibrium is reached, is a nonequilibrium state, which "tends" to change toward the equilibrium. These processes, which involve a free enthalpy decrease, will take place spontaneously. Any nonequilibrium process can be maintained in a stationary way. However, in this case the effects of the spontaneous processes have to be compensated. Biological membranes can be considered as parts of open thermodynamic systems where a high number of events are made possible utilizing either energy demanding metabolic steps or passive spontaneous processes which take place in such a way that the thermodynamic potential is decreased. Living cells have the capability of maintaining such processes, which render unequilibrated ion concentrations and potentials stationary.

Of course, the measurable potential differences between compartments (the actual membrane potential) under normal conditions are not necessarily stationary. Sometimes they may display sudden changes, depending on environmental factors. These changes may rapidly follow the alterations in permeability parameters or those of ion pump activities. Changes in membrane potential may accompany physiological or pathological processes exerting a perturbation on cell metabolism.

B. THE ORIGIN OF ELECTROCHEMICAL POTENTIALS: THE NERNST EQUATION

Let us consider a semipermeable membrane, separating two electrolyte solutions of unequal composition (compartment I and compartment II) and start with the simple case of only one membrane-permeable ion species, e.g., Na^+. The Na^+ ions tend to diffuse along their concentration gradient from the solution of high Na^+ concentration through the membrane into the solution of low Na^+ concentration. However, they cannot reach the bulk of the low concentration solution since they are held back by impermeable anions left behind at the opposite membrane side. Thus, two thin layers of excess positive and negative charges are created at either sides of the membrane surfaces.

In the case of only one membrane-permeable ion species an equilibrium state is established in which the ion diffusion across the membrane is abolished by the created diffusion potential. At equilibrium, the change in the free energy when a mole of Na^+ ions passes from compartment I to compartment II is zero.

$$\Delta G = \mu_c^{II} - \mu_c^{I} = 0 \tag{1}$$

in which the μ_c are the electrochemical potentials, consisting of a chemical component μ and an electrical component $zF\psi$ (with z as the valence of the ionic species, F as the Faraday constant, and ψ as the electrical potential). Substituting the appropriate expressions for the electrochemical potentials we obtain

$$\mu_c^o + RT\ln c_c^I + z_c F\psi^I = \mu_c^o + RT\ln c_c^{II} + z_c F\psi^{II} \tag{2}$$

Solving for the electrical potential difference gives

$$\Delta\psi = \psi^{II} - \psi^{I} = \frac{RT}{z_c F}\ln\frac{c_c^I}{c_c^{II}} \tag{3}$$

The electrical potential difference is, thus, proportional to the logarithm of the ratio of the two concentrations. For cations z is positive and the electrical potential is higher on the more dilute side of the membrane. Equilibrium is attained because the buildup of electrical potential on the dilute side of the membrane raises the electrochemical potential of the solution in the more diluted compartment to that of the more concentrated solution in the other compartment.

If the membranes were only permeable to anions, we would have had the reverse situation. It is to be noted that in each compartment (I and II) the law of electrical neutrality is still valid, because the charge difference (or charge displacement) cannot be detected; it is only manifest as an electrical potential difference across the membrane. Such a potential difference is often referred to as Donnan potential, although it may contain a contribution from the diffusion potential (because it results from an apparent diffusion of ions across the membrane). Equation 3 is often called the Nernst equation. If an external electrical field is applied to a membrane which is only permeable to either negative of positive ions, and the membrane is separating two compartments each containing solutions of the particular ion, the concentrations of the ion at equilibrium are given by the Nernst equation.

C. WHAT IS THE ORIGIN OF DONNAN POTENTIAL?

It is well known that cytoplasm contains a large number of proteins which cannot permeate the plasma membrane. Polyvalent anions, such as proteins, have net negative charge under physiological conditions. The electroneutrality of the cytoplasm is maintained by cations. These are mostly small-sized metallic ions, which can permeate the plasma membrane to a different extent, depending largely on the specific ion channel activities of the membrane. The cytoplasm also contains a significant number and variety of anions which due to their size can also permeate the membrane through channels.

In order to investigate such a system, let us analyze the model, shown in Figure 1.

The extracellular space is represented by the compartment I, and the intracellular space by compartment II. The compartments I and II are separated by a semipermeable membrane. The membrane is permeable to the Na^+ and Cl^- ions and impermeable to the A^{n-} anions which have n unit negative charge.

The system is in equilibrium when the electrochemical potential of the two parts is identical. Let us label the valency of the ith ion with z_i, and the concentration with c_i. The electrochemical potential is

$$\mu_i = \mu_i^o + RT\ln c_i + z_i F\psi \tag{4}$$

where μ_i^o is the standard chemical potential, F V 96500 C, i.e., the Faraday constant, and ψ is the electric potential. The conditions of the equilibrium are as follows

$$\Delta\mu_{Na} = RT\ln\frac{[Na^+]^{II}}{[Na^+]^{I}} + F\Delta\psi = 0 \tag{5}$$

$$\Delta\mu_{Cl} = RT\ln\frac{[Cl^-]^{II}}{[Cl^-]^{I}} - F\Delta\psi = 0 \tag{6}$$

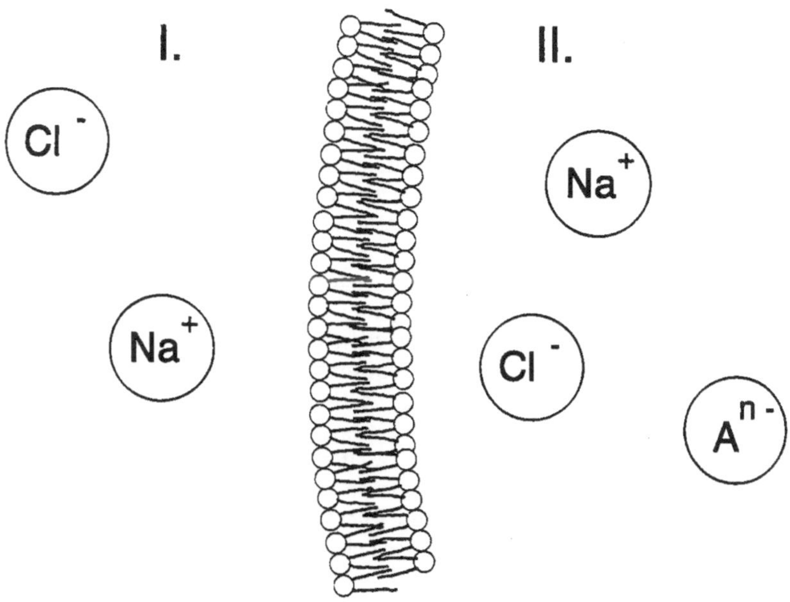

Figure 1. A compartment model of permeable and impermeable ions. The two compartments I and II are separated by a membrane, which can be permeated by the Na^+ and Cl^- ions and are impermeable for the multivalent protein anions (A^{n-}). The equilibrium and the generated membrane potential is described by the Nernst equation. The membrane potential has its origin in the Donnan equilibrium, the diffusion potential of the permeating ions, and, in case the membrane is endowed by this energy demanding capacity, ion pump activity.

Thus, the membrane potential at equilibrium ($\Delta\psi$) is

$$\Delta\psi = -\frac{RT}{F}\ln\frac{[Na^+]^{II}}{[Na^+]^{I}} = -\frac{RT}{F}\ln\frac{[Cl^-]^{I}}{[Cl^-]^{II}} = -\frac{RT}{F}\ln r \tag{7}$$

$$r = \frac{[Na^+]^{II}}{[Na^+]^{I}} = \frac{[Cl^-]^{I}}{[Cl^-]^{II}} \tag{8}$$

The parameter r gives the ratio of the equilibrium concentration of the permeating ions. The electrochemical and electric potential differences are normalized in these cases to the I compartment, representing the extracellular space. For I and II compartments the electroneutrality must be maintained:

$$[Na^+]^{I} = [Cl^-]^{I} = c \tag{9}$$

$$[Na^+]^{II} = [Cl^-]^{II} + na \tag{10}$$

where na is the nonpermeating anion concentration. From the above equations, after some algebra, the

$$r = \frac{na + \sqrt{4c^2 + (na)^2}}{2c} \tag{11}$$

The correlations between concentrations are

$$[Na^+]^{II} > [Na^+]^{I} = [Cl^-]^{I} > [Cl^-]^{II} \tag{12}$$

It is obvious from the definition of r that at $r \neq 1$, one always has membrane potential, which is different from 0. This is called Donnan potential. The $r = 1$ is true only if $a = 0$. If one has nonpermeating anions in the system, $r > 1$. It also involves a negative Donnan potential being generated for a plasma membrane as a consequence of the nonpermeating anions. It also comes from the basic equations that the value of r (and the Donnan potential) increases with the intracellular anion concentration and has an inverse relationship with the extracellular salt concentration. The value of Donnan potential does not (directly) depend on metabolic processes; thus, metabolic inhibitors do not influence its size. This is a basic difference with any other kind of nonequilibrium potential. Further permeating electrolytes, given to a Donnan system, will be distributed according to the Donnan equilibrium and the original ion distribution will be slightly perturbed only.

The mechanism of the Donnan equilibrium and the equilibrium potential difference can also be described as follows: let us take a system with two compartments shown in Figure1. In order to start, let us put a given amount of nonpermeating polyanion to II so that the accompanying ion is Na^+. Into compartment I let us put Na^+Cl^- solution. Because the membrane can be permeated by both Na^+ and Cl^-, the concentration difference will drive Cl^- from I to II. Of course, the concentration difference between I and II cannot be eliminated because of the repulsing force of the polyanions in II. Steady-state conditions assume an equilibrium between diffusion and electric forces. Since the polyanion does not permeate, the separating lipid double layer will have a given surface charge density at each side.

Before attaining steady state, Na^+ ions will also follow the Cl^-, to maintain electroneutrality. This also means that the electric potential field builds up slower, because the resultant transport is mostly NaCl. At dynamic equilibrium such an electric double layer is formed where the charged particle transports equalize each other during the opposing fluxes. The actual amount of ions taking place in this process is practically negligible as compared to the total extra- and intracellular ion concentrations.

D. CAPACITANCE AND CHARGE CONDITIONS IN CELL MEMBRANES

Since Gorter and Grendel, while working in the Department of Physiology of the University of Leiden in 1925, discovered that the basic framework of cell membranes is a double layer of lipid molecules, electric descriptions of the plasma membrane of eukaryotic cells has been frequently compared to capacitors. The lipid double layers can be considered as insulators, separating the boundaries of the aqueous solutions (Figure 2). The actual capacity of such capacitors is extremely large, since the distance between the "condenser plates" is very small (5 to 7 nm). Due to the great variety in cell size (nearly two

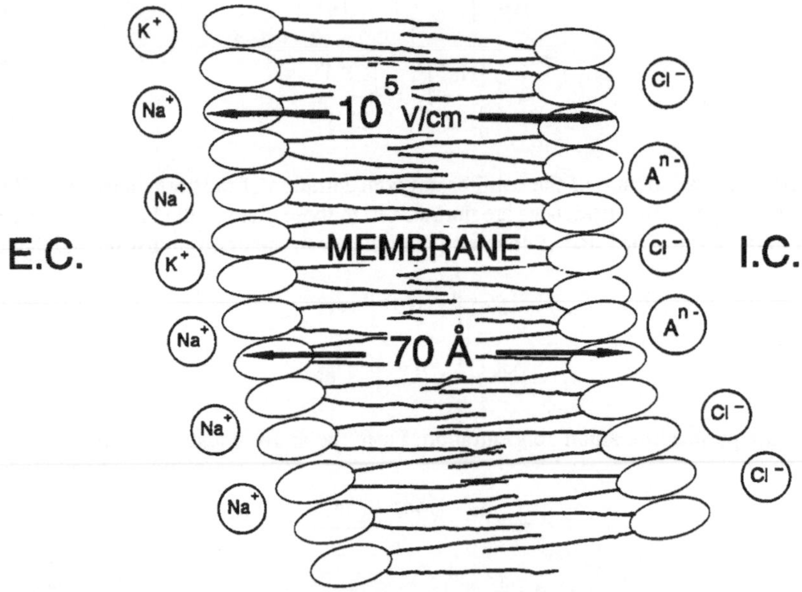

Figure 2. A schematic representation of the extra- and intracellular space, E.C. and I.C., respectively, separated by a membrane.

orders of magnitude!), it is customary to normalize the cell membrane capacitance to the surface area. Thus, the frog-striated muscle fiber has a capacity about 10 μF/cm^2, while the giant axon of squid has 1 μF/cm^2. This latter is the most frequently used experimental cell type in electrophysiology. If we take 90 mV as an average plasma membrane potential, the charge that is necessary to generate this potential difference can easily be calculated:

$$Q = CU = 10^{-5}\, F/cm^2 \times 9 \times 10^{-2}\, V = 9 \times 10^{-7}\, C/cm^2 \tag{13}$$

Thus, the charge density is around 10^{-7} to 10^{-6} coulomb/cm^2. Since the separated charges are in the form of ions (electrolyte solutions), the charge density can also be expressed as moles per area. From the Loschmidt number and the charge of an electron, one mole monovalent ion has

$$6 \times 10^{23} \times 1.6 \times 10^{-19} C \simeq 96500\, C \tag{14}$$

i.e., approximately 10^5 coulombs. Thus, 9×10^{-7} C/cm^2 charge separated at the two sides of frog muscle membrane is equivalent with 9.5×10^{-12} mole ion per squared centimeter. In comparison it is worthy to note that 1 cm^3 intra- or extracellular fluid contains at least 1.55×10^{-4} mole cations and anions. In other words, it is easy to show that the 6×10^{-8} cm (i.e., 0.6-nm) thickness of this fluid has the necessary amount of ions to charge the membrane near 100 mV.[27] This amount of the ionic charges is negligible to the total amount of the charged ions available in cells, having 6- to 10-μm average diameter, not even mentioning large nerve cell with 50- to 70-μm diameter.

The above derivation is valid practically for every living cell, including of course, lymphocytes.

E. THE DIFFUSION MEMBRANE POTENTIAL

When a given concentration of an electrolyte (e.g., an HCl solution, which is more or less a solution of H^+ and Cl^- ions) is overlaid at the top of a higher concentration of the same electrolyte, due to the concentration differences, the positive (H^+) and negative (Cl^-) ions start diffusing toward the lower concentrations. If we consider an electrolyte, such as H^+Cl^-, one has to deal with the higher mobility (larger diffusion constant) of protons as compared to Cl^- ions. The protons will "run" ahead of the chloride ions. Thus, an imbalance in the different charges will be generated. In the transient domain at the lower concentration layer, a relative surplus of H^+ ions will be generated. This part of the layer will have more positive charge than the other, where Cl^- ions lag behind H^+. This potential is called diffusion potential. The size of the potential has a direct correlation with the relative mobility difference between the ions. The charge of the more mobile ion will dominate the lower concentration domains and vice versa. The definition of the ion mobility is the diffusional speed of a particle in a field of unit force. Naturally, diffusion potential is transient, because the concentrations will be equilibrated in a steady-state system. The significance of the measurable diffusion potentials is further decreased by the fact that they are frequently perturbed by convection streams. On the contrary, their significance may be enlarged and stabilized in such systems where layers at different concentration are separated by a selectively permeable membrane. It is well known for biological membranes that their permeability is highly selective for different types of ions. Diffusion potentials across selectively permeable membranes are also transient phenomena if the membrane is permeable for each ion. Of course, besides the different mobility parameters, the concentrations also participate in the generation of diffusion potentials.

If a closed system is considered and the permeability of the different ions is not uniform, one of the variable diffusion potentials approaches the values of the Donnan potential as the system goes toward equilibrium. Thus, finally it equals the Donnan equilibrium potential. Every step toward equilibration can be considered as quasiequilibrium. This is characterized by the electric neutrality of the two compartments, the transient equilibrium between the partition of the concentrations, and ion current. This equilibrium is transient because the sum of individual ion currents is zero, but the currents of the individual ions are not. As the ionic partitions will change with time, the diffusion potential will also be changed.

The degree of transport processes can be characterized with the flux. Flux (J) is the amount of the transported item across unit cross section per unit time. According to the above definition, the flux of particles can be calculated as a product of the particle flux rate and the particle concentration. Flux, drift or flow is generated by some kind of "force". This force (P) is the differential quotient of potential energy function, according to the coordinates of localizations. This gradient has a negative sign, because the system goes toward equilibration, i.e., the driving force decreases. If we have concentration differences (concentration gradients), the driving force of the flux is the differential quotient of the chemical potential. In calculating the particle flux we can further utilize the relationship between the flux rate (v) and mobility (u): $v = uP$.

Thus, the flux

$$J_{diff} = cv = cuP = -cu\frac{d\mu}{dx} \tag{15}$$

If we take into account only the flux along the x axis, the gradient of the chemical potential is

$$\frac{d\mu}{dx} = \frac{d(RT\ln c)}{dx} = RT\frac{d(\ln c)}{dx} \tag{16}$$

If our system has only one salt of monovalent ions we can express the flux of the positive and negative ions alongside the concentration gradient as follows:

$$J^{+}_{diff} = -c^{+}u^{+}RT\frac{d(\ln c^{+})}{dx} \tag{17}$$

$$J^{-}_{diff} = -c^{-}u^{-}RT\frac{d(\ln c^{-})}{dx} \tag{18}$$

A *p* permeability constant which characterizes the passive diffusion across a membrane can be derived from the above equations. Considering the

$$\frac{d(\ln c)}{dx} = \frac{1}{c}\frac{dc}{dx} \tag{19}$$

equation, furthermore a permanent concentration gradient, i.e., where Δc is the

$$\frac{dc}{dx} = \frac{\Delta c}{\Delta x} \tag{20}$$

concentration gradient at the two sides of the membrane and Δx is the thickness of the membrane, the flux can be written in a more simplified form:

$$J = -uRT\frac{\Delta c}{\Delta x} = -p\Delta c \tag{21}$$

The permeability constant of the equation has a meaning, namely, the flux that can be measured at unit concentration gradient.

Any electric potential gradient has an additional effect on charged particles which can be positive or negative, depending on the sign of the charge and its size is $z[(d\psi)/(dx)]$. Here z is the number of unit charge per particles and $(d\psi)/(dx)$ is the electric potential gradient along the x axis. The corresponding contribution to the flux can be calculated in an analogous way as above for concentration gradients. Since in those equations the concentrations are expressed in moles per liter, the electric potential has to be normalized to one mole charge. This for monovalent ions is $F \cong 96500$ C. Thus the ion fluxes can be expressed for positive ions:

$$J^{+} = -c^{+}u^{+}\left[RT\frac{d(\ln c^{+})}{dx} - zF\frac{d\psi}{dx}\right] \tag{22}$$

and

$$J^{-} = -c^{-}u^{-}\left[RT\frac{d(\ln c^{-})}{dx} - zF\frac{d\psi}{dx}\right] \tag{23}$$

for negative ions.

If two compartments are isolated by a membrane the sum of the two low fluxes must be zero; otherwise, charge accumulation may occur. Thus, $J^+ = J^-$. The two basic equations can be substituted and

$$zFd\psi = -\frac{u^{+} - u^{-}}{u^{+} + u^{-}}RTd(\ln c) \tag{24}$$

Integrating this differential equation:

$$zFd\psi = -\frac{u^{+} - u^{-}}{u^{+} + u^{-}}RT\ln\frac{c^{II}}{c^{I}} \tag{25}$$

where the c^I and c^{II} label the salt concentrations in the two separated compartments.

The above type of thermodynamic considerations help us to generate a connection between the characteristic parameters of ion diffusion and those of the diffusion potential which is measurable between the two sides of membranes. Of course, taking into account more than one type of permeating cation and anion, the mathematical analysis will become more complex. Omitting details after a more complicated, but basically similar series of algebra, the final outcome is given by the Goldman-Hodgkin-Katz equation:

$$\Delta\psi = \frac{RT}{zF}\ln\frac{p_{Na}[Na^{+}]^{II} + p_{K}[K^{+}]^{II} + p_{Cl}[Cl^{+}]^{I} + \ldots}{p_{Na}[Na^{+}]^{I} + p_{K}[K^{+}]^{I} + p_{Cl}[Cl^{+}]^{II} + \ldots} \tag{26}$$

It is clear from this equation that particular ions which have higher permeability and unequal distribution will contribute to a greater extent to the generation of a diffusion potential. In the presence of ionophores with specificity for particular ions, the permeability and such contribution can be very great (e.g., ionomycin and Ca^{2+} pair). Of course, nonpermeating ions will not contribute.

Let us consider the simple case when one adds NaCl solution to the left side of a membrane, and the same concentration is added from KCl solution to the right side. Even assuming that the membrane is permeable only for cations, the electroneutrality demands a stoichiometric ion exchange. If the K^+ permeability is greater, the left side will be more positive than the right side due to the microscopic charge separation. At the beginning of the diffusion process, according to the Goldman-Hodgkin-Katz equation:

$$\Delta\psi = \frac{RT}{F}\ln\frac{p_{K}}{p_{Na}} \tag{27}$$

The larger the difference between the two permeability constants, the greater is the potential difference. The ion distribution of K^+ ions keeps a better equilibrium with the diffusion potential than that of Na^+ ions. In the series of quasiequilibriums, singular states can be made stationary, if ion pumps are in the system which can stabilize nonequilibrium concentrations.

Let us consider such a system where we put Na^+Cl^- into the No. I compartment and K^+Cl^- into No. II (Figure 1). Beyond these initial conditions, it is assumed that K^+ and Cl^- ions diffuse rapidly and the membrane permeability for Na^+ ions is much lower. Since we have not included nonpermeating ions into the system, at equilibrium conditions there will not be a potential difference between the two compartments. Before reaching equilibrium, however, we can order a diffusion potential to every intermediate state according to the permeability constants and the actual ion concentration differences. Selecting out an intermediate state and initiating an "active" pump activity which transports, let us say, Na^+ ions from the No. II compartment to No. I so that it stabilizes the Na^+ concentrations of this selected state leads to a pseudoequilibrium. Under such conditions the distributions of K^+ and Cl^- ions will correspond to an apparent Donnan equilibrium, since in this system the Na^+ ions are apparently nonpermeating ions.

$$\frac{[K^+]^{II}}{[K^+]^{I}} = \frac{[Cl^-]^{I}}{[Cl^-]^{II}} = r \tag{28}$$

$$\Delta\psi = \frac{RT}{F}\ln r \tag{29}$$

Despite the apparent and real similarities, the real and pseudo-Donnan systems have several essentially different features. Here we will point out only the following two:

1. In pseudo-Donnan systems the unequal ion distribution and thus the membrane potential also depend on the active pumping. Terminating the pump activity by, for example, metabolic inhibitors, the Donnan-like distribution of the K^+ and Cl^- ions will also be terminated. In true Donnan systems this is not the case because their nonpermeating ions will not become permeating if pump activities are terminated.
2. Pseudo-Donnan systems are more dynamic. In such systems the ion distributions and the diffusion membrane potential are determined by the actual permeability and ion distributions. Since there is no real equilibrium in these systems, the instantaneous changes in permeability immediately result in changes in membrane potential without yet changing ion concentration.

In real Donnan systems the equilibrium is real. The corresponding equilibrium Donnan potential is independent of the permeability of the different ions.

Different pumping activities, transporting, co-transporting, or antiporting ions can be electronically silent or electrogenic. Electronically silent pumps may transport ions with different charges in the same direction or antiport ions with the same sign of charge to opposing directions. As examples, the Na^+ and H^+ or K^+ and H^+ antiports or Na^+Cl^- pump can be mentioned. The first two can be found in lymphocytes or in cells of the stomach wall. The second type is well known, e.g., in the gall-bladder. This silent pump might also influence membrane potential indirectly through changing the ion concentrations in the different compartments. Electrogenic pumps are characterized by net charge fluxes, in opposite direction with the existing potential difference.

This type of pump may transport a particular ion from one compartment to the other or may also transport different ions carrying the same charge in opposing directions in unequal quantities. Owing to the net charge flux such a pump can be considered as generating electric energies. It is obvious that a secondary effect through changing the original ion concentrations which were in equilibrium is also generated by this mechanism.

IV. BIOLOGICAL VARIATIONS ON THE ORIGIN OF PLASMA MEMBRANE POTENTIAL

Transmembrane potential and ionic currents may play a role in signal transduction and differentiation in a number of cells.[2,10-12,18-21,24] A distinct separation between the effects of ion channel activities and membrane potential changes is almost impossible, due to the strong interdependence of these two parameters, which are ubiquitous in living cells. It is a matter of argument whether the contribution of the electrogenic pumps to the membrane potential is significant or not. Most of the data denying this statement comes from the observation that the addition of ouabain (a strophanthin-K derivative, and an arrow poison of South American Indians that can disengage the neuromuscular connections which is also a specific inhibitor of the Na-K dependent ATPase, the Na^+-K^+ pump) to cellular systems does not significantly change their membrane potential. Thus, the contribution of the ion pump activities to the resting potentials are overlooked and few papers or even books claim that they have any significance at all in the generation of transmembrane potential differences. Two relatively recent papers picked up this question and tried to give it a fair treatment. Van Duijn and Vogelzang[28] investigated the potential values of the cellular slime mold *Dictyostelium discoideum*; Bashford and Pasternak[26] investigated the origin of plasma membrane potential of Lettre cells, a highly malignant mutant of the Erhlich ascites tumor cells. The possible counterargument, that these findings might be correct for unusual types of cells, is immediately overcome by the fact that human neutrophils, a numerically dominant type of human leukocytes, have a very similar mechanism of maintaining their plasma membrane potential.

The Na^+-K^+ pump is known to extrude three Na^+ ions for taking up two K^+ ions, simultaneously, at the expense of ATP dephosphorylation. The major argument against such a mechanism is that, had it been a prevalent way to extrude ions from the intracellular space, the steady state-notion of the monovalent ions would have been violated. Many calculations and textbooks claim that the Na^+-K^+ pump contributes no more than 3 to 5 mV or 3 to 4% to the membrane potential of muscle, but also in general to that of other cells. The fact is that Na^+ can also be exchanged for H^+ ions. Thus, cells are "never" in a steady state for H^+ and a steady pumping activity has to compensate for it. Of course, measurements of pumping activities are correct, and under resting conditions the contribution of the pump is not much, beyond the maintenance of the unequal ionic equilibrium. For specific cells with a highly tuned metabolism, this type of generation of plasma membrane potential could be rather commonplace. Lettre cells, as is well known, are prototypes of an aggressively growing cell line with very high metabolic turnover. Thus, it is no wonder that such mechanisms were found in that case. The fact that ouabain does not alter the resting membrane potential of muscle cells is quite natural. A similar finding is true for resting lymphocytes. However, upon stimulation, where the resting potentials are significantly influenced by the metabolism, the effect of Na-K and other pumps can be considered as major contributors to the membrane potential. The evidence provided by Bashford and Pasternak[26] claiming that such effects, i.e., the significant contribution of the pump effects, have to be considered at least in case of Lettre cells is as follows:

1. Evidence that Na^+-K^+ pump can directly set membrane potential
2. The existence of balancing anion leaks
3. The existence of an electroneutral mechanism for Na^+-K^+ and Na^+-H^+ exchange
4. The Lettre cells have the peculiar property that they do not have K^+ or other single cation channels

The membrane potential of these cells was measured by oxonol. Oxonols carry a net negative charge, and the negatively charged dye is excluded from the mitochondria; thus, the values obtained after calibration can be attributed to the plasma membrane potential.

Direct evidence for the contribution of the Na^+-K^+ pump to the plasma membrane potential comes from the experiment, where inhibitors such as ouabain cause rapid depolarization of cells in the absence of measurable changes in intracellular concentrations of Na^+ or K^+. This temperature-dependent and metabolism-driven ATPase activity ultimately maintains the unequal distribution of monovalent cations such as K^+ and Na^+ at the extra- and intracellular side of the plasma membrane.

Further evidence that the above experimental and theoretical analysis has a validity beyond the Lettre cells, which can be considered as an unusual and special kind of tumor cells, is that human neutrophils are completely depolarized by ouabain. Since ouabain extensively inhibits not only the Na^+ extrusion but also the K^+ intake, another alternative to inhibit the Na^+-K^+ pump is the removal of K^+ from the extracellular space. The Na^+-K^+ pump has an absolute requirement of extracellular K^+ for its electrogenic capacity; the depletion of K^+ from the extracellular space could be expected to depolarize the cells. Experimental evidence has shown that this was the case. Furthermore, when the K^+ was re-added to the cells, they were repolarized. An artificial alteration of the Na^+-K^+ pump by increasing the intracellular Na^+ concentration by a Na^+-specific ionophore, nigericin, activates the Na^+-K^+ pump, and, as expected, the cells are hyperpolarized.

Of course, the mechanism is complicated by the electroneutral exchange of Na^+ and K^+, and Na^+ and H^+ ions as well. However, the mechanisms of these "antiports", when one positively charged ion is exchanged for another positively charged one, are not satisfactorily studied yet, and much effort is required to clarify every detail.

Lettre cells have no K^+ channels. Thus, by increasing the extracellular K^+ concentration the potential is not changed. Only the addition of the specific K^+ ionophore, valinomycin, induces a change, depolarization of the membrane potential. In order to judge this fact properly we remind the reader here that valinomycin is a mobile carrier that shuttles K^+ across the plasma membrane along the electrochemical gradient. For "ordinary" cells with a dominant K^+ diffusion potential, the K^+ electrochemical gradient is near zero. That is the actual reason why these cells (like peripheral blood lymphocytes) are only slightly hyperpolarized by valinomycin, which is a highly K^+ ion-specific peptide ionophore.

V. MEASURING TECHNIQUES OF MEMBRANE POTENTIAL

A. A PROBLEM CHILD: THE LYMPHOCYTE

Unlike the conveniently large muscle and nerve cells or particularly thick axons, such as the giant squid axon with its "huge" 1-mm diameter, our targets, mostly lymphocytes, are relatively small-sized cells with an average diameter of 6 to 10 μm. On the other hand, the lymphocytes have a high number of subsets, i.e., a great individual variety, with different functional capabilities. Thus, classical electrophysiological methods usually fail to provide reliable data on ion channel activities and membrane potential, with the only exception of the patch clamp method and its variations, or the very carefully and critically applied measurements with fluorescent dyes, the latter also providing good grounds fot the study of large cell populations on an individual (cell-by-cell) basis.

Basically, three major groups of methodical approaches are available to investigate membrane potential and its variation in lymphocytes.

1. Electrophysiological techniques, like the whole cell current-clamp variation of the patch clamp, employing patch pipettes. For developing the patch clamp technique, around 1976, Erwin Neher and Bert Sakmann were awarded the Nobel laureate in physiology and medicine in 1991.[29-31] Their discoveries led to a better understanding of widespread diseases such as diabetes mellitus, cystic fibrosis, and a number of others, which may not occur that frequently but are lethal.
2. Isotope techniques, i.e., determination of the partition of organic or inorganic cations or anions based on their radioactivity. $^{42}K^+$, $^{86}Rb^+$, $^{24}Na^+$, and $^{45}Ca^{2+}$, $^{28}Mg^{2+}$, or the organic tritiated tetraphenyl phosphonium are frequently applied ions in studying ionic distributions between extracellular and intracellular compartments.[32]
3. Relatively recently, the application of positively (carbocyanine-based cations) and negatively (barbituric acid derivatives, i.e., oxonol-based anions) charged fluorescent dyes has been introduced to estimate membrane potential of either cells in suspension or on a cell-by-cell basis, using flow cytometry. The partition of the fluorescent molecules gives the basis for calculation of the membrane potential and its changes, after the necessary calibrations.[33-38]

B. CRITICAL EVALUATION OF THE MEASURING TECHNIQUES

1. Electrophysiology

The accuracy and resolving power of elegant electrophysiological techniques such as the patch clamp or whole cell clamp methods introduced by Neher and Sakmann[29] (for details, see References 39 and 40), enabling us to investigate the functional states of a single ion channel in plasma membranes, cannot be denied (see Figure 3 for some of the technical details, summarizing briefly the most important assets and major features of patch clamping). In the whole cell configuration membrane potential measurements can be made by clamping the pipette current to 0 (current clamp) with a suitable voltage applied between the pipette and a reference electrode in the bath. Of course, (near-) absolute values in potential changes can easily be studied by electrophysiological measurements.

However, the capability of making measurements on more than a few cells is restricted, owing to the intrinsic constraints of the technique. Larger cell populations or subpopulations can be investigated only with great difficulty, if at all. However, the biological variance among individual cells and the lack of obtained data extendable for certainty over a larger cell population, that can be statistically evaluated, present serious handicaps considering that subpopulations may show different quantitative or qualitative properties.

Recently, fluorescent antibodies have been applied to identify subsets of lymphocytes prior to patch clamping them for electrophysiological studies.[4] Although the authors correctly claim that the applied protocols for staining do not affect channel expression, an influence of the treatment on the finer functional responses of the studied channels cannot be excluded, especially not in those cases, where, in order to increase the visibility of the fluorescent labels, large phycoerythrin molecules are applied.

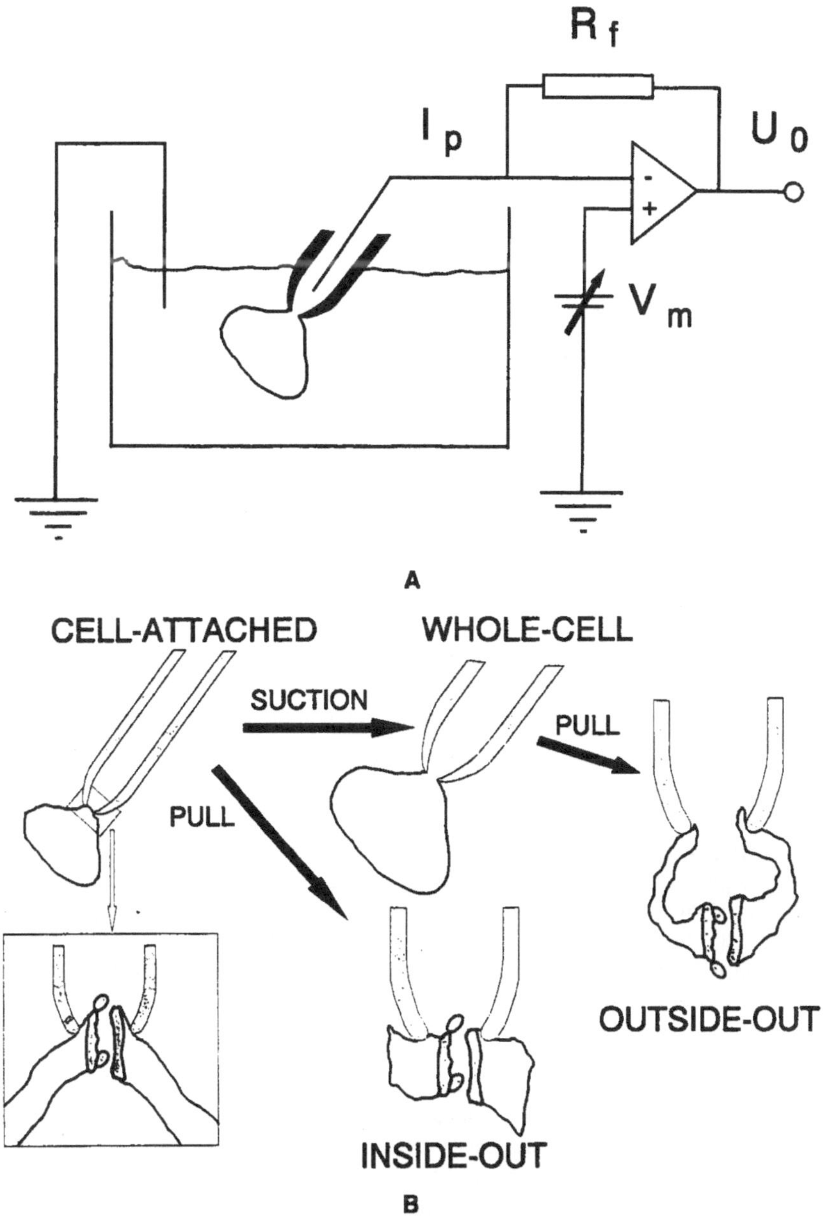

Figure 3. The principles of the patch clamp method. Panel A: The mechanical contact between the membrane and the pipette has high electrical resistance ("giga-seal", 1 to 100 GΩ). That giga-seal insulates the inside of the pipette from the buffer in which the pipette and the examined cells are immersed. In voltage clamp mode the membrane potential can be set to any given value (V_m) and the current passing through the patch pipette (I_p) is measured as the voltage drop on a feedback resistor (R_f): U_0-V_m. Panel B: The schematic drawings show the basic variations of the patch clamp technique.

Nevertheless, introducing fluorescent labeling for the identification of cells by subset-specific monoclonal antibodies prior to or after patch clamping is a further step forward in overcoming the obvious handicap of the patch clamp technique, concerning biological variability.

2. Isotopes

The isotope method is among the most reliable and sensitive ones for cell suspensions. However, these types of measurements carry the unavoidable limitations due to signals from debris, dead cells, and other inequalities originating from the ever-present heterogeneities among cell populations, impairing the accuracy of the measurements.[32]

Nowadays, this approach is less and less applied as other more convenient and faster methods are steadily improved. However, in some specific cases when increased sensitivity is mandatory, the isotopes are extremely useful.

3. Fluorescent Dyes

The fluorescent methods, although they are somewhat less sensitive than those applying isotopes, have the great advantage that cell suspensions can be studied using two fundamentally different methods of data collection and evaluation. Spectrofluorimetry provides data carrying the same inherent limitations of accuracy and resolving power at a single cell level as cation or anion partition measurements based on isotope labeling (Figure 4B). Flow cytometric membrane potential determinations with fluorescent dyes have the unique property that data collection is carried out on a cell-by-cell basis without losing the capability of studying large cell populations in a reasonably short time period (Figure 4A, C, and D). Artifacts originating from debris or nonviable cells can easily be eliminated and, what is more, the differences among subpopulations can be observed directly and immediately parallel to data collection. Collection of membrane potential data in correlation with other parameters is also feasible. Suitably selected second fluorescent dyes can monitor practically any parameter simultaneously with membrane potential on a real time basis, often in a specific and quantitative manner (e.g., enzyme activity, DNA content, subclasses of cells, accessibility of cell surface elements, etc.). Monitoring more than two parameters at the same time can also be dangerous, because unwanted interactions between the applied dyes (excimers, quenching, energy transfer) may occur, without an easy way to control them. On the other hand, the somewhat lower sensitivity of fluorescent dyes increases the danger that the interference of parallel fluorescent labeling with the native state of the cell can be a perturbing factor.

a. Flow Cytometric Evaluation of the Membrane Potential

Membrane potential measurements by flow cytometry have a further advantage over spectrofluorimetric ones since extracellular dye concentration can be neglected in most cases due to technical reasons, namely, the sheath fluid that surrounds the cells is so thin during the measurement that the contribution to the fluorescence signals comes almost exclusively from the cell-bound dye. Thus, even weakly bound ligands can also be studied, because the removal of the excess labels is not a prerequisite of the measurement. Last but not least, working with living cells, the high-power laser excitation of the dyes allows the application of an optimal low dye concentration, thereby minimizing toxicity. A major limitation in the application of fluorescent dyes lies mainly in the scaling problems of the measured fluorescence changes, i.e., the calibration of the obtained data on a millivolt scale. This is an inherent but not unsurmountable problem that has its origin in the indirectness of these methods.

It is interesting to compare the sensitivity of the above methods. Isotopes, as well as fluorescent dyes, can be monitored at as low a level as a few hundred molecules or less. The picoamper current measurements, where subpicoamper values are not easily determined, demand as many as 10^7 ions diffusing through a single channel per second, in order to be detected. However, there is no doubt about the fact that this is a relatively direct measurement, in contrast to the more indirect fluorescence ones. Radioactive cations could put in a claim for the directness of their determination as well, but their detection and the handling of radioactivity renders these measurements more clumsy.

We strongly believe that a combination of these experimental possibilities, such as a combined fluorescent and patch clamp technology, can overcome part of these problems arising with the application of a single technical approach. The sensitive CCD (charge coupled diode) cameras allow simultaneous determination of changes in intracellular fluorescence together with measuring ion channel activities by electrophysiological methods. Thus, new possibilities for combined applications are open even without inventing profoundly new methods.

C. MEASURING MEMBRANE POTENTIALS BY POTENTIOMETRIC FLUORESCENT DYES

The dyes vs. more direct electrophysiological measurements debate is as old as the introduction of these dyes for estimating changes in the membrane potential of cells where the size of the cell makes it difficult to determine the electric activity through inserted electrodes, unlike in muscle or nerve cells, as it was discussed recently by Pratap et al.[41] The patch clamp method solved the problem of applying electrodes to small-sized cells, but could not solve the question of biological variability, as we have seen above, since

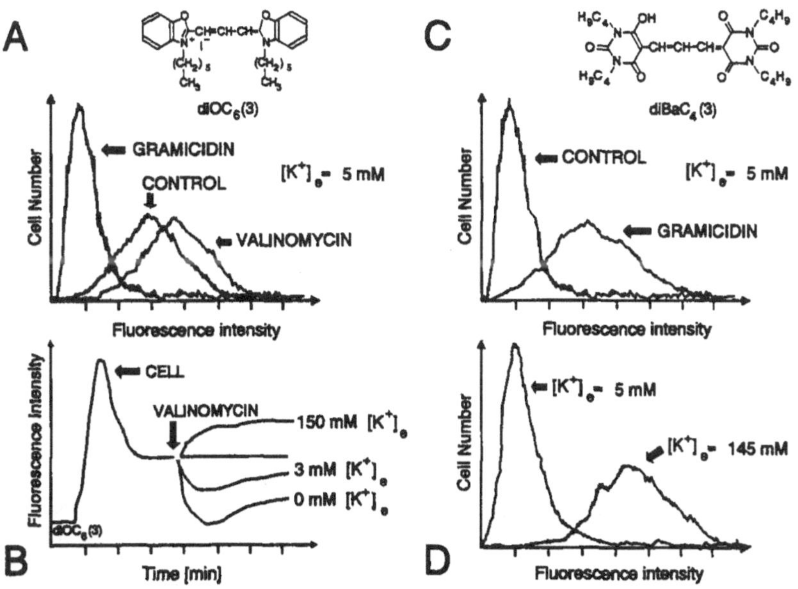

Figure 4. Fluorescent dyes monitor membrane potential changes. Panels A, C, and D show such histograms taken in a flow cytometer that provide data over a large cell population, yet taking measuring points on a cell-by-cell basis. Cells stained with carbocyanine based [diOC$_6$(3)] positively charged dyes (panel A) change their fluorescence intensity parallel with the membrane potential. Using oxonol-based [diBaC$_4$(3)] negatively charged dyes (panels C and D) cells change their fluorescence intensity anti-parallel with the membrane potential. Panel B shows membrane potential measurement of a cell suspension in a spectrofluorimeter. Effect of ionophores and extracellular ion concentration changes on membrane potential is indicated in the figure.

the number of cells which can be investigated per instrument per day is in the very low teens. Of course, the directness of the measurement and its resolving power at an electronically characterizable single channel level cannot be simulated by dyes. Nevertheless, here we will concentrate on the evaluation of potentiometric dyes.

Potentiometric dyes can signal changes in the membrane potential at very different time scales. There are fast and slow dyes, which can serve as potential indicators on the microsecond and millisecond time scales, respectively. It is obvious that this basic property determines boundaries for their application. However, the system vs. indicator interaction problem arises as well, since the same dye may have different response times in different systems. A family of cationic dyes, the carbocyanine compounds (e.g., diOC$_6$3, Figure 4A and B) may respond slowly when the equilibrium of diffusion potentials is measured in erythrocytes, or can change optical properties orders of magnitude faster when responding to millisecond long voltage pulses in planar lipid bilayers.

Recently Pratap et al.[41] revisited this question and gave a thorough analysis of the optical potentiometric indicators which can monitor the transmembrane electrical potential. The response of the dye to changes in the membrane potential (ΔE_m) is defined as

$$R = 100\frac{F^{+a} - F^{-a}}{F^{-a}} \tag{30}$$

where R is the response in the optical (fluorescence) parameter and the $F^{\pm a}$ signs stand for the fluorescence of the cell suspension before and after adding ionophore. The authors compared the type of response of 20 impermeant oxonols in human red blood cells. The majority of the oxonols interacted with the potassium ionophore valinomycin but did not with gramicidin. The negatively charged oxonols showed a decrease of fluorescence with hyperpolarization (the negatively charged dye molecules are expelled from the more negative cell). However, five oxonol type dyes increased their fluorescence with increasingly negative membrane potentials. A comparison of two dyes from the two groups with different responses showed that both dyes displayed potential dependent decreases in their binding capacity. The authors suggested a two-state model with two different quantum efficiencies within the membrane. If the membrane has two localization sites for both dyes and one of the dyes (the one which lowered its fluorescence response upon hyperpolarization) has the same quantum efficiency for both membrane components while the other one (which increases fluorescence upon hyperpolarization) has increased quantum yield when it is in the outer layer of the membrane, the movement of the two dyes may be in the same direction with opposite fluorescence consequences. The authors generally agree that the membrane potential is satisfactorily described by the Goldman equation.[41] Dyes which stimulate chloride conductance will underestimate E_m, which is basically a K^+ equilibrium potential at the given extracellular K^+ concentration.

The majority of the oxonol dyes report ΔE_m in erythrocytes by an on-off mechanism. The few instances where this is not the case always have some specific interaction of the particular system measured and the dye. For example, see the different mechanisms for two equally oxonol based dyes, where one responds with an increase, the other with a decrease of fluorescence upon hyperpolarizing the cells with an ionophore. The one (coded as RGA. 451) which showed a rapid increase in fluorescence with minimal effect on the chloride conductance may be a good indicator of quick changes of ΔE_m in erythrocytes. The on-off process, i.e., the dye translocation between membrane and the surrounding medium (or between two sites within the membrane, where in each site the dye is sensing its surrounding and responds with different quantum yield) could be the most likely mechanism of the response of impermeant dyes. Here the question arises why the authors did not apply flow cytometric membrane potential measurement, where among well-controlled conditions the outer compartment, i.e., the surrounding medium, is practically negligible as a contributing factor. It comes from the simple technical advantage that in the flow cytometer only the illuminated cell will contribute to the measured fluorescence. The "surrounding" is so thin, and the concentration is so negligible in it that the fluorescence contribution, with the exception of extremely high dye concentrations, is within the noise of measurement.

Currently, DiBAC4(3) [bis-(1,3-dibutylbarbituric acid) trimethine oxonol] or briefly oxonol or bis-oxonol is the most frequently applied fluorescent dye, favored for its advantageous properties (Figure 4C and D). It is negatively charged and therefore stays outside the normally polarized cells. This behavior makes it much less toxic than many other dyes. The spectral properties, namely, the excitability and emission characteristics which are very near to those of fluorescein, provide a comfortable range for measurements in any sort of optical device. The high sensitivity, even at 50- to 150-n*M* concentration helps to detect the signals, which monitor potential changes. Naturally, this DiBAC4(3) is also partitioned not only by the electromotive forces, but solubility and concentrations as well. Its incompatibility with such an important K^+ ionophore as valinomycin makes its calibration a little bit inconvenient, but Wilson and Chused gracefully circumvented this problem by applying A-23187, the known Ca^{2+} ionophore, as a major tool for calibration.[42,43] A careful and correct application of oxonol, especially in combination with ion-selective pharmacological agents which have been tested for eventual chemical incompatibility with the dye, gives a relatively easy and quick orientational test for detecting membrane potential changes in lymphocytes or other small cells, where the application of patch clamp alone may not give the desired information on the cell population.

1. A Simple Way to Calibrate Membrane Potentials in Measurements with Fluorescent Potential Indicators

Fluorescent potential indicators can be calibrated to monitor not only a relative plasma membrane potential, but also its real values in a flow cytometer, which offers the advantage of measuring this parameter on a cell-by-cell basis. Recently Trón et al. (personal communication) developed an interesting method which seems to be more direct than the measurements described above.[42,43]

One basic, but definitely affordable, assumption is that bis-oxonol, the negatively charged fluorescent potential indicator, distributes across the cytoplasmic membrane according to the Nernst equation. This premise doubtlessly neglects solubility factors or thermodynamic activities and relies principally on the electrochemical gradient as a major factor, governing the distribution of the dye. However, these factors can be included as corrections or they could really be neglected, depending on the measured cellular entities.

The train of thought is as follows: the measured fluorescence intensity reports on the plasma membrane potential, when bis-oxonol is applied. Although the detected intensity may also depend upon a number of other physical-chemical and spectroscopic parameters (such as aggregation of the dye molecules, binding to specific or nonspecific binding sites, changes in the fluorescence spectral, or quantum efficiency parameters upon binding or aggregation, etc.), it is basically a singular valued function of the free dye concentration in a given cell. Using a monovalent indicator dye, as the oxonol, which follows a Nernst-type distribution, the membrane potential can be readily determined from the Nernst equation (e.g., see Equation 3). This is actually determined by the ratio of the extracellular and intracellular ion concentrations of the indicator dye. A problem arises that the determination of the intra- and extracellular dye concentrations may not be as straightforward as it is desirable.

The negatively charged oxonols have the advantage over the positively charged cyanine dyes that they do not accumulate inside normally polarized cells. Thus, there is no significant decrease of the dye concentration in the suspending medium attributable to cellular uptake. However, the measured fluorescence may not reflect proportional distribution of the dye molecules between the extra- and intracellular spaces, because lipid- or protein-associated dye molecules can change their quantum efficiency. This can be convincingly demonstrated by adding dye at different concentrations to a solution of a particular protein, e.g., bovine serum albumin. Besides the remarkably enhanced fluorescence, there is a slight shift of the emission maximum as well.

Thus, the determination of the intracellular free dye concentration seems to be irrevocably compromised.

Trón et al. (personal communication) overcame this problem applying a few reasonable assumptions, rearranging the Nernst equation and using a calibration curve that can be determined for each particular measurement without difficulties.

The applicability of their method requires the fulfillment of the following assumptions:

1. The oxonol distributes between the extra- and intracellular compartments according to the Nernst equation.
2. The total amount of the intracellular dye is negligible compared to the extracellular throughout the whole experiment.
3. Fluorescence intensity of the dye, measured from a given cell, is solely the function of its intracellular concentration in such a way that changes in the extracellular dye concentration will not alter the cellular fluorescence until the cellular concentration is not altered.
4. A sample of cells with zero transmembrane potential is available for calibration purposes.
5. The cells will not be subject to volume changes during the experiment in any of the investigated samples.

Experimentally, one has to measure the mean fluorescence intensities of normopolarized cells at different dye concentrations in a flow cytometer. The calculated curve fits the experimental data and represents the relationship of the measured signal to the extracellular dye concentration. This value can be converted to intracellular free dye concentration using a rearranged form of the Nernst equation:

$$\frac{[ox]_{ic}}{[ox]_{out}} = K = e^{\frac{F}{RT}\Delta\Psi} \tag{31}$$

where $[ox]_{ic}$ and $[ox]_{out}$ are the free oxonol concentrations in the intracellular and the extracellular space, respectively, and $\Delta\Psi$ is the transmembrane potential.

The exponential term is constant (K) for all samples having identical membrane potential. It is easy to see that the plot mentioned above which provides us with fluorescence values vs. extracellular dye concentration, can also be converted into fluorescence vs. intracellular dye concentration if we express the ratio of those two concentrations from the Nernst equation. Thus, extracellular dye concentration dependence of the measured fluorescence intensities can be converted into cytoplasmic concentration dependence of the indicator dye, making possible the calculation of intracellular dye concentrations from fluorescence readings.

Thereafter, calibrating the curve for fully polarized cells and for those with zero potential, any potential value can be obtained upon comparing a measured cell sample of unknown potential with the plot constructed according to the above very briefly described way. In most works on biological samples, the fluorescent dyes were used to monitor mainly the direction of the changes in membrane potential, and there was only an indirect possibility to calculate membrane potential; the above method, despite its numerous assumptions, may offer us a new perspective.

Comparative and simultaneous calibrations with individual membrane potential measurements will justify or question the reasoning of the above discussion. However, if the data obtainable with the whole cell variation of the patch clamp method in current clamp mode support this interesting possibility, it will be a large step ahead, mainly because flow cytometry permits the study of large cell populations. Of course, the obvious handicap of any dye-based membrane potential measurement that individual channel activities cannot be directly identified and analyzed could not be circumvented yet.

D. VARIATIONS ON THE DETERMINATION OF MEMBRANE POTENTIAL VALUES BY FLUORESCENCE TECHNIQUES

Plasma membrane potential measurements can be based on structural changes in the plasma membrane.[44]

Such structural changes, after suitable calibrations, can be used to determine membrane potential changes. Of course, the possibilities presented here are somewhat indirect approaches, but can be used as versatile methods to measure membrane potential. However, since very little information is available on the membrane potential-induced structural changes in the plasma membrane of living cells, these "clumsy" approaches can provide valuable additional information. Dunnant,[45] Muller,[46] and Balázs and co-workers[44] provided ample experimental evidence that a new dimension and a far deeper insight can be gained into such structure-function correlations by applying sophisticated methodology, which may provide an interface between biophysical and biochemical events of membrane-bound processes.

Fluorescence spectroscopy offers a very broad range of possibilities, but here we only deal with two particular ways of membrane potential measurements.

1. Plasma Membrane Potential can be Determined by Fluorescence Anisotropy and FRET Measurements

The first method — the anisotropy measurement — provides a tool to obtain information on the degree of rotational mobility of fluorescent probes;[47] the second gives us a "magic ruler" to determine proximity relationship of neighboring molecules at an Angstrom-level sensitivity.[48] This method — the Förster-type singlet-singlet resonance energy transfer between suitably selected fluorescence donor acceptor pairs — carries also the great possibility to circumvent the wavelength limitation of the resolving power of visible light by the aid of fluorescence spectroscopy, as it was discussed in detail in preceding sections.

The potentialities of anisotropy measurements become obvious if we consider the changes in the orientation of dipole molecules in changing electric potential fields.[49] Induced and transient dipoles can be generated from a high number of cell membrane-bound proteins, which change their conformation in the potential field across the membrane that is not less than 10^5 V/cm.[27] This high-strength potential field generates the phenomenon of voltage gating (opening or closing) of ion channels, which is one of the major regulatory mechanisms that controls transmembrane ion traffic and a high number of intracellular biochemical processes.

The technique, theoretically, is very simple. A suitably selected probe in the lipid region, or attached to particular proteins which are supposed to undergo large conformational changes upon changing the transmembrane potential field, can be used to monitor the plasma membrane potential (Figure 5A). Of course, the experimental difficulties could be very different, depending on the selected probes, their fluorescence lifetime, and their limiting anisotropy. It is likely that phosphorescence probes can give a better resolution due to their microsecond-millisecond lifetime range that correlates better with the rotational mobility of proteins in plasma membranes.

Fluorescence resonance energy transfer can report on changes in the "thickness" of the membrane or changes in the distance between particular points of proteins oriented perpendicular to the plasma membrane. The thickness of the membrane is likely to undergo changes in different electric potential fields. Furthermore, it is just within the scale that is excellently suited for determining changes of distance with very high sensitivity (Figure 5B and C). The combination of such membrane thickness measurements with controlled alterations of the membrane potential may provide an interesting new tool for studying structural changes induced by the changing potential field. Experiments in our laboratories have shown that conformation of the HLA class I antigen followed the changes of membrane potential in a reversible fashion (Bene et al., personal communication).

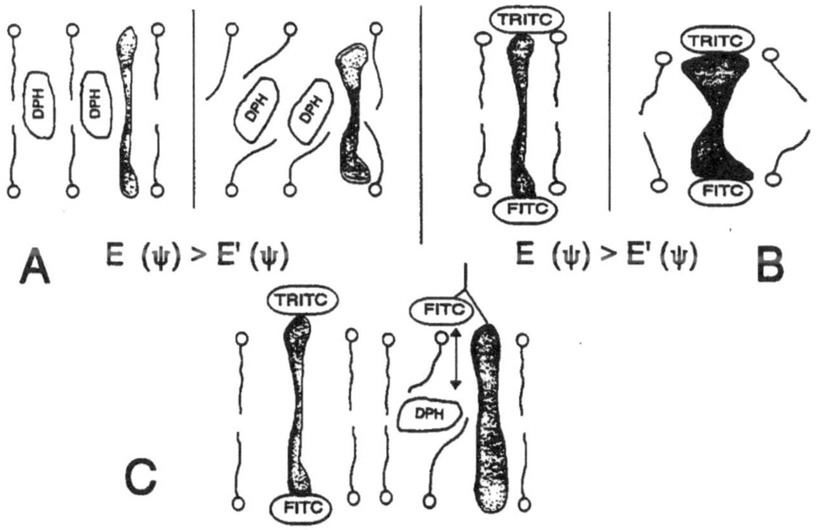

Figure 5. Effect of membrane potential on the structural elements and thickness of plasma membranes. Anisotropy changes of diphenyl hexatriene (DPH) in the lipid region (A), energy transfer measurements between suitably selected fluorescence donor-acceptor pairs across the membrane (B), or lipid probes and protein probes may monitor membrane potential changes through measuring structural alterations (C).

Both anisotropy and energy transfer measurements have been discussed in Chapters 1, 2, and 4 of this book; thus, here we restrict ourselves to presenting only one interesting possibility that combines these two types of measurements in a simple way.

2. Altered Anisotropy of Fluorescent Donor-Acceptor Pairs Monitor Proximity Relationships between the Labeled Cell Surface Elements

Steady-state anisotropy, assuming rotation of spheres under isotropic conditions, is given by the Perrin equation that describes the relationship between fluorescence depolarization of a spherical rotor and the hydrodynamic properties of the surrounding medium:

$$\frac{1}{r} = \frac{1}{r_0}\left(1 + \frac{kT}{V_h \eta}\tau\right) \tag{32}$$

Here r_0, i.e., the limiting anisotropy, is defined as

$$r_0 = \frac{3\cos^2\xi - 1}{5} \tag{33}$$

In Equation 32 r is the actual anisotropy of the fluorescence emitted by the probe, r_0 is the limiting anisotropy, corresponding to the motionless state of the fluorophore according to Equation 33, τ is the fluorescence life time, k is the Boltzmann constant, T is the absolute temperature, while V_h and η are the hydrated molecular volume of the rotating fluorophore and the viscosity of the embedding medium, respectively. The theoretical limit of anisotropy values, r_0, is given by Equation 33, where ξ is the angle between the absorption and emission dipoles of the fluorophore.

Under biological conditions one cannot assume rotation of spheres in an isotropic environment. However, the actually determined apparent steady-state (or time resolved) anisotropy values may carry a large amount of valuable biological information. In a number of cases, not only the sometimes determinable absolute values of particular parameters are of biological interest and importance, but their much more easily determined changes, which can inform us about active or passive participation of certain molecular or other structural elements in given biological functions.

Fluorescence resonance energy transfer efficiency data permit the monitoring of molecular distances between 2 and 10 nm, and their dynamic changes, by optical methods using the visible range of the spectrum in a spectrofluorimeter, fluorescence microscope, image-analyzer, or flow cytometer. In the latter, measurements can be carried out on a cell-by-cell basis without losing the possibility to study large cell populations and combining several parameters, thereby identifying subpopulation with particular properties.

Here we will show that fluorescence anisotropy data can be converted to energy transfer efficiency. The method itself does not offer a great advantage as compared to, for exampe, spectrofluorimetric or any other more direct methods for determining the same parameter. However, beyond that it does offer a technical variation, and such simplifications can be introduced by the aid of it into, for example, flow cytometric energy transfer measurements, or viscosity and temperature-normalized energy transfer measurements,[50-53] which can monitor molecular fluctuations, or membrane potential measurements that it is worthy to describe a concise version of the calculation.

3. Calculation of Energy Transfer Efficiency from Anisotropy Data

Energy transfer efficiency, E, and the fluorescence lifetime of the excited donor in the presence (τ_D^A), and absence (τ_D) of the acceptor dye have the following simple relationship:

$$1 - E_D = \frac{F_D^A}{F_D} = \frac{Q_D^A}{Q_D} = \frac{\tau_D^A}{\tau_D} \tag{34}$$

For determining the same parameter at the acceptors side the expression is somewhat more complicated:

$$\frac{F_A^D}{F_A} = \frac{\tau_A^D}{\tau_A} = 1 + \left(\frac{\varepsilon_D C_D}{\varepsilon_A C_A}\right) E_A \tag{35}$$

where F_A is the fluorescence of the acceptor in the absence and F_A^D in the presence of the donor and τ_A is the lifetime of the acceptor in the absence and τ_A^D in the presence of the donor.

C_A and ε_A and C_D and ε_D are the concentrations and the absorption coefficients of the acceptor and the donor, respectively. The absorption coefficients are considered at the wavelength of the actual excitation.[54]

As we have seen in Equation 32, the Perrin equation also contains the fluorescence lifetime of the rotating fluorophore. When energy transfer takes place, the lifetime of the donor becomes shorter and that of the acceptor may be lengthened. This latter occurs only if the transferred energy reappears as fluorescence of the acceptor molecule, which is not a prerequisite for the energy transfer itself. Determining the limiting anisotropy of the donor and acceptor dyes in separate experiments and measuring anisotropy values of a fluorescence donor molecule (or those of an acceptor) in the presence (r_D^A) and absence (r_D) of acceptor (or donor) molecules, the relationship of the altered lifetimes shown in Equation 35 can easily be obtained by dividing the two separately applied Perrin equations with each other. One equation will describe the anisotropy in the presence of one of the fluorescent dyes; the other in the presence of both of them. After some elementary algebra, it is possible to show that the energy transfer efficiency between the donor and acceptor dyes can be calculated according to Equation 36, if we measure the donor anisotropy. A similar equation is valid if the acceptor anisotropy offers better resolution power or other technical advantages (Equation 37).

$$E_D = 1 - \frac{\tau_D^A}{\tau_D} = 1 - \frac{r_D\left(r_o - r_D^A\right)}{r_D^A\left(r_o - r_D\right)} \tag{36}$$

$$E_A = \left(\frac{r_A\left(r_o - r_A^D\right)}{r_A^D\left(r_o - r_A\right)} - 1\right)\frac{\varepsilon_A C_A}{\varepsilon_D C_D} \tag{37}$$

Thus, a few anisotropy measurements and some elementary calculations can provide us with data, which can be determined in far more direct ways, but those methods require different conditions and individually carry less information. The potentialities of the described combinations and the validity of the simplified theory presented here have to be proven yet by comparing data obtained separately with FRET determined by the classic methods, and FRET values calculated from anisotropy measurements.

VI. CATION CHANNELS

A. VOLTAGE-GATED K⁺ CHANNELS

1. General Properties

The diversity of potassium channels is enormous. "There are more species of potassium selective ion channels than those of any other types of ion channels, with their own voltage dependence, pharmacology, and regulation by intracellular and extracellular factors".[55,56]

Potassium channels occur in practically every eukaryotic cell type and their activity is intimately associated with major cellular functions of the particular cell type in a high number of cases. Thus, it is no surprise that, in 1984, applying patch clamp techniques, the existence of K^+ channels was also discovered in human lymphocytes.[7,57] Circumstantial, i.e., nonelectrophysiological, evidence had been obtained much earlier both with isotope and fluorescence techniques. However, the characterization of the channels demanded the helping hand of the more direct electrophysiological measurements.[10-12] The type of K^+ channel that was identified by patch clamp in lymphocytes belonged to the voltage-gated delayed rectifier type K^+ channels. These channels were discovered much earlier, and were studied extensively, in muscle and nerve cells. Matteson and Deutsch, and simultaneously DeCoursey and co-workers were the first to identify these K^+ channels in T lymphocytes.[7,57] These channels were readily activated by depolarization of the cells. In whole cell configuration the cells were voltage clamped at –80 mV first, and outward directed positive currents were evoked by pulses to more positive potentials than –50 to –40 mV (Figure 6). After a fast rising phase (i.e., activation) the current reaches a maximum value and it declines due to inactivation of the channels. Applying repeated test pulses with short intervals depolarizing to a certain potential, the inactivation accumulates; the maximal amplitude of the current declines.

The K^+ channels of lymphocytes are sensitive to the well-known channel blockers, including tetraethylammonium (TEA), 4-aminopyridine (4-AP), charybdotoxin (CTX), etc. The potency of a channel antagonist can be characterized by the concentration of the drug required for half blocking the K^+ conductance at a given potential.

Based upon the different activation characteristics, the extent of inactivation during sustained (200-ms) depolarization, and the presence of cumulative inactivation and pharmacological sensitivity, at least three

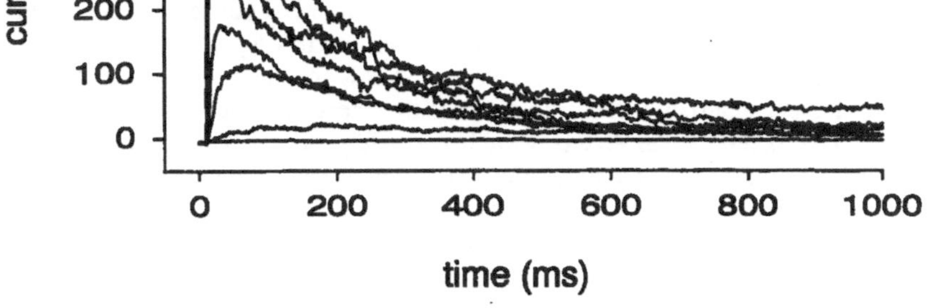

Figure 6. Time course of outward-directed potassium currents in human lymphocytes. After achieving the whole cell configuration the cell was held at –80 mV. K^+ currents were evoked by 1-s long depolarizing pulses to different potentials from –50 to +60 mV. The delay between successive pulses was 40 s. Current records were filtered at 2 kHz and stored in a computer for off-line analysis.

types of these voltage-gated potassium channels can be identified: the *n*, *n'*, and *l* types. The classification follows the terminology that has been introduced for muscle and nerve cells.

Murine immature (CD4 and CD8 positive) cells show the characteristic features of *n*-type potassium channels. The current can be evoked by depolarization to more positive potentials than -40 mV. The outward current inactivates rapidly; it decreases to more than the half of the starting value in 150 ms. The *n*-type channels are cumulatively inactivated during repetitive depolarization, are half-blocked by 10 m*M* TEA, and are sensitive to nanomolar concentrations of CTX. CD4 positive and CD8 negative mature helper-phenotype T cells display one order of magnitude fewer *n*-type channels which is easily detected by the much lower amplitude of the K^+ current at test potentials where current is evoked.

Cytotoxic and suppressor ($CD4^-$ and $CD8^+$) murine lymphocytes show *n'* and *l*-type potassium channels. The *n'*-type channels have the same voltage dependence of activation and CTX sensitivity as the *n*-type channels, but display little or no cumulative inactivation, and little inactivation during a 200-ms long depolarizing pulse. These channels are half-blocked by very high (100 m*M*) concentration of TEA. The *l*-type channels are insensitive to CTX, very sensitive to TEA (0.1 m*M* half-blocking concentration), display no or little cumulative inactivation, and activate at potentials above –10 mV. For recent detailed reviews those of Gardner[58] and Gallin[59] are easily readable by beginners and more advanced researchers as well.

2. Effects of Channel Blockers and other Agents

The effect of blockers of voltage- or Ca^{2+}-activated K^+ channels were tested on cell activation processes, practically simultaneously with the discovery of the existence of these channels in lymphocytes. The (transient) hyperpolarizing effect of mitogens on T cells is easily interpretable through a diffusion potential, induced by an increase of K^+ efflux through channels. Lectins, like phytohemagglutinin (PHA), increased the number of channels open at a given time. This increase, compared to the unstimulated cells, could be as much as tenfold.[60-63] The picture was complicated by the fact that these experiments could not be reproduced by others.[64,65] Instead of a quick increase in channel activities, lectins, phorbol esters, and interleukin-2 could only result in a slow increase in the total K^+ current after many hours of stimulation. This pointed toward an increase in the number of K^+ channels instead of an increase in their activity. Applying channel blockers to activated cells, an inhibition of the major activation steps was reported.[7,61,62,66,67] However, the dose response relationship was different according to various authors.

Recently, a new aspect of this question was investigated by us (Pieri et al., personal communication). Besides the widely applied methods to follow DNA synthesis, some biophysical measurements were combined, wherein more complex information could be attained. The structural changes in the cell membrane were monitored by the kinetics of merocyanine 560 dye uptake. Mitochondrial activation was assessed through implementing rhodamine-123 fluorescence studies. Rhodamine-123 is a fluorescent dye, with advantageous fluorescence and accumulation properties. The latter is specific for mitochondrial membrane potential measurements. This parameter is thought to signal mitochondrial activation processes as well. Mitochondrial mass was determined by nonyl-acridine orange (NAO). The c-myc and c-myb protooncogenes, expressed at different stages of cell activation, and therefore useful as activation markers, were followed by fluoresceinated monoclonal antibodies directed against the respective proteins. Finally, DNA synthesis was followed by ^{3}H-thymidine incorporation and flow cytometric cell cycle analysis. The applied channel blockers were individually added to separate samples to probe their effects on the above-outlined set of parameters. The scale of blockers covered 4-aminopyridine and tetramethylammonium bromide, quinine, verapamil, and an interesting quaternary ammonium compound, bretylium tosylate (BT). The latter is known as a weak K^+ channel blocker and also Na^+ channel opener. Owing to this characteristics, the drug also has the capacity to activate the Na^+/K^+-dependent ATP-ase.[2,18-21,24,68] The extent of the effects varied, depending on the kind of blockers. One thing, however, was clear: not all the measured parameters were equally sensitive to the blockers and their dose dependence was also different. Mitochondrial activation and merocyanine fluorescence of the plasma membrane changed parallel with ^{3}H-thymidine incorporation, giving more complete information about the effect of channel blockers on cell activation processes. However, although in the case of verapamil, a Ca^{2+} channel blocker, the relationship between this and transmembrane signaling triggered by lectins or allo-antigens is clear; the same cannot be stated for K^+ channels.

When the question is posed as to whether a particular mechanism of the cell activation is hampered, the actual type of cell activation pathway must always be identified. We return to this problem in another section of this chapter.

3. Genetic Properties of K^+ Channels

Very recently, mouse K^+ channel genes were isolated. The electrophysiologically identified, *n*, *n*′, and *l* types are well characterized and distinguishable. The *n*-type channels play a role in cell proliferation and activation of mouse and human T lymphocytes. The isolated and closely related mouse K^+ channel genes are the MK1, MK2, and MK3.[69] These seem to belong to the Shaker-related K^+ channel genes. The elegant study of Grissmer et al.[4,5] provided evidence that type *n* K^+ channels are encoded by the MK3 gene.

The mRNA of MK3 was injected to *Xenopus* oocytes, which subsequently expressed the channel and thus provided a versatile model system to be analyzed.[5] These K^+ channels showed the voltage gating, conductance, and pharmacological properties which were similar to those characterizing the voltage-gated *n*-type channels of human and mouse lymphocytes.

The expression of MK3 gene, which contains an uninterrupted exon, was elegantly demonstrated in the EL4 mouse T-cell line by the aid of polymerase chain reaction. The application of probes prepared from the unique 5′ noncoding region of MK3 and MK2 genes was able to show the MK3 and MK2 region on human chromosome 13, and 12, respectively.

The expression of *n* channels varies with different stages of development of T lymphocytes mostly parallel to the rate of cell division. About 200 *n*-type K^+ channels can be found on immature thymocytes. In quiescent T cells this number decreases to one tenth. The activation of mature T cells increases this number to 200 to 400, and these channels keep displaying the electrophysiological and pharmacological properties of type *n* channels.

Despite these detailed and important contributions to our knowledge about K^+ channels in lymphocytes, the mechanism of their participation in cell activation has yet to be answered. Again, it has to be repeated that the degree of the participation of ion channels in different types of activation processes must be highly dependent on the type of channel and the putative role of the particular ion in the signal transducing mechanism. If the channel activities are electrogenic, the longer the duration and the higher the complexity of membrane bound processes, the greater the possibility that they can become principal targets for interference with the signal transducing mechanism.

In order to introduce the reader into a more elaborate and systematic pharmacological treatment of voltage-sensitive channels, one has to refer to the fundamental work of William A. Catterall and co-workers in Seattle.[70] A few excerpts will be taken from his work and discussed in light of some newer data as well.

4. Metabolic Regulation of K^+ Channels

Potassium channels, as we have seen, are ubiquitous and regulate directly or supposedly indirectly a high number of cellular processes, yet the pharmacology of K^+ channels is practically nonexistent, as compared to that of the much less ubiquitous and numerous sodium and calcium channels. This statement is especially true concerning the therapeutical consequences of such modulators, which can interfere with K^+ channel activities. Perhaps one reason for this is the fact that K^+ channels participate in such a number and variety of regulatory processes that it is difficult to find such a specific agent that influences only one of these vastly diverse regulatory effects.

The recently discovered class of K^+ channels that is inhibited by ATP, i.e., their activity is regulated by the intracellular metabolism as reflected by the actual ATP level, promises a broad field of application for new types of ATP-sensitive K^+ channel- (K^+_{ATP} channel) directed drugs. There are activators of the K_{ATP} channels, such as cromakalin (BRL 34915), that can hardly activate the K_{ATP} channels which govern insulin secretion in pancreatic cells, but can activate similar channels in vascular smooth muscle. Thus, cromakalin and perhaps derivatives may have differential effects and may pave the way for new powerful vasodilators. The involvement of K^+ channels, sensitive to ATP, in insulin secretion may forecast a number of serious consequences. Briefly, the regulation of insulin synthesis through ATP-sensitive K^+ ion channels is depicted as follows. The β cells of the Langerhans islands in the pancreas contain these K^+ channels in their membrane and the activity of these channels contributes considerably to the resting membrane potential. The channel suspends its open state either upon addition of carbohydrate secretagoges or intracellularly applied ATP. These processes will lead to depolarization of the cells. However, these signal processes are far from being that simple. An obligatory increase in the $[Ca^{2+}]_{ic}$, that seems to be totally independent of extracellular Ca^{2+} concentration, is also an important factor.

Besides the key role of the K_{ATP} channel, Ca^{2+}, and voltage-activated K^+ channels, nonselective cation channels and delayed rectifying K^+ channels are also prime elements in this regulatory step that influences and controls the glucose metabolism of the whole body. Thus, Miller's suggestion that "glucose-regulated K^+ channels are sweet news for neurobiologists" can be taken at face value.[71]

Due to the potential importance of ATP-sensitive K^+ channels in the general metabolic regulation of the cellular processes, and the perspective in its pharmacological modulation, we will give a concise summary of data available on their biophysical and pharmacological properties.

Basically two different types of ATP-sensitive K^+ channels have been identified in insulin-secreting cells. The real significance of their difference is in their unit conductance, i.e., there is a more (large) and a less efficient one. Most of the investigators restrict themselves to discussing properties of the large K^+_{ATP} channels. The very low frequency of channel opening, which is increased by excising the patch from the cell or adding detergents, i.e., by conditions under which the influence of the intracellular ATP concentration is likely to be overcome, suggests that normally these channels are in a nearly blocked, tonically inhibited, state.

K_{ATP} channel density is not high, being in the range of 5 to 600 in the plasma membrane of one cell. The total number of the channels per cell is about 2400 or even 10,000 according to other calculations. A possible cause of this variance is that either they are inactivated in some patches or their density is very uneven. They are characterized by high selectivity to K^+; for Na^+ ions they are about 150-fold less permeable. Even their Rb^+ conductance, a frequent experimental substitute for K^+, is a fraction of that of K^+, and thus Rb^+ can act as a permeable inhibitor of the channel. The K_{ATP} channels have strong current-rectifying properties. The channels allow K^+ to enter cells and are restricting their efflux. The influx or efflux depend on the ATP concentration.

The intracellular control mechanism depends mainly on the pH and the ATP level. Membrane potential changes and alterations in the intracellular Ca^{2+} ion concentration are relatively ineffective. The pH exerts its effect on the opening frequency at higher pH regions (~7.9) and especially if low concentrations of intracellular ATP are present. It is assumed that the decrease of ATP^{4-} concentration is the real activator of the channel in these experiments. However, the main action of the intracellular ATP is to evoke channel closure. The half-maximal inhibition is around a few 10 μM. Since channel activity is gradually lost whenever the patch is separated from the membrane (this can occur within a fraction of a second or after a few minutes), it was suggested that the deactivation of the channels is connected to cytoskeletal elements. This so-called run down of the channels may also be connected to Ca^{2+}-dependent dephosphorylation processes. In the case of patches, tenths of micromolar concentrations of ATP are enough to seal the channels, while retaining the structural connection demands a higher (0.5 to 1 m*M*) concentration of ATP for inhibition. Absence of ATP and Mg^{2+} activates channel openings and conductance, while in the presence of ATP the absence of Mg^{2+} enhances the effect of ATP. Some models connecting K_{ATP} channel regulation to cAMP-dependent kinase activity fail to explain a number of experimental findings, among others the relationship of ATP and Mg^{2+} and the rapid and reversible action of ATP analogues. Nevertheless, firm experimental evidence supports the hypothesis that low ATP opens these channels and high ATP exerts inhibition. ADP, in the presence of ATP, can open K_{ATP} channels in a dose-dependent fashion. However, higher concentrations of ADP are also inhibitory. The nonhydrolysable ADP analogue, ADP-β-S can also open the channel in the presence of the nonhydrolysable ATP analogue ATP-γ-S. This votes against the involvement of phosphorylation-dephosphorylation processes. The opening effects of both GDP and GTP on the K_{ATP} channels make it complicated to give an explanation based on G protein action. However, recent data suggest that G-proteins may cooperate with K_{ATP} channels in a more complicated way.

The complicated regulatory mechanisms of K_{ATP} channels and their complex role in the regulation of carbohydrate mechanism are beyond the scope of this book. However, it must be pointed out that this relatively late newcomer to the family of K^+ channels may open new ways to enhance the not-too-extended K^+ channel pharmacology, especially with prospective therapeutic applications. The presence of K_{ATP} channels in the central nervous system provides a very useful field to look for possibilities of therapeutic intervention, e.g., in the field of stroke and epilepsy. Cromakalin has been shown to reduce seizure-like activity among experimental conditions and K_{ATP} channel-activating drugs seem to be useful in whole animal models of epilepsy.

5. The Structure of the K^+ Channel

The "Shaker" gene has been identified as a possible gene for encoding K^+ channel activity in *Drosophila melanogaster*, which belongs to the family of fruitflies. The shaker name comes from the very fact that mutants of the *Drosophila* having an abundance of K^+ channels shake their legs in the literal sense. The

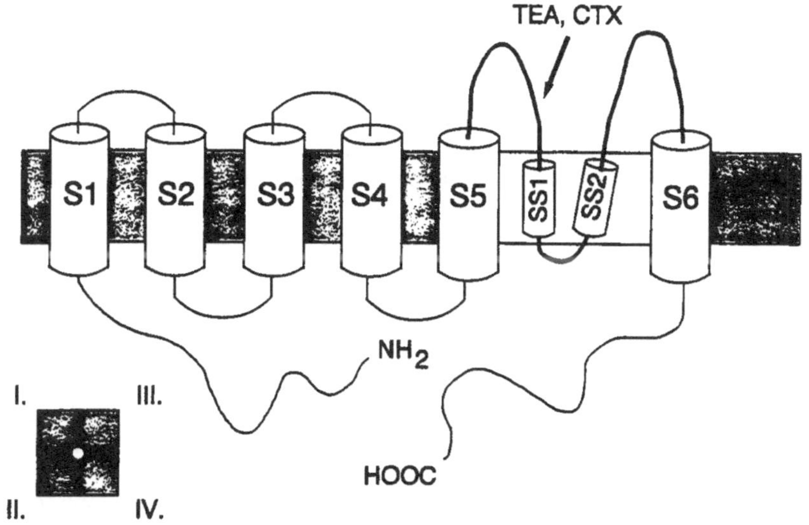

Figure 7. Domain structure and membrane-spanning (α-helical) segments of the potassium channel polypeptide. Numbered cylinders (Sn) represent the major membrane-spanning segments, while subsegments are labeled with SSn. A possible steric configuration of the four K^+ channel domains forming an active channel in the plane of the plasma membrane is shown in the lower left corner. Two major inhibitor binding sites are also indicated.

mRNA of these genes has been injected and expressed in *Xenopus* oocytes and functional K^+ channels have resulted from this procedure. Analyzing the structural properties of this gene, it is interesting to note that the functional K^+ channel (which is a single polypeptide chain) is smaller than the α subunit of the Na^+ channel with its 260-kDa molecular weight or the calcium channel, where each of the α and α_2 subunits are on the order of magnitude of 150 kDa. The K^+ channel polypeptide (Figure 7) is similar to one of the four internally homologous domains of a Na^+ or Ca^{2+} channel.[3] The interspecies homology is nicely demonstrated when one compares the K^+ channel polypeptides of *Drosophila*, mouse, and rat. Thus, the diversity of K^+ channels may mean a diversity in regulatory and auxiliary mechanisms, which are adapted to different functional roles, but basically there is a structural conservatism. The diversity, of course, may also arise from the multiplicity of genes encoding K^+ channels. Alternative gene splicing, as well as a multimeric polypeptide chain structure, can also be responsible for the great variety, not even mentioning the vast possibilities of posttranslational modification. The latter can occur during or after the folding of the polypeptide into the endoplasmic reticulum in the course of its synthesis. Nevertheless, a great variability and heterogeneity is provided and it serves the functional faculties of the channel, which are manifold.

6. Mechanism of Voltage-Dependent Gating

Gating, i.e., regulated opening or closing of ion channels by changing membrane potential, is among the fastest available regulatory processes in the plasma membrane. Key cellular responses in a variety of cells are triggered by membrane de- or hyperpolarizations. However, in general it can be stated that hyperpolarization is most frequently observed interlinked with the relaxation of one or more functions. Depolarization is connected to the activation of excitable cells and hyperpolarization relaxes them. However, this coupling of phenomena seems to be rather general. In airway smooth muscle, the action of relaxants is accompanied by hyperpolarization of the membrane. Hyperpolarization and relaxation are also induced by drugs such as forskolin, theophylline, and dibutyryl-cAMP, indicating that cAMP-dependent phosphorylation may be involved in the development of the electrical response. Similarly, the anti-arrhythmic effects of bretylium thosylate could be attributed to hyperpolarization. The same effect of BT may cause its differential action on two distinctly different types of cell activations. Cyclosporin A, an indispensable drug in solid organ transplantation, also affects membrane potential, very likely through activating the Ca^{2+}-activated K^+ channels and thereby having the unlikely characteristic of being both a depolarizing and a hyperpolarizing agent, depending on the duration of its effect and the dose.[72] This suggestion is somewhat weakened by the controversy in the occurrence rate and functionality of such channels in T and B lymphocytes, as it will be discussed.[59,73]

Practically every channel has a kind of voltage regulation, and the gating of a channel is based on the alteration of the electrical gradient across the plasma membrane in such a degree that the active channels may stop or the inactive ones may begin their function. Hyperpolarization "generally" closes while depolarization opens channels. The electrical gradient across the membrane is on the order of 10^5 V/cm. This offers an extremely strong force field, which is eminently suitable for the formation of transient dipoles. These types of dipoles are generated by the field; however, of course, one cannot exclude the reorientation of permanent dipoles from the molecular variations. The mobility of amino acid residues carrying net positive or negative charges or mobile π electron islands of protein molecular orbits can be subject to such movement. When textbook drawings of membrane-spanning regions of channel proteins are studied, one has to keep in mind that in many cases the membrane-spanning sequences are deduced from the hydrophobicity and length of the specific amino acid segments. They are far from always supported by digestion experiments of relevant membrane loops from inside and outside the membrane. However, these possibilities are not real handicaps in understanding the general mechanism of voltage gating.

By definition, it means the translocation of a part of the transmembrane protein (or its charge) in such a way that the channel function is made operative. The frequently assumed all-or-nothing-type gating is somewhat discredited by the very fact that, even with the great advantage of patch clamping, one needs to have a transfer of 10^7 monovalent ions per second through a single channel in order to measure picoamper level currents. It is likely that the time delay between partial and full opening states of channels must be very short, but the rise time of the current is unlikely to be infinitely short.

In contemporary physiology, voltage gating refers almost exclusively to processes where ion traffic is involved across membranes, utilizing pores, formed from specific proteins, the ion channels. Looking for the intricate molecular mechanism of voltage gating, one has to take into account the widely accepted possibility that transient dipoles can be formed from ion channel proteins. Furthermore, weak van der Waals-type intermolecular interactions can be highly influenced by the oscillatory fluctuations of transmembrane potential. Thus, it is more than likely that enzyme actions and the regulation of intramembrane intermolecular interactions must be influenced by membrane potential changes. An indirect evidence to support this view is that those signal-transducing mechanisms are more sensitive to membrane potential manipulations wherein a higher number of membrane-bound molecular processes are involved.

We have obtained more direct experimental evidence for the general validity of the above assumption with flow cytometric energy transfer experiments. Inter- and intramolecular proximity changes have been detected in the HLA class I antigen of human lymphocytes upon changing the membrane potential either toward lower or higher values. Significantly enough, the change in proximity of the monoclonal antibody-labeled moieties of the antigen was opposite upon depolarization as compared to that after hyperpolarizing the membrane. This experiment allows us to suggest a membrane potential-controlled dynamic change of integral membrane proteins which naturally regulates functional properties of a particular protein exposed to such physical force.

The mechanistic models, i.e., mobilizing distinct parts of membrane proteins during the gating or channel-activating process, are not always justified. Although they are nicely visualizable models, one has to keep in mind that removal of a single positive or negative charge from the pathway of charged ions in a narrow, practically virtual tunnel can open or lock the channel lumen.

However, there is an interdependence between channel activity and voltage gating. The principal reason is that an alteration in the membrane potential assumes some variations in the "resting" ion currents under physiological conditions.

Thus, voltage gating depends on the channel opening and closing with respect to their duration, rate, and kinetics. There are Ca^{2+} channels of the L (i.e., dihydropyridine) type, where, for example, strong depolarization can modify the channel from its "normal" gating pattern into a mode of gating, characterized by long opening and high open probability. The rate constants for conversions between gating modes can be slower than "normal" channel opening and closing rates. At the same time, the equilibrium between modes is almost as steeply voltage dependent as channel activation and deactivation at more negative potentials.[74] It is no wonder that Ca^{2+} channels have such stringent and manifold regulation. Ca^{2+} has the steepest electrochemical potential across the membrane (10^4-fold concentration difference plus the attracting coulombic forces) and any change in the intracellular Ca^{2+} concentration may influence a number of Ca^{2+}-dependent functions. Eminently among them are the activation or inhibition of phosphorylation processes. These types of regulatory steps are further complicated by those inhibitors and activators which modulate the equilibrium of phosphorylation and dephosphorylation processes in the cell. An example is that Ca^{2+} concentration-dependent K^+ channel activity is cross-regulated by the presence of protein kinase A and phosphatases (or rather the phosphatase inhibitor ocadaic acid) in myocytes.[75]

a. The Sliding Corkscrew Model of Voltage Gating

Intensive work has been carried out on the molecular details of voltage-dependent gating. One of the most accepted models is the sliding helix model. The overall concept is that the S4 segments of those ion channels where this important subunit has been identified undergo a sliding movement on the external surface of a cylinder. The primary amino acid sequence of the subunit carries positively charged amino acid residues such as arginine. The positive charge carrier residues are separated from each other by nonpolar residues. The positive charges are neutralized by fixed negative charges on other (non-SH) transmembrane segments of the channel. The transmembrane segment is held in (or at least greatly helped to be held in) position by the intracellular negativity. When depolarization occurs, a well-ordered sequence of subunit conformational changes will occur, since the coulombic forces between the positive and negative charges are not strong enough to maintain the original structure. The voltage-driven conformational changes position the individual domains in a concerted way to such a position that finally an open ion channel is formed. The helical structures of the S4 subunit will move about 5 Å outward relative to their original position and will also undergo a spiral (corkscrew-like) rotational motion. The angle of this rotation, according to computer-simulated motions, is near 60°. All these movements together will result in a charge separation. An unpaired negative charge will be on the cytosolic side of the plasma membrane, while the extracellular side will have a net positive charge. The intramembrane positive and negative charges are in paired form in both conformations, i.e., before and also after depolarization. This may be understandable, because unpaired charges may easily make the ion passage impossible. Summing up the complicated processes, one can simply claim that membrane depolarization can make a high number of different ion channels open, and this process is generally termed activation. The prototype of the voltage-gated channel is the simplest one, the potassium channel, but more complicated and larger channel structures of the sodium and calcium channels share the same motives. The central element is a polypeptide with four repeating units of homology (repeats I to IV), and each of these contain the actual voltage-sensing region, the above-described S4 segment.[76]

7. Evolutionary Considerations

An evolutionary aspect of these mechanisms is that, while K^+ channels are widespread in many unicellular organisms, Ca^{2+} channels are found mostly above this level, in higher protozoa and Na^+ channels seem to be the latest on the evolutionary tree: these are only in higher multicellular organisms. The development of the higher organisms selected such genes which carried the encoding properties for Ca^{2+} and Na^+ channels. It is interesting to note that Ca^{2+} seems to be the most ubiquitous regulatory ion with its easy complex forming capability without making it too tight, i.e., irreversible complexes. Despite this, K^+ ion channels seem to be more ancient than Ca^{2+}. The similarity of the most important features of these channel peptides and the practically common mechanism for voltage gating also point toward a common ancestor gene.

In the briefly described sliding corkscrew model of voltage-regulated ion channel gating the importance of charged amino acid residues seems to override everything else. Yet, the significant experimental evidence recently obtained by Catterall and co-workers[70] supports that a change in a single nonhydrophilic residue may explain a high number of differences in channel conductance. Although there are experiments demonstrating that changes in the position of positively charged amino acids, the exchange of a single residue, may have substantial consequences, the possibility for unrevealed roles of conformational changes bound to positioning other, noncharged amino acid residues cannot be underrated.

B. CALCIUM-ACTIVATED POTASSIUM CHANNELS

Since the original discovery by George Gárdos in 1958, that Ca^{2+} influx regulates K^+ channels in erythrocytes, the presence of calcium-activated K^+ channels has been recognized in practically every cell type.[12,77-80] There is some controversy in the case of lymphocytes. The presence of Ca^{2+}-activated K^+ channels in lymphocytes was suggested more than a decade ago by Rink and Deutsch,[81] but their characterization and electrophysiological confirmation in the different subclasses of lymphocytes was published only very recently and mostly for B lymphocytes. Wilson and Chused[42,43] provided evidence, using carefully elaborated flow cytometric membrane potential measurements, and applying the negatively charged oxonol dye that T lymphocytes and monocytes have a quinine-sensitive, intracellular Ca^{2+} increase-induced K^+ efflux. The calcium ionophore, A23187 was used to increase the $[Ca^{2+}]_{ic}$. Again, Ishida and Chused[73] studied the presence of Ca^{2+}-activated K^+ channels in T and B lymphocytes. In T cells, they found a saturation of the activation of calcium-sensitive potassium channels at the 200-n*M* intracellular Ca^{2+} level; however, they failed to find Ca^{2+}-activated K^+ channels in B cells.

Since the above authors have claimed that they found this type of channel in T and not in B cells, contradictory reports have surfaced, using the more unambiguous electrophysiological technique, the patch clamping. So far the patch clamp experiments have shown the presence of calcium-activated K^+ channels only in human B cells and rat thymocytes.[82,83] The few successful experiments with T cells were carried on Jurkat cells, a T lymphoma line, and with the perforated patch clamp method, where the membrane of the cells was punctured by nystatin.[84] Recently Gallin[59] suggested that failures to detect this important type of the regulatory ion channels might have been due to a "washout" of the cell interior in whole cell patch clamp configuration. Applying the negatively charged oxonol dye in flow cytometric measurements, we found the calcium-activated potassium channels present in spleen cells of nude mice, which are undoubtedly B cells.

Applying the same flow cytometric techniques, we had no difficulties to find reliable (although indirect fluorescence) evidence for the presence of Ca^{2+}-activated K^+ channels in all kinds of T cells from human peripheral blood and also in T cell lines. Accepting the saturation concentrations of the intracellular Ca^{2+} determined in T lymphocytes for these channels by Ishida and Chused[73] as 200 n*M*, and assuming that at least one Ca^{2+} atom is necessary for their activation, the easily calculated upper limit of the number of these channels for a lymphocyte plasma membrane is around 500.

In human peripheral blood lymphocytes, which are mostly for T cells, we also found an interesting voltage modulation effect on Ca^{2+}-activated K^+ channels[85,86] (Figure 8). In order to administer a controlled amount of Ca^{2+} into the cells, human peripheral blood lymphocytes were treated with the calcium ionophore ionomycin without extracellular Ca^{2+}. After determining the membrane potential distribution over the cell population by equilibrating the cells with 100 n*M* oxonol, the addition of 100 μ*M* Ca^{2+} immediately decreased the dye uptake of the cells, indicating a hyperpolarization of the plasma membrane (Figure 9A). This signal was explained as the contribution of the diffusion potential of Ca^{2+}-activated K^+ channels to the resting potential. A small fraction of the cells did not respond. However, this nonresponding population was also observed when the cells were depolarized by high extracellular potassium. When the extracellular Ca^{2+} was increased to the physiological 1 m*M* among conditions similar to the above, it caused, in a high number of cases, depolarization in the majority of lymphocytes (Figure 9B). A similar phenomenon was observed in patch clamp experiments by Bregestovski et al.[87] The effect they found was attributed to a Ca^{2+}-induced decrease in the number of depolarization-activated K^+ channels. However, if we calculate the Ca^{2+} influx-initiated depolarization on the basis of the Goldman equation and the ionomycin-mediated increase in permeability, it turns out that the steady Ca^{2+} flow can itself have a depolarizing effect, besides inhibiting K^+ channel activity. An increase in the level of ionomycin in the cell membrane significantly enhanced the depolarization that occurred even at lower concentrations of extracellular Ca^{2+}. It seems that the amount of ionophore, the calcium concentration, and some other physical parameters such as temperature and the responsiveness of the Ca^{2+}-activated K^+ channels together will determine the response of the cell that can be monitored by the changes in membrane potential.

In our experiments, we found such parameter sets quite easily, wherein the same concentrations of the ionophore and extracellular calcium, which depolarized the peripheral blood lymphocytes in phosphate buffered saline, could cause a hyperpolarization upon a previous slight depolarization of the cells (Figure 9C). The slight depolarization of the cells prior to addition of the Ca^{2+} was an absolute prerequisite to elucidate the phenomenon. The time kinetics of this activation of the Ca^{2+}-sensitive K^+ channels has yet to be determined in patch clamp experiments, since the flow cytometric membrane potential measurements generally have a 1 to 2 min time lag between two samples; therefore, the hyperpolarization must occur within this time. This technical handicap compromises the time resolution of the experiments. Thus, obviously this period is only the upper limit, and further experimentation will determine the actual time kinetics of the above-described phenomenon.

A higher degree of depolarization prior to addition of ionomycin and extracellular Ca^{2+} to trigger the Ca^{2+}-sensitive K^+ channels, provided that the increase of the extracellular K^+ was between 73 and

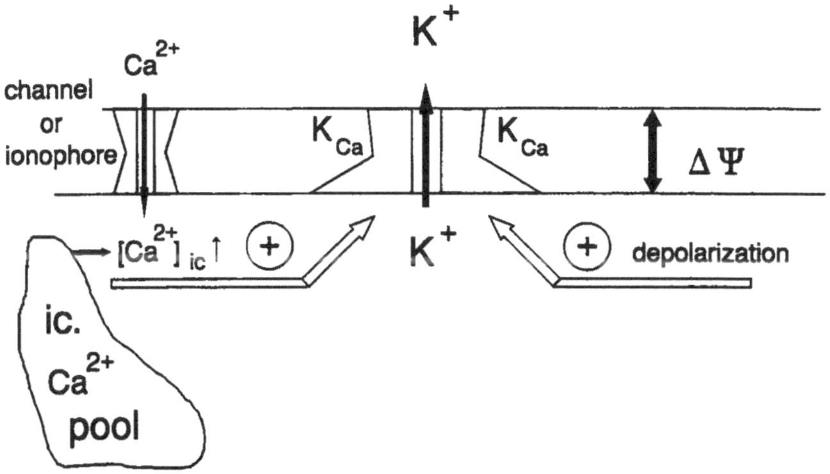

Figure 8. Effect of depolarization and release (or influx) of Ca^{2+} ions on the calcium-activated potassium channels. Calcium-activated potassium channels can be triggered by membrane depolarization and by increase in the cytosolic-free calcium concentration. The subsequent rise in the K^+ permeability of the membrane results in a hyperpolarization of the plasma membrane.

140 m*M*, resulted in a further depolarization upon addition of even modest Ca^{2+} doses (100 to 500 μ*M*). These doses generally induce hyperpolarization at 35-m*M* extracellular K^+ concentration.

It was suggested earlier that depolarization of the plasma membrane has an inhibitory effect on divalent cation influx induced by ionomycin; however, depolarization does not influence the equilibrium concentration, just the influx rate.[88] Since in our experiments the hyperpolarization of previously depolarized lymphocytes was not followed by depolarization, the lower Ca^{2+} influx rate alone does not explain the phenomenon. The first conclusion to be drawn, regarding the regulation of the Ca^{2+}-activated K^+ channels in lymphocytes was that these channel functions are also voltage modulated.

A very similar mechanism was observed by Sarkadi et al.,[79] who practically at the same time when we made our observations on the potassium channels have shown that membrane depolarization selectively inhibited receptor (CD3)-operated Ca^{2+} channels in human (Jurkat) T lymphocytes. Interestingly enough, they found an effect pointing to the opposite direction, i.e., to an inhibition of the Ca^{2+}-activated K^+ channels upon depolarization. Nonetheless, we believe that both phenomena belong to the same category of regulatory events.[79] Of course, we have to keep in mind that Sarkadi and co-workers investigated the voltage sensitivity of the receptor-operated calcium influx through monitoring the hyperpolarization caused by the activated calcium-sensitive potassium channels. Meanwhile, in our case the calcium channels were circumvented by applying ionomycin. Thus, in our measurements the combined dependence of this particular type of potassium channel on voltage and internal calcium concentration was in the focus of interest.

1. The Role of Transient and Induced Dipoles

It is generally accepted that the functional behavior of membrane proteins is highly dependent on their conformation.[11] The transient and induced dipoles in the membrane, generated by transmembrane potential, are reversibly spanning proteins and other molecules across the membrane. This type of electric regulation is reflected, for example, in the voltage gating of channel activities and some of our direct measurements as well. The potential-dependent conformational changes of HLA class I antigens have been convincingly demonstrated in flow cytometric experiments, wherein the resolving power of fluorescence resonance energy transfer between fluorescently tagged monoclonal antibodies directed to different epitopes of the antigen was exploited (Bene et al., personal communication). Thus, the accessibility of chemical groups, epitopes, or regulatory sites at the cell surface can certainly be altered by changes in the membrane potential, as it was also shown by Balázs et al.[44] in experiments detecting different numbers of cell surface SH groups in depolarized and normopolarized cells. The facilitated responsiveness of Ca^{2+}-activated K^+ channels under modest depolarization was clearly shown in our case

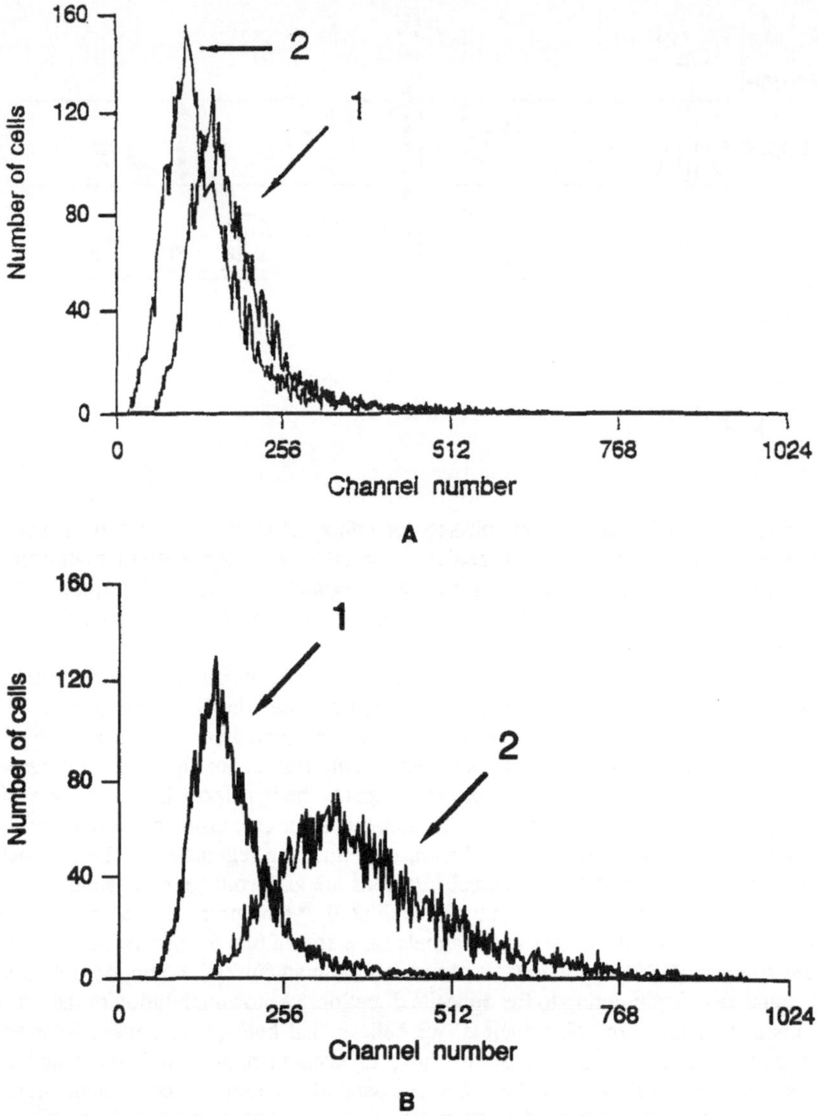

Figure 9. Triggering of Ca^{2+}-activated K^+ channels by adding calcium and ionomycin. Panel A: histogram 1 shows the membrane potential distribution over a human peripheral blood lymphocyte population. Histogram 2 shows the hyperpolarizing effect of 0.1 m*M* extracelular Ca^{2+} concentration added in the presence of calcium-specific ionophore (2 μ*M* ionomycin). Histograms were obtained with diBaC_4(3), a negatively charged fluorescence dye in a flow cytometer. Panel B: histogram 1 shows the membrane potential distribution over human lymphocytes. Histogram 2 represents depolarized cells, obtained after adding 2 m*M* extracellular Ca^{2+}. Other details were identical to those of described above. Panel C: a substantial depolarization (histogram 1) reverted the depolarizing effect of 2 m*M* Ca^{2+} (histogram 2).

above. The receptor-operated Ca^{2+} influx, probably inhibited in the case of Jurkat cells may have a mixed background, as the binding of anti-CD3 antibody to the corresponding antigen alone did cause a partial depolarization according to the work of Balázs et al.,[44] not even mentioning its cell activating capability.

Here the following model can be offered which may provide a general explanation for a number of known experimental facts: "Transient and induced dipoles may provide a significant contribution to the regulation of intramembrane molecular interactions through changing molecular conformations. This can be modified through fine modulations of membrane potential by external trigger signals or metabolic processes. The recently discovered similarities between the cells of nervous and immune systems, and the

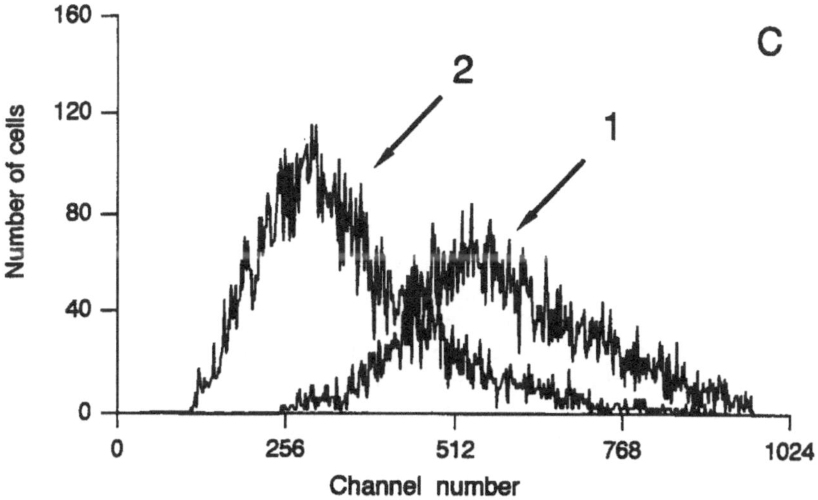

C

ever accumulating evidence that proximity and dynamic behavior of membrane proteins and their lipid-domain dependent conformational changes can significantly influence, up or down regulate membrane functions, may also, at least in part, refer to the above mechanism."[17-21] Thus, due to the high sensitivity of the above regulation, we feel justified to claim that only a narrow potential window of the plasma membrane is optimal for those intermolecular interactions which dominate membrane ion channel and also enzyme activities during the initiation and propagation of transmembrane signaling at the level of plasma membrane. While studying the stimulatory signals generated by antigen or mitogen, using mostly Ca^{2+} influx-dependent signal transducing pathways, one has to consider the physiologically very high extracellular Ca^{2+} concentration, although it is likely that natural signals will initiate a smaller and better controlled Ca^{2+} influx than a calcium specific ionophore. However, taking together the above-discussed events and the changes in membrane potential accompanying cell stimulation, a potassium channel with combined voltage and calcium regulation seems to be more sensitive and adaptable to the ever-changing extracellular conditions than a channel sensitive to calcium alone.

Cross regulation of channels are not novel in muscle and nerve cells. The activity of the so-called maxi BK channels is modulated in the presence of $[Ca^{2+}]_{ic}$. Slight depolarization increases the activity of the channel over the 0.01 to 100-μM $[Ca^{2+}]_{ic}$ range. The percentage of time the channels are open for each condition was measured when the membrane potential was changed from –50 to +50 mV. The BK channels of muscle and nerve cells are opened by Ca^{2+} at the intracellular membrane surface. The mean open duration increases when $[Ca^{2+}]_{ic}$ is increased. Studies on the kinetic properties of BK channels suggest that openings can occur with 1, 2, 3, or even more Ca^{2+} ions bound to the channel, with the apparent mean duration of the openings increasing with the number of bound ions.[89] Among the many possibilities, the following scheme is also used to describe the gating of channels in a simple way:

$$\begin{array}{c} C \overset{Ca^{2+}}{\rightleftharpoons} C(Ca^{2+}) \overset{Ca^{2+}}{\rightleftharpoons} C(Ca^{2+})_2 \overset{Ca^{2+}}{\rightleftharpoons} C(Ca^{2+})_3 \\ O(Ca^{2+}) \overset{Ca^{2+}}{\rightleftharpoons} O(Ca^{2+})_2 \overset{Ca^{2+}}{\rightleftharpoons} O(Ca^{2+})_3 \end{array} \quad (38)$$

where C and O mean the closed and open state of the channel, respectively.

2. Biochemical Control of Calcium-Activated Potassium Channels

The practically ubiquitous Ca^{2+}-activated K^+ channels are under stringent control of a variety of regulatory processes. Recently, it has been reported that neuropeptides such as somatostatin are capable of stimulating these channels.[90]

The importance of this observation is that this type of channel behaves similarly in a number of different cell types, and thus these findings may point to a general regulatory mechanism. Furthermore, phosphorylation-dephosphorylation processes are frequently instrumental in activating cellular events, but in the majority of cases phosphorylation activates something and dephosphorylation slows down or terminates the activated processes. Here, we have an obvious example of the opposite effect.

The effect of somatostatin, a neuropeptide, is known to reduce voltage-gated Ca^{2+} currents and has an opposite effect on inward-rectifying K^+ currents. These effects are thought to be G-protein mediated ones, since they are blocked by pertussis toxin, but can be maintained in the presence of exogenous cAMP. Under specific conditions, which stabilized membrane currents for hours without disturbing the concentration of divalent cations and small metabolites (e.g., with nystatin-permeabilized patches), Ca^{2+} currents were identified to be originating from dihydropyridine sensitive L-type channels. These channels were also inhibited by various somatostatin concentrations, but the inhibition by somatostatin was not complete concerning the peak current. From additional fluorescence measurements, it was suggested that these phenomena, namely that dihydropyridine had to be given at a concentration that completely blocked Ca^{2+} currents to inhibit secretion as effectively as somatostatin, directed the attention to the K^+ channels.

With a patch pipette containing high K^+ concentration, a sustained outward K^+ current was observed with the same voltage steps and these K^+ currents completely obscured the underlying Ca^{2+} flux. Maximally effective concentrations of somatostatin stimulated this potassium current at all voltages. The cobalt sensitivity of the calcium current and the parallel inhibition of the effect of somatostatin indicated that Ca^{2+}-activated K^+ channels might be involved in the observed sustained hyperpolarization by somatostatin. The observations were made on rat pituitary tumor cells, which have secretory properties and, as such, express large conductance potassium channels as well as Ca^{2+}-activated K^+ channels (BK type). Detailed electrophysiological clarification of channel conductances and application of pharmacological agents helped to reveal that the effect of somatostatin on membrane excitability of metabolically active pituitary tumor cells is a large increase in the opening frequency of high conductance (120 to 180 pS) charybdotoxin and tetraethylammonium sensitive, Ca^{2+}, and voltage-activated K^+ channels. Here we have to refer back to our finding discussed earlier that voltage gating, or at least modulation, of the Ca^{2+}-activated K^+ channels, observed in lymphocytes, may be a general phenomenon. The above data of White et al.[90] supports our view. Interestingly enough, (as it has been published by Sarkadi et al.[79]), the Ca^{2+} channel side of processes involving Ca^{2+}-activated potassium channels is also voltage regulated. However, the direction of this regulation has an opposite effect, i.e., partial depolarization decreases the Ca^{2+} influx. In their thorough analysis, White et al.[90] concluded that the neuropeptide somatostatin stimulates the calcium- and voltage-dependent large potassium channels through dephosphorylating a cAMP-dependent phosphorylation site on the BK channel proteins.

An open alternative could be that a similar process takes place in a closely associated regulatory molecule. Since membrane depolarization requires a cAMP-dependent phosphorylation step for activating DHP-sensitive calcium channels, the same dephosphorylation process can be responsible for the inhibitory effect of somatostatin on these channels as well. Further evidence along this line came from the specific and potent phosphatase inhibitor, ocadaic acid. This inhibited the serine and threonine protein phosphatases and also inhibited the effect of somatostatin when it was added together with the neuropeptide into intact cells. Further interesting information was that somatostatin stimulated tyrosine phosphatase activity in pancreatic tumor cells. Despite these very well elaborated experiments revealing a part of a complicated mechanism, it is highly probable that many more experiments have to be done before we can come closer to a better understanding of the coupling mechanisms of biochemical and biophysical events in the cell membrane.

A very elegant and promising recent experiment of Tanabe et al.[76] produced chimeric calcium channels, where one or more of the four repeats of the skeletal muscle dihydropyridine receptor were replaced by repeats derived from the cardiac equivalent of the same receptor. It was found that repeat I alone determined whether the chimeric calcium channel showed skeletal muscle-type slow or cardiac muscle-type rapid activation. Thus, the future seems to plead for combined applications of different technologies.

3. An Involvement of Ca^{2+}-Activated K^+ Channels in the Membrane Effects of Different Alcohols

Lipid solubility is not the sole criterion for the inhibition of Ca^{2+}-activated K^+ channels by alcohols.[91,92] Although no lesser personality than the Nobel Laureate Linus Pauling attempted to explain the mechanism by which general anesthetics generate anesthesia, the mechanism of their action is largely unknown.

Pauling, in 1963, attempted to explain the anesthetic potency of these important drugs by the changes in the physicochemical properties of the lipid bilayer, caused by the dissolution of these drugs in membranes.[93] The perturbed structure of the lipid bilayer was utilized in the explanation of down regulation in nerve conductance, which consequently could lead to anesthesia. Relatively recent evidence, obtained mostly in the 1980s, supports the hypothesis that general anaesthetics may directly interact with the hydrophobic region of membrane proteins. Membrane spanning protein loops are theoretically deduced from studying the average hydrophobicity of amino acid sequences of primary structures of transmembrane proteins. A high number of membrane-spanning polypeptide chains must contain hydrophobic regions which are supposed to traverse the membrane. Such a peptide chain can traverse a membrane region more than once. It is no wonder that membrane-bound lipid-soluble anesthetics, having a hydrophobic nature, can interact with these peptide moieties. Unfortunately, to provide unequivocal evidence for the existence of a direct interaction is not easy. Probing an alteration in the lipid structure may itself cause a functional change in an indirect way as well. This question has a great importance for at least two reasons. One obvious reason is the effects of anesthetics and an urgent need to reveal their mechanism. The second, having not lesser importance, is that the most general and misused drug in the world is alcohol. A better understanding of its effect may help mankind to get rid of the harmful consequences of drinking alcohol.

Tas et al. tried to gather more direct evidence on the effect of different alcohols at the cellular or membrane level.[92] The authors utilized a recently characterized ion channel, the Ca^{2+}-activated K^+ channel of rat glioma C6 cells. Ca^{2+}-activated K^+ channels have been characterized in the last 34 years since their discovery by George Gárdos in erythrocytes and later by others in practically all types of cells.[78]

This channel has the capability of quick regulation or counterregulation of the plasma membrane potential in the majority of cells. These Ca^{2+}-activated potassium channels regulate not only the K^+ concentration of the cells but also participate in controlling transmembrane signaling. Tas et al.[92] studied the effects of the volatile anaesthetics, such as halothane, enflurane, isoflurane, and methoxyflurane. In order to have a parameter that might differentiate between effects on Ca^{2+}-activated K^+ channels, namely their inhibition, and the lipid solubility of the examined agents in the plasma membrane, the following experiments were carried out. A Ca^{2+} ionophore was applied to trigger Ca^{2+}-sensitive K^+ channels through elevating intracellular free Ca^{2+} concentration. The Ca^{2+}-activated K^+ channels are known to increase the influx of ^{86}Rb isotope, which serves as an exchange cation for the K^+ ions that are lost during the potassium efflux. The K^+ diffusion potential, which generates higher negativity in the intracellular compartment, is responsible for the enhanced ^{86}Rb isotope influx. This effect can be inhibited by K^+ channel inhibitors such as charybdotoxin.

N-alkanols have a lipid solubility, and also anesthetic potency, increased by approximately a factor of three upon extending the particular n-alkanol applied with one additional carbon atom. The principal goal of these very important experiments was to look for a putative correlation among lipid solubility, inhibition of the Ca^{2+}-activated K^+ channel and the anesthetic effect of alcohols. Since the Ca^{2+}-activated K^+ channel opens up directly upon activation at the intracellular end of the K^+ channel by Ca^{2+} ions, any effect which could influence the Ca^{2+} influx would also complicate the explanation. The availability of fluorescent calcium chelators such as quin-2, indo-1, and fura-2 made it a routine investigation to look for the effectivity of the calcium ionophore in the presence of alcohols with different chain lengths. Using fura-2 loaded rat glioma C6 cells there was no difference in the ionomycin-mediated shuttling of Ca^{2+} into the cells whether the alcohols were present or not. Neither stimulatory nor inhibitory effects of the alcohols have been detected at the measurable intracellular Ca^{2+} level. The major second messengers of the transmembrane signaling system were also tested and it was found that increases in cAMP, cGMP, or protein kinase C activity did not affect the function of Ca^{2+}-activated K^+ channels in the investigated glioma cell line.

The 50% inhibitory concentration of different alcohols on Ca^{2+}-activated K^+ channels, their anesthetic potency in tadpoles, and the membrane/buffer partition coefficients of the alcohols, i.e., their lipid solubility, were correlated. The anesthetic potency of alcohols having different numbers of carbon atoms from 2 to 8 correlated extremely well with their lipid solubility. The IC_{50} (i.e., the 50% inhibitory potency) of alcohols for the Ca^{2+}-activated K^+ channels showed a very interesting behavior when it was represented as a function of their lipid solubility. For one, it was found that from pentanol to octanol the IC_{50} for Ca^{2+}-activated K^+ channel inhibition is much higher than the anaesthetic potency in tadpoles, as it was reported by Alifimoff et al.[94] For 1-hexanol the difference was a factor of 8 for the IC_{50} of Ca^{2+}-activated K^+ channel, and for 1-heptanol it was a factor of 5. This was the first well-established experimental evidence that alcohols may interact directly with the hydrophobic parts of intra- (or rather *trans*) membrane proteins, beyond the trivial alteration of the lipid environment and the protein lipid domain interaction. Modifying the alcohols in such a way that did not affect their solubility (for example, introducing more OH groups) did alter their effect on the Ca^{2+}-activated K^+ channels. This influence had obviously no correlation with the lipid solubility of the alcohols in question. The observation that even a change in the number of OH groups could drastically alter the effect of an alcohol on integral membrane proteins and their functions while showing no correlation with its lipid solubility has a far reaching consequence. The pioneering work of these authors confirmed experimentally the more than 10-years-old prediction in a general sense that alcohols may have a selective effect on membrane proteins depending on the structure of the protein and the alcohol as well.[91,92,94] The fact that lipid solubility of alcohols is (at least partially) inconsequential concerning their particular effects on a functional membrane protein suggests that alcohols interact directly with membrane proteins. Structural features of the inhibitory site on the protein would in this case modulate the inhibitory potential of the alcohols having different numbers of C atoms in their backbone and diverse positioning of their OH groups. Thus, a rather stringent sterical matching of anesthetic alcohols and channel activities can be suggested. Some other similar experiments show that other enzyme activities might also be inhibited relatively independently from the general anaesthetic properties. Such an enzyme is the luciferin-luciferase system. It seems that other membrane bound ion-transporter mechanisms can also be altered independently from the lipid solubility of alcohols. The $Na^+/K^+/Cl^-$ cotransporter of glial cells[91] was inhibited by lower n-alkanols through competition with the Cl^- ion in binding to the protein. A competitive inhibition always suggests a direct interaction with the protein structure, even if a competitive inhibition, determined by kinetic methods, does not necessarily mean the binding of the inhibitor to the same site as an apparently (i.e., kinetically) competing substrate.

Taken together the data presented in this section clearly suggest that clinically applied concentrations of general anaesthetics affect a number of sensitive cellular functions. Furthermore, besides anesthetics, alcohol consumption not only slows down the reaction time of muscles and the nervous system, but can also alter a number of cellular functions in the cells of the immune and endocrine systems. The ever more convincing notion that anesthetics may have target sites at more than one membrane protein having a key metabolic function can be true for practically each and every mammalian cell. Thus, besides regular clinical anesthesia, alcohol drinking is also capable of temporarily, or in chronic cases perhaps on a long-term basis, inhibiting important cellular functions and causing disharmony between the nervous, endocrine, and immune systems. Of course, on a short-term basis, the transient disadvantages can be compensated, but carrying such a load is certainly beyond the hope of repair in chronic alcoholic cases.

C. NA+ CHANNELS IN LYMPHOCYTES

Sodium channels are extremely important in excitable cells as the voltage-dependent sodium channel carries the main component of the Na^+-inward current when a propagating action potential is evoked in excitable tissues. Capitalizing on the specific, high-affinity toxins and the power of the whole armory of current molecular biology, Na^+ channels belong to the best-studied and -understood elements of excitable cells. Noda and co-workers[95,96] and other groups determined not only their primary structure but also, with high probability, the membrane-spanning regions. The α, β_1, and β_2 subunits are 260, 36, and 33 kDa, respectively and the β_2 is attached to the α by disufhide bonds (Figure 10).

There are many receptor sites associated with the Na^+ channels which can inhibit them (tetrodotoxin, amiloride, scorpion toxins, etc.), while others can cause a persistent activation. Out of the α, β_1, and β_2 subunits the α and β_2 subunits can be covalently labeled from the outside of the cells. All three subunits are richly glycosylated to such an extent that 30% of their molecular weight comes from carbohydrates. A cAMP-dependent protein kinase can phosphorylate the α subunit, proving that it penetrates the membrane. A possible model of the channel is shown in Figure 10.

The cDNAs encoding the largest (α) subunits of sodium channels had been cloned from eel electroplax, rat brain (four different kinds: I, II, IIA, and III), rat skeletal muscle, and *Drosophila melanogaster*. The rat IIA and II differ from each other by only seven amino acid residues. One very useful way to study these channels and their electrophysiological and other functional properties is to inject the channel mRNA into *Xenopus* oocytes and then study the properties of the expressed channels. Just to illustrate the fine details already known for this channel, it is enough to mention that quite recently the changing

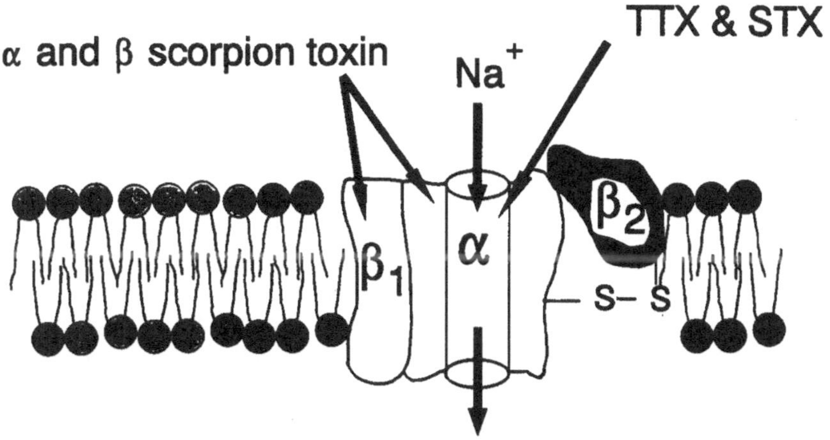

Figure 10. A possible model of the Na^+ channels with the indication of specific inhibitor binding sites.

of a single amino acid, a Phe, to Leu at position 860 of domain-II was enough to evoke a dramatic change in voltage-gating properties. Seemingly no changes have been involved or rather expected, since a neutral amino acid was changed for another neutral one in the membrane-spanning domain of the α helix. The α subunits of rat brain and skeletal muscle could also form functional sodium channels when expressed in frog oocytes. However, some inactivation properties of the channels were altered. Each α subunit contains four internally homologue domains, each of which has been predicted to contain either 6 or 8 transmembrane segments. A general trend points toward such efforts that aim to identify structural elements that function in the critical channel properties of activation, inactivation, and ion selectivity.

The major voltage sensor of the four homologue domains is the S4, having a 20 amino acid residue segment, where every third position contains positive charges, with mostly nonpolar residues in between. It is of extreme interest that voltage-dependent Ca^{2+} and K^+ channel encoding cDNAs have sequences with striking similarity to the S4 motif of the Na^+ channel α subunit. Thus, it is a likely hypothesis that nature invented a similar mechanism for all voltage-gated channels. Voltage dependence of channel activation was altered when mutations were introduced to change positively charged residues of the S4 motif.[97]

Recently, we have shown that, in contrast to the correctly identified voltage-gated Na^+ channel appearing in only 3% of human and murine lymphocytes, there was an almost ubiquitous Na^+ influx triggered by a particular ligand, bretylium tosylate, which activated Na^+-K^+-ATPase and hyperpolarized the plasma membrane. A 10- to 15-mV or higher depolarization proved to be necessary prior to addition of the ligand to achieve the hyperpolarization. (More details will be given in a later section.)

If Ca^{2+} and K^+ channels are endowed with a similar S4 motif like the α chain of voltage-gated Na^+ channels in excitable cells, it is unlikely that this new channel had a different sequence for the voltage-gating region, whatever its structure in other respects might be. If the ligand gating is also a prerequisite to activate its mechanism, as we observed, the role of the ligand could be the correct positioning of the S4 motif across the membrane, in order to activate its voltage sensitivity. The inverse possibility is that a previous potential decrease is required to facilitate or trigger the effect of the ligand.

D. THE CA^{2+} CHANNEL

Classified according to biophysical and pharmacological properties, not less than four types of Ca^{2+} channels have been found in excitable cells, and taking into account the current trends there is not much reason to believe that nonexcitable cells, such as those of the immune, secretory, or endocrine, etc. systems, have less than that. The four major subtypes are labeled P, T, N, and L. The latter type of Ca^{2+} channels, providing sustained currents, are known to be inhibited by three different families of Ca^{2+} channel antagonists: phenylalkylamines, dihydropyridines, and benzothiazepines, of which the best known representatives are verapamil, nifedipine, and dilthiazem, respectively. The binding sites of these Ca^{2+} channel antagonists are allosteric receptor sites in a true sense, i.e., the structure of the antagonist binding peptide is changed upon binding the drug and this steric alteration is conferred also upon other components of the channel. The binding sites are different for each drug and can be identified on the α_1 subunit of the Ca^{2+} channel receptor complex.

Photoaffinity labeling work of Catterall and co-workers contributed much to the identification of such sites and possible functional roles.[70] Recent typical work of this group will be utilized in the discussion below.

First, using the available methodological approach, the cDNA technique contributed much to the discovery of important subunit structures, and their homology to other cation channel types. For example, the important drug-binding subunit of Ca^{2+} channels, the α_1, has structural homology with the functionally most important subunits of Na^+ and K^+ channels from rabbit skeletal and heart muscle. The hydrophobicity of the primary sequence allows the assumption that six transmembrane segments are available (from S1 to S6). Despite the known structural details of the α_1 subunit, the actual drug-binding sites are not really identified beyond a good approximation. The drugs of all three chemical classes of Ca^{2+} channel blockers bind to the intracellular part of the channel. This is easily verifiable through applying penetrating and nonpenetrating chemical variations of these drugs. Patch clamp measurements also offer an elegant possibility to apply drugs from the inside or the outside of the membrane patches. The future trend definitely points toward the necessity of determining the actual drug-binding sites. The determination of the allosteric mechanism of their action will certainly greatly contribute to the understanding of channel gating and the intracellular physiological regulation of ion channels in general and of voltage-regulated Ca^{2+} channels in particular. Fortunately, photoaffinity labeling techniques allow us to label drug-binding sites with radioactive drug analogues of, for example, verapamil. Such studies support the view that Ca^{2+} channels have a similar subunit structure in a number of excitable tissues. Proteolysis with a series of endoproteases can excise smaller pieces of the α_1 subunit. The peptide sequence-specific antibodies afford, if not the precise identification, at least the determination of the sequence interval that binds the peptide.

This general strategy was applied recently to identify a phenylalkylamine binding region within the α_1 subunit of skeletal muscle Ca^{2+} channels.[98] The phenylalkylamine receptor specific verapamil analogue (N-methyl^{3}H) LU49888 allowed the group of Striessnig to determine (by the aid of immunoprecipitation with several sequence-specific antibodies) that the site of labeling was between Glu 1349 and Trp 1391. Thus, the boundaries of the verapamil-binding site were determined in the α_1 segment of the Ca^{2+} channel, although the accuracy of 42 amino acid residues was not too high, yet this result is the closest to reality. "This segment of α_1 contains transmembrane S6 helix of domain IV and the beginning of the long intracellular C terminal tail."[98] Thus, it was proposed that the intracellular end of helix S6 of domain IV and the amino acids in its vicinity at the intracellular part may play a role in the formation of the phenylalkylamine receptor site. Phenylalkylamines regulate the open channels from the inside, which implies a physiological possibility for intracellular formation of a channel configuration that leads to opening, i.e., a pore for Ca^{2+} traffic, regulated from inside. This argument firmly supports that a transmembrane pore can be formed by the cytoplasmic end of the transmembrane segment in S6. It was further speculated "that Asp 1382 and Asp 1385 which would be located on adjacent turns on the same side of a helix may play a critical role in both binding of phenylalkylamines and in coordination of Ca^{2+} ions entering and exiting from the transmembrane pore on the cytoplasmic side."[98]

These types of experiments are definitely more convincing than even a generally highly esteemed site-directed mutagenesis. The fact that a neutral amino acid change in segment II S4 dramatically altered the gating properties of the voltage-dependent sodium channel (which has basically an analogous functional behavior and voltage sensor structure to K^+ and Ca^{2+} channels) warns us against drawing a hasty conclusion concerning a general model of voltage gating.[99] More direct information can be obtained about actively involved segments of the voltage-gated channels that form their transmembrane pores from studies that localize receptor sites for drugs having a channel blocking action. Important amino acid residues which form high-affinity receptor sites externally for extracellular blockers of K^+ and Na^+ channels have also been identified in or around the S6 segment. The conclusion was that these segments on the extracellular side of S6 are the extracellular sites for Na^+ and K^+ channels, i.e., the actual sites for Na^+ and K^+ ions to enter the transmembrane pore. These types of experiments may provide a key to the pharmacology of the future. However, the observation of Auld et al.[99] undoubtedly points out that the logic of "Nature made" biological machinery and that of even the most modern directions of research are not always overlapping. Theoretically nonfunctional parts of a macromolecule may carry such important stabilizing features or hidden functionality that is far from being obvious at first glance. A better understanding of the primary protein folding at the level of the endoplasmic reticulum may be a great help in deciding which parts of a protein play a key role in their function. However, the strong conservatism of ion channel structures in the evolutionary tree suggests that there is not much room for "unimportant" or less-functional parts of these polypeptide structures. The more likely event is that we have not recognized a particular function yet that is carried by a given moiety or fold. This philosophy justifies those refined and intricate investigations which aim to understand the functional role of every molecular detail of biological macromolecules. On the other hand, a strong emphasis must be put on the fact that macromolecular structures are perfectly functional only in their native environment.

Two major types of Ca^{2+} channels have been identified so far, namely, the voltage- and ligand- (mostly hormone) operated ones. In the case of excitable cells the number and types of Ca^{2+} channels are classified either electrophysiologically or pharmacologically. There are Ca^{2+} channels which mediate long-lasting (L-type) Ca^{2+} ion currents. These can be further subdivided according to their sensitivity to organic Ca^{2+} antagonists: dihydropyridines, phenylalkylamines, and benzothiazepines. Frequently applied representatives of these groups (nifedipine, verapamil, and dilthiazem, respectively) were mentioned before. The binding sites for these antagonists are well-defined, allosterically regulated receptor sites. Again, a conservatism in the receptor and channel "jungle" can be pointed out, since neurotoxins for Na^+ channels have binding sites similar to these allosteric regulatory ones of L-type Ca^{2+} channels. The subunit structure of this L-type channel solubilized in digitonin from skeletal muscle has a structure similar to that of the Na^+ channel. The α, α_2, β, δ, and γ subunits have 175-, 143-, 54-, 24-, and 30-kDa approximate molecular weights, respectively. They can be more glycosylated (α_2, δ, γ) or less endowed with carbohydrates (α, β), associated in a disulfide-linked complex (α_2, δ) with the α, and have hydrophobic domains to a different extent. The α and β subunits can be phosphorylated by a cAMP-dependent protein kinase, which serves as a regulatory mechanism of ion conductance. These data are soundly confirmed by immunochemical and biochemical analysis as well.[70]

VII. KEY WAYS TO UNDERSTAND ION CHANNEL REGULATORY MECHANISMS

A. PARTICIPATION AND INFLUENCE OF ION CHANNEL ACTIVITIES IN DIFFERENT SIGNAL TRANSDUCTION PATHWAYS

A prototype of the cellular signal transduction pathways is that of the activation of lymphocytes. Although T and B lymphocytes have distinctly different functions, the transmembrane signaling, the set of those intramembrane and intracellular processes, triggered by an extracellular signal, which finally activate their specific function has many common features. The major trigger signals for T cells are the specific lymphokines, which can be secreted by monocytes or the T lymphocytes themselves in a self-stimulatory way. Lectins, antigens, or particular types of cell surface antigens binding monoclonal antibodies and tumor promoters such as phorbol esters, calcium ionophores, and a number of combinations of the above are in the long line of stimulatory signals. B cells can also be triggered by IL-2, the interleukin that first had, and not by accident, the name of T-cell growth factor, and also a number of specific and nonspecific signals, which have various relevance to the physiological triggering of cell proliferation.

1. The Role of Ca^{2+}

The large transient changes in intracellular Ca^{2+} concentration are very extensively studied in mitogen- or allo-antigen-induced activation of lymphocytes.[10-12] The modern era of fluorescent Ca^{2+} chelators synthesized by Tsien and co-workers at Berkeley, CA, the quin-2 and later the far more sensitive indo-1 and fura-2 dyes made it possible to follow changes of free Ca^{2+} concentration in the cytosol with relatively little harm to cellular functions.[8,9]

The existence of voltage-gated Ca^{2+} channels in lymphocytes was justifiably assumed a long time ago, but the electrophysiological evidence was only presented in 1986 and 1987 by Kuno et al. and Kuno and Gardner,[12,100,101] who reported that inositol 1,4,5-trisphosphate or mitogen-induced Ca^{2+} influxes were voltage insensitive.

Fukushima and Hagiwara[102-105] determined the existence of voltage-gated Ca^{2+} channels in a B lymphoma line. This was done by the whole cell clamp variation of patch clamp. Single channel currents were not detected, probably because of the scarcity of the channels in the membrane.

Thus, data obtained so far presented evidence for the existence of ligand- and voltage-gated Ca^{2+} channels in B lymphocytes, and indirect evidence supports that both kinds of Ca^{2+} channels are operative in T lymphocytes as well. However, even nowadays fluorescence methods are more widely applied to study Ca^{2+} channels in lymphocytes than electrophysiological ones. The principal reason is that lymphocytes are not easy targets for patch clamping; furthermore, flow cytometric membrane potential measurements or pharmacological ion channel analysis can overcome the problem of biological variability, as we discussed before.

Of course, beyond the established role of Ca^{2+} as a second messenger in lymphocyte activation, and its other actions like activating secretion, muscle contraction, etc., there are many phases of cellular events when its presence is mandatory. However, even in the case of its second messenger role in T lymphocyte activation, one has to specify the trigger signals before voting for Ca^{2+} as a second messenger. Data are presented in other sections, according to which the presence of Ca^{2+} is not a prerequisite for signal transduction in T lymphocytes, namely, when interleukin-2 is the activation signal, in contrast to antigens, or antibodies directed to the T-cell receptor and CD3 system, or mitogenic lectins.[106-108]

2. Effects of Ionophores

Sometimes Ca^{2+} channels are simply circumvented by ionophores to study the secondary consequences of Ca^{2+} influx. Thus, lymphocytes can, for example, be activated by A23187 and the phorbol ester TPA. The combination of ionophore and TPA leads to a proliferative response of cultured bovine lymphocytes, prepared from lymph nodes. Different combinations of mitogenic lectin Con-A, TPA, and A23187 were also tested. Ishida and Chused[73] concentrated on the Ca^{2+}-activated K^+ channels, by a similar method, and established their presence in T but not in B lymphocytes. It is interesting to note that patch clamp experiments which are thought to provide more reliable evidence failed to demonstrate these channels in T lymphocytes and determined their existence in B lymphocytes.[59] Pharmacological experiments, i.e., the testing of all three major Ca^{2+} channel blocker families using verapamil, nifedipine, and diltiazem, proved to be efficient in pointing toward the existence of voltage-gated Ca^{2+} channels in lymphocytes. These drugs were able to inhibit PHA-induced T cell activation in a dose-dependent fashion. Parallel to this effect, there was a marked decrease observed in the $^{45}Ca^{2+}$ influx upon PHA activation.[109] The blocking of Ca^{2+} channels also hindered the appearance of the anti-Tac monoclonal antibody-defined subunit of the complex IL-2 receptor, as well as the IL-2 synthesis itself. These experiments were carried out on normal human lymphocyte populations.

A proposal that voltage-dependent K^+ channels may mediate Ca^{2+} permeability was quickly rejected experimentally by Gelfand and co-workers.[110-112] Neither K^+ channel blockers, nor depolarization of the membrane, which was expected to open K^+ channels, had any significant influence on the increase of intracellular Ca^{2+} concentration. Fukushima and Hagiwara[105] provided evidence for just the opposite process. They succeeded in presenting evidence that monovalent cations, such as Na^+ and K^+, were able to generate currents, which were identified as currents through calcium channels in mouse neoplastic B lymphocytes. This seems to be more plausible than a Ca^{2+} flux through K^+ channels. Since channel specificity mainly depends on the size and charge of the ion, a channel specialized to carry currents of a larger and more heavily charged cation may be permissive to smaller ions such as K^+ and Na^+. Na^+, although a smaller atom, particularly for this reason has a larger hydrate shell and diffuses about two orders of magnitude slower than K^+ ions. A recently found putative voltage- and ligand-gated Na^+ channel, which has yet to be characterized in greater detail, but which undoubtedly allows the passage of Na^+ ions, may be identical to a special type of Ca^{2+} channel.[2] However, pharmacological evidence together with whole cell patch clamping (e.g., amiloride sensitivity, Ca^{2+} channel blocker insensitivity) suggest that these channels are a novel type of lymphocyte Na^+ channel and not disguised Ca^{2+} channels used as pathways for Na^+ ions in the fashion described by Fukushima and Hagiwara.[105]

There is much controversy in the literature about the coupled Ca^{2+} influx and release of Ca^{2+} from the intracellular stores. However, there is a consensus in more recent data that both voltage- and ligand-gated Ca^{2+} channels are present in lymphocytes, but their cooperation and significance is not yet known in full detail.[12]

3. Cell Activation and the Complex Role of Different Calcium Concentrations

Concerning the biological role of calcium and the known consequences of changes in its concentration during lymphocyte activation, it has been suggested that IL-2 production is likely to be Ca^{2+} sensitive. Recent studies, denying the role of $[Ca^{2+}]_{ic}$ increase in alternative signal-transducing pathways in lymphocyte activation somewhat change this picture.[106-108] Early steps in transducing activation (proliferation and/or differentiation) signals, regardless of their complexity, point toward the necessity of activating kinase-type enzymes, which can phosphorylate specific proteins that are supposed to relay the signals further toward the nucleus, in order to initiate new gene expression, etc. Ample evidence has been presented that many different pathways exist, but only two major steps of phosphorylation or, rather, phosphorylating enzyme systems are known.

The protein kinase C family phosphorylates upon activation by inositol phosphates and diacylglycerol. This process is interlinked with an increase of intracellular calcium concentration. The protein kinase C phosphorylates seryl and threonyl residues of different proteins. Mitogens, allo-antigens, or monoclonal antibody-induced signals are processed this way.

Another group of kinases is the tyrosine kinase family, which, in particular cases, seems to have a less-complicated transmembrane signaling mechanism. In the case of interleukin-2-dependent signal processing, the situation might be the simplest and most efficient. In the complex IL-2 receptor, there is a subunit that is supposed to be a major element of the high-affinity binding site, and that has a molecular weight of about 75 kDa in human lymphocytes. Similar structures have also been isolated from mouse systems.[113-115] Unlike the p55 (Tac) protein that also binds IL-2 and is assumed to cooperate with the p75 during tight lymphokine binding, the p75 has a long peptide tail in the cytosol. This peptide tail, although itself has no tyrosine kinase activity, is supposed to have a tyrosine kinase in the immediate vicinity and the activation of this is largely independent of both $[Ca^{2+}]_{ic}$ increase and the inositol phosphate cascade. Nevertheless, the biochemical mechanism involved in IL-2-induced signal transduction is not resolved yet.[114-119] The p55, or, according to the present nomenclature, the IL-2R α chain, has an affinity for IL-2 with $K_d = 10^{-8}$ *M*. This dissociation constant is the lowest for IL-2 as compared to those of the p75 (IL-2R β chain, $K_d = 10^{-9}$ *M*), and the tight binding site ($K_d = 10^{-11}$ *M*), which is very likely formed from the α and β chains. The α chain has a cytoplasmic domain formed only from 13 amino acid residues, containing several positively charged amino acids. The cytoplasmic domain of the β chain contains a much longer peptide tail, which consists of 286 amino acids with serine-, acid-, and proline-rich regions. However, neither the IL-2Rα nor the IL-2Rβ express the consensus sequence for the ATP binding site of tyrosine kinase, i.e., the Gly-X-Gly-X-X-Gly pattern. This amino acid sequence is mandatory for the known tyrosine kinases to bind the ATP that provides its γ-phosphate group for phosphorylating the tyrosine residues. Thus, it is clear now why we have to cope with the present-time theoretical suggestion that an unknown peptide, carrying the tyrosine kinase activity, must be in the vicinity of the IL-2R α and β chains and triggered by the IL-2 binding of the above two receptor sites.[115]

In order to characterize the complexity of the problem and provide evidence that the history of the multisubunit IL-2R is not completed yet, it is worthwhile mentioning that Sugamura announced the cloning of a third subunit of the IL-2R in 1992 (personal communication). The logical conclusion that the β subunit must have a direct contact to a yet-unidentified protein kinase activity was quite recently confirmed. Two independent groups provided evidence for the possible participation of a $p56^{lck}$ tyrosine protein kinase of the Src tyrosine protein kinase family in the phosphorylation processes, triggered by the interaction of IL-2 and the IL-2 receptor. Thus, it could be the missing link in this particular signal transducing pathway.[120,121] The structural features of the multisubunit IL-2R and the TCR-CD3 complex are schematically shown in Figure 11.

It is interesting to note that the β chain of the IL-2R, which has an intermediate binding capacity of the IL-2, is sharing a number of structural features with the erythropoietin receptor that includes a significant sequence homology. Based upon this analysis, the IL-2Rβ chain seems to be a member of a new family of growth and differentiation factor receptors (see Reference 115), the hematopoietic receptor superfamily "that is distinct from G protein-associated hormone receptors, receptors expressing a tyrosine kinase in their cytoplasmic domains, as well as those that belong to the immunoglobulin superfamily."[115] The reason we analyzed these features in such fine detail is that in another section we suggest that certain receptors triggering tyrosine kinase activity may be less sensitive to structural and potential changes in the membrane than those involving Ca^{2+} influx and inositol phosphate cascade.

Of course, this discussion does not mean that the two major phosphorylating mechanisms cannot be combined in signal transduction. Participation of tyrosine kinase activation is strongly supported by facts in antigen-, antibody-, or lectin-induced proliferation as well as in signal processing-inducing differentiation, where the T cell-receptor and CD3 antigen complex is involved. T cell specific members of the Src tyrosine kinase family, *lck* and *fyn*, are supposed to phosphorylate proteins before PKC activation. However, a detailed analysis has shown that those tyrosine residues which were phosphorylated in TCR-CD3 complex-operated activation processes were distinctly different from those involved in IL-2-triggered transmembrane signaling.[115-125] Now the most likely mechanism is that the $p59^{fyn}$ tyrosine protein kinase phosphorylates the ζ subunit of the TCR-CD3 receptor complex and the $p56^{lck}$ of the same

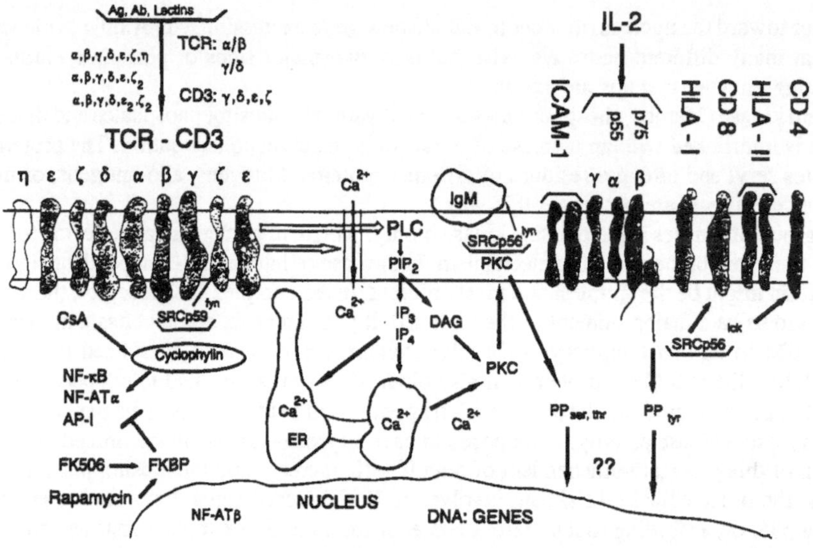

Figure 11. The multisubunit system and some of the surrounding molecules of two important receptor system of human T lymphocytes. The composite drawing shows molecules which are not supposed to occur together on the same cell surface. The intracellular part of the drawing indicates some of the signal-transducing pathways and also inhibitors. The number of the subunits reflects a current state of the research and may not be the "final" version. The symbols are as follows: Ag = antigen; Ab = antibody; TCR-CD3 = T cell receptor-CD3 complex, the subunits are marked with greek letters; PLC = phospholipase C enzyme; PKC = protein kinase C enzyme; PIP_2 = phosphatidyl inositol 4,5-bisphosphate; DAG = diacylglycerol; IP_3 = inositol 1,4,5-trisphosphate; IP_4 = inositol 1,3,4,5-tetrakisphosphate; ER = endoplasmic reticulum; CsA = cyclosporin A; ICAM-1 = intercellular adhesion molecule 1; IL-2 = interleukin-2, the subunits are marked with greek letters.

Src kinase family, as we have seen above, is triggered by the IL-2 activated IL-2 receptor. This latter is also able to phosphorylate the CD4 and CD8 antigens, while the $p59^{lyn}$ will phosphorylate the IgM molecules.

4. Oscillations in Ca^{2+} Concentrations

Rhythmic activity of excitable cells is a known phenomenon. The cardiac cycle, brain activity, or the rhythmic contraction of the muscle of shellfish are extensively studied. Since the difference between excitable and nonexcitable cells is less and less due to the extensive studies applying both patch clamp and fluorescence techniques to study ion channel activities and membrane potential changes in a wide variety of immune and endocrine cells, research efforts to find oscillatory phenomena in this field gained higher priority and impetus.[126] The technical means have been provided by the pioneering work of Dr. Tsien at Berkeley, and the very rapidly developing and spreading image analysis methods. The obvious principal target for these investigations was the oscillation of free calcium concentration and the related biochemical and membrane events.

Free Ca^{2+}, as a major second messenger of transmembrane signaling, was long suspected to have an oscillating pattern of intracellular concentration. The reason is the steepest known concentration gradient between extra- and intracellular space and the quick release from and rebinding possibility to intracellular stores of this particular ion. The principal cell types of the immune system, insulin-secreting pancreatic β cells and a number of *par excellence* excitable cells, are known to have functions stringently regulated by Ca^{2+} concentration changes. The patterns of oscillations also show a wide variety. These are induced, modulated, transient, permanent, receptor controlled, and/or second messenger-controlled oscillations, as summarized in a concise fashion by Berridge's recent minireview in 1990.[127]

When we speak about the oscillations of ion concentrations, we understand that it means a controlled fluctuation of the concentration of a particular ion (e.g., Ca^{2+}) or metabolite in a well-defined cellular compartment. The prototype of such oscillations, that of free Ca^{2+} concentration is governed by two principal elements: the plasma membrane and the endoplasmic reticulum (ER).[127] The first regulates the entry of Ca^{2+}, which is extremely quick due to the 10,000-fold concentration gradient across the plasma membrane; the second can store and release Ca^{2+} in and from its storage vesicles. All other regulatory processes which can influence or trigger the relevant channel activities or induce a quick release of Ca^{2+} from the endoplasmic reticulum are mostly modulatory. We do not make (and essentially do not see) much difference between endoplasmic reticulum and calciosomes as both of them have the same origin together with the majority of other intracytosolic organelles.

Recently, Irvine[128] studied the interrelationship between the ER pool and the Ca^{2+} channels in the plasma membrane, utilizing the discontinuous "quantal" release of Ca^{2+} from the stores and the structural features of the inositol 1,4,5-trisphosphate (IP_3) receptor. Quantal release of Ca^{2+} from the internal stores has been suggested, based upon the dose-dependent release of Ca^{2+} from the intracellular stores upon hormone or IP_3 stimulation. Irvine suggested that the intracytoplasmic domain of the IP_3 receptor can be instrumental in modulating Ca^{2+} channel activities both in the ER and the plasma membrane. This regulation was described by a model of conformational coupling of the Ca^{2+} release and entry and the IP_3 receptor. This regulated influx of Ca^{2+} is necessary to maintain the oscillation of Ca^{2+} concentration.

The actual Ca^{2+} oscillations are experimental facts and have been observed in a number of different cell types responding to trigger signals. The pattern of oscillations can phenomenologically be described either as sinusoidal oscillations or transient oscillations (also called baseline spiking). In terms of oscillating reactions, the sinusoidal type oscillations are the "real" ones. If we extend the term oscillation over the most different kinds of fluctuations in intracellular Ca^{2+} concentration, a much broader family of this phenomena is obtained. The sinusoidal types of oscillations are generally superimposed on an increased level of intracellular calcium concentration and can be considered as a well-regulated effect to maintain the necessary high concentration of intracellular calcium. Transient oscillation means transient increases in the cytosolic Ca^{2+} concentration from a constant resting level.

Acetylcholine, phorbol esters, and a number of other agonists have the capacity of triggering oscillation which, using biocybernetical terminology, may reflect the lack of an immediate response of feedback regulation. The negative feedback regulations which maintain a certain concentration have the general feature of such mechanisms, where the maintenance of the quantitative relationships is important.

A decrease in the concentration of the agent to be regulated signals the mechanism, which comes to the rescue by increasing the level of the agent. However, the increase itself will inhibit the effect causing the increase. Berridge[127] suggested several models for the generation of transient oscillations. In the "receptor-controlled" models, the oscillation of the IP_3 level results in a periodic increase of $[Ca^{2+}]_{ic}$, while the "second messenger-controlled" models suggest that second messenger interactions at the level of the ER are responsible for the periodic release and uptake of Ca^{2+}.

In the first model of the receptor-controlled Ca^{2+} oscillation the diacylglicerol/protein kinase C pathway feeds back into the receptor-G protein-phospholipase C transduction process to inhibit further hydrolysis of phosphatidyl inositol 4,5-bisphosphate and production of IP_3 (Figure 12a). Protein kinase C exerts its inhibitory effect by phosphorylating the putative G proteins responsible for the stimulation of phospholipase C. Thus, the IP_3 concentration itself oscillates.

Another model of the receptor-controlled Ca^{2+} oscillation proposes a positive feedback between the production of IP_3 and the increase in cytosolic Ca^{2+} concentration (Figure 12b). The initial small rise in $[Ca^{2+}]_{ic}$, evoked by IP_3, stimulates the phospholipase C enzyme and the resulting increased amount of IP_3 releases more Ca^{2+}. The outcome of this amplification is a pulse of IP_3 and a fast rise in $[Ca^{2+}]_{ic}$. The calcium signal then begins to decline due to the depletion of the IP_3-sensitive Ca^{2+} store and the fast uptake of Ca^{2+} into mitochondrial stores. The consequent decrease of $[Ca^{2+}]_{ic}$ terminates the positive feedback, and the IP_3 level also decreases. The cooperative action of IP_3 and the above positive feedback might be responsible for the sharp peaks, the so-called Ca^{2+} spikes.

The two-pool model for the second messenger-controlled oscillation is based on an IP_3-sensitive and an IP_3-insensitive intracellular pool of Ca^{2+} (Figure 13). The IP_3-insensitive pool is capable of accumulating Ca^{2+} from the cytosol. Reaching a certain load (overloading of the pool) triggers the release of Ca^{2+} from this store into the cytosol. This phenomenon is referred to as calcium-induced calcium release (CICR). To drive this oscillator, an input of Ca^{2+} is necessary. This input may come from an IP_3-induced release of Ca^{2+} from the IP_3-sensitive stores, or (since the capacity of the IP_3-sensitive pool is finite) influx of external Ca^{2+} across the plasma membrane. The relation of Ca^{2+} entry and the state of filling of the IP_3-sensitive store can be explained by the model of Irvine.[128] He speculated that the IP_3 receptor can undergo a conformational change which can influence both the calcium channel within the IP_3-sensitive store and

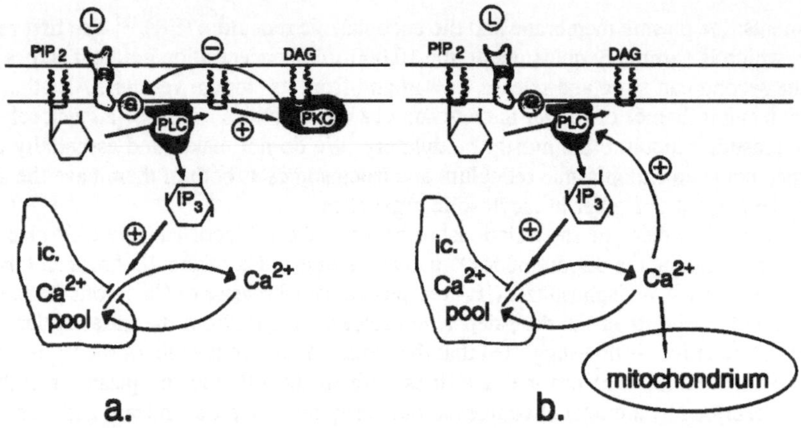

Figure 12. Receptor-controlled Ca^{2+} oscillation in the cytoplasm. The symbols are as follows: L = ligand; R = receptor; G = G-protein; PLC = phospholipase C enzyme; PKC = protein kinase C enzyme; PIP_2 = phosphatidyl inositol 4,5-bisphosphate; DAG = diacylglycerol; IP_3 = inositol 1,4,5-trisphosphate; The a and b parts represent two models described in the text in more detail.

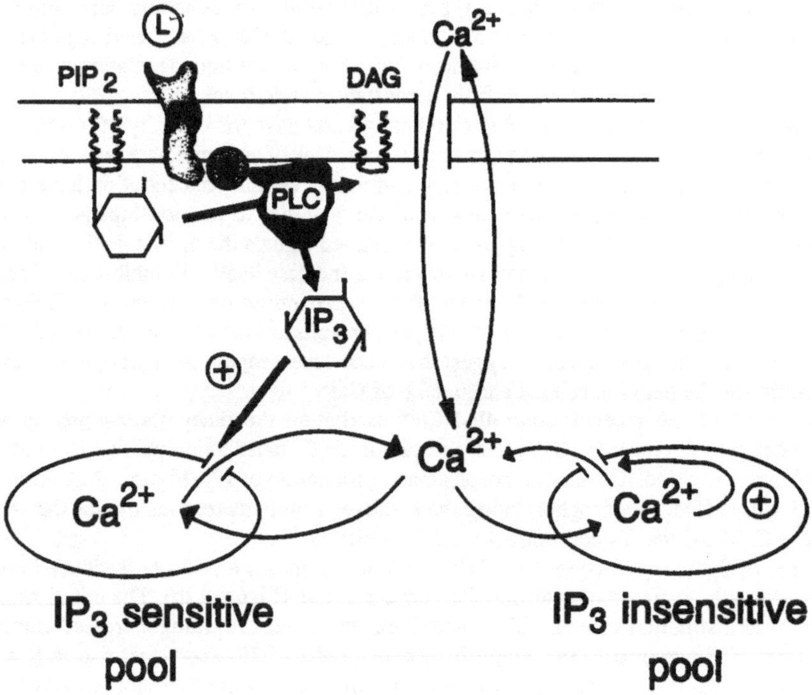

Figure 13. The two-pool model of the second messenger-controlled Ca^{2+} oscillation. Symbols are the same as in Figure 12.

that within the plasma membrane, as we mentioned it earlier. Likely candidates for the IP_3-sensitive pool are the calciosomes; the IP_3-insensitive pool might be some component of the ER.

The stable Ca^{2+} oscillations are probably governed by the triggering agonist (norepinephrine, histamine, vasopressin, angiotensin, etc.), the negative and positive feedback loops of second messengers, the capacity and number of Ca^{2+} channels in the plasma membrane, and the different pools in the ER.

Mitogens are also capable of inducing oscillations of cytosolic Ca^{2+} and transmembrane Ca^{2+} current in human leukemic T cells.[129,130] The induced calcium oscillation depends on the mitogen-regulated calcium channels in the plasma membrane. Although the gating of these calcium channels is voltage independent, the calcium influx is governed by the electrical driving force, which is thought to be established primarily by K^+ conductance. Fluorescence video imaging techniques and high-affinity toxins were used to study the role of K^+ channels in modulating mitogen-induced calcium signaling. Human leukemic T cells (Jurkat E6 1) were loaded with the fluorescent calcium chelator, the acetoxymethylester of the fura 2 dye, and bathed in physiological solution containing a membrane potential sensing dye, bis-oxonol. The intracellular calcium concentration was monitored through exciting the fura dye selectively at two wavelengths and the membrane potential was simultaneously detected by exciting bis-oxonol at a third wavelength while emission for both dyes was collected at $\lambda > 510$ nm.

Stimulation with phytohemagglutinin, a potent T-cell-activating mitogen, elicited parallel oscillations of intracellular Ca^{2+} and membrane potential with a cycle-period of 100 to 200 s. Jurkat cells express primarily two kinds of K^+ channels that could contribute to this response: voltage-gated K^+ channels that are blocked by charybdotoxin and calcium-activated K^+ channels which are sensitive to apamin. Charybdotoxin or apamin alone had little or no effect on intracellular calcium concentration. In contrast, simultaneous addition of the two toxins caused membrane depolarization and inhibited intracellular calcium oscillations in approximately 70% of the cells undergoing oscillations at the time of treatment. These results suggest that K^+ channels modulate calcium signaling in activated T cells by influencing the membrane potential and thus regulating the driving force for calcium entry. The lack of any effect for either of the blockers alone indicates that voltage-gated and calcium-activated K^+ channels are each capable of fulfilling this function individually.[130]

The slightly more detailed description of these recent data can be justified by the importance of the cross-regulation of Ca^{2+} and K^+ channels under such conditions when cell activation takes place.

B. CATION CHANNELS ARE NOT NECESSARILY SPECIFIC FOR PARTICULAR IONS

There are nonspecific cation channels in lymphocytes. Fukushima and Hagiwara studied calcium channel activities in a mouse splenic B myeloma cell line.[105] Their major finding was that these cells were capable of transporting other ions, namely K^+ and Na^+, across their calcium channels. They applied the whole cell variation of the patch clamp technique throughout their investigations.

Monovalent cations are known to carry currents through calcium channels under specific conditions:

1. The membrane potential has to be shifted to large positive values.
2. The external divalent cation concentrations have to be low.

Detection of such currents of monovalent cations through Ca^{2+} channels is not easy because of the frequent coexistence of voltage- and ligand-dependent ion channels together with Ca^{2+} channels in most of the cell membranes. Thus, the contribution of such channels to the measured currents has to be suppressed pharmacologically and also with substitution of ions. The most disturbing contribution may come from the voltage-gated K^+ current (of course, through the coexisting voltage-gated potassium channels). Things are further complicated by the fact that different types of Ca^{2+} channels can also frequently occur in the plasma membrane of the same cell. Different Ca^{2+} channels are not likely to be permissive to the same monovalent cations in the same way. In the first half of the 1980s, Fukushima and Hagiwara discovered a voltage-gated type of Ca^{2+} channel in cells fused from the mouse B myeloma line and mouse splenic B lymphocytes.[105] These seemed to be excellently suited to carry out studies wherein the properties of the voltage-gated Ca^{2+} channels, including their capacity to allow the passage of other ions, were to be investigated. Kinetic analysis of both the inward and outward whole cell current demonstrated the activation of one type of channel between –120 and +100 mV, the voltage-dependent Ca^{2+} channel. This unique property made it possible to obtain Ca^{2+} channel currents without being modulated by currents from other types of channels. Briefly, the results of Fukushima and Hagiwara showed that at reduced external Ca^{2+} concentrations an inward current arose due to the passage of monovalent cations and outward current was detected, carried by monovalent cations, at large positive potentials.

The analysis to be summarized here in a concise form showed that both inward- and outward-directed currents were separated by the same mechanism. Three different holding potentials were applied. When –108 mV was the holding potential, inward-directed currents became observable at –65 mV and reached a peak at –35 mV. The more positive the potential, the faster the rise to peak and the smaller the decay time was. Around +25 mV, the inward-directed Ca^{2+} current reversed (reversal potential). The extracellular Ca^{2+} concentration was 2.5 m*M* and the patch electrode contained the so-called internal Na^+ solution with 155 m*M* Na^+. The outward currents were increased in peak amplitude progressively up to the applied +59 mV. When the holding was changed to −78 mV, there was an almost uniform 40% decrease in peak currents, but the time course and the reversal potential remained the same. When the holding potential was decreased to −58 mV, the time-dependent current practically disappeared during 150-ms pulses that were more positive than −50 mV. Thus, in this case the channel was inactivated. The inward current was always carried by Ca^{2+}, since there was no change in the peak amplitude of the membrane current upon replacement of the external Na^+ with Cs^+ or TEA^+, or interchanging Na^+ with K^+. The size of the inward current was increased with increasing Ca^{2+} concentrations in the bath, and was also carried by other divalent cations, except for Mn^{2+}. The outward current had to be Na^+, since it was the only major internal cation species in the pipette during the whole cell patch clamping. On the other hand, internal and external Na^+ concentrations were nearly the same, thus the equilibrium potential for Na^+ was supposed to be near 0 mV. Higher concentrations of extracellular Ca^{2+} shifted the opening toward more positive potentials and it was established from the reversal potentials, which were also shifted, that the proportion of Na^+-permeable channels depended on both the extracellular Ca^{2+} concentration and the membrane potential values. The reversal potential changed with the Nernst slope of a Ca^{2+} electrode.

Results with different monovalent cations indicated a reversal potential for Na^+ very near 21 mV. For K^+ the same value was near 28 mV. Thus, K^+ was less permeant than Na^+. It is an interesting finding because of the identical charge and the fact that the hydrated Na^+ ion is significantly larger than the hydrated K^+ ion. It is possible that these monovalent cations move across channels without their hydrate shell. In this case the dehydration process demands extra energy. The other possibility is that they move together with their hydrate shell, but beyond the size and charge there must be some other characteristics which influences ion movements across membrane channels. Of course, one has to consider that in this case both Na^+ and K^+ are using Ca^{2+} channels. However, great similarities among voltage-gated ion channel features and the mechanism of voltage gating makes it difficult to explain these facts.

C. G-PROTEIN REGULATION OF ION CHANNELS, RECEPTORS, AND TRANSMEMBRANE SIGNALING

G proteins are a superfamily of proteins having the characteristic capability of binding GTP or, more generally, as we will see, guanosine phosphates. GTP binding is accompanied by structural and conformational changes. When the GTP is dephosphorylated, it is released from the protein that can go back to the original structural state and the cycle can start again. The most operative species of this family are those built from three subunits, designated α, β, and γ (Figure 14). Guanosine phosphates bind to the α subunit in the form of GTP. The subunit, which is also endowed with GTPase activity, has a weakened association to the β and γ subunits upon GTP binding. The dissociated α subunit is the activated G-protein component in most of the cases, although data are available showing that the liberated βγ dimer can also activate ion channels, e.g., in cardiac muscle. The released α subunit, before dephosphorylating itself, will diffuse to specific places where it can influence enzyme activity, receptor functions, or ion channel gating. After a few minutes-long deactivation process, the α subunit-GTP complex is converted into α subunit-GDP again, and the wandering α-GDP will be reassociated to the βγ subunits. G-proteins are apt to act through known second messenger systems, such as cAMP and cGMP, arachidonic acid, inositol polyphosphate degradation products, and diacylglycerols. A frequent consequence of G-protein activation is an altered enzyme activity. A certain type of classification labels them as G_s, G_f, G_o, or different (1, 2, 3) G_i G-proteins. The G_s (stimulator G-protein) stimulates the adenylate cyclase enzyme through hormone receptors. Besides these types of G-proteins, there are small molecular weight ones, which form a superfamily and have their origin in the ras protooncogene. Of course, these small molecular weight G-proteins do not have a subunit structure.

One of the best-studied G-proteins, transducin, regulates sensory phenomena and plays a very important role in the mechanism of vision (Figure 15). Stryer et al.[131] detected the participation of a protein, transducin, which is joined by GTP upon the light activation of rhodopsin. The activated α subunit-GTP complex, after dissociation from the βγ part moves with either two- or three-dimensional diffusion to a cGMP-dependent phosphodiesterase and activates it. Thus, the communication between rhodopsin and phosphodiesterase is maintained by α transducin. The G_o protein is the most abundant type in the brain. The function of the G-proteins is intimately linked to ion channels. That is the actual reason why a third type (after the voltage and ligand) of ion channel gating, i.e., G-protein gating, was

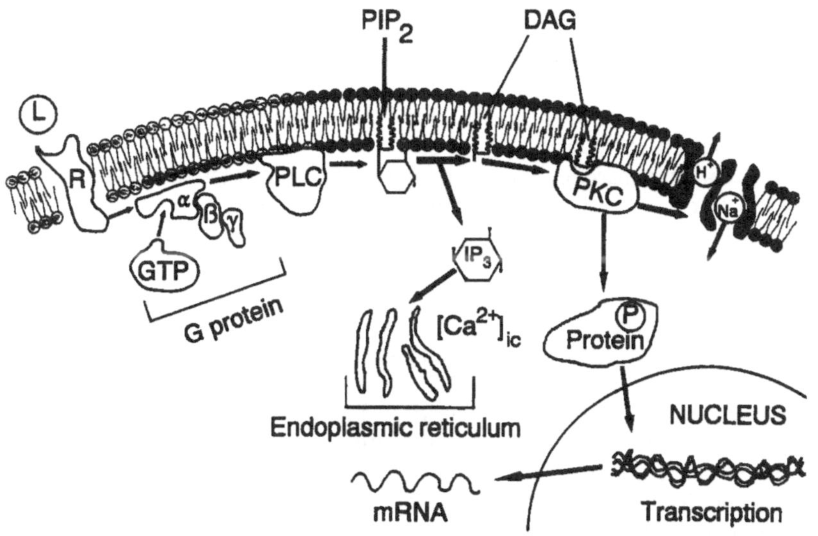

Figure 14. The interaction of the G-protein with the ligand binding receptors and the phospholipase C enzyme. The symbols are as follows: L = ligand; R = receptor; GTP = guanosine triphosphate; α, β, γ are the subunits of the G protein; PLC = phospholipase C enzyme; DAG = diacylglycerol; PKC = protein kinase C enzyme; PIP_2 = phosphatidyl inositol 4,5-bisphosphate; IP_3 = inositol 1,4,5-trisphosphate.

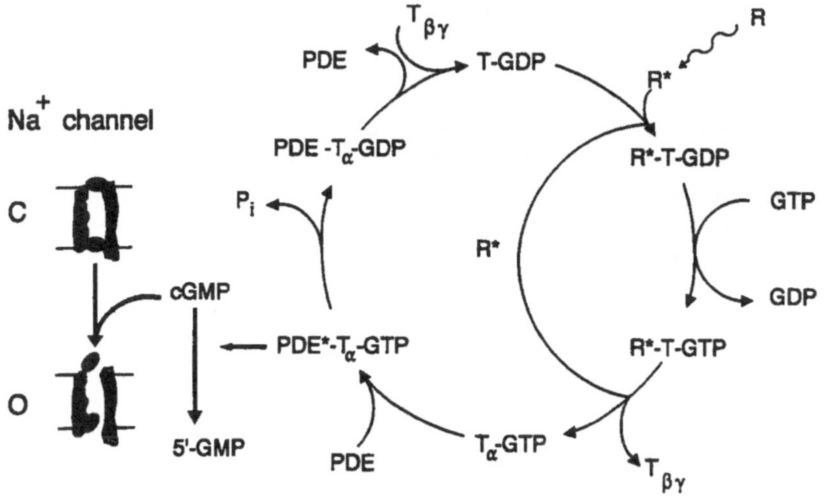

Figure 15. Cyclic scheme of the effect of transducin. T_α, T_β, and T_γ are the subunits of the first discovered G protein. R = rhodopsin; GTP and GDP are the tri- and diphosphorilated forms of guanosine nucleotides; PDE = phosphodiesterase enzyme; X* = activated molecule. The left-hand side shows the open (o) and closed (c) Na^+ channels.

introduced. The G_s-type G-proteins gate Ca^{2+}, Cl^-, and K^+ channels. The α transducin (G_t) activates the cGMP phosphodiesterase, which results in a decrease of the cGMP level and the consequent closure of Na^+ channels in retinal rods. The G_o- and G_i-type G-proteins also regulate Ca^{2+} and K^+ channels. High-affinity guanine nucleotide analogues such as GTP-γ-S, a thio-guanine analogue, and toxins specific for G proteins, such as the cholera and pertussis toxins, helped to elucidate the mechanisms of their action.

One of the obvious handicaps of physiological investigations, where the studied mechanisms cannot be understood without biochemical details, is that the structure has to be destroyed in order to gain access to molecular details. However, the destruction of the structure may impair a number of interactions which are observable only in the original structure. Biochemists, whose contribution to our understanding of contemporary biology simply cannot be overestimated, are frequently hampered by the problem of not understanding complicated structural-functional relationships through destroying the finer connections first. There is a beautiful example where this problem can be elegantly circumvented and living mechanisms studied in the unharmed integrity of the system.

Uhl and co-workers[132] summarized data on the "work" of G-proteins in the photoreceptor cell, where transducin mediates the communication between photoexcited rhodopsin and cGMP-specific phosphodiesterase. A real time assay for the activation and deactivation of transducin was carried out utilizing those light-scattering properties which accompany molecular changes in the functionally intact rod outer segment. One of the major photoreceptor organelles, the rods, will swell and contract when they are illuminated. The light scatter signals show an alteration, undergoing isotropic change when photoreception is taking place. During the absorption of the $h\upsilon$ photon, transducin binds to the activated rhodopsin. After this quick process, due to the GDP $\rightleftarrows$ GTP exchange, the GTP-binding α subunit reaches the phosphodiesterase enzyme either by two-dimensional diffusion along the membrane or by a three-dimensional movement, i.e., a "hopping". These dissociation and association processes can be connected to the changes in the scattering power and provide a dramatic insight into the molecular details of a physiological function. Time kinetics of the changes in light scattering allows us to visualize the dissociation and association processes of an intact G protein system.

The dissociation of G-proteins generates potentially two-signal transducing (or at least modulating) elements: the α subunit GTP complex and the $\beta\gamma$. As briefly mentioned above, the dimeric component may also participate in activating channels. An example was described for the muscarinic K^+ channel of cardiac myocytes. Picomolar concentrations of $\beta\gamma$ dimer were enough to activate an approximately 35- to 40-pS K^+ channel, when they were applied to inside-out patches of neonatal rat atrial membranes.[133] Since $\beta\gamma$ dimers have been shown to activate membrane-bound phospholipase A_2 enzyme in retinal rods, the question has arisen whether this action on the K^+ channels was a primary or a secondary one. Arachidonic acid (Ach) is known to be a product of the action of phospholipase A_2 on phospholipids of the membranes. Ach is then quickly metabolized into further products, which may act as second messengers of ion regulation. Assuming such an indirect mechanism, the inside-out rat atrial patches were treated by affinity-purified antibodies to phospholipase A_2. This treatment completely prevented the activation of K^+ channels by the $\beta\gamma$ dimer even at micromolar concentrations. Since the antibody treatment could not block the effect of GTP or GTP-γ-S binding subunit when Ach was present in the pipette, the possibility arose that the α-GTP subunit can also activate these channels. Thus, these experiments provided a beautiful example of the complexity of biological regulations.[132]

The decomposition products of arachidonic acid[134] released from the cell membrane are responsible for activating receptor-dependent and independent processes, i.e., stimulation in a variety of cells. Ach is converted to prostaglandins and leukotrienes by different oxygenases, which are biologically very active and can modulate platelet aggregation, smooth muscle contraction, and also neural excitation. The leukotrienes were shown to activate K^+ channels of atrial myocytes. Data of Kurachi et al.[134] have shown that arachidonic acid decomposition products can stimulate membrane channels in combination with G-protein action. Since the enzymatic degradation of arachidonic acid is not necessarily receptor dependent, this procedure can act in a receptor-independent way.

D. MACROMOLECULES ARE NOT ONLY ENDOCYTOSED BUT MAY USE CHANNELS FOR CROSSING MEMBRANES

Simon and Blobel[135] reported recently the existence of a protein-conducting channel in the membrane of the endoplasmic reticulum. In order to put their work in the right frame, justified by its significance, following the major trend of their interesting paper, here we attempt to summarize these extremely important findings as follows.

The proteins, freshly synthetized at the surface of ribosomes, are quickly integrated into the membrane of the endoplasmic reticulum if they have enough hydrophobicity. Their orientation can be predicted and depends on the signal-anchor sequence in eukaryotic transmembrane proteins, which are synthesized on the rough endoplasmic reticulum.[136] Analysis of integral membrane proteins having ion channel function showed that membrane-spanning loops and sequences of the peptide chain can be predicted from the primary sequence with high probability. It is the hydrophobicity of the primary sequence that decisively dominates the integration of proteins into the membrane.

Since the endoplasmic reticulum and the ribosomes are sterically closely related, the accommodation of *denovo* synthesized proteins in lipid regions of the membrane can occur even during the synthesis. The "tail" end of the peptide can be manufactured while the "head" is already integrated into the lipid environment. Reconstitution experiments into lipid vesicles support the view that proteins span the lipid layers in a nonrandom, very well-determined fashion. Blobel and Sabatini and a number of other authors suggested the signal sequence, or signal hypothesis, to classify the steps of integrating proteins into the membrane.[135] The membrane-specific signal sequence is supposed to interact with a sequence-specific membrane receptor, which recognizes this sequence. If the protein is water-soluble, the translocation of the peptide chain may occur across a protein channel-like structure. While the first part of the signal sequence hypothesis has been supported, there was no evidence convincing enough for the existence of protein-conducting channels.

The breakthrough came in 1991, when pancreatic rough microsomal vesicles were fused to one side of a planar lipid bilayer, which separated two compartments where the salt concentration could easily be exchanged.[135] Under these conditions electrophysiological measurements made it possible to follow changes in conductance while manipulating the transportable proteins by puromycin treatment. The clearance of nascent proteins which plugged the free passage of ions from the putative protein channels by puromycin dramatically increased the conductance of the channel. The opening of the protein-plugged channels occurred in stepwise jumps and a single channel conductance of 220 pS was determined. The channel, once open, remained so and active for many minutes. When a higher than 220 pS jump in conductance occurred, it was a multiple of this "unit" and could be resolved as two jumps with a refined analysis. The role of the protein is only the gating of these channels and the conductance of the channels is a function of the ions passing through.

At this point, before further describing the major findings of this very significant experiment, a few critical remarks may be pertinent. The mechanism disclosed by these measurements questions the name "protein-conducting" channel. In reality, it has nothing to do with protein conductance. Proteins can be accommodated in such an environment of specific membranes so that after their removal a hole, if one prefers, a temporary channel, is left behind. However, here the conductance is a *par excellence* ion conductance and as such, a nonspecific one, since the steric vacancy leaves a wide-enough channel allowing the trespassing of different kinds of ions along the electrochemical gradient. The precise conductivity values (220 pS) and the stepwise "single channel" currents are neatly measured, yet one can challenge the name of protein-conducting channel. The channel, from the electronic side, may be classified as a nonspecific ion-conducting maxichannel. Considering protein transportation in a vectorial, i.e., unidirectional way, that is perpendicular to the membrane, another approach is also possible. Proteins can be integrated into the (or certain specific) lipid membranes in a removable way. The removal (by puromycin or any other, yet-unidentified way) makes the protein transport across membranes possible. This is undoubtedly a new mechanism of protein transportation and the work of Simon and Blobel is a remarkable one.[135]

Further analysis showed that the vectorial removal of the puromycin peptidyl complex has a salt concentration dependence, and at high salt concentrations the channels are closed.

This specific protein translocating mechanism revealed that the so-called protein-conducting channel was not freely permeable to ions when it was occupied by the polypeptide to be translocated. Since the translocating peptides are allowed to loop in order to move across the membrane, the inability of small ions to pass simultaneously may be a surprise. However, the solvated ions have a considerably thick hydrate shell, sometimes with a diameter of near 10 Å and shedding this shell demands energy. Nevertheless, subthreshold passage of ions cannot be excluded. All the more so, the authors correctly state that the measured conductance at –50 mV, the applied potential difference between the two compartments, corresponds to the passage of 68 million ions per second. Thus even 10^5 ions can "sneak" through the membrane, along the properly positioned peptide chain, undetected. Removal of the ribosomes resulted in channel closing. The conclusion was that this may hint that ribosomes are directly connected to the subunits of the channel. This is correct, but an alternative possibility is that the high ionic strength that was necessary to remove the ribosomes of the pancreatic microsomes might also be enough to close the channel by a direct effect on the channel-forming, yet-unidentified structures. The finding that the protein-conducting channel closes after the passage of each translocation event suggested that it must be opened when a new peptide is ready for translocation. Blocking of "ordinary" ion channels was also observed when a flexible peptide chain occupied their lumen. A similar phenomenon, where impermeant ions keep a channel open, but at the same time block the passage through it, was also observed, though generally this is not possible.

The detailed mechanism of this exciting series of observations deserves further thorough studies to clarify the physiological gating of the channel. The possibility that other maxi ion channels may be able to smuggle at least small peptides across membranes may open a new perspective in the understanding of secretory mechanisms of cells. The understanding of the physiological regulation of such systems will be a great step ahead in this field. Nevertheless, the direction has been shown and the first steps made.

E. MULTIDRUG RESISTANCE AND TRANSMEMBRANE MOLECULAR TRANSPORT

The above paragraph presented convincing evidence that, besides charged ions or small metabolites, macromolecules can also be transported across plasma membranes. Considering such a molecular transport, we implicitly stress the importance and seek for the mechanisms of the inward-directed processes. As Higgins and Gottesman reviewed recently,[137] multidrug resistance poses an opposite problem in clinical practice and is supposed to be responsible for a very high number of death cases each year. This particularly important problem arises from a mechanism that is acquired by several patients treated against cancer with supposedly effective drugs. In a significant number of cases and types of cancer, the patients develop a resistance not only against the actual chemotherapeutic agent they were currently treated with, but they become cross-resistant to a multitude of drugs. This phenomenon is called multidrug resistance. Multidrug resistance is generally associated with high expression levels of a particular protein in cancer cells. This phosphoglycoprotein has 170-kDa mol wt and, when its gene (MDR1) is transfected to drug sensitive cell lines, a multidrug resistance develops parallel to the expression of this protein which is operated similarly to the ATP-binding cassette carrying proteins. Thus, we are encountered again by an ATP-operated pump mechanism that pumps out molecules from the cell interior at the expense of ATP decomposition.

However, the models which are essentially consonant with a macromolecular transport across the membrane without a significant involvement of the lipid phase might have to be changed. In contrast to a generally accepted hydrophilic pathway of the drug-eliminating mechanism, now a new idea has been introduced, suggesting that drugs can also be smuggled through the lipid phase of the membrane. The drugs can be intercalated into the lipid bilayer and flip over to the extracellular side. This spontaneous flip takes place at a very low rate. The 170-kDa phosphoglycoprotein may serve as a "flippase" enzyme, transporting drugs from the cellular interior through the lipid phase by simply enhancing, with an enzyme-like action, the flip-over efficiency of the drugs, accommodated by the inner sheet of the lipid bilayer. This interesting possibility emphasizes that Nature invented far more than one way for eliminating or endocytosing molecules from or into the cell interior. Of course, the individually and successfully growing cancer cells, which protect themselves from poisonous drug effects, will not serve the survival of the whole organism. However, the experimental possibility to study this and similar mechanisms *in vitro* may provide a profitable way to understand and also overcome such transport pathways.

VIII. LYMPHOCYTE STIMULATION, MEMBRANE POTENTIAL, AND OTHER ION CHANNEL ACTIVITIES

Cell activation, by definition, means those processes when internal or external signals make the cells proliferate, differentiate, secrete, or initiate any other cell function. In excitable cells such as muscle or nerve cells, activation induces an excited state that enables these cells to carry out their specific functions they are differentiated for. In these cells the activation processes are intimately connected to changes in the membrane potential and particular ion channel activities. The so-called nonexcitable cells were found to be endowed with almost the same, or at least very similar, sets of ion channels. Thus, they also have the capability to change their plasma membrane potential in different functional states. Nevertheless, not much is known about the details of these changes and their connection to cell activation processes. Here again, lymphocytes will serve as a good example of the so-called nonexcitable cells.

The values of membrane potential measured in unstimulated lymphocytes are in rather good agreement, despite the high number of diverse, indirect techniques applied.[12] Measurements with triphenylmethyl phosphonium ($TPMP^+$), tetraphenyl phosphonium (TPP^+), carbocyanine, or oxonol dyes gave values between −10 and −70 mV for human peripheral blood cells (mostly T cells) or tonsil cells (B cells). Similar values were obtained for murine (mouse and rat) spleen and thymus cells. These values were confirmed by a number of patch clamp measurements as well. Applying the whole cell variation of patch clamp in current clamp mode we sometimes found a little higher values, sometimes near –80 mV. The potential values are sensitive to the preparation of cells. The more gentle the preparation, the higher the measured potential value. It is also generally accepted that the membrane potential of unstimulated lymphocytes is primarily a K^+ diffusion potential. This view was supported, among others, by the potassium-specific mobile ionophore valinomycin, which shuttles K^+ ions along the electrochemical gradient. In the presence of valinomycin, only a small hyperpolarization was observed, indicating that the resting potential of lymphocytes is very close to the K^+ potential. However, the measured potential values do not fit the Nernst equation well. Despite this fact, which may point toward the contribution of other ions, the majority of investigators failed to find a significant dependence of membrane potential on Na^+ or Cl^- ions. Of course, it does not necessarily mean that there is no such contribution.

Cell-activation associated membrane potential changes have been reported by many authors, but there is a great diversity in the size, and even the direction, of potential changes in these data.

Hyper- and depolarization were equally reported, but neither the method nor the exact timing, activator, or cell types were normalized. Thus, the comparability of available data is doubtful.[10-12] Our recent measurements showed a slight hyperpolarization upon lectin- (phytohemagglutinin or concanavalin A) induced stimulation of human peripheral blood lymphocytes. It occurred within a few minutes (1 to 10 min) and was monitored by bis-oxonol fluorescence in a flow cytometer.[24]

One technical reason for contradicting earlier and latter measurements could be that the positively charged carbocyanine dyes applied first (up to the mid 1980s) had to be equilibrated with the cells before starting the measurement. Thus, the cells were exposed to toxic effects for 10 to 20 min before the actual measurements could be started. Furthermore, upon penetrating the cytosol, these dyes also monitored the mitochondrial potential, which is generally even higher than the plasma membrane potential, and were technically apt to measure depolarization more easily than hyperpolarization.

The bis-oxonol is negatively charged, stays outside the cytoplasm of fully polarized lymphocytes, penetrates only into the outer regions of the plasma membrane, and thus its toxicity can hardly be a factor in the case of completely viable cells. Its handicap could be that cross-reacting with the stimulating agent itself may signal false hyper- or depolarization, since, the lower the oxonol fluorescence, the more negative the membrane potential. Therefore, the actual reason for a change of fluorescence, when we measure membrane potential by potentiometric fluorescent dyes, has to be thoroughly investigated in each individual case.

Patch clamp measurements on activated lymphocytes are obviously not trivial, mostly due to the altered membrane parameters, not favoring the seal formation with patch pipettes.

Nevertheless, despite all these problems there are studies which investigated not only the changes in membrane potential during stimulation, but also the influence of artificial changes in membrane potential upon cell stimulation. Gelfand and co-workers conducted a valuable series of investigations from the mid-1980s, studying the role of ion fluxes and the changes of membrane potential and also the effects of these changes on cell activation processes.[109-112,138,139] At this point, it is mandatory to mention that these investigations were conducted under circumstances where cell activation was triggered by mitogens. It is obvious today that more than one cell-activating transmembrane and intracellular signal-transducing pathways are available. Gelfand and colleagues themselves contributed much to clarifying this problem.[108-112,138-141] The selection of investigated pathways depends mostly on the nature of the trigger signal. Thus, the conclusions drawn from the above-analyzed experiments are mostly relevant for lectin and antigen-antibody- (CD3 antigen and T cell receptor) induced activation processes, since lectins are known to bind to and activate cells through the TCR-CD3 receptor complex.[142]

Activation of lymphocytes (i.e., eliciting a built-in response of the cells to more or less specific outer or sometimes internal signals) has a number of different pathways. These pathways mean the signal-transducing mechanisms, i.e., the series of events that an extracellular signal evokes in leading to proliferation and/or differentiation of a lymphocyte. Data are available for both T and B lymphocytes. The best-known signal-transducing pathway is the phosphatidyl inositol 4,5-bisphosphate (PIP_2) cascade, which finally activates protein kinase C[143-145] (Figures 11 and 14). Alloantigen, anti-CD3 or anti-T cell receptor antibodies, or lectins, which bind specific or nonspecific elements of the cell surface, trigger a phospholipase C enzyme activity which leads to numerous breakdown products of PIP_2.[142] The accompanying Ca^{2+} influx, a concomitant intracellular calcium release and the end products of the inositol cascade, i.e., diacylglycerol (DAG), inositol 1,4,5-trisphosphate (IP_3), and 1,3,4,5-tetrakisphosphate (IP_4), although with different efficiency, activate a protein kinase C enzyme. The cytosolic enzyme is sequestered to the plasma membrane upon its activation and phosphorylates proteins. These phosphoproteins play an important role in activating specific genes, starting proliferation and differentiation. However, more and more data have been accumulated suggesting that some lymphokines activate target lymphocytes through a different pathway. Mills et al.[108] and Valge et al.,[107] independently of each other, provided evidence that IL-2 signaling (activation of T lymphocytes by a lymphokine, a peptide with a molecular weight of about 15 kDa, which seems to be a more natural way) does not demand protein kinase C activation. The first group used protein kinase C-deficient mutant cells and the other depleted this enzyme, yet IL-2 could induce proliferation in both cases. Tigges et al.[106] gave convincing experimental evidence that B cell proliferation and differentiation can be triggered by a single receptor when IL-2 is added, and neither Ca^{2+}, inositol phosphate products, nor cyclic nucleotide and protein kinase C play any role in this specific signal transduction process.

In 1991, evidence was presented by Fung et al.[124] that, although the IL-2 receptor itself does not carry a subunit sequence with tyrosine kinase activity in the intracytosolic domain of its tail, it has a tyrosine protein kinase-activating capacity that has a distinctly different pattern from that of the antigen- (or mitogen) induced tyrosine phosphorylation. This tyrosine kinase activity was physically associated with the β subunit of the human interleukin-2 receptor. This putative tyrosine kinase is very likely identical to the $p56^{lck}$ tyrosine protein kinase of the Src tyrosine kinase family[120,121] (Figure 11).

Recently, it was demonstrated, in the case of mouse B cells, that surface immunoglobulin cross-linking can activate a tyrosine kinase pathway that is independent of protein kinase C. This study established that surface immunoglobulin-mediated B-cell stimulation induces not only a PIP_2-dependent but also a tyrosine kinase-dependent biochemical pathway of activation that can occur in the absence of detectable PIP_2 hydrolysis. The evidence that protein tyrosine kinases are associated with growth regulation in many cellular systems is substantial, and in the study of Brunswick et al. this was clearly demonstrated for the B cell antigen receptor as well.[146]

Taking together the above paragraphs, it is obvious that more than one signal-transducing pathway is available. Thus, when we study the effect of signal-transducing processes on membrane potential or, conversely, the effect of changing membrane potential on cell activation, we have to specify the transducing pathways under investigation. Membrane potential changes are easily induced by changes in the ion transport processes. Such processes may influence with higher probability those signal transduction pathways that utilize more membrane bound steps. The same direction in the change of membrane potential may generate an impairment of one type of signal-transducing mechanism and, at the same time, facilitate another.

Gelfand and co-workers[112] described that an isotonic increase in the extracellular K^+ concentration which fully depolarized lymphocytes prevented alloantigenic or PHA-induced cell stimulation. They observed that the intracellular Ca^{2+} release was not satisfactory for the PHA signal when the cells were depolarized prior to the addition of the lectin.

Crabtree[147] in his excellent review paper: "Contingent Genetic Regulatory Events in T Lymphocyte Activation" categorized the T lymphocyte "activation" molecules using an analogy to viral gene expression. After an efficient trigger signal that exceeded a given threshold these activation steps could be classified as

1. Immediate events
2. Early events
3. Late events

Immediate events occur independently of protein synthesis. Early events require protein synthesis but precede cell division, and late events occur only after cell division. This classification helped him in organizing about 67 events into a time table (Figure 16) that shows the appearance of molecules becoming detectable at different stages of cell activation. Figure 16 also shows the localization of the molecular elements and the time after the activation signal that is necessary for detectability. The time table alone does not help in understanding these complex events, but Crabtree's paper contains a high number of carefully drawn conclusions which are worthy of being analyzed separately, before we try to discuss the role of membrane potential and ion channel activities in cell activation processes. It is amazing how long an exposure to the original trigger signal is necessary for the final outcome of the activation even when the known earlier consequences are already in progress. In the great "jungle" of individually studied phenomena many questions can be put forward, but among them the most inconvenient types are those which are interested in the causal relationship of the appearance of "reporter" molecules and particular activation steps, not even mentioning those aimed at the final understanding of the whole process of

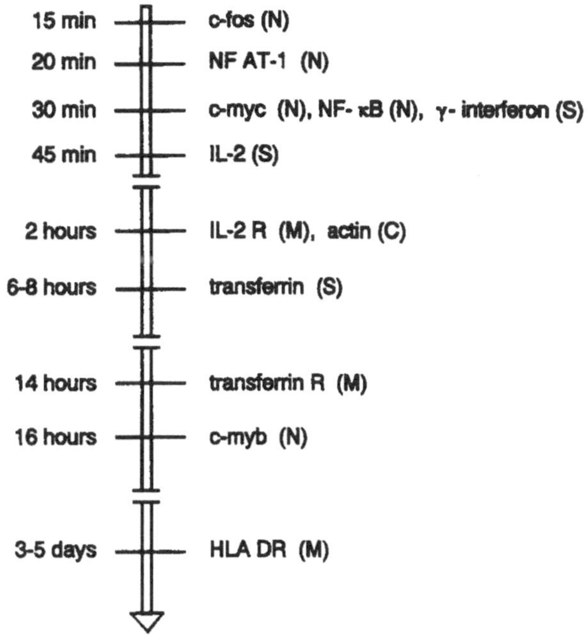

Figure 16. Different molecular factors appear at different stages of cell activation. The time indicated on the left side of the arrow shows the elapsed time after initiating an effective activation signal. M, C, and N refer to the membrane, cytoplasm, and nucleus as localization sites, respectively, while S is in brackets when the molecule is secreted. IL-2 = interleukin-2; IL-2R = interleukin-2 receptor; NF AT-1 = nuclear factor AT-1; NF κB = nuclear factor κB.

activation. The major problem is that when we face such questions, the most essential condition, i.e., the kind of activation pathway (or pathways) we are looking at, is often forgotten. Interleukins (IL-1, IL-2, etc), these leukocyte-activating factors of peptide nature, which can have high affinity binding receptors, may be able to send a signal through the membrane in a more direct way, i.e., through steps that are less membrane dependent than those signals which are induced by antigen-antibody reactions or lectins.

The best-studied (phosphatidyl inositol and Ca^{2+}-coupled) pathways are not necessarily the most physiological ones. The real common feature of practically every kind of signal transduction in lymphocytes is the protein phosphorylation step. Relatively little is known about the fate of the phosphorylated proteins, but certainly they must, directly or indirectly, participate in initiating the involvement of the genetic apparatus in cell proliferation and differentiation. The dependence of the phosphorylation processes on tyrosine kinase or protein kinase C-type enzymes seems to have significance only from regulatory aspects at the membrane level. The obligatory expression of the protooncogenes that many times, or almost always, is accompanied by the activation processes may or may not have a causative relationship with the activation itself.[148-150] It has been shown that inhibition of the activation process does not necessarily eliminate the expression of c-*fos*, c-*myc*, or c-*myb*.

Normal human fibroblasts were studied in this respect and it was shown that quiescent fibroblasts could be activated with serum. When both serum and alpha interferon (IFN-α) were administered, growth was inhibited.[151] The cells did not normally go into S phase and were significantly retarded by the end of a 1-week-long incubation period. However, this growth inhibition was not preceded by down regulation or delay in the induction of c-*myc* and c-*fos* at the mRNA level, as Northern blot analysis revealed. Thus, at least for normal fibroblast growth, induced by serum and inhibited by IFN-α, the inhibition of growth is not necessarily dependent on down regulation of either c-*myc* or c-*fos* protooncogene transcription. Furthermore, INF-α may act to inhibit cell growth either through a posttranscriptional effect on cellular protooncogenes, which would be necessary for proliferation, or through induction of other, as-yet unrecognized gene(s) associated with growth arrest. In order to point out the complexity of the question, we have to cite that inhibition of growth-related protooncogene expression by IFN was reported in several malignant cell lines and precursor cells of the hematopoietic system.[151]

A. KINETIC ANALYSIS OF K^+ ION CHANNEL FUNCTION IN LYMPHOKINE-ACTIVATED KILLER (LAK) CELLS

Ion channel activities have a role in transmembrane signal transduction, e.g., signal propagation in nerve cells is mediated by modulating ion channel functions. Lymphocytes express hormone receptors, which can be liganded by a high number of neuroendocrine products. Thus, one can conclude that such effectors, neuropeptides, and hormones (e.g., adrenocorticotropic hormone (ACTH), glucocorticoids, acetylcholine, etc.), which are circulating in the blood, may have access to such lymphocyte receptors which respond through changing ion channel activities, thereby regulating the immune system at the single cell level. On the other hand, primary and secondary lymphoid organs are directly innervated by the nervous system. Taken together, we may claim that there is such a thing as a neuroendocrine-immuno cross-regulation. We put the stress on cross-regulation, because lymphokines (e.g., IL-2) are known to influence K^+ channel activities; thus, beyond the neuroendocrine control of immune functions, the same immune functions may also have an influence even on the central nervous system.[152] Of course, the blood-brain barrier somewhat mitigates the feedback, but does not absolutely rule it out. One of the better known neuroendocrine products, 5-hydroxytryptamine (5HT), which is also known as serotonin, has a number of effects on different cells. Serotonin modulation of the immune mechanism is of great interest since it increases NK (natural killer) cell lytic activity *in vitro* and modulates tumor growth *in vivo*.

LeFever and Liepius studied the cytolytic activity of LAK cells in the presence of 4-aminopyridine (4AP), a voltage-gated K^+ channel blocker; 5HT; the serotonin antagonist quipazine or in the presence of quinidine, a blocker of Ca^{2+}-activated K^+ channels; and found that killer activity of LAK cells was inhibited in a dose-dependent fashion in the presence of these effectors.[153] Evidence was presented that 4AP or quinidine had its effects on the LAK cells themselves and not on the target cells lysed. Thus, it was suggested that LAK cell activity involves intact K^+ channel activation, and neuroendocrine products can also influence or modulate LAK cell functions.

An important feature of the experimental and theoretical analysis of LeFever and Liepius is that they borrow the formalism of enzyme kinetics to describe the sequence and connections of the lytic activity exerted by the various effector cells of the immune system.[153]

These steps are classified by them as

1. The binding of effector and target cells
2. The lytic programming of target cells by effector lymphocytes
3. The lysis of the programmed target cells

A general formal model of their kinetic analysis is as follows:

$$E + T \overset{k_1}{\rightleftharpoons} ET \overset{k_2}{\rightarrow} ET_P \overset{k_3}{\rightarrow} E + T_P \overset{k_4}{\rightarrow} T_L \tag{39}$$

Here E and T are effector and target respectively, while T_P and T_L stand for programmed and lysed target cells. The k_i rate constants are similar to those in enzyme kinetics. Similarly to a Michaelis-Menten-type classic kinetic analysis, the velocity of target cell lysis can be described as

$$V = \frac{V_{max}(T)}{K_m + (T)} \tag{40}$$

accordingly, V_{max} is the possible maximum rate of lysis, under the given conditions and at optimally high target cell concentration. T is the concentration of the target cells and K_m is the target cell concentration to achieve half maximal velocity of lysis, i.e., $1/2V_{max}$. LeFever and Micha provided evidence that K_m can be used to determine the number of actual LYTIC LAK effector cells (E) in the studied lymphocyte population. A concise description of the work of LeFever and Liepius on the application of their kinetic method to study the participation of K^+ channels in LAK cell activity will be given below.[153]

Their first question was, during the analysis of the effects of K^+ channel blocking 5HT and serotonin antagonists, to show the step of the lytic process where K^+ channels were involved. To meet their goal, a formal Michaelis-Menten-type kinetic analysis was given, also using the Lineweaver-Burk double reciprocal plots. The plot enabled them to determine V_{max} and K_m values. A modified version of this analysis (Hane's plot) overcomes the mathematical problems present in the traditional Lineweaver-Burk plot. The inhibition of lytic activity significantly decreased V_{max}, which was inversely related to the concentration of ion channel blockers. Thus, the maximum rate of target cell lysis decreased in a dose-dependent fashion. On the other hand, K_m was not changed and remained constant throughout the experiments. The percent decrease in lysis could even reach 100%, i.e., no lysis occurred above a certain inhibitor concentration. The most effective drugs were quipazine and quinidine which resulted in a more than 50% inhibition at 0.5 m*M*, and 100% at 1-m*M* concentrations. Of serotonin, and especially of 4AP, more than an order of magnitude higher concentration was necessary to obtain the same effect.

Decreased V_{max} and unaltered K_m values pointed toward a noncompetitive inhibition. LAK cells retained their viability and the effect of channel blockers was exerted through the reduction of LAK effector cell lytic activity. Their final conclusion was that the time necessary to program target cells for lysis was prolonged. Expressed in terms of reaction kinetics, V_{max} and k_2 were decreased and K_m remained constant. Separate studies indicated that the binding of effector and target cells was not impaired even when channel blockers inhibited the lytic activity by 90%. Thus, the prolonged programming time was not actually due to a shortened encounter time with target cells. Hence, as they suggest, we have to look for a post-binding step in the mechanism, during the generation or release of the lytic signal.

Taken together, the important data of LeFever and Liepius demonstrated that K^+ ion channels are involved in the lytic function of LAK cells and that such neuroendocrine hormonal products such as serotonin also exert a control over the process.[153] Kinetic analysis proved to be very useful in obtaining a bridge-head to step toward clarifying the mechanism of this effect. The effector-target cell binding was not impaired and lytic programming might involve K^+ channel activity in the period following target cell binding. The maximal rate of lysis was significantly impaired. The above kinetic analysis provided a good parameter, K_m, for a reliable estimation of the number of LYTIC LAK cells in a lymphocyte population. Preincubation experiments with K^+ channel blockers showed that their action was reversible and did not result in a permanent alteration in the lytic function of LAK cells, and continuous presence of the blockers was necessary to maintain and maximize their inhibitory effect. They postulated that ion channel function was involved in the generation or delivery of the lytic signal from LAK cells to the target tumor cells, and that this function bears the responsibility for triggering the target cell injury process. Serotonin may have a role in the neuroendocrine modulation of immune responsiveness *in vivo*. Localized stress and/or inflammation, releasing serotonin, may decrease the activity and effectiveness of LAK cells in protecting us against tumor cells. Experimental evidence was given that serotonin increases the rate of growth of murine fibrosarcomas. These findings draw our attention to an important system of cross regulation also regarding the therapeutic effectivity of LAK cells.

IX. CHANGES IN DYNAMIC RESPONSIVENESS OF CELL MEMBRANE ELEMENTS DURING AGING

A high number of physiological functions decline either quantitatively or qualitatively with aging. A particular interest has been paid to those declining processes that are connected to the immune response. Immunosenescence has been studied at both cellular and subcellular levels and recently reviewed by a special volume of the *Annals of the New York Academy of Sciences* and also by Thoman and Weigle.[154,155] However, as our knowledge on the immunological role of individual immunocompetent cells is developing, our view of immunological aging has to be revisited. Aging itself, apart from the chronological aspect, is an imprecisely defined process, which refers to the gradual loss or changing of abilities either inherited or acquired throughout the progression of lifetime. There are remarkable individual differences between different strains, as well as great individual variations of the entities of the same strain thereby making the definition of a clearly established indicator parameter extremely difficult. The same problem concerns subindividual parameters as well.

The great individual variations clearly indicate that the chronological age itself, although it gives some orientation, does not necessarily determine the actual stage of aging for a particular individual. The harmful effect of increasing age on the human and animal immune systems has been clearly documented. Our attention here will be restricted to lymphocytes. Data concerning the aging of lymphocyte subpopulations are conflicting. The "numerical" parameters, such as particular antigen densities of peripheral blood lymphocytes, were determined by several authors through fluorescently tagged monoclonal antibodies for almost the total commercially available panel. The contradicting data sometimes report a slight numerical decrease; sometimes a normal density of these antigens in the membrane of lymphocytes when comparing those from elderly to those from young indicate that real significant changes had not been observed. Thus, the confirmed alterations in the process of aging, such as decreased responsiveness of competent lymphocyte populations to activating stimuli, may come from a perturbation of regulatory processes or activation steps and not from trivial quantitative differences of the cell surface parameters.

Some regulatory parameters of lymphocytes may respond more sensitively to the aging process. Therefore, the parameters we analyzed were mostly regulatory ones and were also connected to the lymphocyte plasma membrane, a major theater of the immune processes.

The plasma membrane is the primary level where all alloantigenic or any other extracellular stimuli encounter the cells. Besides, in the last 1½ decades, it emerged that the plasma membrane is the main theater of signal transduction between the outside and the inside of cellular systems in general and of lymphocytes in particular. There is no doubt that any serious and permanent alteration of plasma membrane properties, may it be either structural or functional, could significantly contribute to the aging processes. However, a number of cell surface parameters investigated so far in connection with the immunological decline failed to show, as we mentioned earlier about antigen density, a correlation with chronological aging. Furthermore, not only the proliferative response to lectins and alloantigens, but the generation of cytotoxic effector cells, delayed-type hypersensitivity, or primary and secondary antibody responses are decreased or even lacking in aged humans or in murine models.[155]

The advent of more refined techniques in lymphocyte membrane physiology, bringing along patch clamping, flow cytometric membrane potential measurements, image analysis, etc., made it possible to monitor on an objective, cell-by-cell basis such parameters as individual channel activities and changes in the membrane potential with great accuracy. On the other hand, in the case of flow cytometry even individual variations can easily be overcome by extending our investigation of an individual parameter over a large cell population within a short time. Despite all these excellent and highly sophisticated hardware possibilities and emerging new trends, we are quick to admit that the ultimate reason behind the changes in the cell-mediated immune response during aging is far from being understood. Still, it becomes more and more likely that it has a multifactorial background.

Thus, the more key parameters are studied, the higher is the likelihood that some of the decisive factors can be determined. Without such deeper knowledge of the effects of aging on immunological responsiveness, we are just shooting in the dark without even seeing the targets. In the last decade, the lack of large individual changes in gross parameters induced everybody to look for more refined parameters. The work of several authors[18,156,157] points in this direction. The pioneering work of several groups[158-164] has determined that lipid dynamics decreases with advancing age in rat and mouse lymphocytes. Lateral mobility of rat hepatocyte membrane proteins was studied by Zs.-Nagy and Kitani in a series of experiments and a gradual decline of the measured diffusion constants was observed with aging.[165-174]

A. CHANGES IN PLASMA MEMBRANE POTENTIAL AND TRANSMEMBRANE SIGNALING DURING AGING

Here we will focus our interest mostly on those experiments which are connected with the changes in transmembrane signaling during aging, and consequently are correlated with changes in plasma membrane potential and dynamic events in the cell membrane. Although other authors have not found a difference in the cell surface density of antigens monitored by the fluorescently labeled monoclonal antibodies anti-CD3, -CD4, -CD8, -CD11, and OKM-1, we revisited this question under extremely well-defined quantitative conditions, but in vain. Some differences observed so far, which may be a reason for the contradictory data in the literature, may arise from the fact that parameter sets were examined among different conditions. Most probably the cells of the elderly group do not have different volume and do not carry a higher density of antigens. The membrane area of lymphocytes from elderly is thought, by some, to be somewhat smaller, but most probably this is not the case. Measurements may determine that their volume is smaller but this could be accompanied by the same surface area of plasma membrane. Elderly populations are known to lose a certain amount of water, as is generally expected in the case of the majority of different cell types during aging. Elderly cells may shrink, displaying a different (smaller) scatter signal in a flow cytometer, although they retain the plasma membrane antigens at the same level. Since volume parameters indicated by the scatter signal are regularly extrapolated to a sphere-like membrane surface, the reason for an apparently higher antigen density is obvious. In our case the only small yet significant difference between young and elderly populations occurred in the number of anti-CD8- and OKM-1- positive cells. This is consistent with the long-known decreased cytotoxicity and macrophage activation in elderly peripheral blood mononuclear cell populations. Conflicts in data and their interpretation may arise from, and reflect, the physiological heterogeneity of the elderly and also the differences in the choice of the elderly population studied. In our case, the subjects were selected by more severe criteria than those used for the SENIOUR protocol.[157] The higher rate of OKM-1-positive cells correlated well with the fact that an increase was observed in the peripheral blood monocyte count of this population. The effector functions of monocytes were found to be decreasing with aging. Meanwhile, the increased phagocytic activity may be explained, as an adaptation process, by the increased number of macrophages.

Treating peripheral mononuclear cells of the young and elderly with monoclonal antibodies against four antigens applied to the search for putative differences in their cell surface densities was further extended. The effect of OKT-3, OKT-4, OKT-8, and OKM-1 monoclonal antibodies on the phosphatidyl inositol breakdown, which is essential for certain types of transmembrane signaling, was investigated. Monoclonal antibodies against the differentiation antigens CD4, CD8, and CD11a did not induce the inositol phosphate cascade in lymphocytes. It is difficult to claim with absolute certainty, but in a few cases these antibodies not only failed to induce inositol phosphate breakdown, but we even found a lower titer of inositol phosphates after the treatment of lymphocytes from both young and elderly. OKT-3 antibody stimulated the formation of IP_3, IP_1, and IP_2 less effectively in elderly than in young. However, this difference between the stimulating power of the differentiation antigens was not exceeding 50, 40, and 40% for IP_1, IP_2, and IP_3, respectively, with the data for young taken as 100%. The potentiality of activating phospholipase C, which is responsible for the breakdown of phosphatidylinositol 4,5-bisphosphate, can also be considered as a measure of protein dynamics in the plasma membrane. The G-protein phospholipase C complex can partially be responsible for these effects. On the other hand, decreased availability of the substrate cannot be excluded from among the circumstances that can hinder the final outcome of the enzyme reaction.

Since changes in the plasma membrane potential may be responsible for alterations in the responsiveness of integral membrane proteins and activation processes not only in excitable (muscle or nerve) cells but also in lymphocytes, we investigated such changes and their possible correlation with aging. There was no difference between the resting potential of lymphocytes from peripheral blood of the old and young. However, when depolarizing concentrations of KCl were added to the extracellular space, this treatment proved to be less effective for samples from the elderly. Since the negatively charged bis-oxonol dye was applied in a flow cytometer to determine membrane potential values, the smaller fluorescence indicated the more negative potential on a cell-by-cell basis. This method highly decreases the toxic effects of the dye on normal cells, because the dye will enter only those cells which are depolarized, i.e., at least partially impaired. The preparation process of peripheral blood lymphocytes might impair the supposedly more fragile cells from the elderly. This effect may falsify the measured data, because in this case only the more viable elderly lymphocytes keep their integrity for the measurement. Care was taken to control the yield of the preparation of lymphocytes from both the young and the elderly. A carefully applied cell preparation method gave the same yield for both groups. Thus, any differences obtained could not be attributed to an artifact.

Although transmembrane potential is not supposed to exist if differences of the intra- and extracellular ion concentrations are equalized, lymphocytes from elderly patients showed a significant remaining resting potential at as high as 150-m*M* extracellular potassium concentration. The reason for this "resistance", or lowered responsiveness, may be connected to the lower permeability of the plasma membrane and a remaining ion gradient across the membrane.

Taken together with those data where Ca^{2+}-activated K^+ channels also showed a decreased responsiveness in plasma membranes of lymphocytes from the elderly,[18] it seems likely that, besides the increased lipid viscosity of these membranes, the integral proteins spanning the membrane and carrying important functions may also lose their flexibility in either a direct or an indirect way. Recently it was found that voltage regulation of Ca^{2+} activated K^+ channels may overcome the altered response of these channels in the plasma membrane of peripheral lymphocytes from the elderly.[18]

X. DISEASES INVOLVING OR CAUSED BY ION CHANNEL DEFICIENCIES

Recently more and more diseases are connected in a causal or, as to our present knowledge, phenomenological way to impaired ion channel activities. In this section a few of such better-known diseases will be discussed, with special references to the causal or symptomatic nature of the involvement of ion channels in their pathogenesis.

An increasing number of syndromes proved to be autoimmune diseases with a high number of very different and seemingly unrelated phenomena. The general importance of ion channel activities of cell membranes, regardless of their excitable or nonexcitable nature, is underlined by the availability of more and more data showing that autoimmune diseases and also those of genetic origin have in the background of the colorful syndromes some regulatory or other functional problems of ion channel disorders. Without aiming to describe here a full series of such diseases, some particular, but characteristic examples will be given, where ion channel functions or their regulations are involved. For example, data are available showing that the regulatory problems of a single Cl^- ion channel in the aerial pipe (bronchial tube) inner tissues may be the sole cause of the cystic fibrosis.

A. THE LAMBERT-EATON MYASTHENIC SYNDROME

The Lambert-Eaton myasthenic syndrome, a so-called paraneoplastic disorder, is based on an autoimmune reactions to the voltage-gated calcium channel. The altered functions and regulatory properties of the voltage-gated K^+ ion channels in autoimmune diseases, such as systematic lupus erythematosus (SLE), diabetes mellitus, and experimental allergic encephalomyelitis, are just about to be a little better understood. However, an emerging problem is the determination of their causative role in the development of these diseases. In the case of the Lambert-Eaton syndrome, this problem seems to be solved as evidence is accumulating that the syndrome causing heavy myasthenic symptoms is due to the presence of auto-antibodies directed against a particular plasma membrane protein. This protein was identified as the presynaptic voltage-gated Ca^{2+} channel in the neuromuscular junctions. The Lambert-Eaton syndrome is also known as a paraneoplastic disorder, since it is often associated with small cell lung carcinoma, a vicious type of lung cancer. The syndrome emerges in about 60% of the patients with small cell lung carcinoma, but their causal relationship is an open question yet. The serum of patients with Lambert-Eaton syndrome can be used to prepare immunoglobulin. This immunoglobulin injected into mice can transfer a high number of abnormalities observed in the syndrome. Since neuromuscular junction cannot function properly in the absence of functionally intact voltage-operated Ca^{2+} channels, it is no wonder that the first symptoms of the Lambert-Eaton syndrome are muscular weakness and a number of other corollary signs pointing toward the disorder of the autonomous nervous system, such as constipation, sexual impotence, or dry "mucous membranes". Electromyographic studies revealed that peripheral muscle will respond even to supramaximal excitation with very weak action potentials. The post-tetanic potentiation of electrophysiological activity is characteristic of the disease.

Due to the very high co-occurrence rate of small cell lung carcinoma, it was suggested that the autoimmunity might be tumor related.[175] However, this type of lung carcinoma can occur separately, not related to Lambert-Eaton syndromes. Even experimental evidence provided by Lambert himself in 1982 voted against the possibility that this particular type of carcinoma produces auto-antibodies against the voltage-operated calcium channel.[175] The presence of organ-specific auto-antibodies and also the co-occurrence with diseases known to accompany autoimmune diseases (e.g., hyperthyroidism) pointed toward autoimmunity. The therapeutic evidence comes when patients with Lambert-Eaton syndrome are treated with plasmapheresis, thereby depleting circulating antibodies or with immunosuppressive treatment, or the combination of these, and the treatment brings about a regression of clinical symptoms of the disease. The success of transferring the disease with the serum of Lambert-Eaton syndrome patients to mice provided strong evidence for the autoimmune origin of the problems. Subcellular immunohistochemical techniques detected the passively transferred immunoglobulin molecules of patients with Lambert-Eaton syndrome at such regions of the nerve terminal plate membrane where the acetylcholine release is supposed to occur.

The obvious autoimmune nature of the Lambert-Eaton disease has not entirely eliminated the possibility of a causative connection of small cell lung carcinoma and the syndrome. The interesting finding that voltage-gated Ca^{2+} channels are present in the carcinoma cells also revived the investigations in this field.

Cultured lung carcinoma cells also proved to possess the voltage-operated calcium channels, as Ca^{2+} influx inhibitable by calcium channel blockers was observed in K^+ depolarized cells *in vitro*. These obviously voltage-gated Ca^{2+} channel functions were significantly inhibited by immunoglobulin prepared from patients with Lambert-Eaton syndrome. It was hypothesized that the high number of voltage-gated calcium channels in the cancer cells, which are *per se* different immunologically from normal ones, but with high probability for cross reactions, may participate in the buildup of the Lambert-Eaton syndrome. However, not all the patients have lung carcinoma. It can be argued that there is more than one way to develop autoimmunity against voltage-operated calcium channels.

In order to find which type of calcium channels are those inhibited by the immunoglobulin of patients, experiments were carried out with pharmacological agents relevant to decide the question. Here we take for certain only the positive identification, because negative effects always provide less sturdy evidence. The L-type dihydropyridine- (nifedipin) sensitive channels, which belong to the class of the most frequently occurring voltage-gated Ca^{2+} channels, have been identified beyond doubt. N-type ω-conotoxin sensitivity in some cases suggests that these channels may also be involved. Molecular biological investigations, cloning cDNAs for the α_1 and α_2 subunits hint that voltage-gated calcium channel subunits from patients with Lambert-Eaton syndrome may not be identical to the normal ones.

Molecular immunological identification of the channel-forming subunit bearing the responsibility in developing autoimmunity may substantially contribute to the handling of this very inconvenient and deadly disease at the therapeutic level.

B. AN IMPORTANT EXAMPLE OF ANIONIC CHANNELS: THE Cl^- CHANNEL

Anionic channels are relatively neglected, and less intensively investigated than cationic ones, although their functional importance is commensurate with that of the cationic channels. They participate in the stabilization of membrane potential in a number of cells, eminently in muscle and also nerve cells. An important function of the chloride channels is the regulation of cell volume and also that of the transepithelial transport. A more intensive research was initiated by the discovery that deadly diseases such as cystic fibrosis have been connected with disfunction of Cl^- channels. A mutation in the cystic fibrosis gene product can block the cyclic AMP activation of epithelial Cl^- channels. This genetic defect, as we have seen, leads to such a perturbation of Cl^- channel regulation that the airways cannot be properly moisturized and the consequent repeated inflammations of the pulmonary system will kill those unfortunate persons who are suffering from this form of regulatory perturbation. Recently, a type of chloride channel of *Torpedo marmorata* was isolated and the primary structure was determined.[176]

This remarkable achievement came relatively late, as compared to similar ones with cationic channels (e.g., voltage-gated Na^+, K^+, or Ca^{2+} channels).[70] Nevertheless, it has to be greatly acknowledged, since identification and purification of Cl^- channels are not helped by high-affinity ligands unlike in the case of K^+ or Na^+ channels. Only the postsynaptic glycine receptors and those of γ amino-butyric acid were available, which helped the cloning of ligand-gated Cl^- channels.

The Cl^- channels are not only present in the plasma membrane of most cells and participate in various functions, but, similarly to cationic channels, there is a large variety of different Cl^- channels. Due to this fact it is assumed that Cl^- channels cloned and identified as ligand-gated ones may even belong to a different subclass. Cl^- channels have been isolated from kidney, trachea, as well as gall bladder cells. The most successful isolation and identification has been carried out with the Cl^- channel of the electric ray. Earlier isolation attempts failed mostly because of the nonspecific binding of inhibitors. Jentsch et al. used the advantage that the electric organ of *Torpedo california* is a rich source of voltage-gated Cl^- channels.[176] As they claim, this channel is thought to ensure the high conductance of the noninnervated membrane of the electrocyte necessary for efficient current generation caused by Na^+ influx through the acetylcholine receptor of the innervated membrane. The team applied an expression-cloning approach. The application of the method was made available by the lucky situation that the Cl^- channel from *Torpedo* electric organ, in contrast to other Cl^- channels, can easily be expressed in *Xenopus* oocytes. Jentsch et al.[176] cloned the complete cDNA, which coded this type of Cl^- channel. They showed that it was a single protein subunit, having a 89-kDa molecular weight, and containing several putative membrane-spanning domains which can be expressed in *Xenopus* oocyte plasma membrane resulting in the formation of functional Cl^- channels.

RNA from *T. marmorata*, injected into *Xenopus* oocytes, was sufficient to increase the Cl^- conductance of the oocyte plasma membrane fivefold by the third day after injection. It is interesting to note that the newly expressed channel is not calcium dependent, which provided an excellent tool for differentiation between the newly expressed and the endogenous Cl^- channels of *Xenopus* oocyte. The authors found that a 10-kilobase (kbase) single mRNA was enough for functional expression of Cl^- channels in the oocytes. Three clones with cDNA inserts, having sizes from 2.7 to 3.2 kbases, were isolated from a randomly primed cDNA library using suitable clones as probes. A partial sequence analysis showed that they contain the complete coding region. Transcribing all three original clones into mRNA and injecting them into *Xenopus* oocytes resulted in a large Cl^- conductance. This was similar to that observed with injection of the total *Torpedo* electric organ RNA, indicating that a single polypeptide chain encoded by this cDNA was sufficient to initiate a Cl^- channel activity.

Electrophysiological and ion substitution experiments proved that the increase in ion conductance was contributed really by chloride ions. Hyperpolarizing conditions activated the channel and induced an outward directed, rectifying current, similar to the one elicited by the injection of the total RNA. Stepping from a holding potential of -100 mV toward more positive potentials initiated small currents first, which then increased with larger depolarization. This leads to an overall outward rectification of currents after a hyperpolarization-induced activation. Again, the chloride conductance expressed from total electric organ RNA showed the same complex voltage dependence.

From the cDNA-derived RNA the amino acid sequence of a functioning chloride channel was deduced. The deduced amino acid sequence was analyzed for local hydrophobicity. The calculated molecular weight of the protein was 89 kDa, and the average hydrophobicity suggested as many as 13 membrane-spanning domains. Here, we wish to emphasize again that hydrophobicity alone does not necessarily mean that a particular loop of the polypeptide chain really spans the membrane. In this case, for example, the number of domains, which had the high enough number of amino acid residues in order to span the whole thickness of the membrane, was only nine. These domains had not less than 20 hydrophobic amino acids each. Two more loops had the theoretical possibility of crossing the membrane twice. These latter, closely spaced domains are separated by short sequences, containing proline or glycine residues which may provide a turn in the polypeptide chain. N-linked glycosylation sites, a consensus phosphorylation site (position 600 in the long hydrophilic stretch after domain 12), which can easily be phosphorylated by cAMP-dependent protein kinase, were identified in the amino acid sequences. Based on the determined primary structure and assuming that all hydrophobic segments are really spanning the membrane and the N terminus is located in the cytoplasm, the authors prepared a tentative model. The C terminus of the molecule, which has a potential glycosylation site, was put to the extracellular side of the membrane. The other glycosylation site (this is also only a potential one) was neglected and all 13 possibilities in the primary sequence to span the membrane were utilized as real membrane-spanning domains. Some membrane-spanning domains contain positive or negative charges (Arg, Lys, Hys). The authors offer these regions as selective filters which may direct the Cl^- ions across the membrane confining them to a narrow enough pathway. The beauty of this work is that, after a high number of unsuccessful attempts, these authors proved that a single polypeptide chain was sufficient for the *Torpedo* chloride channel. The analytical investigations and molecular biology experiments in *Xenopus* oocytes were expertly done.

Electrophysiological and biophysical data showed that the Cl^- selectivity of the channel was voltage dependent. A slow activation by hyperpolarization and, when activated, an increasing conductivity with depolarization characterized the channel.

Data also suggested that the Cl^- channel is likely to be homolinear, or, as is more commonly known, a double-barreled channel. Crosslinking studies may further support this suggestion. The isolation technique of this channel, maintaining the presence of the potential cAMP-dependent phosphorylation site, gives hope that this team may be successful in isolating and further characterizing the epithelial cAMP-regulated Cl^- channel, which is strongly suggested to be involved in cystic fibrosis. The cDNA probe developed in the above-analyzed experiments can be instrumental in this endeavor. It is interesting to note here that, while many cationic channels have closely resembling sequences, this chloride channel does not seem to belong to that superfamily of ion channels. As we have seen before, the members of a large gene family of potassium channels which belong to the so-called Shaker gene family of the *Drosophila melanogaster*, generally have six putative membrane-spanning domains. Voltage-gated Na^+ and L-type (dihydropyridine sensitive) Ca^{2+} channels have four repeats of six similar domains. The overall voltage gating mechanism of these channels has more and more accurate models; the above-analyzed chloride channel is unlikely to belong to this family. Even cGMP-gated (G-protein regulated) cation channels of the retinal rod photoreceptor could be classified into the large superfamily of voltage-gated (cation) channels. With some fantasy, the contradiction between the conservation of the voltage-gated cation channel mechanism during evolution and this Cl^- channel might be eased slightly. However, the so-called S4 domain, which dominates the gating mechanism as the key subunit of the cationic superfamily is completely missing here. Nature might have found more than one way for voltage gating.

In biochemical sciences there is a plethora of protein sequences, and many functionally closely related proteins have been found utilizing the same building stones or modules. The epidermal growth factor module is a good example of this.[177] Biological diversity has been achieved by a complex process of linking simple units. Proteinases in cascade systems such as blood clotting have epidermal growth factor and "kringle" modules or serine proteinase domains. Although these findings strongly support the logical view that Nature favored well-constructed mechanisms during evolution, more than one mechanism equally fit for survival can be imagined. There is a large number of orphan sequences as well (the TOMBs, i.e., temporary one-member bunches) which may point in this direction. Nevertheless, especially in more complex cellular systems, such as the central nervous system, it is likely that some uniformity of flexible structure-function relationships may promote plasticity. The plasticity of ion channels, i.e., that channel expression and function is modifiable and strongly influenced by the function triggered by extra-

or intracellular signals, may demand such conservation of functional elements. The voltage-gated cation channels, which can be classified into a large superfamily, where a particular domain (S4) plays a dominant role in the gating of ion flux upon changing the transmembrane potential, provide an excellent example of this. The opposing data, obtained very recently with a particular type of anion channel from *T. marmorata* do not necessarily contradict this elegant and very profitable direction of protein structure-function analysis. On one hand, there are quite a few different chloride channels. This freshly analyzed one could be a first member of another large superfamily of anionic voltage-gated channels which have to be discovered yet. Thus, this assumption offers another possible solution, which might explain the contradiction between the structure of the anionic and cationic channels, showing that the S4 domain may not have such singular importance as is currently suggested. It is noteworthy that it may provide a better exercise field to study such questions in lymphocytes if the low number of channels per cell can technically be overcome.

1. Porin Channels in Lymphocytes

As has been mentioned, Cl^- channels can have great importance in genetically determined diseases such as cystic fibrosis. Thus, it is no wonder that research in this field has been significantly intensified. Recently, Thinnes and co-workers significantly contributed to our understanding of the function of a human chloride channel, identified and isolated from human B lymphocytes.[178-180] This voltage-dependent anion channel, the porin 3IHL, forms a part of the chloride channel complex, which is observed in different cells and may also be affected in cystic fibrosis. The name, porin 3IHL, refers to the physical parameter that it is a 314-Da protein from human lymphocytes. Thinnes et al. described this porin late in 1989 for the first time in the plasmalemma of a eukaryotic cell line, where porins have only been found in the outer mitochondrial membrane.[178] The expression of porins in the plasmalemma of transformed human B lymphocytes is demonstrated by cytotoxicity and indirect immunofluorescence techniques with living and fixed cells, respectively. For this latter purpose a rabbit antisera was used by the authors, which was directed against the purified porin 3IHL and also against the free or acetylated synthetic peptides of its 19 N-terminal amino acids.

This channel-forming membrane protein proved to be a basic polypeptide. The single polypeptide chain is composed of 282 amino acids. It is not a glycoprotein, but its N terminal was found to be acetylated. As is expected for porins, the amino acid sequence shows 56% hydrophilic (i.e., charged) amino acids. In addition the 18 N-terminal amino acids of this porin 3IHL can be arranged into an amphiphilic α-helix, as in the case of other porins. It shows a relative conservation in the primary structure so that *Neurospora crassa* and *Saccharomyces cerevisiae* mitochondrial porins are approximately 29 to 24% identical with the 3IHL porin. A much higher degree of identity was found comparing at least partial data on mitochondrial porins from rat kidney and bovine heart, showing a sequence identity of about 90% to the human B lymphocyte porin analyzed by Thinnes et al.[178-180]

After determining the primary structure of this integral plasma membrane protein of human B lymphocytes, the purified porin 3IHL was reconstituted into lipid bilayer membranes, and its electric parameters were further studied. It was found that porin Cl^- channels built into lipid bilayer membranes formed voltage-dependent channels. The 4,4′diisothiocyanatostilbene-2,2′-disulfonate (DIDS), a potent inhibitor of chloride transport, applied at 100-μM concentration for 5 min, significantly altered the distribution and also the voltage dependence of these porin channels in the lipid bilayers. The porin channels showed a small irregular response in the presence of DIDS instead of having distinct steps. Furthermore, the voltage dependence of channels was abolished by DIDS. These properties together suggested that the voltage-dependent anion channel porin 3IHL might be a homologue of a similar porcine channel molecule observed recently by Breuer.[181] This was confirmed by two-dimensional gel electrophoresis experiments.

Some differences between the properties of the porin 3IHL channels and other maxichannels were revealed by patch clamp experiments, especially concerning anion selectivity. However, it has to be taken into consideration that electrophysiological characterization of the porin 3IHL channel was always achieved in reconstitution studies using an almost homogeneous preparation of the channel molecule. On the other hand, other maxi-chloride channels were always measured in the presence of other protein molecules, which obviously could influence their electrophysiological responsiveness.

It was suggested by Thinnes et al.[179] that porin 3IHL might be the part of the chloride channel which is deregulated in cystic fibrosis, due to a genetic deficiency. This deficiency is manifested in the lack of transcription of a regulatory protein that is mandatory for normal regulation of chloride transport and the consequent moisturization of the wind-pipe epithelial cells. As is known, the lack of this mechanism leads to a repeated occurrence of pneumonia cases, finally causing a death of these patients in the early teens. According to Thinnes,[178-180] another striking similarity between eukaryotic chloride transport and chloride channels involved in cystic fibrosis is that in both cases a possibly cell specific regulation of the channels by cytoplasmic modulators is postulated. These modulators are impaired by the genetic deficiency in this deadly disease. The porin 3IHL was shown to exist in B lymphocytes as well as in epidermal and endothelial cells. The protein, constituting the chloride channel, is always accompanied by antigens of the HLA DR, i.e., MHC class II, molecules in the plasma membrane. The MHC molecules are often in the steric vicinity of important regulatory molecules and receptors. Not much is known about the functional role of this neighborhood, although the nonrandom coexistence may not be without physiological significance.[182-188]

C. DISEASES INVOLVING MALFUNCTIONS OF K^+ CHANNELS

Physiological characterization and description of important cell membrane constituents, especially ionic channels, sometimes go almost parallel with the identification of their metabolic role or an impairment of this function in diseases. Similarly to what we have seen in the case of the Lambert-Eaton myasthenic syndrome, where an autoimmune mechanism blocked a voltage-gated calcium channel, thereby impairing the physiological calcium transport or in the case of cystic fibrosis, where a regulatory gene of a chloride channel (maybe a porin) is changed, some other channels might also be involved in the development of autoimmune diseases. Indeed, K^+ channels were also suspected to participate actively in, or at least monitor the development and stages of, a number of — mostly autoimmune — diseases.[4,5]

Arthritis, systematic lupus erythematosus, encephalitis, and certain types of diabetes mellitus are on the list of diseases where K^+ channel impairment may play a putative role. Thus, it is quite natural that eminent researchers of this topic paid attention to the possibility that disfunction of the most frequently occurring cation channel in lymphocytes may also have a role in the development of autoimmune diseases. An experimental approach to this important topic was taken by Grissmer et al.[4,5] Experimental polyarthritis can be relatively easily generated by immunizing animals with type II collagen for several weeks. The polyarthritis developed upon this treatment is usually characterized by cellular infiltration of the subsynovial tissues, synovial hyperplasia and, finally, by developing periostitis, which destroys the cartilage of the joints. This picture is practically identical to the general symptoms of human rheumatoid arthritis. A similar arthritis was induced by repeatedly injecting into the joints monoclonal antibodies which were developed against different cells of the cartilage by Dr.Tibor Glant in Debrecen, Hungary, and Chicago.[189] When not antibodies but type II collagen is injected, a T cell dependence is prevalent because this collagen is a T cell-dependent antigen. The CD4 antigen, characterizing the helper/suppressor T cells, plays an interesting role in this process, since monoclonal antibodies directed against this antigen retard the development of the arthritis and also anti-CD4 therapy decreases the clinical symptoms of not only the collagen arthritis, but also a number of other allergic diseases. The same authors have recently shown that in a number of diseases where the allergic origin can be assumed with great probability, the number of voltage-gated K^+ channels of the *l* type is greatly increased.[4,5]

In a study of Grissmer et al. the question was addressed whether the enhanced appearance of these channels in autoimmune diseases is a result of a generalized immune response *in vivo*, and whether these *l*-type channels can be found in collagen arthritis as well.[5]

CD4 and CD8 negative and Thy 1,2-positive cells from mice with collagen arthritis were investigated to see whether they possessed a higher than normal number of *l*-type K^+ channels. The mode of recognition and the signal differentiating between *n*- and *l*-type K^+ channels was discussed in an earlier section.

It was clearly demonstrated that the diseased cells expressed a very high number of *l*-type K^+ channels. Surprisingly, when other T cell subsets were compared in a similar way, it was found that either the $CD4^+$ and $CD8^-$ or the $CD4^-$ and $CD8^+$ cells retained their normal responsiveness to electrophysiological stimuli. The conclusion of the authors was that the abnormal appearance of *l*-type K^+ channels was confirmed in the double-negative ($CD4^-$, $CD8^-$) cells. Meanwhile, other T cells had the usual amounts of *n* type or the cytotoxic $CD4^-$, $CD8^+$ cells of the n' and *l*-type potassium channels, their number being between 10 and 20 per cell as measured by the whole cell clamp variation of the patch clamp. This enhanced expression of the *l*-type K^+ channels in double-negative T cells did not seem to be a generalized immune response *in vivo*, but rather associated exclusively with the symptoms of autoimmune diseases. The enhancement of channel expression is not only a signal of activation, since quiescent cells are expressing the above-mentioned low number of cells and, e.g., the mitogen activation increases the number of voltage-gated K^+ channels by a factor of approximately 20. The expressed new channels show the electrophysiological and biophysical properties of the *n* type. The exclusive abundance of the *l*-type channels in the $CD4^-$ and $CD8^-$ (double negative) cells seems to correlate well with the autoimmune diseases. However, no correlation was observed between the appearance of these *l*-type channels in double-negative cells and the stage of the autoimmune disease.

In a systematic work, Grissmer et al. have shown that collagen arthritis, SLE, type I diabetes, and chronic experimental allergic encephalomyelitis are sharing a number of immunological features in which this electrophysiological difference can always be found.[4] The authors, describing the phenomenon suggest its correlation with the pathogenesis of chronic autoimmune diseases. The *l*-type channels may be related to a yet-uncharacterized activation pathway which has no relationship to mitogenic or simple antigen activation. The causal relationship with autoimmunity still has to be proven. The actual normal occurrence rate of the double-negative cells (<5% of T cells) was not enhanced by the disease. Knowing the mechanism behind these observations at the genetic and membrane level may significantly contribute to the general understanding of autoimmune diseases. It is no wonder that the most easily accessible channels, the potassium channels were studied from this aspect.

However, a similar participation of other key regulatory channels, determining membrane potential levels and dynamic molecular events in the plasma membrane, may have important functions in the pathogenesis of autoimmune and also other diseases as well. Na^+/H^+ exchange, the ATP-dependent Na^+-K^+ATP-ase, Ca^{2+}-activated K^+ channels are eminent candidates for further investigation. Data obtained so far clearly indicate the future of a cellular pathology that will be accessible only by those methods, which have the capacity to monitor dynamic functional events. The importance of pathomorphology is enormous, despite its static character that reveals a "frozen" steady state of microscopic morphology. Physiological and biophysical methods advocate a pathology at the level of molecules, without losing the information coming from the intact cell, in contrast to biochemistry, which in most of the cases is forced to work from cellular homogenates. This statement, of course, does not hold for the beautiful experiments of site-directed mutagenesis; however, those are not purely biochemical techniques anymore.

XI. PHARMACOLOGICAL MECHANISMS OF SOME IMPORTANT DRUGS MAY BE CONNECTED TO CHANGES IN THE PLASMA MEMBRANE POTENTIAL

There exist a practically endless list of pharmacological agents and toxins which directly or indirectly influence ion channel activities in muscle and neurobiology with narrower or broader specificity and consequently the membrane potential. From a pharmacological point of view it is worthy to note that the abundance of effectors and toxins which are available to modulate the function of a particular ion channel or the membrane potential does not necessarily correlate well with the number or potential importance of even widespread therapeutic agents. K^+ ion channels are perhaps the best-characterized family of ion channels with very specific agents directed to different subspecies of their large families. However, the number of calcium channel agonists and antagonists or even those of Na^+ channel activity modulators are much greater.[70]

Another field of not-negligible interest is that not infrequently the mode of action of important or sometimes indispensable drugs is not fully understood, despite extensive and widespread investigations. These compounds of usually well-known chemical structure frequently have some rather neglected or even overlooked "side" effects also on ion channels and/or plasma membrane potential. Since the real mechanism of their mode of action has not yet been resolved, despite the ever-accumulating amount of data available on biochemical and molecular details, a new approach of the problem seems to be mandatory. Analyzing the great haystack of biochemical and molecular biological data looking truly important, and of course they really are, one factor must not be forgotten: the plasma membrane (with its ion channels, and also membrane potential and other intricate structures) is *per se* the first and earliest theater of any drug-cell encounter. Thus, effects initiated here may trigger or at least modulate any latter effects, including even contingent genetic regulatory functions.[147] Even steroid hormones, which have well-characterized intracytosolic receptors that translocate upon hormone binding into the nucleus, have recently been shown not to have negligible effects at the level of the plasma membrane.[190,191] As an

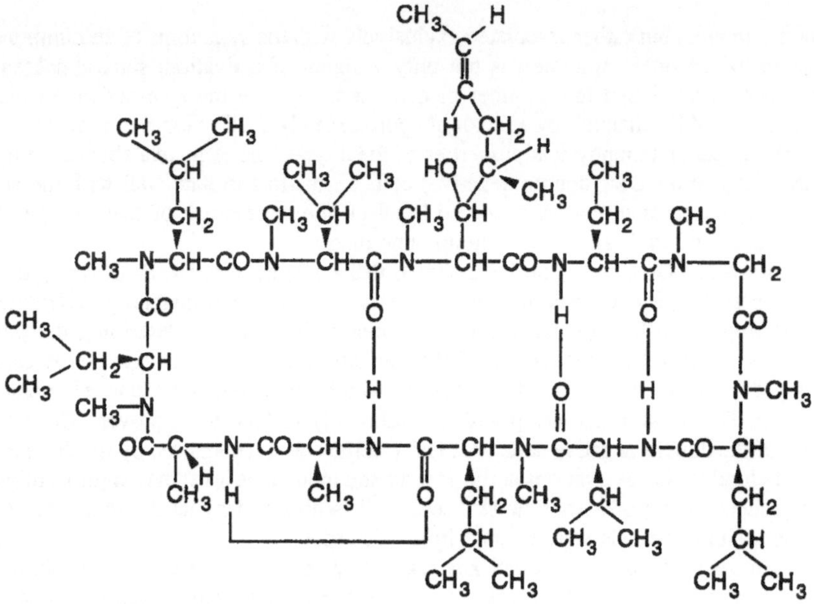

Figure 17. The molecular structure of cyclosporin A. The structure does not exclude an ionophore-like effect, as is suggested.

example of the above-outlined problems, the effects of cyclosporin A will be discussed in detail. Cyclosporin A (CsA) is an extremely important drug, known for nearly two decades without accurate knowledge of its mode of action, and also known to have membrane effects.

CsA is an antibiotic with a cyclic endecapeptide structure which is shown in the Figure 17. The immune suppressive effect of this drug was discovered by Borel,[192,193] and its stereospecific synthesis was elaborated by Wenger.[194] Since the discovery of the therapeutic effect of this compound, isolated from fungi found in the Norwegian mountains, several variations of the active and inactive configurations of the original chemical structure have been synthesized. The great interest in this drug (or rather drug family) is highly justified by its extremely potent medical effectivity. It can selectively inhibit humoral and cellular immunity. The inhibition of the latter is indispensable in solid organ (kidney, heart, liver, etc.) transplantation and also in the treatment of autoimmune diseases. An obvious handicap of the original drug was that the range of its effective therapeutic dose overlapped with the toxic dose interval. The unfortunate paradox effect that handicapped the application of the drug in kidney transplantation was its nephrotoxicity. A few years ago Hiestand et al.[195] reported about new derivatives with much less severe nephrotoxic side effects. Taking into account that this is a most important key drug in organ transplantation, it may sound strange that the mechanism of its action has not been clarified so far.[192]

A new drug, FK-506, with similar potential impact on transplantation was recently found by a Japanese scientist.[196-198] However, there was no obvious difference found in the biological actions of FK-506 and CsA, although they have distinctly different chemical structure. Here we provide only a short excerpt of the many and thorough investigations on CsA, before focusing our interest on the analysis of its membrane effects.

One of the highlights of the many rigorous investigations into the mode of action was that of Krönke et al.[199] and parallel with them that of Elliot et al.[200] Both groups concentrated on the possible impairment of lymphokine synthesis and showed that the IL-2 synthesis of CsA-treated cells was blocked at the mRNA level. However, the synthesis of IL-2 receptors and their expression in the plasma membrane remained intact and the receptors were responsive to IL-2 in lymphocyte cultures of BALB/c and C57BL/67 mouse spleen cells. The obvious reason for targeting the investigations on the lymphocytes has been that CsA inhibits T cell activation (cellular immunity and also cytotoxic effects), and leaves B cell stimulation relatively unhampered (humoral immunity, i.e., antibody production). Many investigations revealed a vast forest of different effects on the mitochondrial or lysosomal functions; however, none of them proved to be unequivocally linked to the so-important immune-suppressive effect. Receptors for

CsA were supposed to be in the cell membrane. Radiolabeled CsA on mouse thymoma cells indicated that the drug was associated with a specific intracellular protein fraction, termed cyclophilin.[201]

CsA may also inhibit the Ca^{2+} binding function of calmodulin through binding to this Ca^{2+}-specific protein. Investigations of the preferential uptake of dansyl-labeled CsA by different subsets of human peripheral blood lymphocytes revealed a dim and a light population of cells, classified on the basis of their dansyl-CsA content.

In 1985 we observed an effect of CsA on the plasma membrane of human and murine lymphocytes. We pursued this line and in the light of the above-outlined great variety of effects of CsA at very different levels of cell metabolism, where none of the listed data could be linked to the mode of action of CsA with a high enough certainty, our experiments and the derived conclusions offered a simpler approach. We suggested that an early effect of CsA on the cell membrane of lymphocytes may be a common denominator for its different activities.[15,16,18-22,202-204]

Although our data to be outlined here were soon confirmed and we also expanded the scope of these types of investigations, they may not have higher significance than any of the many futile attempts to provide a unified explanation for the effects of CsA.[138] The significance of our approach has been revitalized by some most recent investigations.[4] These new investigations pointed out a possible involvement of K^+ ion channel activities in a number of autoimmune and also some other diseases. Some of the data have already been mentioned, dealing with the regulation of K^+ channels and the consequences of their different perturbations. If autoimmune diseases can be connected to altered ion channel activities, then our earlier findings that under specific conditions CsA depolarized or depending on its dose hyperpolarized human and murine lymphocytes have to be re-evaluated and revisited. This is all the more so, because the causal involvement of K^+ ion channel activities in T cell activation is more and more generally accepted. On the other hand, there is a difference, very likely only in the quantitative sense, between the occurrence of particular K^+ ion channels in T and B lymphocytes. Below we briefly summarize our findings, assumptions, and suggestions about the mode of action of this drug. Thereafter will try to analyze our available data and also that of other authors in light of the hypothesis that the membrane effect of CsA can be instrumental in its immunosuppressive effect.

A. THE MEMBRANE HYPOTHESIS OF THE MODE OF ACTION OF CsA

Upon encountering the cell membrane, CsA, having a hydrophobic character, merges into the lipid region of the membrane. Of course, this hydrophobic character does not prevent its translocation into the cytoplasm, especially along the endoplasmic reticulum. Nevertheless, at the first theater of the encounter, the plasma membrane, it causes a dose-dependent hyper- or depolarization of the membrane.[15,16,22,72] Data will be presented which show that its molecular structure allows, and measurements support, that it can act as a weak Ca^{2+} ionophore.[72] A modest increase in intracellular Ca^{2+} can activate calcium-sensitive K^+ ion channels in the membrane and the subsequent diffusion potential generated by the activated K^+ channels causes hyperpolarization. Depolarization will be observed when a higher dose of CsA is added (about tenfold of the therapeutic dose range). Such depolarization of lymphocytes by CsA, also in combination with established calcium ionophores, like ionomycin or A-23187, was described and discussed in detail by us.[15,22] This could be the first and maybe only a trigger effect of the drug that can produce a number of other effects later on. Of course, since CsA also penetrates the cell interior, further effects, independent of the outlined very first one, may and can modulate the development of a full-scale immunosuppression.

One crucial point is the way CsA can pharmacologically differentiate between T and B cells, i.e., cellular and humoral immunity. Ishida and Chused[73] provided a detailed, fluorescence technology-based analysis suggesting that T cells do, while B lymphocytes do not, have Ca^{2+}-activated K^+ ion channels in their plasma membranes. Since their experiment, a number of data have surfaced claiming that Ca^{2+}-activated K^+ ion channels were present in B lymphocytes or lymphoid cell lines as well. Electrophysiological, i.e., patch clamp, experiments (which were expected to confirm the more circumstantial fluorescence data on the existence of these channels in lymphocytes) were also reported. However, while fluorescence spectroscopic membrane potential measurements indicated the presence of Ca^{2+}-activated potassium channels on T lymphocytes, patch clamp measurements failed to confirm this finding. The only exception was the Jurkat cell line analyzed by clamping nystatin perforated patches. On the contrary, B lymphocytes yielded easier and displayed this particular type of K^+ channels.[59] Our data from flow cytometric membrane potential measurements and patch clamp as well suggest that quantitative differences may exist between the two major immunocompetent cell types in this respect (Gáspár et al., personal communication). Accepting this view, the importance of the trigger signal through the Ca^{2+}-activated K^+ channel becomes far more crucial.

Thus, the membrane hypothesis on the mode of action of CsA, in the final form, is as follows: CsA inhibits T cell activation through hyper- (or in case of higher doses with de-) polarization of the plasma membrane. Both sustained hyper- and depolarizations have been reported as processes inhibiting cell activation.[24,112] Since B cells are supposed to have a lower density of this particular type of channel, the Ca^{2+}-activated K^+ channel in their plasma membrane, the effects of CsA are far less expressed, i.e., humoral immunity is not much impaired. Any other effects which develop later in the time course of the action can be considered not unimportant, but secondary or tertiary, which may not be relevant in the explanation of the effect of the drug.

At this point we would like to emphasize that this is certainly not a claim that the only and absolute decisive effect of CsA has been found. Rather, it is a new point of view which may raise the interest of future researchers to investigate biophysical and biochemical regulatory processes in their complexity. Whatever the importance of the above findings may be, one cannot neglect the fact that a single gene mutation, which, let us say, encodes the proper regulatory protein of the chloride channel and thereby controls the suitable moisturing of the air pipes, thus preventing the developing of cystic fibrosis, can cause such a deadly disease. However, one cannot neglect the other side, namely, that effects on such regulatory processes may be evoked at the membrane level as well. If we perturb an allosteric regulation of an ion channel with a specific drug, we may generate a similar unfortunate or fortunate (i.e., pathophysiological) misregulatory effect. In the case of CsA, the mode of action is likely to have more than one focal point. A similar title of a paper by Gelfand et al.[138] is truly expressing the status of our knowledge: "Cyclosporins inhibit lymphocyte activation at more than one site."

Now, in the analysis below, we will present the experimental background.

1. The Effect of CsA on the Membrane Potential of Lymphocytes

The effect of CsA on the membrane potential has not been a chance observation. Since the early 1980s, the two-dimensional pattern formation of cell surface elements (the domain-like structure of lipids, nonrandom codistribution of cell membrane elements, association and dissociation of complex receptor structures) and its dynamic interaction with cell function was a working hypothesis.[17-23]

Since CsA was known to differentiate between the two major classes of immunocompetent cells, the T and the B lineage, by inhibiting the activation of the first and leaving relatively unperturbed the second, our interest was focused on the mechanism of the action of the drug. Many effects of CsA were already known when we started considering the possible membrane effect of CsA. In collaboration with Dr. Aszalós at the Food and Drug Administration, we applied $DiOC_6(3)$, a carbocyanine-based fluorescent dye at that time easily available, to lymphocytes in a flow cytometer to look for the possible changes in their plasma membrane potential. CsA is hydrophobic, and usually is dissolved in dimethylsulfoxide or alcohols when added to *in vitro* cellular systems. Thus, the assumption that it may have specific or nonspecific effects in the plasma membrane seemed to be highly probable.

The first experiments with CsA immediately showed a decrease in the fluorescence of the investigated human peripheral blood and also mouse splenic and thymic and rat thymic lymphocytes.[15,22] The positively charged cationic dye had to be equilibrated for about 20 to 30 min with the cells. After this, CsA was applied in about five times higher concentrations than the pharmacological dose used in transplantation therapy or mixed lymphocyte-culture experiments. The philosophy behind the increased dose was that we wanted to see if there were any immediate effects in the membrane and wanted to enlarge them, in order to make monitoring easier. The membrane potential decreased and this depolarization was dose dependent and well observable even at the upper limit of the pharmacological dose range (0.5 μg/ml). Gelfand et al.[138] confirmed our findings both with oxonol- and carbocyanine-based (i.e., anionic and cationic) dyes as well. They supported our observations concerning the directions of the membrane potential change upon addition of CsA.

Since it was published at that time that insulin pretreatment concentrated the immune suppressive effect of CsA, we tested the effect of adding two units of insulin on CsA-induced depolarization.[22] Insulin eliminated the depolarizing effect of the CsA. On the other hand, when an increase in the Ca^{2+} concentration, induced by the calcium ionophore A-23187, was studied on the same cell types (T and B cells of human and murine origin), the separately depolarizing dose of CsA caused a decrease in the depolarizing effect of Ca^{2+} and ionophore. Now it is quite obvious that depolarization modulates the effect of Ca^{2+}

ions.[72] A more specific Ca^{2+} ionophore, ionomycin, showed a similar countereffect, the same sort of direct or indirect antagonism with CsA.

The lipid structure was made responsible for these observations, and to pursue this line, electron spin resonance experiments with 5-doxylstearic acid, a spin probe, were implemented to get a closer insight into the mode of action of CsA. ESR spectra obtained from the membranes of intact cells spin labeled with 5-doxylstearic acid show contributions mostly from the spin labels of restricted motion with negligible contribution from free moving spin labels. The spin label concentration was kept low enough in order to avoid spin-spin interactions. Results clearly showed the effects of CsA on the mobility of phospholipids in the plasma membrane. Parallel and perpendicular components of the motionally averaged nitrogen hyperfine tension, order parameter, and polarity factor, the usual spectral parameters of these types of measurements, were compared before and after CsA treatment. The results provided evidence that CsA pretreatment significantly decreased the mobility of the phospholipids in the plasma membrane, as was well expressed in the significant change of the order parameter from 0.6 to 0.626. Since these measurements were done immediately after the addition of CsA to spin-labeled cells, a rapid equilibrium between CsA and cellular membrane components can be assumed. This finding might still be worth mentioning since, for example, natural killer cell-mediated cytotoxicity is altered by an increase in membrane fluidity.

In a later study on the early stage of drug-cell interaction more attention has been paid to the effect of CsA on the intracellular calcium level. The investigations were made in the mid-1980s, thus the first member of a series of fluorescent calcium-chelating agents, the quin-2, was applied. The measurements were carried out with cell suspensions in a spectrofluorimeter, and changes in intracellular Ca^{2+} concentration were calculated after calibration. The samples were kept at 37°C, and the buffer (phosphate buffered saline = PBS) simulated the ion composition of the normal extracellular space, with the exception that 10 m*M* Hepes was applied to keep the pH at 7.3 throughout the experiment. Human peripheral blood lymphocytes were measured this way. Addition of 12.5 µg/ml CsA in DMSO increased the fluorescence by a factor of 1.5. The calculated increase in the intracellular Ca^{2+} concentration was more than 100 n*M*. At the end of the experiment, calculations were verified by addition of 0.6 µ*M* ionomycin, which brought the intracellular Ca^{2+} concentration to equilibrium with the extracellular one.

It was an interesting observation that monoclonal antibodies also had some effect on the membrane potential changes triggered by the CsA. The OKT-3, OKT-11, OKTa1, OKT-4, and OKT-8 antibodies were tested. Preincubation with the last two antibodies directed against the CD4- and CD8-positive cells provided a slight but significant protection against the depolarizing effect of CsA. Since the CD3-directed OKT-3, among other antibodies, has a much higher number of binding sites and CD3 positivity is almost universal for human peripheral blood lymphocytes, the phenomenon observed with OKT-4 and OKT-8 might deserve more attention than it has received so far. It is interesting to note that anti-CD3 antibodies alone also had a depolarizing effect on human peripheral lymphocytes, but there was no synergism with the effect of CsA. Since CsA decreased the intracellular K^+ concentration in a dose-dependent way as revealed by ^{42}KCl experiments, the possibility arises that CD4 and CD8 antibodies might be in the immediate vicinity of K^+ channels. This hypothesis was not further pursued at that time.[16]

Since insulin *in vitro* decreased the effectivity of CsA in changing the membrane potential, two other important hormones, known to affect cell metabolism and activation of lymphocytes, have been further tested in combination with CsA. Interleukin-2 and interferon-γ (IFN-γ), which are known to influence the maturation of lymphocytes, were selected. While 10^3 units of human recombinant IFN-γ abolished the depolarizing effect of 10 µg/ml CsA, the recombinant IL-2 (50 units) enhanced the depolarizing effect of the same amount of the drug. The actual interpretation of these kind of data is not simple. However, they are not entirely untreatable in the light the background data. CsA is also known to inhibit the synthesis of IFN-γ.[203] Since IFN-γ showed a strong interaction with CsA at the membrane level by preventing the depolarizing effect of CsA on human peripheral blood mononuclear cells, both direct and indirect interactions are possible. However, the strong immediate interaction with one of the earliest effects of CsA (i.e., changes of the membrane potential) suggests a direct interaction with the early effects as well as with the final outcome.

The opposite effect of IL-2 in comparison with that of the IFN-γ gives a valuable control for the above observations, because it helps to exclude the possibility of an artifact caused by the nonspecific effect of the peptide. However, the human peripheral blood lymphocytes were not primed for IL-2, yet we obtained a facilitation of the depolarizing effect of the CsA. A direct interaction with the dye cannot be the explanation because in that case IFN-γ would have the same effect. The binding of IL-2 (or at least its presence) alone did somewhat decrease the membrane potential; however, it was less than its due share based on experiments when it was added together with CsA. The actual existence of a low amount of IL-2 receptors was verified with FITC-anti IL-2R antibody, although their level was really very low. It cannot be excluded that weak, nonspecific binding of IL-2 to cell surface elements also contributed to the effect. IL-2-primed cells, in separate experiments, carried out with oxonol as potentiometric dye showed a slight, but well-defined hyperpolarization. This also points toward the nonspecific effect of IL-2 in the cited cases. Ionophores specific for K^+ ions, such as valinomycin, did not interact with CsA directly and decreased the effect of CsA on the membrane potential. This is in agreement with data that CsA facilitates K^+ release from the cells.[16]

2. Ca^{2+}-Activated K^+ Channels may be Responsible for the Effects of CsA on the Membrane Potential

The apparent contradiction that in murine cells the effect of A-23187 and ionomycin on the membrane potential was somewhat decreased in the presence of CsA, while in human lymphocytes calcium ionophores and CsA acted in an additive way can satisfactorily be interpreted by data obtained in our laboratory by Mátyus et al.[85] The phenomena are likely to depend on the voltage regulation of Ca^{2+}-activated potassium channels. According to their observations high calcium concentration (1 to 2 m*M*) in the extracellular space and addition of Ca^{2+} ionophores cause depolarization. On the other hand, modest amounts of Ca^{2+} (below 0.1 m*M*) activate calcium-sensitive K^+ channels, and hyperpolarize the lymphocytes through the diffusion potential which is generated by the outward-directed flow of K^+ ions. As we have seen before, depolarization of cells before adding Ca^{2+} ionophore could overcome the inhibition of Ca^{2+}-activated K^+ channels by micromolar concentrations of Ca^{2+}. This phenomenon had not been observed earlier; however, Bregestovski et al. have also seen the potential decreasing effect of calcium influx, and it was attributed to a calcium poisoning of the calcium-activated potassium channels.[87] Experiments on human cells were carried out with higher ionophore and CsA concentrations.

In our present view, CsA can be considered a weak Ca^{2+} ionophore. This is also supported by the temperature-dependent effect of CsA on the membrane potential (unpublished observations). The release of ^{42}K ions and increase (upon addition of a low amount of CsA, within a short time) or decrease (higher amount of CsA after longer incubation) of membrane polarization can all be interpreted in light of the different excitabilities of Ca^{2+}-activated K^+ channels by low or high amounts of Ca^{2+} ions. In a more recent investigation by Vereb et al. the effect of CsA on the membrane potential was further studied also using, besides oxonol, the fluorescent indo-1 Ca^{2+}-chelating agent, which is significantly more fluorescent than the earlier applied quin-2.[72] In order to make the investigations more focused on the T and B cells in a comparative fashion, the human lymphoblastoid T cell line, HUT-78, and the B lymphoblastoid line also of human origin, P3HR-1, were studied. The "biphasic" effect (hyper and depolarization) of CsA on the membrane potential was observed with the anionic bis-oxonol in a flow cytometer on HUT-78 cells. Immediate effect of CsA was hyperpolarization, and a 10- to 60-min preincubation of samples with CsA and subsequent membrane potential determinations showed a dose-dependent depolarization. In the human B lymphoma line JY, the hyperpolarization was not observed, only the depolarization. Since B cell lines either do not have,[73] or, if they do, supposedly have only a lower number of calcium-activated potassium channels as compared to T lymphocytes, the results seem to be consistent with data of the current literature.[59,72,73] Calcium measurements with indo-1 and quin-2 fluorescent calcium indicators showed basically the same pattern of action on HUT-78, P3HR-1, and JY cells or rat thymocytes. The only noticeable difference was that the increase in the intracellular calcium concentration was transient on a minute time scale.

3. Further Data on the Mechanism of CsA

Importantly enough, Gergely and Aszalós studied the effect of CsA on the anti-CD3 monoclonal antibody and lectin-Con A-induced activation and the membrane potential of human T lymphocytes.[202] Their philosophy was that the early effects of CsA have already been studied on human and murine lymphocytes;[15,16,22,72,138] however, the changes in plasma membrane potential occurring in the later stages of activation had not yet been studied *in vitro*. Therefore, the field of investigation was also extended over CD4-positive cells of CsA-treated transplant patients.[204] They investigated the differences in cell proliferation induced by anti-CD3 monoclonal antibody and Con A and the effect of CsA on a number of subsequent measurable parameters and the membrane potential. CsA equally suppressed anti-CD3 and

Con A triggered T cell proliferation in human peripheral mononuclear cells. An enhancement of the suppression was observed, when the cells, after 30 min of incubation with anti-CD3 antibodies, supposedly became free of the unbound or weakly bound antibodies. A similar treatment did not alter the Con A-induced proliferative response. The authors suggested that very likely more than 30 min were required for the antibody to crosslink with T cells and macrophages, which was thought to be essential for effective proliferation. It was actually already shown that an elimination of anti-CD3 antibodies and the phorbol ester TPA even after 4 h of activation may result in a loss of IL-2 production. The authors supported the finding of other authors that a slight decrease was observable in the expression of the p55 subunit of the IL-2 receptor complex. The externally added (10 U/ml) recombinant IL-2 decreased the inhibitory effect of CsA on the proliferation, but this effect was varied depending on the donor of the investigated human peripheral blood mononuclear cells. The proliferation enhancing effect of macrophages was quantitatively decreased by 1 to 5 μg/ml CsA pretreatment.

Activation of cells by both lectin and the anti-CD3 treatment decreased the membrane potential. The decrease in plasma membrane potential was running parallel with the proliferating ability of the cell population. The degree of the repolarizing effect of CsA was different for the lectin and the antibody. However, this could be caused by a direct interaction of CsA, Con A, or anti-CD3 with oxonol. Thus, in such cases, the quantitative evaluation was difficult. The finding, however, suggests that the transmembrane signal transduction at the level of membrane potential was reversed for both stimulatory agents in the presence of CsA.

Both Con A and anti-CD3 activate the inositol phosphate system, increase intracytosolic Ca^{2+}, and, finally, the effective signal transduction requires PKC activation. The authors suggest that under their experimental conditions CsA had a similar effect on the K^+ channels of lymphocytes as already described earlier in collaboration with us.[16]

B. OTHER IMMUNOSUPPRESSANTS AND IMMUNOPHILINS

The immune suppressor FK 506, which is structurally unrelated to CsA but has a very similar immune-suppressing effect on T cells and its structural analogue, rapamycin, were studied recently. The data pointed out that two distinct signal-transducing pathways are inhibited in T lymphocytes by complexes formed between immunophilins, a putative immunosuppressant-binding protein family and either FK 506 or rapamycin. Rapamycin and FK 506 change the resting plasma membrane potential level. What makes it more interesting is that FK 506 interferes with the antigen-receptor induced signals, while rapamycin bound to its immunophilins interferes with IL-2 induced signals. Bretylium tosylate, a quaternary ammonium ion, which increases membrane potential through activating the Na^+-K^+ pump, inhibits the antigen receptor and lectin-induced transmembrane signaling, while enhances the IL-2 induced signals. The whole effect depends on a number of different parameters: (1) the addition of the immune-suppressant together with the stimulatory agents, (2) the time that elapses between the addition of the suppressant and the activator, (3) their respective doses, and (4) the kind of the activation signal that is supposed to be triggered by the stimulatory agent in question.

A 17-kDa putative intracellular protein receptor for CsA has been suggested by Handschumacher et al.[201,205] and Harding et al.[206] in the mid 1980s. This protein was code named cyclophilin A, and had high affinity to immunosuppressive analogues of CsA and a significantly lower affinity to its inactive analogues. The FK 506 is also associated with a specific group of proteins in the cytosol that were simply labeled FKBPs being FK-binding proteins. Both immunophilins act as peptidyl-prolyl *cis-trans* isomerases, or rotamases.

A recent review of McKeon suggested that receptor-drug complexes may serve as functional mediators of immunosuppression.[207] However, the intracellular immunophilins can lose their rotamase activity without suppressing T cell activation. Thus, the possibility has arisen that the drug-receptor complex can act as a single effector. It was observed that cyclophilin C bound a higher molecular weight intracellular protein component (77 kDa) in the absence of CsA, while a 55-kDa protein migrated together with the cyclophilin-CsA complex. A very promising set of data on the binding of CsA to cyclophilin A and C as well as on FK 506-FKBP complexes showed that both types of complexes were associated with calmodulin and the calcium- and calmodulin-dependent phosphatase 2B. This latter is also known as calcineurin. While neither the drugs (CsA or FK 506) nor the respective specific immunophilins alone had any effect on these molecules, the drug-receptor complexes inhibited their phosphatase activity also *in vitro*. These findings may provide the key to the ultimate mechanism of these immunosuppressive agents, originally found in fungal toxins. The fact that rapamycin and FKBP complexes are likely to suppress separate signal-transducing pathways does not contradict the above statement. On the contrary, it is just underlining the significance of different signal transferring routes.

These very recent data have not diverted us in the least extent from attributing an important role to the immunosuppressants at the level of the plasma membrane.

The plasma membrane as the first theater of the drug-cell encounter can evidently modulate the effects of many drugs. We provided data and also present a model later in this discussion suggesting that, without diminishing the importance of experimental data dealing with the high number of intracytosolic biochemical effects of immunosuppressants, especially those of the recently discovered interactions with phosphorylation processes, the plasma membrane effects of these changes alone can be responsible for a great deal of the immunosuppression.[207]

XII. THE FUNCTIONAL ROLE OF LIGAND-INDUCED MOVEMENTS IN TRANSMEMBRANE SIGNALING

The ligand-induced movement of cell surface receptors seems to be a primary event in the transduction of several extracellular signals that control cell growth and differentiation. Beyond this generally accepted view, however, a relatively low number of solidly established experimental facts are available to support this view. Zidovetzky et al.[182,184,186] were among the first who studied such phenomena with a novel physical method.

They determined that rotational diffusion of epidermal growth factor complexed to cell surface receptors reflected rapid micro-aggregation and endocytosis of occupied receptors. This study applied a time-dependent phosphorescence anisotropy decay measurement that is described in chapter 2 of this book. More recently van Belzen et al.[188] determined that interaction of EGF receptors with the cytoskeleton are also related to receptor clustering.

Other proteins such as the $p21,^{ras}$ which is ubiquitously expressed in mammalian cells and is thought to be involved in a growth-promoting signal-transducing pathway are also linked to redistribution in the membrane upon signal processing. In B lymphocytes, surface immunoglobulin receptors upon cross-linking by antibodies first form small aggregates, patches, which will be collected into a single assembly, called a cap, on a minute scale. This cap is located at one pole of the cell (this is called capping). This process is the trigger signal for the activation of B cells. Immunofluorescence investigations in microscopes revealed that $p21^{ras}$ co-caps with surface immunoglobulin molecules in mouse splenic B lymphocytes. This redistribution (or rather coredistribution of the receptors) does not occur when Con A causes a similar but less specific capping phenomena.[208] Con A is known to bind to proteins, with specificity only, to the oligosaccharide moiety: the α-mannopyranozide rings. Such experiments were able to point toward the possibility that $p21^{ras}$ may be a component of the signaling pathway, very likely through G-protein coupling, initiated specifically by the capping of surface immunoglobulins together with $p21^{ras}$ in B lymphocytes.

Despite the scarcity of available data, it is obvious from the above-cited examples that such kind of investigations are prerequisites for the better understanding of transmembrane signal-transducing events. Biochemical facts can give us the chemical background and sometimes the sequence of the molecular steps, but, in order to have a physiological picture, we have to fit those data into the cellular structure and the dynamic behavior of cell organelles and their molecular components.

XIII. THE EXISTENCE, TYPES, AND POSSIBLE FUNCTIONAL ROLES OF CELL SURFACE PATTERNS

Membrane potential and the closely interdependent ion channel activities are naturally highly influenced by the molecular interactions of cell membrane protein elements. A possible contribution from the lipid moieties of the plasma membrane cannot be excluded, although the lipid environment can also be simply considered as the surrounding of the protein molecules. However, this latter view or rather model is a far cry from the real understanding of the actual direct participation of lipid elements and their domain structure in the protein-protein interactions in cellular membranes in general and in the plasma membrane in particular.

The great success of the Singer-Nicolson fluid mosaic membrane model that has been justified indeed, overshadowed some facts for a considerably long period.[209]

The authors themselves had never thought about their model as an absolute dogma, i.e., that every membrane constituent was completely independent in its relative movement. Co-capping phenomena, successful cross-linking experiments of functionally cooperating elements, like receptor subunits in the membranes signaled very early that cell surface elements, although they were independently mobile in general, did not have an absolute freedom in their relative mobility.[210,211]

In an earlier suggestion by us it had been predicted that a quasi-two-dimensional pattern of protein and lipid elements existed in the plane of the plasma membranes.[14] Furthermore, it was also suggested that any significant change in this pattern might have functional consequences.[14,17-19] Cells, like lymphocytes, are capable of responding to external signals in their specific ways, and these capabilities include practically all kinds of the known immunological defense mechanisms. Specific responses, being highly complex phenomena, are generally preceded by a well-defined number of steps, partially confined to molecular events in the cell membrane. Such activation processes are far less clear at the level of cell nuclei.

Among the activation phenomena in the plasma membrane, the best understood is redistribution of receptors or other membrane constituents. These redistributions frequently occur inside, or rather in the plane of the plasma membrane. Ligands, like epidermal growth factor, having bound to their respective receptors, often induce local aggregation processes, which take place relatively slowly. At the same time the plasma membrane, altered by the ligand binding and the triggered activation processes, offers a different target to the external world, with a new pattern of ligand-binding capacity and thereby reactivity.[18,19,21] In order to obtain topological information of such alterations in these patterns at the molecular level, one has to introduce refined methods exploring distance relationships, i.e., the proximity of cell surface elements at a molecular (preferentially at nanometer or Angstrom) scale.

Beyond the possibility for chemical cross-linking of nearby elements, a novel biophysical method was introduced in the late 1970s and the early 1980s, the flow cytometric energy transfer measurement or FCET.[10,186,210-219] The method is described in detail in earlier chapters; thus, there is no need even for a brief summary here. However, we can claim that, since the mathematical and physical problems of cell surface proximity measurements have been solved, the method was successfully applied to obtain precise data on cell surface pattern changes in a number of cases, especially in combination with other biophysical methods.[86,220,221] A consistent application of the FCET method by us and also by others revealed that many protein elements are really interdependent in the cell membrane. The murine H2-K^k antigen and the Con A receptors showed a nonrandom codistribution without having any covalent molecular connection.[210] The existence of a new "subunit" of the IL-2 receptor complex was confirmed by FCET as well as by fluorescence recovery after photobleaching techniques.[17-21,215,220] The new subunit later proved to be identical with the intercellular adhesion molecule, the ICAM-1.[115,222]

The fluorescence photobleaching recovery studies provided data on the co-mobilities of the ICAM-1 molecules, the Tac antigen and the HLA class-I antigen.[220] An unexpected molecular vicinity of the HLA class I and II antigens was recently reported, providing additional data on the dimeric nature of some HLA class II subspecies and the monomeric distribution of the HLA class I molecules.[187] The ligand binding-induced co-localization of the CD4 and T-cell receptor-CD3 complex was elegantly described in CD4-transfected mouse cells by applying the FCET.[223] Similar, rather sophisticated ways of studying cell surface phenomena are more and more readily available, supporting our view that a functionally very important dynamic, two-dimensional pattern facilitates intercellular communications and also reflects metabolic states of the cells.[224-227]

Now it is firmly established that co-localization of protein entities in the plasma membrane are not only beyond a chance co-occurrence, but rather predetermined at the level of protein synthesis on the surface of the rough endoplasmic reticulum.[136]

A. TYPES OF CELL SURFACE PATTERNS: A POSSIBLE REGULATION OF ION CHANNEL CONFORMATIONS[228]

Membrane elements of the cell surface have different meaning for the environment and reflect the metabolic state of the cell interior. New membrane elements are expressed in greater quantity when, for example, the cells are activated (i.e., triggered to proliferate, differentiate, or secrete), and these elements may arise from *de novo* synthesis as well as from sequestration of otherwise cytosolic elements into the membrane, e.g., the protein kinase C enzyme.

The basic sequence and localization of the protein-elements is given in an almost final form in the endoplasmic reticulum, where the protein subunits with different hydrophobicity will be located accordingly in the lipid structure.[136] Lipid vesicles from the endoplasmic reticulum will float to the plasma membrane and fuse with it. As compensation for this process, there is a vesicle migration toward the cell interior, separated from the plasma membrane. This process is generally termed endocytosis, wherein the internal parts or contents of the vesicles might be digested (processed) in lysosomes; other parts may be recycled to the membrane. A steady-state coupling at very different time scales exists between the two processes. The lack of equilibrium causes either a growing or a shrinking of the cell surface.

Thus, an ever-changing pattern of the cell plasma membrane is formed which shows a domain-like structure especially concerning the different lipid molecules forming the domains which have a wide range of different hydrophobicity and protein-accommodating capacity. The integral proteins of the membranes (and transmembrane ion channels *per se* have to belong to this class of membrane proteins) can move relatively freely as the Singer-Nicolson fluid mosaic membrane model predicts. The limitations, at least partially, arise from the preferential lipid domain islands, which accommodate particular integral proteins.

Cytoskeletal connections may put another constraint on the free mobility, and co-localizations with other membrane elements may also limit the individual mobilities to co-mobilities. Such joint mobilities are not necessarily "inborn" in the membrane structure, but can also be evoked by ligand binding and, for example, subsequent phosphorylation. Fixing of a substantially large charged group to any protein will certainly alter its conformation, allosteric regulation, intermolecular interactions, or even its preferential localization in a lipid domain.

Fromherz, in a series of very elegant theoretical papers, analyzed the possibility that dissipative membrane structures have to be formed from integral membrane proteins under the compulsory influence of membrane potential fluctuations.[229-231] The framework is provided by the fluid mosaic nature of the lipid bilayer and the membrane-bound proteins. Proteins are characterized by diffusion coefficients, in the case of ion channels also by channel conductance and charges. The proteins have a lateral mobility due to the diffusion. Fromherz's model considers only those proteins which have ion channel structures, i.e., specific ion conductivities inside of these molecular tunnels. The protein density and the membrane potential are the two macroscopic variables which generate a self-organization of the ion channels. Self-organization occurs through "charged channel condensation". The ion fluxes generate electric currents which induce local membrane potential gradients. Since ion channels are electrically charged proteins, while they diffuse, they move in an applied electric field. Thus, there is a similarity between this movement and the electrophoresis. Taken together the basic elements of Fromherz's conception, they are as follows:

1. The Singer-Nicolson fluid mosaic membrane model
2. The electrical circuit model of the active ion channels
3. The molecular model of ion channels
4. The electrophoresis model

By skillfully coupled application of Smoluchowski's equation describing the lateral mobility of the proteins and Kelvin's cable equation that describes the local electric circuits, patterns of ion channels can be generated. Stationary and oscillatory patterns have been generated, which depend on the average density of charged channels, the level of their reversal potential, and, of course, the size of the membrane. These patterns are far from electrochemical equilibrium under such conditions which can occur in membranes of living cells. Beyond the Saffman-Delbrück model of biological membranes,[232,233] which was mainly an attempt to describe the thermodynamic properties of the lipid bilayer, this important work of Fromherz is the first model that predicts the compulsory existence of membrane patterns, without assuming any chemical cross-links, i.e., trivial pattern formations among membrane elements.[229-231]

Thus, singular physical forces can be satisfactory to generate nonrandom codistributions of membrane proteins without an interference by the cytoskeleton or even the solvating properties of the lipid domains. However, the pattern generation predicted by the model resulted in rather "macroscopic" patterns as compared to the fine complex receptor structures, which are known to display nonrandom and functional codistributions, for example, in lymphocyte membranes.

The last two decades have expanded our knowledge on the structuredness of the plasma membrane components and now we can be convinced that patterns formed by the sheer physicochemical forces can be maintained in equilibrium.[234,235] Even steady states can be maintained, which are far from equilibrium if a transmembrane ion gradient is present as a source of the energy. Richard and Miller[228] described a beautiful example of steady-state coupling of ion channel conformations to a transmembrane ion gradient.

Thus, nowadays the existence of patterns in the cell membrane in the sense of a relatively fixed two- (or even three-) dimensional co-localization of plasma membrane protein elements seems to be commonplace. It was about a decade ago when we claimed the necessity of and also later elaborated a method for studying such molecular patterns as characterizing elements of plasma membranes which reflect metabolic or activation states and are generated and maintained both by extracellular and intracellular mechanisms.[14,19,21,210,211,217,218]

Even those scientists, whose experiments gave the final impetus to the formulation of the Singer-Nicolson model, point out that many membrane protein-bound functions (especially receptor functions) are interconnected with molecular associations.[183,236,237] A well-elaborated example is given by Zidovetzky et al.[182,184,186] showing that, when epidermal growth factor binds to its receptor, this event-initiated association is an essential part of signal transduction. Complex receptor structures, such as that of the IL-2, have such subunits, which have auxiliary functions such as the adhesion of the cell, along with the signal-transducing p55 and p75 complex which has the capacity to bind IL-2, although, only the latter (the p75 subunit) is believed to transmit the signal through triggering a putative-associated protein that is supposed to have tyrosine kinase activity. Neither of the actual IL-2 binding proteins have the ATP binding amino acid sequence that is a prerequisite for kinase activities. The actual model for signal transduction has been described earlier. Here the important information for us is that the ICAM-I and the complex receptor of IL-2 functionally and sterically are co-localized and have also a co-mobility.[215,220,222] These and similar formations support the suggestion that different patterns of proteins exist in the cell membrane and have functional significance.

B. TRANSMEMBRANE POTENTIAL DIRECTLY INFLUENCES THE DYNAMIC BEHAVIOR OF CELL MEMBRANE LIPID AND PROTEIN COMPONENTS

As it has been pointed out, physical as well as biochemical events in plasma membranes, such as changes in ion channel activities, structure of phosphatidyl inositol 4,5-bisphosphate, membrane permeability to different ions, membrane potential and lipid fluidity, conformational changes of protein receptor association, and dissociations triggered by external or internal stimuli, may play essential roles in information transfer through the cell membrane. Although we have a reasonably great deal of information on the proximity and mobility of membrane proteins and some lipids during transmembrane signaling, less is known at least in general terms about the lipid constituents.[238,239] Lipids and proteins are confined to their ever-present dynamic interactions in the plane of and perpendicular to the membrane. The internalization processes and *de novo* membrane synthesis which are the major factors in the determination of the number and localization of lipid and protein mosaic structure work mostly in the plane of the membrane. Little attention has been paid to the lipid protein interactions concerning their vertical translocation. The membrane lipid fluidity can modulate the expression and availability of some membrane proteins.[240] The exposure of proteins on the cell surface can also be affected by changing the membrane potential.[44–46,240,241]

Changes of potential gradients across the membrane of phospholipid vesicles and living cells are accompanied by appreciable modifications in membrane structure such as changes in membrane rigidity, thickness, conformational changes of proteins, or accessibility of reactive chemical groups. Some data will be presented here, which show that changes in the membrane potential of living cells, mostly lymphocytes, would be connected to definite changes in the structure and dynamics of well-defined lipid regions of the plasma membrane. A few examples will also be presented in this section to point out that membrane potential changes can definitely change particular receptor structures, which have very specific roles in transmembrane signaling.

The relationship between transmembrane potential and lipid dynamics in the cytoplasmic membrane of mouse thymus cell were investigated with fluorescent methodology. Parallel to lipid viscosity measurements with three different fluorescent lipid probes (1,6-diphenyl-1,3,5-hexatriene, DPH, 1-[4-(trimethylammonium)]-6-phenyl-1,3,5-hexatriene, TMA-DPH, and 12-(9-anthroyloxy)stearic acid, 12AS), monitoring different structural regions of the lipid domain structure through changes in their optical anisotropy, membrane potential measurements were carried out with bis-oxonol.[240] The fluorescence anisotropy of the probes was measured as a function of temperature at two different values of the membrane potential. The following data suggest that changes in the transmembrane potential are accompanied by local alterations in membrane lipid dynamics, and their extent is temperature dependent in an interval that is physiologic for most of the mammalian cells.

Although the membrane potential could be changed in a stepwise fashion between 10 and 60 mV in 10-mV steps through an isotonic increase of the extracellular K^+ concentration, in order to magnify the expected changes, the two extremes, that is, the fully depolarized cells and those which had not lost their original potential level were actually compared. In the comparison of intact and depolarized cells the temperature dependence of steady-state fluorescence anisotropies of the three dyes, DPH, TMA-DPH, and 12AS, embedded into the plasma membrane, were studied. Since these dyes are accommodated by different lipid regions of the membranes, they made possible the monitoring of physical properties of different domains.

Here a few words are pertinent about the definition of lipid fluidity. This term, which has been strongly criticized many times, and not without reasons, is a phenomenologic one with all the problems of averaging particular properties and the lack of detailed, fine-resolution power. However, its application cannot be objected if the real limited significance of such measurements is kept in mind, and data are not overinterpreted.

Thus, lipid fluidity is a phenomenologic term widely applied to phospholipid bilayers and also to natural membranes with respect to the physical state of the phospholipid matrix. Accepting the fluidity concept with the restriction that, although the bilayer will be considered as an isotropic system, we are aware of its anisotropy. The measurements were evaluated for a homogeneous hydrocarbon fluid, and only the bulk lipid structure will be characterized by "microviscosity" (η) or "anisotropy" parameter, defined by $[(r_0/r) - 1]^{-1}$. This anisotropy value provides a quantitative, but averaged parameter of membrane fluidity. The resolution power is restricted to the rotational diffusion of the fluorescent probe embedded in the analyzed region. The mobility of the reporter molecule reflects all those complicated intermolecular interactions which can occur between the dye and the surrounding molecules. The information content of the steady-state anisotropy can be further increased by time-resolved measurements, where two or three major components of the complex mobility of the probe may provide additional variables for analysis. Two major components can be the dynamic component which measures the rate of rotation of the probe, and the static component which indicates the extent of this motion (i.e., the angular range of probe rotation). This latter reflects the molecular packing of the fatty acid chains (order) within the bilayer. Therefore, changes in the values of steady-state anisotropy of lipid probes may result from either changes in the microviscosity of the probe environment or from alterations in the local lipid packing density. Steady state anisotropy data obtained with TMA-DPH and 12AS reflected no major difference in the structural or physical state in the probed lipid membrane regions of the intact and depolarized temperature range, i.e., between 18 and 40°C. In contrast to these findings, DPH reflected some alteration of the bulk lipid structure of depolarized cells as compared to the intact ones when the temperature is above 28°C (Figure18). The findings can be explained by the different localization and orientation of the probes in the membrane and/or their different kinds of motion. Space-filling models show that the length of the molecules of DPH and TMA-DPH is similar to the length of the fatty acyl chains of the lipids and, of course, the dye molecules have a lipophilic nature. It seems plausible that the long carbon chains of TMA-DPM and 12As are largely packed parallel to the phospholipid chains and they are anchored at the polar-apolar interphase due to their ionic charges. The DPH molecules are likely to be accommodated in the region near the center of the bilayer and preferentially oriented perpendicular to the surface. The DPH molecules were localized in the plasma membrane of living cells even after 3 h after the labeling, as our microscopic investigations revealed.[240] This was also indicated by the fact that similar temperature dependence of emission anisotropy of DPH was obtained with normal and depolarized cells, independent of whether the given temperature range was covered by decreasing or increasing the temperature. The indication came from the measurement of the emission anisotropy of DPH performed at 20 and 38°C, which showed no time dependence up to 3.5 h. Although the possibility could not be excluded that DPH molecules had been translocated into the membranes of cell organelles, according to the above results the contribution of such fluorophores to the total fluorescence can be neglected. Unlike 12-AS and TMA-DPH, the DPH molecules are not anchored in the bilayer; they have a vertical motional freedom as well. This fact can explain why DPH did show a distribution along the membrane normal, inside the lipid bilayer. The membrane microviscosity varies along the membrane normal. Thus, the DPH molecules which are located at different distances from the center of the lipid bilayer can experience different motional freedom that results in different steady-state anisotropy values. Accordingly, the emission anisotropy of DPH is an ensemble average. This average is further altered by the possible lifetime distribution of the dye molecules. Since the lifetime decreases with the increase of the dielectric constant of the environment, the DPH molecules situated closer to the membrane surface can be characterized by shorter lifetimes due to the gradient of the dielectric constant across the membrane. Recently it has been pointed out that an increase in the mobility of DPH along the membrane

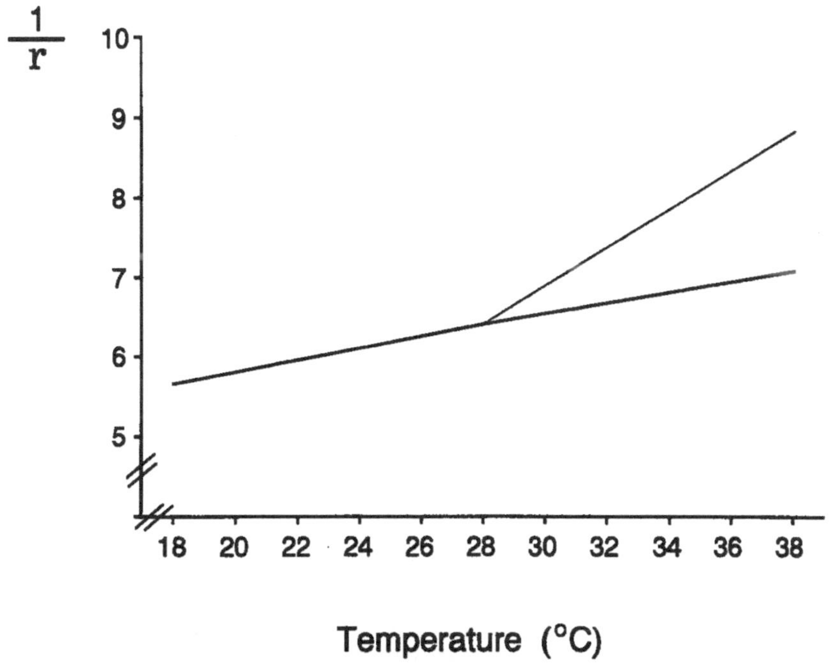

Figure 18. Reciprocal values of diphenylhexatriene anisotropy in plasma membranes of human lymphocytes plotted against the measuring temperature. There was a change in the bulk lipid structure above 28°C, as is indicated by the broken (dashed) line.

normal may also result in a decrease of the average lifetime. Of course, in this case it has been assumed that a dye molecule can sample every position, characterized by different lifetimes along the membrane normal during its excited state lifetime. This phenomenon can again introduce a change (i.e., an increase) of the average anisotropy. Based upon the above short analysis two ways are offered to interpret the findings with DPH anisotropy in connection with the plasma membrane potential.

(1) Assuming that the DPH distribution is the same for the intact and depolarized cells over the whole temperature range one can conclude that intact cells seem to have a "looser" membrane structure above 28°C than the depolarized cells. Then, concerning the measured anisotropy, the looser membrane structure of intact cells (above 28°C) could result in many effects as compared to the depolarized cells. The higher level of water penetration may cause a decrease in the average lifetime of the excited state of DPH molecules due to the increase of the local dielectric constant. Another factor, the increased level of DPH mobility again would decrease the average lifetime. These two effects would increase the average anisotropy. Then, according to our own measurements, the increased range of molecular mobility, regarding both the range and rate of the motion, would have to dominate the anisotropy difference between the intact and depolarized cells.[240]

(2) Regarding those data sets which were obtained with TMA-DPH monitoring the membrane structure near the membrane surface, i.e., the polar apolar interphase, the application of the above considerations leads to the obvious conclusion that the increased mobility should be exactly counterbalanced by the decrease of lifetime (due to the change in the local dielectric constant) over the whole temperature range investigated so far. This would result in the same emission anisotropy for both intact and depolarized cells. We believe that this is a remote possibility which becomes even less likely in the view of the findings with 12-AS, which also detected no difference between the depolarized and intact cells over the whole temperature range.

Accordingly, the most reasonable interpretation of the above data is that depolarization introduces (either directly or indirectly) such changes in the membrane structure which are localized to the central core region of the lipid bilayer. It was assumed that above 28°C the intact cells, compared to the depolarized ones, had a plasma membrane with higher mobility of the phospholipid chain ends in the core region. This allowed the DPH molecules situated close to the "membrane core" of intact cells to experience a higher mobility as compared to those embedded in the depolarized membrane. As a consequence, DPH distribution will also be more centered around the middle part of the bilayer in the intact, i.e., fully polarized cells. Since the DPH molecules have a longer lifetime close to the center than those located closer to the surface, this difference in the localization alone can increase the average lifetime and therefore the average value of the reciprocal anisotropy, i.e., 1/r. In this interpretation all differences, regarding the distribution of DPH molecules, their mobility, and also the average lifetime, would result in the same type of deviations observed between the anisotropy values of depolarized and intact cells. In this case even a slight change in the middle of the bilayer can induce a well-detectable alteration in anisotropy values.

Finally, in conclusion one can say that a change in the potential gradient across the plasma membrane very likely results in an alteration of the mobility of those DPH molecules which are located in the central region of the bilayer. This change can either be due to a direct effect, characteristic of the bilayer structure itself, or even more likely it can be realized through conformational changes of particular membrane proteins. This latter case implies that changes in the membrane structure have been induced by conformational changes of membrane proteins and they are centered around and located within not too far distances from membrane proteins, i.e., in the boundary lipid regions. Some authors observed an enhanced membrane rigidity upon creation of transmembrane potential in lipid membrane vesicles. The effect was monitored through anisotropy measurements with DPH. An opposite tendency was observed by us using mouse thymus cells, i.e., the depolarization of the membrane led to an increased apparent rigidity.[240] Probably, some potential dependent and protein-mediated changes in the lipid structure could be responsible for the observed differences.

The relationship between the transmembrane potential and the lipid structure and dynamics in thymocytes seems to have correlation to the transmembrane signaling processes. However a generalization of the observed particular effects demands more extensive investigations. The paper of Lakos et al. and the sizeable literature cited in it represents a current cross section of these problems.[240] Such investigations will obviously lead us toward a better understanding of rearrangements of cell surface elements associated with membrane potential changes.

C. LATERAL MOBILITY OF ION CHANNELS

Ion channels are not necessarily immobilized in the plasma membranes of different cells. At present relatively little is known about the functional significance of the mobile ion channels. However, there are a few experiments supporting the idea that ion channels, like the majority of cell membrane elements, have a restricted mobility. Weiss et al.[242] studied the mobility of lectin receptors and two different types of cation channels in the sarcolemma of frog skeletal muscle by fluorescence recovery after photobleaching. While one of the lectin receptors, the wheat germ agglutinin (WGA) showed a measurable lateral mobility in 53% of the cells, with a diffusion constant $D = 6.4 \times 10^{-11}$ cm^2/s, the fluoresceinated succinyl-Con A-labeled receptors were immobilized. Practically there was no measurable diffusion constant of this class of receptors. These experiments served only as positive and negative controls for mobility parameters in the skeletal muscle of frog *Rana temporaria*. An interesting part of this important work focused on the mobility of two cation channels. The mobility of voltage-dependent Na^+ and delayed rectifier K^+ channels have been determined with the loose patch clamp method, coupled with photodestruction of the channels. The channels were identified by the patch clamp, and the photodestruction was enforced by application of UV light. The specificity of the measurement was provided by the so-called through-the-pipette photodestruction of the channels. The most effective wavelength was around 289 nm. The energy density requirement for Na^+ channels was around 0.5, for K^+ channels approximately 1J/cm^2. The same dose of irradiation did not alter the mobility level of the WGA; thus, it was suggested that the UV irradiation did not cause the immobilization of ion channels directly. Na^+ channels, similarly to the immobile Con A receptors, displayed a $D \leq 10^{-12}$ cm^2/s diffusion constant. On the contrary, calculations from the recovery of potassium currents suggested that about 25% of these channels had a diffusion constant $D = 5 \times 10^{-11}$ cm^2/s. The authors concluded that mobile and immobile membrane proteins coexisted in the sarcolemma of frog skeletal muscle and that both voltage-dependent Na^+ and K^+ channels were nearly immobilized.[242] These interesting series of experiments raise the question whether the degree of immobilization is the same or might be altered when ion channels are activated during signal transduction events. Nonexcitable cells may also behave like muscle cells, but this is not an absolute requirement. One feels it an interesting and very promising endeavor to extend the above far-from-trivial investigations over the field of nonexcitable cells, such as lymphocytes.

D. MEMBRANE POTENTIAL, ACCESSIBILITY, AND ELECTRON MICROSCOPIC VISUALIZATION OF CELL SURFACE ELEMENTS

A completely different type of approach studied the direct influence of membrane depolarization on the accessibility of cell surface elements. In order to avoid the possibilities of artifacts arising from the method applied for depolarization, the Ca^{2+} ionophore A23187 was used in Ca^{2+} or EGTA-containing buffer to obtain depolarized and intact cells. An easy target on the cell surface to look for changes in protein conformation is the titration of the accessibility of the SH groups. These groups are characteristic of proteins. In general, they can even be classified into two major groups regardless of the localization of their protein carriers: readily available and masked SH groups. Masking means that some of these hydrophilic groups are buried in the interior of the protein and will be readily accessible only upon conformational changes of the proteins. An excellent reagent for studying such possibilities is the 5,5′dithiobis-2-nitrobenzoic acid (DTNB) or Elmann's reagent.[44] Titration of cell suspensions of human and murine lymphocytes showed that the accessibility of cell surface SH groups was substantially enhanced in depolarized cells. The results were similar, whether the depolarization was achieved by ionophore or OKT-3 mAb, directed against the CD3 receptor.[44]

Our data were basically in a good agreement with and very similar to those of Müller et al.[46] and Dunant et al.,[45] who applied a very elegant method to study protein localization in pre- and postsynaptic membranes during electric signal transduction. The membranes were deep-frozen within microseconds in a special apparatus during the different stages of the transmission of the evoked electric potential pulse. The concomitant freeze fracture analysis under electron microscope clearly showed the difference of protein distribution in the plain of the presynaptic membrane. The number of the protein "heads" were highly enhanced and their number gradually decreased per unit area of the membrane.[46] In the case of postsynaptic membranes, out of the observed two types of intramembrane particles, the globular ones with a diameter of 8.8 nm were the most frequently found entities (about 80%) in the unstimulated cells. Elongated intramembrane particles have a diameter about twice that of globular particles. Transmission of a single nerve impulse was accompanied by a marked decrease in the number of the first type and by an increase in the number of the elongated particles. This momentary alteration of the postsynaptic membrane during transmission of a single nerve impulse is probably linked to activation of the acetyl-choline nicotinic receptors.[45] Thus, it is evident from data obtained with a completely different type of cells and also a different method that changes in the transmembrane potential are accompanied by vertical transitions of the integral membrane proteins.

A more target-specific approach revealed that conformation changes in the structure of the HLA class I antigen of the human major histocompatibility complex also takes place upon depolarizing human peripheral mononuclear cells through isotonic increase of the extracellular K^+ ion concentration. Flow cytometric energy transfer measurements were carried out between the monoclonal antibody-labeled β_2 microglobulin component of the HLA class I antigen and the much bulkier framework, labeled by another monoclonal antibody, the W6/32, with a specificity to epitopes of HLA class I A, B, and C molecules. The model of this measurement was discussed in detail in the earlier chapters, describing the technical details of the FCET experiments more specifically. Here we summarize the data only briefly. The altered energy transfer efficiency distribution demonstrated that, upon depolarization, the molecules changed either their relative orientation or the actual molecular distance. We think that this is the first direct evidence that a particular type of membrane proteins changes its conformation upon depolarizing the plasma membrane (Bene et al., in preparation).

XIV. DYNAMIC BEHAVIOR OF CELL MEMBRANE MACROMOLECULES MAY SIGNIFICANTLY CONTRIBUTE TO THE ION-TRANSPORT ACROSS MEMBRANES

As we have seen, unequal cationic and anionic distributions can be maintained at the two sides of biological membranes by a high number of very different factors. These parameters and factors, like the Donnan potential, or diffusion potential can maintain such systems even far from equilibrium. Of course, this has a considerable requirement for chemical potential and may occur in living as well as physical-chemical systems. However, the so-called pumping mechanisms, where unequal ion distributions are further increased, or ion and molecular transports are sustained "uphill", i.e., against an opposite electrochemical potential, is solely attributed to living systems. Of course, the basic thermodynamic laws are not violated in these cases either, since such transport processes are energetically supported at the expense of chemical energy previously accumulated in the living systems, for example, during anabolic processes of metabolism. Metabolic processes provide chemical energy (potential) through the decomposition (digestion) of molecules, synthetized beforehand in botanical or zoological systems. Thus, these "expensive" energy-demanding transport processes have their ultimate energy source from the radiation energy of the sun, which provides the energy backup for the whole biosphere. Typical examples for these pumping mechanisms are those of the Na^+ and K^+ ion antiport, where three Na^+ ions are transported from the intracellular to the extracellular space, coupled to the transportation of two K^+ ions from the extracellular space into the cell interior. This is a net electrogenic pump, generating electric potential difference. A similar mechanism is the transport of Ca^{2+} ions out of cells against a very steep concentration gradient. Among normal circumstances the intracellular Ca^{2+} concentration is 10,000-fold less than the 1- or 2-m*M* extracellular calcium concentration. Both of these well-known examples have in their energetic background the ATP hydrolysis carried out by the actually ion-transporting proteins. The mechanism of coupling between ATP hydrolysis and, for example, Ca^{2+} transport is strongly dependent upon the molecular dynamics of the transporting proteins. The mechanism of coupling is also strongly influenced, if not eventually determined by the spatial relationship of the phosphorylation and ATP-binding domains of the Ca^{2+}-ATPase to the Ca^{2+} channel that is involved in the translocation of the calcium. One of the major problems of contemporary science is the actual coupling mechanisms of these important cell biological events at the molecular level. Two models are available, which equally rely upon the structural characteristics of the Ca^{2+}-ATPase. The two models of the structure and the transport coupling of the Ca^{2+}-ATPase are very different and in reality we do not know much about the real situation; the only thing that can be done is to investigate the inherent consistencies or inconsistencies of the models. Since the Ca^{2+}-ATPase (similarly to the Na^+-K^+-ATPase) is embedded into lipid membranes, a direct experimental approach, for example, at the resolution level of X-ray crystallography is far from being trivial. One of the two mechanism predicts a conformational coupling for the utilization of the free energy of ATP hydrolysis for Ca^{2+} transport. This assumption predicts a mechanical coupling at the molecular level between the ATP binding and Ca^{2+} transport sites which are physically well separated. The other mechanism, which is called "ligand conduction", postulates a direct interaction between the γ-phosphatidyl group of ATP and the transported Ca^{2+}. The first hypothesis is actually utilized in most of the cases in schemes based upon the $E_1 \rightarrow E_2$ conformational transmissions, where at least two conformationally different states of the Ca^{2+}-ATPase must exist.

The (at least) two different conformational states which have sound experimental evidence for their existence have differing affinities and vectorial specificities for their substrates,[243] such as ATP, P_i, and Ca^{2+}. The powerful methods of modern molecular biology, i.e., chemical modification, site-directed mutagenesis, and biophysical measurements (e.g., application of intramolecular energy transfer), contributed much to the understanding of a considerable amount of these details. The ATP-binding domain is located in the cytoplasmic moiety of the Ca^{2+}-ATPase enzyme and the tight binding sites of the Ca^{2+} reacting with the ion with high affinity are supposed to be in the interior of the lipid bilayers. Since these two sites are thus separated by a considerable distance (20 to 40 Å), the conformational coupling seems to be the most convenient way to connect these two events in a functional way. The structural information on these steps of the process is far from being complete as the probes of biophysical investigations have tentatively described binding sites and the obligatorily assumed conformational changes do not alter the secondary structure of the Ca^{2+}-ATPase molecule to a great extent.

Martonosi and his group in a systematic work aimed at localizing the structural changes connected with the $E_1 \rightarrow E_2$ transition of Ca^{2+}-ATPase and its three-dimensional implications applied a number of crystallographic, immunological, and chemical modification techniques.[243,244] Although a three-dimensional crystallographic structure and the Angstrom level changes during Ca^{2+} transport have not yet been revealed, most of our knowledge on the relationships between a structure-function relationship of the Ca^{2+}-ATPase is due to their pioneering work. Here we will concentrate only on their spectroscopic data, since these fit our goal, to present an example, where dynamic motions of plasma or endoplasmic reticulum membrane proteins are essential from functional point of view.

Static distances between particular binding sites of fluorophores, fulfilling the requirement for Förster-type fluorescence energy transfer, have been determined by this method.[10,11,48,131,221,224-227,245,246] However, these distances are subject to considerable uncertainty, and can and should be used only as approximate indicators of the relative locations of the various binding sites within a protein and, here in particular, the ATPase molecule. As we have seen earlier, the application of Förster's theory and the appropriate measuring techniques to determine distances among cell surface elements is a rather imprecise, approximate method unless a number of stringent requirements are fulfilled. These are the determination of the orientation of fluorophores, the knowledge of the binding sites, etc. However, when we intend to determine changes in distances or in fluctuation pattern of a molecule at a nanometer scale, variations of this method may reveal otherwise unattainable properties of molecules. If the labeling can be carried out in a site-specific way even dynamic behavior of particular molecules can be measured. A brief outline of this method follows.

The Förster-type resonance energy transfer senses the molecular fluctuations due to the negative 6th power of the distance relation between donor and acceptor molecules.[245] The problem is that the fluctuations of macromolecules which could have been nicely monitored in nanosecond time scale is not synchronized. Thus, the disordered fluctuation cannot be observed due to the averaging effects of the unsynchronized molecular motions. However, at different temperatures the fluctuation frequencies and amplitudes will certainly change. These changes are also masked by the parallel alterations in the quantum yield of the fluorescent dyes which could otherwise monitor the changing fluctuation pattern. Somogyi et al.[52] published a relatively simple and very useful method that solved the problem of the changing fluorescent quantum efficiencies by normalizing the energy transfer values to the donor's fluorescence at each temperature. This procedure eliminated the perturbing effect of the temperature on the measuring device and the new normalized energy transfer parameter had a direct relationship with the fluctuation of the protein. These ideas were utilized in the determination of the fluctuation changes of bacterial RN-ase T1 molecule and the Ca^{2+} ATP-ase.

The Ca^{2+} is transferred across the membrane in association with the Ca^{2+}-ATPase. Thus, the environmental heat, i.e., thermal fluctuation of critical regions of the enzyme molecule, perhaps influenced in specific ways through phosphorylation by ATP may play a significant or even a key role in Ca^{2+} translocation. A similar idea was proposed some time ago by Huxley in relationship to myosin cross-bridge movements, by us in several publications[247-252] on the role of protein fluctuations in enzyme action, and by Martonosi and also others on the Ca^{2+} transport in sarcoplasmic reticulum.[244,253] In order to investigate our idea that enzyme action can at least be modulated if not driven by thermal motions, the temperature-dependent changes in normalized energy transfer efficiency were theoretically analyzed and also experimentally applied.[52,247,249,251] Jóna et al.[53,244] utilized the method of temperature-dependent and normalized energy transfer, where the transfer efficiency was normalized, according to our theory to the quantum efficiency of the donor fluorescence at every measured temperature point, to determine fluctuating distances between different site-specific donor acceptor systems in the Ca^{2+}-ATPase molecules of the sarcoplasmic reticulum. Energy transfer from a danzyl compound to fluorescein (from AEDANS to FITC) was measured in the E_1 conformational state of the enzyme and when the $E_1 \rightarrow E_2$ transition was triggered by sodium vanadate (Na_3VO_4). Similarly, two other transfer pairs were applied to measure distances among other particular points of the enzyme, the transfer from FITC to Nd^{3+} and from AEDANS to Pr^{3+}. The measurements were repeated at different temperatures between 10 and 37°C. The temperature dependence of the normalized energy transfer efficiency was highly dependent on the localization of the donor acceptor pair in the enzyme molecule. Thus, the authors' conclusion was that a flexibility gradient emerged within the Ca^{2+}-ATPase molecule.

The region of the molecule monitored by the AEDANS-Pr^{3+} donor-acceptor pair proved to be rather insensitive to a temperature change of approximately 28°C. A change in this temperature interval enhanced the fluctuation of the RNase T_1 enzyme by nearly 40% as it was determined earlier.[52] The region of the Ca^{2+}-ATPase molecule which connects the ATP binding site with the Ca^{2+} binding one, labeled by FITC and Nd^{3+}, respectively, increased its fluctuation by 25% and there was also a near 20% enhancement of fluctuation in the E_2 state transition by vanadate. In contrast to the latter, the same donor-acceptor pair showed only half of the temperature-dependent fluctuation enhancement in the E_1 state of ATPase. At elevated temperatures the proteins increase their thermal extensions which may also reflect their more intensive fluctuations.[251,254] Since the rate of the Ca^{2+} transport is extremely sensitive to the temperature, a coupling between enhancement of this transport and structural fluctuations in the protein molecule may be assumed. Thus, dynamic motions in the structure of the Ca^{2+}-ATPase molecules are thus righteously connected to their function.

Another parameter that can also contribute to the structural flexibility and responsiveness of a membrane-bound enzyme is the membrane potential.[11,18-21,44-46,240,255-258] This statement is also supported by the experimental finding with membranes of artificial vesicles that the rigidity in the lipid phase was significantly enhanced by an increase in the transmembrane potential.[259] The "uphill" pumping of Ca^{2+}

from the intracellular (internal) compartments generates membrane potential. Thus, we can postulate that the ATP-dependent Ca^{2+} transport is electrogenic. Quite a few examples are known in which the changes in the remarkably high transmembrane potential fields spanning membranes can alter protein conformations. A reversible alteration, for example, was determined in the structure of Ca^{2+}-ATPase through observing quantum efficiency changes in covalently bound FITC and the intrinsic tryptophan fluorescence.[259] The altered membrane potential values produced by electrogenic ion pumping mechanisms can exert either negative or positive feedback effects. The former is the more likely, since that can serve as a regulatory, stabilizing brake in biological systems. Recently, similar effects have been observed by FCET measurements, when proximity parameters of the subunits of HLA class I antigen were altered upon changing plasma membrane potential (Bene et al., personal communication). The membrane potential changes also had a strong influence on the rate of formation and stability of the E_1- and E_2-type Ca^{2+}-ATPase.[244]

Tsong[260] applied repeated high-voltage pulses to vesicles formed from sarcoplasmic reticulum. He observed that upon this treatment the Ca^{2+} uptake of these vesicles was significantly enhanced. In light of these data described above which claim a different conformational stability of the E_1 and E_2 forms of the Ca^{2+}-ATPase, an electroconformational coupling was suggested by Martonosi et al.[244] However, an analysis also by the same authors determined that the effects of high-voltage pulses were likely to increase the Ca^{2+} permeability of the membrane, through facilitating the passive diffusion of calcium into the vesicles. This final analysis, of course, did not eliminate the validity of the observation that the E_1 and E_2 conformations showed different responsiveness in an electric potential field. The strong electric field across membranes must exert a stringent control on the dynamics and flexibility of those structural domains of proteins where mobile π electron islands can be subjects of induction of transient dipoles or in those cases where net negatively or positively charged amino acid residues occur. A similar mechanism is extensively studied in the mechanism of the voltage gating of ion traffic. Much less attention has been devoted to the possibility that cell surface receptors and ligand binding sites or important enzymes (their conformational regulation) can also be influenced by electric fields. No doubt that observations which were made on vesicles of the endoplasmic reticulum are very likely monitoring the physiological responsiveness. However, in the plasma membrane, the cell surface pattern can be significantly altered by changes in the transmembrane potential.[14,15,17-21,240,255,261,262]

Taken together, the above-presented analysis on the dynamic behavior of an important membrane protein element, the Ca^{2+}-ATPase, which significantly contributes to the regulation of Ca^{2+} ion distribution at the two sides of the membrane, underlines the fact that flexibility and fluctuation freedom of the cell membrane-embedded molecules are not uniform in all directions but have a natural structural constraint. This was shown elegantly by temperature-dependent energy transfer measurement between well-defined structural subdomains of the Ca^{2+}-ATPase. On the other hand it was analyzed that, although the electric potential field does influence the different conformations of the protein to a different extent, observed functional alterations may be caused by other effects of the (fluctuating) potential field. This directs our attention to the reality that in such complex systems as biological membranes effects of external physical or chemical (liganding) perturbations can lead to changes in flexibility, mobility, or even proximity of separate binding sites. However, the actual functional consequences of such effects have to be individually determined. Conclusions cannot be drawn from circumstantial evidence for certainty without a further, many-sided approach to the same question. On the other hand, the likelihood that in a complex system any substantial change in the stringent control of the structural flexibility and conformation will result in altered functional responsiveness will grow with the degree of complexity. Principal targets of such regulation or perturbation are not necessarily only the protein elements of the plasma membrane. The high mobility and relatively disordered state of lipid elements were confirmed by direct measurements and also by the latest models of the structure and mobility of lipid bilayers.[263-267] However, the lipid layers are rather isolating elements which separate compartments filled with water-soluble elements. Important biological roles and particular functions were more or less confined to oligomeric proteins and their subunits, mostly to integral (transmembrane) proteins. Receptor functions, enzyme activity, ion pumping properties, and ion channeling across membranes are prerequisites of proteins.[268-271]

Only in the last 10 to 15 years have biologists paid more attention to lipid degradation products, which in a phosphorylated form may serve as second messengers. Phosphorylation and dephosphorylation seem to be a major biochemical means of triggering or deactivating a potential effector molecule. Phospholipids and their changes in their phosphorylation degree may surprise us in the future as equally important elements in regulating cell surface mobility, proximity, and function, more so because the altered electric charge distribution in the plane and perpendicular to the membranes *per se* has to influence mobility parameters in the available electric fields.

A. RECEPTORS AND CYTOSKELETAL ELEMENTS

Interaction of receptors with the cytoskeletal elements can directly be related to the clustering of receptors in the cell membrane as was shown for epidermal growth factor.[182,186] An opposite effect was also observed, when depolymerization of microtubules altered membrane potential, and affected the motional freedom of membrane proteins.[272] It has been known for a long time that the microtubular system can directly influence plasma membrane dynamics, such as microviscosity and fluidity.[272,273] A direct structural association of microtubules or tubulins with cell membrane components has also been observed.[272] However, the interaction of the microtubules, and the cytoskeletal system in general, is rather complex with the elements of the cell membrane, and the actual mode of interactions has been identified only in a few cases. Following the line of observations made by Aszalós[272] that the integrity of microtubules is not indifferent to the motional freedom of the lipid components of the membrane, it was a quite logical extension of this work to look for the motional freedom of protein moieties of the membrane and another parameter of dynamic behavior of the membrane elements, such as the membrane potential. The results of Aszalós et al. clearly indicated that microtubules influenced the motional freedom of membrane proteins of Chinese hamster ovary cells either directly or indirectly, and beyond this they can alter the membrane potential of intact viable cells.[272] The Chinese hamster ovary cells were selected for these investigations on the basis of the availability of mutants with altered responsiveness to drugs, deteriorating microtubule system, and those having different tubulins. The measurements were carried out with flow cytometric membrane potential determinations and the mobility of protein elements of the membrane was followed by electron spin resonance (ESR) of protein binding probes. The measurements were based on the observation that the motional freedom of ESR membrane lipid probes increased if the microtubule system was depolymerized by drugs known for this action.

The ESR probes directed to protein SH groups showed that not only the lipid but also the protein elements seated in the lipid domains of membranes increased their motional freedom. A parallel decrease in the plasma membrane potential was also demonstrated, although in this case the effect is still waiting for electrophysiological confirmation since the altered membrane-cytoskeleton interactions can also expel or accommodate membrane potential-sensing dyes in the plasma membrane through changing partition coefficients. The likeliness that a real membrane potential change was observed, nevertheless, is not negligible since the disrupted cytoskeletal connections objectively alter mobilities in the lipid domains and also generally that of membrane proteins. Thus, there is a great probability that channel activities and ion transport, which regulate cell membrane potential, were also affected.

XV. DYNAMIC PHYSICAL INTERACTIONS OF PLASMA MEMBRANE MOLECULES GENERATE CELL SURFACE PATTERNS AND REGULATE CELL ACTIVATION PROCESSES

A considerable amount of evidence accumulated during the last two decades on the dynamic behavior of molecular elements of the plasma membrane in general.[10,11,13,260] Lymphocytes have been a prime (but not exclusive) target of such investigations due to their functional importance and also the role of their plasma membrane in cell activation, antigen presentation, intercellular recognition processes, and several other, mostly immunological key functions.[17,240] Dynamic molecular events of plasma membranes have been defined as rotational, translational (lateral), and segmental movements of molecules or their domain-like associations in the plane of and perpendicular to the membrane.[10,11,14] Mobility and proximity changes, assembly, aggregation, patching, capping, sequestration processes, channel activities, or flip-flop motions in the lipid domains all refer to certain restrictions on a putative free mobility in membranes or other organelles or even cell free systems.[240,274-297] General features of free and restricted mobilities in the plasma membrane of lymphocytes have been reviewed in the second half of the 1980s.[10,11]

The principal role of evoked changes (ligand induced, generated by membrane potential changes, or triggered by any other physical, chemical, or biological means) in these molecular activities during cell stimulatory processes or intercellular interactions can hardly be questioned.[15-21,32,88,100,101,240,298-341]

The major goal of this last section is to present a brief account and draw a conclusion from the proceeding sections on the importance, variety, and measurability of dynamic events governing cell activation, one of the key cellular processes. Our attention will mainly, but not exclusively, be focused on lymphocytes, the key cellular elements of the immune system.

A. PROXIMITY OF CELL SURFACE PROTEINS

A nonrandom codistribution of plasma membrane integral proteins has been known for long and supported by co-capping phenomena of seemingly unrelated membrane antigens.[13] A possible connection between macromolecular dynamics in plasma membranes and transmembrane signaling was assumed by Damjanovich et al. in 1981.[14] Furthermore, it was suggested that the existence of a dynamic pattern-like assembly of membrane-bound protein molecules may have a coded message, reflecting metabolic state and readiness for interactions.[14,19,342]

Chen et al.[343] applied the fluorescence resonance energy transfer (FRET) method to determine proximity relations between fluoresceinated and rhodamine-conjugated Con A molecules and thus their respective receptors in a flow cytometer. The high FRET efficiency between the appropriately labeled Con A molecules indicated that in the membrane of transformed Friend erythroleukemia cells the Con A receptors were microclustered even at 4°C.

Jenis at al.[344] observed time-dependent average distance changes of Con A receptors present on HL-60 promyelocytic leukemia cells. However, a real breakthrough concerning proximity relationships of membrane-bound antigens was achieved only when a new measuring technique of FRET had been applied, utilizing dual wavelength excitation and an algorithm that enabled us to calculate the absolute value of the energy transfer efficiency in a more elaborate and controlled way and on a single cell basis in flow cytometers in a series of experiments.[10,186,210-213,215,217,218,345]

The first application of the FCET revealed a nonrandom codistribution of $H2K^k$ antigens and Con A receptors, despite simultaneous observations indicating that Con A receptors (an inhomogeneous membrane-protein family where the specificity is carried by the α-mannopyranoside sugar ring, attached to the protein moieties) were capped independently of the $H2K^k$ antigens.[210] Detailed analysis of these experiments[10,213,214,217] and an extension of the applicability of the method for instruments having only single laser excitation possibility[214] or the application in an image analyzer[246] provided solid support for the versatility of the FCET method. Thus, the FRET method, supported by other methods as well, made available the measurement of proximity relationships of cell surface antigens on a cell-by-cell basis.[18,86,186,215,220,221,345-349] One should emphasize the importance of the FCET measurements in contrast to the relatively easily applicable microscope-based similar method, since the latter ones would definitely lack the advantages of determining proximity distributions over a large cell population. The population analysis is the only way by which one can overcome the problems generated by the biological variabilities, that would otherwise many times compromise the actual values and even quality of measured data.[85,86,221]

Another successful application of FCET as well as fluorescence recovery after photobleaching (FRAP) experiments looking for comobilities of cell surface elements decided in favor of a possible association between interleukin-2 (IL-2) receptors and class I and II HLA molecules.[215,220] The IL-2 receptor has a complex nature. The FCET method helped to determine a nonrandom codistribution of a putative 95-kDa subunit that was also expressed on human T cells and seemed to be an activation antigen. Thus, besides the 55-kDa Tac peptide and the 75-kDa peptide which are supposed to bind IL-2 and upon association with each other form the tight binding site for this important interleukin, a third subunit of the complex was suggested.[215-220] Our first communication on this subject had already reported that the HLA class I antigens had a distinct proximity relationship to the Tac peptide.[215] However, these data were misinterpreted that time as a "chance encounter" of the Tac peptide and the HLA class I antigen, that is, we believed that the measured energy transfer between the IL-2 receptor and the HLA class I was due to the high level of HLA class I antigen density in the cell membrane of the particular cell line (HUT-102B2) wherein this parameter was measured. The extended work that applied a different method, lateral diffusion measurements, also resulted in a distinct comobility of the Tac peptide and the HLA class I antigen.[220] In this work it was explicitly stated that the Tac peptide and the HLA class I antigen must be in close vicinity. Sharon et al.[348] demonstrated by immune precipitation that IL-2 molecules can be cross-linked with HLA class I antigen. Their conclusion that a molecular interaction could exist between the HLA class I antigen and the IL-2 receptor was identical with our earlier suggestion.[220] Harel-Bellan et al.[346] confirmed the above conclusion using a single laser excitation variation of our FCET method, measuring proximity of FITC- and TRITC-labeled bound IL-2 molecules and rhodamine-conjugated anti

HLA class I antibodies. The observed strong energy transfer between FITC-IL-2 and TRITC anti-HLA class I antibody suggested a close proximity between low-affinity IL-2 binding sites and HLA class I heavy chain.

Interestingly enough, a functional significance of the 95-kDa putative subunit of the IL-2 receptor was established by Burton et al.[222] They described that the protein recognized by OKT 27 and 27b monoclonal antibodies was identical with the intercellular adhesion molecule 1 (ICAM-1). The association of this ICAM-1 with the multichain high-affinity IL-2 receptor "may facilitate the paracrine IL-2 mediated stimulation of T-cells expressing IL-2 receptors by augmenting homotypic T-T cell interaction, by receptor-directed focusing of IL-2 release by helper T cells, and by focusing IL-2 receptors of the physically linked cells to the site of lymphocyte function-associated antigen-ICAM-1-IL-2 receptor interaction."[222] The major histocompatibility antigens are frequently surrounded by a number of receptor molecules, although the functional significance of such co-distribution patterns is not clear.[182] Nevertheless, co-localization of membrane proteins, determined via FCET, and fluorescence photobleaching comobility studies contributed significantly to our understanding of the multisubunit IL-2 receptor complex.[115]

Recently, Liegler et al.[345] have presented interesting data on the proximity of H-2 antigens and the insulin receptor by fluorescence energy transfer measurements in a flow cytometer. Murine T cell lines were transfected with human kidney insulin receptor and high-energy transfer was measured between FITC-conjugated human insulin receptor-directed antibodies and rhodamine-conjugated monoclonal antibodies, with specificities to H-2K^s, H-2D^q, and H-2D^q, L^q antigens. A more vigorous research into this direction can further extend our understanding about the biological significance of such molecular patterns, described so far, mainly, but not exclusively, by application of the FCET method.[345]

Mittler et al.[349] observed physical association between T-cell receptor and CD4 antigen in a murine T-cell hybridoma that was transfected by the gene for the CD4. The CD4 antigen that was successfully expressed in the plasma membrane was translocated next to the T-cell receptor only upon induction by antigen receptor ligation.

Gorvel et al.[347] successfully applied the FCET method to follow conformational changes of rabbit amino-peptidase N in the plasma membrane domains of erythrocytes. Thus, conformational changes of a specific membrane protein could be followed in a large population of cells on a cell-by-cell basis. To our knowledge, this was the first attempt to measure such a type of molecular biological changes for enzymes in combination with their distribution over a sizeable cell population. Further steps into this direction may open a new branch of cell biology at molecular level.

Recently Szabó et al.[246] elegantly implemented another variation of the fluorescence energy transfer technology, the photobleaching fluorescence energy transfer for epitope mapping in a laser scanning microscope, the ACAS 570. The sensitivity of this method surpasses that of the flow cytometric measurements, mainly due to the larger viewing angle of the microscopes, in general, and the fact that the energy transfer efficiency can be detected on the donor's side, in particular, while in case of flow cytometry, for technical reasons, mostly the fluorescence enhancement of the acceptor's side is utilized in calculating the transfer efficiency. However, we will emphasize again that a microscope-based method cannot provide a reliable analysis at a large-enough population level for parameters which are inevitably subjects of biological variability. The method itself, a beautiful innovation, the digital imaging of FRET in combination with fluorescence photobleaching and its application in cell biology, was originally introduced by Jovin and Arndt-Jovin.[225-227] Nevertheless, Szabó and colleagues successfully mapped distance relationships between elements of the TCR receptor and CD4 epitopes. A successful decoding of the biological meaning of two-dimensional receptor patterns demands much effort and all different kinds of methodical approaches have their particular value, resolution, and contribution potentialities.

B. CELL SURFACE PATTERNS

The existence of cell surface patterns, i.e., nonrandom codistribution of membrane-bound antigens, was suggested and observed first by Boyse et al. with his "blocking test".[261] He determined a supramolecular pattern of different alloantigens in the plasma membrane of lymphocytes. The Lyt-2 was associated with H-2D, Lyt-1 with H-2K, and H-2D with TL. Flaherty and Zimmerman[262] introduced new variations of Boyse's method, such as blocking of a second antibody, Fc receptors, Cb3 rosettes, or phagocytosis. They also observed a mutually induced mobility of the TL and H-2D antigens.

Damjanovich et al.[14] suggested a generalized occurrence of such two-dimensional patterns and started developing the FCET method to explore such supramolecular formations. Relatively recently, Fromhertz has provided an elegant theoretical analysis as well as data, obtained by computer simulations, wherein, through applying the cable equation for the ion channels, and utilizing the lateral diffusibility of molecular elements in the plasma membrane, nice pattern-like distributions of membrane proteins emerged.[229-231]

Szegedi et al.[350] demonstrated that in autoimmune diseases the antigen density, i.e., a certain type of pattern, in the plasma membranes of lymphocytes could be significantly altered.

A phenomenological classification of cell surface patterns was attempted by Damjanovich and Pieri, utilizing the available rather scarce data.[18] A brief summary of the classification is presented here, including some more recent new data, collected since the above publication.

1. *Nonrandom codistribution of the membrane elements in the plasma membrane.* This class might include numerous such formations; however, we suggest to include only those topological arrangements wherein particular lipid domains preferentially accommodate functionally unrelated protein elements in close proximity due to their physical-chemical properties. Con-A receptors and H-2K^k antigens may serve as prototypes.[210] No functional correlation is known between the co-localized entities of this group, and they also retain some independent mobility such as separate capping.
2. *Evoked redistribution in the plane of the membrane.* T cell receptor and CD4 physical association was induced by antigen receptor ligation by Mittler et al.[349] A murine T cell hybridoma line was transfected with CD4. The physical association between the expressed CD4 antigen and the T cell receptor was observed by flow cytometric resonance energy transfer only after inducing the cells by monoclonal antibodies directed against the T cell receptor complex. The functional significance of such induced redistribution of membrane proteins is very plausible, although it has not been clarified yet.
3. *Evoked redistribution of proteins and lipids in a plane perpendicular to the membrane.* Constant endocytosis of coated pit-captured external agents and the membrane itself is, of course, continuously compensated by *de novo* synthesis of the membrane. The orientation of the eukaryotic membrane-spanning proteins is mostly determined at the level of the endoplasmic reticulum by their hydrophobic moieties.[136] However, this process is not necessarily without an oscillation in the accessibility of epitopes or other functional groups in the plane of the plasma membrane. Spectacular examples were presented by Balázs et al.,[44] Müller et al.,[46] and Dunant et al.[45] when depolarization of lymphocytes or postsynaptic membranes significantly altered accessibility of cell surface SH groups or time-dependent freeze-fracture technique. In this latter case, the subsequent electron microscopic investigations showed the altered cell surface pattern of proteins. The antigen presentation can also be classified to this group of cell surface patterns. Synthesis of new dyes can significantly extend our experimental possibilities in this field.[351]
4. *Assembly of functionally related subunits by ligands.* A likely example is the IL-2 receptor. The tight binding site is likely to be formed only when IL-2 is present.[114,115,352,353] This class of assemblies is suggested only for those cases when functionality cannot be questioned. Thus, complex receptor structures are major candidates for this group.
5. *Conformational change-induced functional alteration of membrane-spanning proteins or charged oligosaccharide moieties of lipids.* Age-related changes, enzymatic digestion, adhesion, or intercellular interaction may cause such deformations or destructions of particular parts of cell surface molecules.

Electroconformational couplings may also cause such changes in the cell surface pattern, beyond those already mentioned in connection with the perpendicular molecular localization in the previous paragraph.

Implicitly, several authors referred to such cell surface patterns as a possible new code system that regulates the communication of the cell and the extracellular world. Not much effort has been made so far to decode the unmistakable message of the cell surface. Of course, in this case the problem is far more complicated than it was when Nirenberg solved the alfabeth of the genetic code. The genetic code has a direct and in a way simple relationship to the amino acid sequences of the proteins. The cell surface code is probably more complicated, because we have to determine the coded message before we start looking for a decoding algorithm. It is known that an incompletely differentiated, but already committed cell can be transformed by its environment, making it identical with the surrounding cell types. This transformation is very likely mediated by the cell surface patterns of nearby cells, but intercellular chemical, hormonal, etc. interactions cannot be excluded or taken for granted. A character of the code of the cell surface definitely depends on the types and the two- or three-dimensional distribution of its receptors and other even nonprotein molecular elements. These receptors behave like sensors (eyes and ears of the cell) toward the outer world. Cell-mediated cytotoxicity is a well known example of such "communications". The extreme complexity of transmembrane signaling pathways, the diversity of different complex receptor systems which finally generate cellular responses from inside (secretion, activation, differentiation, expression of new membrane elements, intercellular interactions, etc.), gives us an idea about the difficulties on the long and hard way we have to take, until we learn the real algorithm to decode the messages of the cell, encoded by cell surface structures. In a way, the code may not be purely chemical, in contrast to the genetic code. The charge distribution, the steep and remarkable transmembrane potential, mobility, and proximity changes in the cell membrane during cellular activities, and in one word the "pattern" refers to a complex appearance of chemical and physical steps which must occur in a concerted way.

At present the majority of such discussions belongs to the realm of philosophy, but emerging new techniques and conceptions can help to put such a series of arguments on a solid and scientific basis.[14,17-21,229-231,255]

C. FLUCTUATING PROTEINS IN THE CELL MEMBRANE

FRET can be utilized to monitor dynamic changes of individual molecular elements of membranes in a more direct way than those mentioned above. Somogyi et al.[249] have shown that FRET efficiency that varies with temperature of the measurements could be controlled or even utilized to determine molecular fluctuations. The actual energy transfer, determined at a well-defined temperature value, was made independent of the influence of the temperature by normalizing the transfer efficiency to the fluorescence of the donor molecules. This simple way enabled the authors to follow the changes in the fluctuation pattern of proteins due to the changing temperature. RN-ase T_1 Ca^{2+}-dependent membrane ATPase molecules have been investigated so far, utilizing this very promising yet not very frequently applied method.[53,249] An elegant application of this method — as we have seen before — revealed that the calcium-dependent ATPase molecule had different fluctuational freedom in its subunits. It was suggested that this fluctuation may play a role in the calcium transport across the membrane.[244]

D. TRANSMEMBRANE SIGNALING AND MEMBRANE POTENTIAL CHANGES

Much attention has been paid to cell activation processes in the last decade. It was unequivocally established that the information transfer across plasma membranes involves essentially biochemical and biophysical processes. These processes might be instrumental in the understanding of such mechanisms as the elucidation of several cellular functions such as secretory, synthetic, gene activation, or proliferative steps. According to an early suggestion of Cone,[255] variations in the transmembrane potential level can serve as a basic mechanism of mitotic control. A better understanding of these functions may significantly contribute to the available diagnostic tools of pathological functions reporting on malignantly transformed cells as well. Although a significant amount of important information is available on the biochemical pathways and different cell activation processes, which are highly dependent on the nature of the trigger signals (mitogen or antibody-antigen reaction-induced responses as compared to the interleukins or hormone induced ones), far less information is available on the contribution of physical events to the selection of transducing pathways and the actual responses.[10-12,18,127,144] Among physiological or even pathological conditions, the earliest events during signal transduction in the plasma membrane are direct consequences of the signal itself, regardless of its physical or chemical nature. When transmembrane signaling is induced via a receptor-operated pathway, the first event is the ligand-receptor encounter and binding. Devotedly to our consistent goal to study principally the physical events evoked by signal-transducing mechanisms at the level of the plasma membrane, we focus our interest here upon membrane potential changes during signal transduction in lymphocytes. This can be justified by the common knowledge that membrane potential changes must precede all voltage-gated ion transport processes, which seem to be instrumental in elucidating the responses of the stimulated cells.[10-12,130]

Electroconformational coupling and membrane protein function have been studied theoretically as well as experimentally, reporting on the changes in the essential proteins' structural and functional responsiveness upon the changing electric potential field.[260] Examples, beyond those of Balázs et al.,[44] Muller et al.,[46] and Dunant et al.[45] mentioned above, were also given by many authors. Depolymerization of microtubules altered membrane potential and affected the motional freedom of membrane proteins in intact Chinese hamster ovary cells.[272] Mátyus et al.[85] and Sarkadi et al.[79] observed that regulation of the Ca^{2+}-activated K^+ channels is substantially influenced by the actual membrane potential values. Depolarization decreases the calcium influx,[79] while it sensitizes the intracellular Ca^{2+} binding site of the K^+ channel in such a way that calcium doses which usually inhibit the responses of K^+ channels can activate them at the lower membrane potential values.[85] Such cooperation of ligand and voltage may be more frequent phenomenon than has been suspected.

Pieri et al.[2] observed that bretylium tosylate (BT), a quaternary ammonium ion, known for a weak potassium channel inhibiting and sodium channel opening property, rapidly hyperpolarized practically all kinds of lymphocytes, regardless of their T or B origin. The apparent contradiction with physiology was immediately solved when the activation of the Na^+-K^+-ATPase was found to be responsible for the relative hyperpolarization. The hyperpolarization had an absolute requirement of a slight depolarization before the addition of the BT.[2] It is interesting to note that acetylcholine, a well-known mediator of the nerve system can also hyperpolarize certain types of cells.[354] Since the sodium influx upon addition of BT was solidly established, a ligand- and voltage-gated sodium channel was assumed to regulate electrogenic Na^+-K^+ pump activity in human, mouse, and rat lymphocytes. This mechanism was independent of the Na^+/H^+ exchange,[68] although the proton and sodium antiport and other co-transport processes have a multiple regulatory role in lymphocytes and their activations.[298,355,356] A real novelty of the finding with the application of the BT is that only 3% of the T lymphocytes have been thought to have Na^+ channel activity. On the other hand, the BT-initiated Na^+ influx proved to be a requisite of practically all kinds of lymphocytes. Electrophysiological evidence obtained with patch clamp method also supported the existence of this new kind of Na^+ channel in lymphocytes.[357]

BT inhibited the PHA or antigen receptor-induced cell activation processes in a dose-dependent fashion, when BT was added simultaneously with the trigger signal. Human peripheral lymphocytes were blocked in G_0 and G_1 phase by the presence of BT. The removal of BT lifted the block and the proliferation continued even without readdition of lectins. The sustained hyperpolarization of the cells was suggested as a principal cause of the inhibitory action.[24] A remarkable effect of the BT was found when the same doses significantly enhanced the proliferative responsiveness of lymphocytes after simultaneous treatment of IL-2 primed cells with BT and IL-2 (Pieri et al., personal comunication).

Nevertheless, it is generally accepted now that IL-2 has a distinctly different signal-transducing pathway than that of the lectin or TCR-CD3-operated one.[106-108] We suggest that those signal-transducing processes which are more exposed to membrane potential changes of the plasma membrane are more sensitive to BT.[358] The enhancement of the proliferation can be interpreted as a consequence of the Na^+-water-metabolite cotransport, initiated also by the BT. Indeed, it seems that IL-2 receptor-dependent signal transduction is less sensitive to membrane processes and is independent of the Ca^{2+} and the inositol phosphate cascade.[106-108,114,115] Of course, the above model and discussion do not mean that different signal-transducing pathways do not influence each other in general;[359-364] however, in this particular case, even a crosstalk between the two pathways was more or less excluded.[106]

E. IMMUNOSUPPRESSANTS

Immunosuppressive agents, such as CsA, block the signal-transduction pathways of T lymphocytes, thereby preventing host vs. graft reactions of the activated T lymphocytes.[192,193,203] This property, together with the relative insensitivity of humoral immunity (B-cell functions) toward them, make their application indispensable in solid organ transplantation. Although a high number of sophisticated effects of the immunosuppressants (cyclosporins, rapamycin, FK 506) are well known, the actual mode of action by which they can differentiate between cellular and humoral immunity is yet unsolved. Thus, "in addition to the fundamental interest in the molecular basis of immunosuppressive action, the clinical importance of CsA has also spurred vigorous research activity on the structural basis of its function."[365] Here we feel proper to emphasize again that the cyclosporins and similar compounds have a well-defined biphasic membrane effect beyond their better-known biochemical interactions. There is a dose dependent de- or hyperpolarization of the plasma membranes of lymphocytes.[15,16,22,72,138] These effects seem to be the earliest ones, since any other effect assumes a previous encounter of the drug with the membrane.

Recently the old observation of Handschumacher's group[201,205,206] that CsA had an intracytosolic receptor, called cyclophilin, has been further studied by mostly by highly sophisticated X-ray crystallographic and NMR methods.[365,366] Other immunosuppressants have also been linked to intracytosolic receptors, other immunophilins which proved to be different from that of the CsA. A current biochemical theory for the action of this family of drugs attaches their action to the inhibition of nuclear gene-activating factors.[367-369] The immunosuppressant is supposed to bind to its respective receptor, and this complex prevents the translocation of a cytosolic subunit of the nuclear factor, NF-AP, into the nucleus, without suppressing the synthesis of a nuclear subunit of the same factor.[368,369] However, the factor is active only if the translocation of the cytosolic subunit occurs and it can form a complex with its nuclear counterpart. Distinct immunosuppressants may inhibit different nuclear factors.

Taken together our present knowledge and the easily foreseeable future perspectives, we believe that the membrane-bound events of cell activation processes cannot be overvalued. The particularly high importance of membrane potential, proximity, and mobility of cell surface elements and an emerging view of two-dimensional molecular patterns of the receptor complexes in the plasma membrane are yet to be better understood.

REFERENCES

1. **Lewis, R. J. and Cahalan, M. D.,** Plasticity of ion channels: parallels between the nervous and immune systems, *Trends Neurosci.*, 11, 214, 1988.
2. **Pieri, C., Recchioni, R., Moroni, F., Balkay, L., Márián, T., Trón, L., and Damjanovich, S.,** Ligand and voltage gated sodium channels may regulate electrogenic pump activity in human, mouse and rat lymphocytes, *Biochem. Biophys. Res. Commun.*, 160, 999, 1989.
3. **Jan, L. Y. and Jan, Y. N.,** How might the diversity of potassium channels be generated?, *Trends Neurosci.*, 13, 415, 1990.
4. **Grissmer, S., Hanson, D. C., Natoli, E. J., Cahalan, M. D., and Chandy, K. G.,** $CD4^-CD8^-$ T cells from mice with collagen arthritis display aberrant expression of type *l* K^+ channels, *J. Immunol.*, 145, 2105, 1990.
5. **Grissmer, S., Dethlef, B., Wasmoth, J. J., Godlin, A. L., Gutman, G. A., Cahalan, M. D., and Chandy, K. G.,** Expression and chromosomal localization of a lymphocyte K^+ channel gene, *Proc. Natl. Acad. Sci. U.S.A.*, 87, 9411, 1990.
6. **Rich, D. P., Anderson, M. P., Gregory, R. J., Cheng, S. H., Paul, S., Jefferson, D. M., McCann, J. D., Klinger, K. W., Smith, A. E., and Welch, M. J.,** Expression of cystic fibrosis transmembrane conductance regulator corrects defective chloride channel regulation in cystic fibrosis airway epithelial cells, *Nature*, 347, 358, 1990.
7. **DeCoursey, T. E., Chandy, K. G., Gupta, S., and Cahalan, M. D.,** Voltage gated K^+ channels in human T lymphocytes: a role in mitogenesis? *Nature*, 307, 465, 1984.
8. **Tsien, R. Y., Pozzan, T., and Rink, T. J.,** T-cell mitogens cause early changes in cytoplasmic free calcium and membrane potential in lymphocytes, *Nature*, 295, 68, 1982.
9. **Tsien, R. Y., Pozzan, T., and Rink, T. J.,** Measuring and manipulating cytosolic Ca^{2+} with trapped indicators, *Trends Biochem. Sci.*, 9, 263, 1984.
10. **Szöllősi, J., Damjanovich, S., Mulhern, S. A., and Trón, L.,** Fluorescence energy transfer and membrane potential measurements monitor dynamic properties of cell membranes: a critical review, *Prog. Biophys. Mol. Biol.*, 49, 65, 1987.
11. **Matkó, J., Szöllősi, J., Trón, L., and Damjanovich, S.,** Luminescence spectroscopic approaches of cell surface dynamics, *Q. Rev. Biophys.*, 21, 479, 1988.
12. **Grinstein, S. and Dixon, J. S.,** Ion transport, membrane potential and cytoplasmic pH in lymphocytes: changes during activation, *Physiol. Rev.*, 69, 417, 1989.
13. **Alberts, B., Bray, D., Lewis, J., Raff, M., Roberts, K., and Watson, J. D.,** *Molecular Biology of the Cell*, 1st ed., Garland, London, 1983, 255.
14. **Damjanovich, S., Somogyi, B., and Trón, L.,** Macromolecular dynamics and information transfer, *Adv. Physiol. Sci.*, 30, 9, 1981.
15. **Damjanovich, S., Aszalós, A., Mulhern, S. A., Balázs, M., and Mátyus, L.,** Cytoplasmic membrane potential of lymphocytes is decreased by cyclosporins, *Mol. Immunol.*, 23, 175, 1986.
16. **Damjanovich, S., Aszalós, A., Mulhern, S. A., Szöllősi, J., Balázs, M., Trón, L., and Fulwyler, M. J.,** Cyclosporin depolarizes human lymphocytes: earliest observed effect on cell metabolism, *Eur. J. Immunol.*, 17, 763, 1987.
17. **Damjanovich, S., Balázs, M., Trón, L., Szöllősi, J., and Somogyi, B.,** Protein dynamics and function. An overview, *J. Mol. Catal.*, 47, 763, 1988.
18. **Damjanovich, S. and Pieri, C.,** Electro-immunology: membrane potential, ion-channel activities and stimulatory signal transduction in human T lymphocytes from young and elderly, *Ann. N.Y. Acad. Sci.*, 621, 29, 1991.

19. **Damjanovich, S., Szöllősi, J., Balázs, M., Mátyus, L., Gáspár, R., Krasznai, Z., Pieri, C., and Trón, L.**, Dynamic physical interactions of plasma membrane molecules generate cell surface patterns and regulate cell activation processes, *Immunobiology*, 185, 337, 1992.
20. **Damjanovich, S., Pieri, C., and Trón, L.**, Ion channel activity and transmembrane signalling in lymphocytes, *Ann. N.Y. Acad. Sci.*, 605, 205, 1992.
21. **Damjanovich, S., Szöllősi, J., and Trón, L.**, Transmembrane signalling in lymphocytes, *Immunol. Today*, 13, A12, 1992.
22. **Mátyus, L., Balázs, M., Aszalós, A., Mulhern, S. A., and Damjanovich, S.**, Cyclosporin depolarizes cytoplasmic membrane potential and interacts with Ca^{2+} ionophores, *Biochim. Biophys. Acta*, 886, 353, 1986.
23. **Pieri, C., Recchioni, R., Moroni, F., Falasca, M., and Damjanovich, S.**, Parameters to monitor aging with a possible perspective for intervention; an immunological approach, *Arch. Gerontol. Geriatr.*, 12, 231, 1991.
24. **Pieri, C., Recchioni, R., Moroni, F., Marcheselli, F., Falasca, M., Krasznai, Z., Gáspár, R., Mátyus, L., and Damjanovich, S.**, A sodium channel opener inhibits stimulation of human peripheral blood mononuclear cells, *Mol. Immunol.*, 29, 517, 1992.
25. **Neumcke, B.** Membrane potentials, in *Biophysics*, 1st ed., Hoppe, W., Lohmann, W., Markl, H., and Ziegler, H., Eds., Springer-Verlag, Berlin, 1983, chap. 12.3.
26. **Bashford, C. L. and Pasternak, C. A.**, Plasma membrane potential of some animal cells is generated by ion pumping, not by ion gradients, *Trends Biochem. Sci.*, 11, 113, 1986.
27. **Setlow, R. B. and Pollard, E. C.**, *Molecular Biophysics*, Addison-Wesley, Reading, MA, 1962, 469.
28. **Van Duijn, B. and Vogelzang, S. A.**, The membrane potential of the cellular slime mold Dictyostelium discoideum is mainly generated by an electrogenic proton pump, *Biochim. Biophys. Acta*, 983, 186, 1989.
29. **Neher, E. and Sakmann B.**, Single-channel currents from membrane of denervated frog muscle, *Nature*, 260, 799, 1976.
30. **Neher, E.**, Ion channels for communication between and within cells, *Science*, 256, 498, 1992.
31. **Sakmann, B.**, Elementary steps in synaptic transmission revealed by currents through single ion channels, *Science*, 256, 503, 1992.
32. **Sagi-Eisenberg, R. and Pecht, I.**, Membrane potential changes during IgE mediated histamine release from rat basophilic leukemia cells, *J. Membr. Biol.*, 75, 97, 1983.
33. **Waggoner, A. S., Wang, C. H., and Tolles, R. L.**, Mechanism of potential dependent light absorption changes of lipid bilayer membranes in the presence of cyanine and oxonol dyes, *J. Membr. Biol.*, 33, 109, 1977.
34. **Waggoner, A. S.**, Dye indicators of membrane potential, *Annu. Rev. Biophys. Bioeng.*, 8, 47, 1979.
35. **Waggoner, A. S.**, Fluorescent probes for analysis of cell structure, function and health by flow and imaging cytometry, in *Applications of Fluorescence in the Biomedical Sciences*, 1st ed., Taylor, D. L., Waggoner, A. S., Lanni, F., Murphy, R. F., and Birge, R. R., Eds., Alan R. Liss, New York, 1986, chap. 3.
36. **Brauer, T., Hulser, D. F., and Strasser, R. J.**, Comparative measurements of membrane potentials with microelectrodes and voltage sensitive dyes, *Biochim. Biophys. Acta*, 771, 208, 1984.
37. **Cohen, R. L., Muirhead, K. A., Gill, J. E., Waggoner, A. S., and Horan, P. K.**, A cyanine dye distinguishes between cycling and non-cycling fibroblasts, *Nature*, 290, 593, 1981.
38. **Shapiro, H. M., Natale, P. J., and Kamentsky, L. A.**, Estimation of membrane potentials of individual lymphocytes by flow cytometry, *Proc. Natl. Acad. Sci. U.S.A.*, 70, 5728, 1979.
39. **Hamill, O. P., Marty, A., Neher, E., Sakmann, B., and Sigworth, F. J.**, Improved patch clamp techniques for high resolution current recording from cells and cell free membrane patches, *Pflugers Arch.*, 391, 85, 1981.
40. **Sakmann, B. and Neher, E.**, Eds., *Single Channel Recording*, Plenum Press, New York, 1983.
41. **Pratap, P. R., Novák, T. S., and Freedman, J. C.**, Two mechanisms by which fluorescent oxonols indicate membrane potential in human red blood cells, *Biophys. J.*, 57, 835, 1990.
42. **Wilson, A. H. and Chused, T. M.**, Lymphocyte membrane potential and Ca^{2+} sensitive potassium channels described by oxonol dye fluorescence measurements, *J. Cell. Physiol.*, 125, 72, 1985.
43. **Wilson, A. H., Seligmann, B. E., and Chused, T. M.**, Voltage sensitive cyanine dye fluorescence signals in lymphocytes: plasma membrane and mitochondrial components, *J. Cell. Physiol.*, 125, 61, 1985.

44. **Balázs, M., Matkó, J., Szöllősi, J., Mátyus, L., Fulwyler, M. J., and Damjanovich, S.,** Accessibility of cell surface thiols in human lymphocytes is altered by ionophores or OKT3 antibody, *Biochem. Biophys. Res. Commun.*, 140, 999, 1986.
45. **Dunnant, Y., Garcia-Segura, L. M., Muller, D., and Párducz, A.,** Momentary alteration of the postsynaptic membrane during transmission of a single nerve impulse, *Proc. Natl. Acad. Sci. U.S.A.*, 86, 1717, 1989.
46. **Muller, D., Garcia-Segura, L. M., Párducz, A., and Dunnant, Y.,** Brief occurrence of a population of presynaptic intramembrane particles coincides with transmission of a nerve pulse, *Proc. Natl. Acad. Sci. U.S.A.*, 84, 590, 1987.
47. **Thulborn, K. R. and Sawyer, W. H.,** Properties and locations of a set of fluorescent probes sensitive to the fluidity gradients of the lipid bilayer, *Biochim. Biophys. Acta*, 511, 125, 1978.
48. **Stryer, L.,** Fluorescence energy transfer as a spectroscopic ruler, *Annu. Rev. Biochem.*, 47, 819, 1978.
49. **Ameloot, M., Hendrickx, H., Herreman, W., Pottel, H., Cauwelaert, F., and Van der Meer, W.,** Effect of orientational order on the decay of the fluorescence anisotropy in membrane suspensions, *Biophys. J.*, 46, 525, 1984.
50. **Chan, S. S., Arndt-Jovin, D. J., and Jovin, T. M.,** Proximity of lectin receptors on the cell surface measured by fluorescence energy transfer in a flow system, *J. Histochem. Cytochem.*, 27, 56, 1979.
51. **Gelfand, E. W., Mills, G. B., Cheung, R. K., Lee, J. W. W., and Grinstein, S.,** Transmembrane ionfluxes during activation of human T lymphocytes: role of Ca^{2+}, Na^+/H^+ exchange and phospholipid turnover, *Immunol. Rev.*, 95, 59, 1987.
52. **Somogyi, B., Matkó, J., Papp, S., Hevessy, J., Welch, G. R., and Damjanovich, S.,** Förster type energy transfer as a probe for changes in the fluctuations of protein matrix, *Biochemistry*, 23, 3403, 1984.
53. **Jóna, I., Matkó, J., and Martonosi, A.,** Structural dynamics of the Ca^{2+}-ATP-ase of sarcoplasmic reticulum. Temperature profiles of fluorescence polarisation and intramolecular energy transfer, *Biochim. Biophys. Acta*, 1028, 183, 1990.
54. **Cantor, C. and Schimmel, R.,** in *Biophysical Chemistry*, Vol. 2, 1st ed., W. H. Freeman, San Francisco, 1980, chap. 8-2.
55. **Cook, N. S., Ed.,** *Potassium Channels: Structure, Classification, Function and Therapeutic Potential*, Ellis Horwood, Chichester, England, 1990.
56. **Kaczmarek, L. K.,** A book review on *Potassium Channels: Structure, Classification, Function and Therapeutic Potential*, Cook, N. S., Ed., *Trends Neurosci.*, 14, 534, 1991.
57. **Matteson, D. R. and Deutsch, C.,** K^+ channels in T lymphocytes: a patch clamp study using monoclonal antibody adhesion, *Nature*, 307, 468, 1984.
58. **Gardner, P.,** Patch clamp studies of lymphocytes activation, *Annu. Rev. Immunol.*, 8, 231, 1990.
59. **Gallin, E.,** Ion channels in leukocytes, *Physiol. Rev.*, 71, 775, 1991.
60. **Chandy, K. G., DeCoursey, T. E., Cahalan, M. D., and Gupta, S.,** Ion channels in lymphocytes, *J. Clin. Immunol.*, 5, 1, 1985.
61. **Chandy, K. G., DeCoursey, T. E., Cahalan, M. D., and Gupta, S.,** Electroimmunology: the physiologic role of ion channels in the immune system, *J. Immunol.*, 135(Suppl. 787), 1985.
62. **Chandy, K. G., DeCoursey, T. E., Cahalan, M. D., McLaughlin, C., and Gupta, S.,** Voltage gated potassium channels are required for human T lymphocyte activation, *J. Exp. Med.*, 160, 369, 1984.
63. **Chandy, K. G., DeCoursey, T. E., Fischbach, M., Talal, N., Cahalan, M. D., and Gupta, S.,** Altered K^+ channel expression in abnormal T lymphocytes from mice with the lpr gene mutation, *Science*, 323, 1197, 1986.
64. **Deutsch, C. J., Krause, D., and Lee, S. C.,** Voltage gated potassium conductance in human T lymphocytes stimulated with phorbol ester, *J. Physiol.*, 372, 405, 1986.
65. **Schlichter, L., Sidell, N., and Hagiwara, S.,** Potassium channels mediate killing by human natural killer cells, *Proc. Natl. Acad. Sci. U.S.A.*, 83, 451, 1986.
66. **Lee, S. C., Sabath, D. E., Deutsch, C., and Prystowsky, M. B.,** Increased voltage-gated potassium conductance during IL-2 stimulated proliferation of a mouse helper T lymphocyte clone, *J. Cell Biol.*, 102, 1200, 1986.
67. **Legrue, S. J.,** Interleukin-2 stimulus-response coupling is calcium independent, *Lymphokine Res.*, 6, 1, 1987.
68. **Trón, L., Pieri, C., Márián, T., Balkay, L., Emri, M., and Damjanovich, S.,** Bretylium causes a Na^+-K^+ pump activation that is independent of Na^+/H^+ exchange in depolarized rat, mouse and human lymphocytes, *Mol. Immunol.*, 27, 1307, 1990.

69. **Chandy, K. G., Williams, C. B., Spencer, R. H., Aguilar, B. A., Ghanshani, S., Tempel, B. L., and Gutman, G. A.**, A family of three mouse potassium channel genes with intronless coding regions, *Science*, 247, 973, 1990.
70. **Catterall, W. A.**, Structure and function of voltage sensitive ion channels, *Science*, 242, 50, 1988.
71. **Miller, F. J.**, Glucose-regulated potassium channels are sweet news for neurobiologists, *Trends Neurosci.*, 13, 197, 1990.
72. **Vereb, Gy., Panyi, Gy., Balázs, M., Mátyus, L., Matkó, J., and Damjanovich, S.**, Effect of cyclosporin A on the membrane potential and Ca^{2+} level of human lymphoid cell lines and mouse thymocytes, *Biochim. Biophys. Acta*, 1019, 159, 1990.
73. **Ischida, Y. and Chused, T. M.**, Heterogeneity of lymphocyte calcium metabolism is caused by T cell-specific calcium-sensitive potassium channel and sensitivity of the calcium ATP-ase pump to membrane potential, *J. Exp. Med.*, 168, 839, 1988.
74. **Hille, B.**, *Ionic Channels of Excitable Membranes*, 1st ed., Sinauer Associates, Sunderland, MA, 1984.
75. **Kume, H., Takai, A., Tokumo, H., and Tomita, T.**, Regulation of Ca^{2+} dependent K^+ channel activity in tracheal myocytes by phosphorylation, *Nature*, 341, 152, 1989.
76. **Tanabe, T., Adams, B. A., Numa, S., and Beam, K. G.**, Repeat I of the dihydropyridine receptor is critical in determining calcium channel activation kinetics, *Nature*, 352, 800, 1991.
77. **Gárdos, G.**, The function of calcium in the potassium permeability of human erythrocytes, *Biochim. Biophys. Acta*, 30, 653, 1958.
78. **Sarkadi, B. and Gárdos, G.**, Drug actions on potassium fluxes in red cells, in *The Red Cell Membrane*, Raess, B. U. and Tunnicliff, G., Eds., Humana Press, Clifton, NJ, 1989, 369.
79. **Sarkadi, B., Tordai, A., and Gárdos, G.**, Membrane depolarization selectively inhibits receptor-operated calcium channels in human T (Jurkat) lymphocytes, *Biochim. Biophys. Acta*, 1027, 130, 1990.
80. **Tatham, P. E., O'Flynn, K., and Lynch, D. C.**, The relationship between mitogen induced membrane potential changes and intracellular free calcium in human T-lymphocytes, *Biochim. Biophys. Acta*, 856, 201, 1986.
81. **Rink, T. J. and Deutsch, C.**, Calcium-activated potassium channels in lymphocytes, *Cell Calcium*, 4, 463, 1983.
82. **Mahaut-Smith, M. P. and Schlichter, L. C.**, Ca^{2+} activated K^+ channels in human B lymphocytes and rat thymocytes, *J. Physiol.*, 415, 69, 1989.
83. **Mahaut-Smith, M. P. and Mason, M. J.**, Ca^{2+} activated K^+ channels in rat thymic lymphocytes: activation by concanavalin A, *J. Physiol.*, 439, 513, 1991.
84. **Grissmer, S., Lewis, R. S., and Cahalan, M. D.**, Ca^{2+}-activated K^+ channels in human leukemic T cells, *J. Gen. Physiol.*, 99, 63, 1992.
85. **Mátyus, L., Pieri, C., Recchioni, R., Moroni, F., Bene, L., Trón, L., and Damjanovich, S.**, Voltage gating of Ca^{2+} activated potassium channels in human lymphocytes, *Biochem. Biophys. Res. Commun.*, 171, 325, 1990.
86. **Mátyus, L., Takács, L., Pohubi, L., Balázs, M., Szöllősi, J., Matkó, J., Trón, L., and Damjanovich, S.**, Dynamic behaviour of cell surface receptors as revealed by laser excited fluorescence spectroscopy, in *Light in Biology and Medicine*, Vol. 2, Douglas, R. H., Moan, J., and Rontó, Gy., Eds., Plenum Press, New York, 1991, 833.
87. **Bregestovski, P., Redkozubov, A., and Alexeev, A.**, Elevation of intracellular calcium reduces voltage-dependent potassium conductance in human T cells, *Nature*, 319, 776, 1986.
88. **Fasolato, C., Pirro, P., and Pozzan, T.**, Receptor-mediated calcium influx in PC12 cells. ATP and bradykinin activate two independent pathways, *J. Biol. Chem.*, 265, 20351, 1989.
89. **Blatz, A. L. and Magleby, K. L.**, Calcium-activated potassium channels, *Trends Neurosci.*, 10, 463, 1987.
90. **White, R. E., Schonbrunn, A., and Armstrong, D. L.**, Somatostatin stimulates Ca^{2+} activated K^+ channels through protein dephosphorylation, *Nature*, 351, 570, 1991.
91. **Tas, P. W. L., Kress, H. G., and Koschel, K.**, General anaesthetics can competitively interfere with sensitive membrane proteins, *Proc. Natl. Acad. Sci. U.S.A.*, 84, 5972, 1987.
92. **Tas, P. W. L., Kress, H. G., and Koschel, K.**, Lipid solubility is not the sole criterion for the inhibition of a Ca^{2+} activated K^+ channel by alcohols, *Biochim. Biophys. Acta*, 1023, 436, 1990.
93. **Klotz, I. M.**, Water, in *Horizons in Biochemistry*, Kasha, M. and Pullman, B., Eds., Academic Press, New York, 1962, 545.

94. **Alifimoff, J. K., Firestone, L. L., and Miller, K. W.,** Anaesthetic potencies of primary alkanols: implications for the molecular dimensions of the anaesthetic site, *Br. J. Pharmacol.*, 96, 9, 1989.
95. **Noda, M., Shimizu, S., Tanabe, T., Takai, T., Kayano, T., Ikeda, T., Takahashi, H., Nakayama, H., Kanaoka, Y., Minamino, N., Kangawa, K., Matsuo, H., Raftery, M. A., Hirose, T., Inayama, S., Hayashida, H., Miyata, T., and Numa, S.,** Primary structure of electrophorus electricus sodium channel deduced from cDNA sequence, *Nature*, 312, 121, 1984.
96. **Noda, M., Ikeda, T., Kayano, T., Suzuki, H., Takeshima, H., Kurasaki, M., Takahashi, H., and Numa, S.,** Existence of distinct sodium channel messenger RNAs in rat brain, *Nature*, 320, 188, 1986.
97. **Stuhmer, W., Conti, F., Suzuki, H., Wang, X. D., Noda, M., Yahagi, N., Kubo, H., and Numa, S.,** Structural parts involved in activation and inactivation of the sodium channel, *Nature*, 339, 597, 1989.
98. **Striessnig, J., Glossmann, H., and Catterall, W. A.,** Identification of a phenylalkylamine binding region within the α1 subunit of skeletal muscle Ca^{2+} channels, *Proc. Natl. Acad. Sci. U.S.A.*, 87, 9108, 1990.
99. **Auld, V. J., Goldin, A. L., Krafte, D. S., Catterall, W. A., Lester, H. A., Davidson, N., and Dunn, R. J.,** A neutral amino acid change in segment "II. S4" dramatically alters the gating properties of the voltage dependent sodium channel, *Proc. Natl. Acad. Sci. U.S.A.*, 87, 323, 1990.
100. **Kuno, M., Goronzy, J., Weyand, C. M., and Gardner, P.,** Single-channel and whole-cell recordings of mitogen-regulated inward currents in human cloned helper T lymphocytes, *Nature*, 323, 269, 1986.
101. **Kuno, M. and Gardner, P.,** Ion channels activated by inositol 1,4,5,-trisphosphate in plasma membrane of human T-lymphocytes, *Nature*, 326, 301, 1987.
102. **Fukushima, Y. and Hagiwara, S.,** Voltage gated Ca^{2+} channel in mouse myeloma cells, *Proc. Natl. Acad. Sci. U.S.A.*, 80, 2240, 1983.
103. **Fukushima, Y., Hagiwara, S., and Henkart, M.,** Potassium current in clonal cytotoxic T lymphocytes from the mouse, *J. Physiol.*, 351, 645, 1984.
104. **Fukushima, Y., Hagiwara, S., and Saxton, R. E.,** Variation of calcium current during the cell growth cycle in mouse hybridoma lines secreting immunoglobulins, *J. Physiol.*, 355, 313, 1984.
105. **Fukushima, Y. and Hagiwara, S.,** Currents carried by monovalent cations through calcium channels in mouse neoplastic B lymphocytes, *J. Physiol.*, 358, 255, 1985.
106. **Tigges, M. A., Casey, L. S., and Koshland, M. E.,** Mechanism of IL-2 signalling: mediation of different outcomes by a single receptor and transduction pathway, *Science*, 243, 781, 1989.
107. **Valge, V., Wong, J. G. P., Datlof, B. M., Skinskey, A. J., and Raer, A.,** Protein kinase C is regulated for responses to T cell receptor ligands but not to interleukin-2 in T cells, *Cell*, 55, 101, 1988.
108. **Mills, G. B., Girarad, P., Grinstein, S., and Gelfand, E. W.,** IL-2 induces proliferation of T lymphocyte mutants lacking protein kinase C, *Cell*, 55, 91, 1988.
109. **Gelfand, E. W., Cheung, R. K., and Grinstein, S.,** Role of membrane potential in the regulation of lectin induced calcium uptake, *J. Cell. Physiol.*, 121, 533, 1984.
110. **Gelfand, E. W., Cheung, R. K., Mills, G. B., and Grinstein, S.,** Mitogen trigger a calcium independent signal for proliferation in phorbolester treated lymphocytes, *Nature*, 315, 419, 1985.
111. **Gelfand, E. W., Cheung, R. K., and Grinstein, S.,** Mitogen induced changes in calcium permeability are not mediated by voltage gated potassium channels, *J. Biol. Chem.*, 261, 11520, 1986.
112. **Gelfand, E. W., Cheung, R. K., Mills, G. B., and Grinstein, S.,** Role of membrane potential in the response of human T lymphocytes to phytohemagglutinin, *J. Immunol.*, 138, 527, 1987.
113. **Herrmann, T. and Diamantstein, T.,** The mouse high affinity IL 2 receptor complex I. Evidence for a third molecule, the putative γ-chain associated with the α- and/or β chain of the receptor, *Immunobiology*, 175, 145, 1987.
114. **Waldmann, T. A.,** The structure, function and expression of interleukin-2 receptors on normal and malignant lymphocytes, *Science*, 23, 727, 1986.
115. **Waldmann, T. A.,** The interleukin-2 receptor, *J. Biol. Chem.*, 266, 2681, 1991.
116. **Greene, W. C. and Leonard, W. J.,** The human interleukin-2 receptor, *Annu. Rev. Immunol.*, 4, 69, 1986.
117. **Shackelford, D. A. and Trowbridge, I. S.,** Ligand-stimulated tyrosine phosphorylation of the IL-2 receptor beta chain and receptor-associated protein, *Cell Regul.*, 2, 73, 1991.
118. **Tsudo, M., Kozak, R. W., Goldman, C. K., and Waldmann, T. A.,** Demonstration of a non-Tac peptide that binds interleukin-2: a potential participant in a multichain interleukin-2 receptor complex, *Proc. Natl. Acad. Sci. U.S.A.*, 83, 9694, 1986.

119. **Weiss, A., Shields, R., Newton, M., Manner, B., and Imboden, J.,** Ligand-receptor interactions required for commitment to the activation of the interleukin-2 gene, *J. Immunol.*, 138, 2169, 1987.
120. **Horak, I. D., Gress, R. E., Lucas, P. J., Horak, E. M., Waldmann, T. A., and Bolen, J. B.,** T-lymphocyte interleukin-2 dependent protein tyrosine kinase signal transduction involves the activation of $p56^{lck}$, *Proc. Natl. Acad. Sci. U.S.A.*, 88, 1996, 1991.
121. **Hatakeyama, M., Kono, T., Kobayashi, N., Levin, S. D., Perlmitter, R. M., and Tanaguchi, T.,** Interaction of the IL-2 receptor with the Src-family kinase $p56^{lck}$: identification of novel intermolecular association, *Science*, 252, 1523, 1991.
122. **Klausner, R. D. and Samelson, L. E.,** T cell antigen receptor activation pathways: the tyrosine kinase connection, *Cell*, 64, 875, 1991.
123. **Saltzman, E. M., White, K., and Casnellie, J. E.,** Stimulation of the antigen and interleukin-2 receptors on T lymphocytes activates distinct tyrosine protein kinases, *J. Biol. Chem.*, 265, 10139, 1991.
124. **Fung, M. R., Scearce, R. M., Hoffman, J. A., Pfeffer, N. J., Hammes, S. R., Hosking, J. B., Schmandt, R., Kuziel, W. A., Haynes, B. F., Mills, G. B., and Greene, W.,** A tyrosine kinase physically associates with the beta-subunit of the human IL-2 receptor, *J. Immunol.*, 147, 1253, 1991.
125. **Merida, I., Diez, E., and Gaulton, G. N.,** IL-2 binding activates a tyrosine-phosphorylated phosphatidylinositol-3-kinase, *J. Immunol.*, 147, 2202, 1991.
126. **Rink, T. J. and Jacob, R.,** Ca^{2+} oscillations in non-excitable cells, *Trends Neurosci.*, 12, 43, 1989.
127. **Berridge, M. J.,** Ca^{2+} oscillations, *J. Biol. Chem.*, 265, 9583, 1990.
128. **Irvine, R. F.,** "Quantal" Ca^{2+} release and the control of Ca^{2+} entry by inositol phosphates — a possible mechanism, *FEBS Lett.*, 263, 5, 1990.
129. **Lewis, R. S. and Cahalan, M. D.,** Mitogen-induced oscillations of cytosolic Ca^{2+} and transmembrane Ca^{2+} current in human leukemic T cells, *Cell Regul.*, 1, 99, 1989.
130. **Lewis, R. S., Grissmer, S., and Cahalan, M. D.,** Role of K^+ channels in regulating calcium signalling in leukemic T lymphocytes, *Biophys. J.*, 59(abstr.), 13a, 1991.
131. **Stryer, L.,** Cyclic GMP cascade of vision, *Annu. Rev. Neurosci.*, 9, 87, 1986.
132. **Uhl, R., Wagner, R., and Ryba, N.,** Watching G-proteins at work, *Trends Neurosci.*, 13, 64, 1990.
133. **Kim, D., Lewis, D. L., Graziadei, L., Neer, E. J., Bar-Sagi, D., and Chapman, D. E.,** G-protein $\beta\gamma$-subunits activate the cardiac muscarinic K^+ channel via phospholipase A_2, *Nature*, 337, 557, 1989.
134. **Kurachi, Y., Ito, H., Sugimoto, T., Shimizu, T., Miki, I., and Ui, M.,** Arachidonic acid metabolites as intracellular modulators of the G protein gated cardiac K^+ channel, *Nature*, 337, 555, 1989.
135. **Simon, S. M. and Blobel, G.,** A protein conducting channel in the endoplasmic reticulum, *Cell*, 65, 371, 1991.
136. **Hartmann, E., Rapoport, T. A., and Lodish, H. F.,** Predicting the orientation of eukaryotic membrane spanning proteins, *Proc. Natl. Acad. Sci. U.S.A.*, 86, 5786, 1989.
137. **Higgins, F. and Gottesman, M.,** Is the multidrug transporter a flippase?, *Trends Biochem. Sci.*, 17, 18, 1992.
138. **Gelfand, E. W., Cheung, R. K., and Mills, G. B.,** The cyclosporins inhibit lymphocyte activation at more than one site, *J. Immunol.*, 138, 1115, 1987.
139. **Gelfand, E. W., Cheung, R. K., and Grinstein, S.,** Characterization of the role for calcium influx in mitogen-induced triggering of human T cells. Identification of calcium-dependent and calcium-independent signals, *Eur. J. Immunol.*, 16, 907, 1986.
140. **Mills, G. B., Cheung, R. K., Grinstein, S., and Gelfand, E. W.,** IL-2 induced lymphocyte proliferation is independent of increases in cytosolic free calcium concentrations, *J. Immunol.*, 134, 2431, 1985.
141. **Mills, G. B., Stewart, D. J., Mellors, A., and Gelfand, E. W.,** IL-2 does not induce phosphatidylinositol hydrolysis in activated T cells, *J. Immunol.*, 136, 3019, 1986.
142. **Chilson, O. P. and Kelly-Chilson, A. E.,** Mitogenic lectins bind to the antigenic receptor on human lymphocytes, *Eur. J. Immunol.*, 19, 389, 1989.
143. **Berridge, M. J. and Irvine, R. F.,** Inositol triphosphate, a novel second messenger in cellular signal transduction, *Nature*, 312, 315, 1984.
144. **Berridge, M. J. and Irvine, R. F.,** Inositol phosphates and cell signalling, *Nature*, 341, 197, 1989.
145. **Parker, P. J., Coussens, L., Totty, N., Rhee, L., Young, S., Chen, E., Stabel, S., Waterfield, M. D., and Ullrick, A.,** The complete primary structure of PKC. The major phorbol ester receptor, *Science*, 233, 853, 1986.

146. **Brunswick, M., Samelson, L. E., and Mond, J. J.,** Surface immunoglobulin crosslinking activates a tyrosine kinase pathway in B cells that is independent of PKC, *Proc. Natl. Acad. Sci. U.S.A.*, 88, 1311, 1991.
147. **Crabtree, G. R.,** Contingent genetic regulatory events in T lymphocyte activation, *Science*, 243, 355, 1989.
148. **Gamble, D. A., Schwab, R., Weksler, M. E., and Szabó, P.,** Decreased steady state c-myc RNA in activated T cell cultures from old human is caused by a smaller proportion of T cells that transcribe the c-myc gene, *J. Immunol.*, 144, 3569, 1990.
149. **Gewirtz, A. M., Anfossi, G., Venturellio, D., Valpreda, S., Sims, R., and Calabretta, B.,** G_1/S transition in normal human lymphocytes requires the nuclear protein encoded by c-myb, *Science*, 245, 180, 1987.
150. **Venturelli, D., Travali, S., and Calabretta, B.,** Inhibition of T cell proliferation by a MYB antisense oligomer is accompanied by selective down regulation of DNA polymerase α expression, *Proc. Natl. Acad. Sci. U.S.A.*, 87, 5963, 1990.
151. **Yaar, M., Peacocke, M., Cohen, M. S., and Gilchvest, B. A.,** Dissociation of proto-oncogene induction from growth response in normal human fibroblast, *J. Cell. Physiol.*, 145, 39, 1990.
152. **Nistico, G. and De Sarro, G.,** Is interleukin 2 a neuromodulator in the brain?, *Trends Neurosci.*, 14, 146, 1991.
153. **LeFever, A. and Liepius, A.,** Kinetic analysis of K^+ ion channel function in lymphokine-activated killer (LAK) cells, *Immunopharmacol. Immunotoxicol.*, 12, 23, 1990.
154. **Pierpaoli, W. and Fabris, N.,** Eds., Physiological senescence and its postponement, *Ann. N.Y. Acad. Sci.*, 621, 1, 1991.
155. **Thoman, M. L. and Weigle, W. O.,** The cellular and subcellular basis of immunosenescence, *Adv. Immunol.*, 46, 221, 1989.
156. **O'Leary, J. J. and Hallgren, H. M.,** Aging and lymphocyte function: a model for testing gerontological hypotheses of aging in man, *Arch. Gerontol. Geriatr.*, 12, 199, 1991.
157. **Varga, Zs., Bressani, N., Zaia, A.-M., Bene, L., Fülöp, T., Leövey, A., Fabris, N., and Damjanovich, S.,** Cell surface markers, inositol phosphate levels and membrane potential of lymphocytes from young and old human patients, *Immunol. Lett.*, 23, 275, 1990.
158. **Pieri, C., Recchioni, R., Moroni, F., Marcheselli, F., Falasca, M., and Piantanelli, L.,** Food restriction in female Wistar rats. I. Survival characteristics, membrane microviscosity and proliferative response in lymphocytes, *Arch. Gerontol. Geriatr.*, 11, 99, 1990.
159. **Pieri, C., Falasca, M., Moroni, F., Marcheselli, F., and Recchioni, R.,** Food restriction in female Wistar rats. III. Thermotropic transition of membrane lipid and 5′-nucleotidase activity in hepatocytes, *Arch. Gerontol. Geriatr.*, 11, 117, 1990.
160. **Pieri, C., Moroni, F., Recchioni, R., Marcheselli, F., Falasca, M., Antonicelli, R., and Damjanovich, S.,** Aging impairs membrane potential responsiveness as well as opening of voltage and ligand gated Na^+ channels in human lymphocytes, *Arch. Gerontol. Geriatr.*, 14, 145, 1992.
161. **Rivnay, B., Globerson, A., and Shinitzky, M.,** Viscosity of lymphocyte plasma membrane in ageing mice and its possible relation to serum cholesterol, *Mech. Ageing Dev.*, 10, 71, 1979.
162. **Rivnay, B., Bergman, B., Shinitzky, M., and Globerson, A.,** Correlation between membrane viscosity, cholesterol, lymphocyte activation and ageing in man, *Mech. Ageing Dev.*, 12, 119, 1980.
163. **Witkowski, J. and Micklem, H. S.,** Decreased membrane potential of T lymphocytes in aging mice: flow cytometric studies with a carbocyanine dye, *Immunology*, 56, 307, 1985.
164. **Witkowski, J. and Micklem, H. S.,** Transmembrane electrical potential of lymphocytes in ageing mice. Flow cytometric analysis of mitogen stimulated cells, *Mech. Ageing Dev.*, 62, 167, 1992.
165. **Zs.-Nagy, I., Ohta, M., Kitani, K., and Imahori, K.,** An automated method for measuring lateral mobility of proteins in the plasma membrane of cells in compact tissues by means of fluorescence recovery after photobleaching, *Microscopie*, 41, 12, 1984.
166. **Zs.-Nagy, I., Kitani, K., Ohta, M., Zs.-Nagy, V., and Imahori, K.,** Age-dependent decrease of the lateral diffusion constant of proteins in the plasma membrane of hepatocytes as revealed by fluorescence recovery after photobleaching in tissue smears, *Arch. Gerontol. Geriatr.*, 5, 131, 1986.
167. **Zs.-Nagy, I., Kitani, K., Ohta, M., Zs.-Nagy, V., and Imahori, K.,** Age-estimations of rats based on the average lateral diffusion constant of hepatocyte membrane proteins as revealed by fluorescent recovery after photobleaching, *Exp. Gerontol.*, 21, 555, 1986.

168. **Zs.-Nagy, V., Zs.-Nagy, I., Kitani, K., Ohta, M., and Imahori, K.,** Possibilities and limitations of the smearing technique for cellular studies in compact tissues. A transmission electron microscopic analysis of liver smears, *Tissue Cell*, 20, 229, 1988.
169. **Zs.-Nagy, I., Ohno-Iwashita, Y., Ohta, M., Zs.-Nagy, V., Kitani, K., Ando, S., and Imahori, K.,** Effect of perfringolysin O on the lateral diffusion constant of membrane proteins of hepatocytes as revealed by fluorescence recovery after photobleaching, *Biochim. Biophys. Acta*, 939, 551, 1988.
170. **Kitani, K., Zs.-Nagy, I., Kanai, S., Sato, Y., and Ohta, M.,** Correlation between the biliary excretion of ouabain and the lateral mobility of hepatocyte plasma membrane proteins in the rat. The effects of age and spironolactone pretreatment, *Hepatology*, 8, 125, 1988.
171. **Zs.-Nagy, I., Kitani, K., and Ohta, M.,** Age-dependence of the lateral mobility of proteins in the plasma membrane of hepatocytes in C57BL/6 mice. FRAP studies on liver smears, *J. Gerontol. Biol. Sci.*, 44, B85, 1989.
172. **Zs.-Nagy, I., Ohta, M., and Kitani, K.,** Effect of centrophenoxyne and BCE-001 treatment on the lateral diffusion constant of proteins in the hepatocyte membrane as revealed by fluorescence recovery after photobleaching in rat liver smears, *Exp. Gerontol.*, 24, 317, 1989.
173. **Zs.-Nagy, I., Ohta, M., and Kitani, K.,** Effect of idebenone treatment on the lateral diffusion constant of proteins in the hepatocyte membrane as revealed by fluorescence recovery after photobleaching in rat liver smears, *Arch. Gerontol. Geriatr.*, 11, 243, 1990.
174. **Kitani, K., Ohta, M., Kanai, S., Sato, Y., Nokubo, M., and Zs.-Nagy, I.,** Age-induced restricted mobility of proteins in the hepatocyte surface membranes. A possible determinant of membrane transport functions, in *Liver and Aging 1990*, Kitani, K., Ed., Elsevier, Amsterdam, 1990, 305.
175. **Vincent, A., Lang, B., and Newsom-Davis, J.,** Autoimmunity to the voltage gated calcium channel underlies the Lambert-Eaton myasthenic syndrome, a paraneoplastic disorder, *Trends Neurosci.*, 12, 496, 1989.
176. **Jentsch, T. J., Steinmeyer, K., and Schwarz, G.,** Primary structure of torpedo marmorota chloride channel isolated by expression cloning in Xenopus oocytes, *Nature*, 348, 510, 1990.
177. **Patthy, L.,** Modular exchange principles in proteins, *Curr. Opinion Struct. Biol.*, 1, 351, 1991.
178. **Thinnes, F. P., Gotz, H., Kayser, H., Benz, R., Schmidt, W. E., Kratzin, H. D., and Hilschmann, N.,** Identification of human porins. I. Purification of a porin from human B lymphocytes (porin 31HL) and the topochemical proof of its expression on the plasmalemma of the progenitor cell, *Biol. Chem. Hoppe Seyler*, 370, 1253, 1989.
179. **Thinnes, F. P., Schmid, A., Benz, R., and Hilschmann, N.,** Studies on human porin. III. Does the voltage dependent anion channel "Porin 31HL" form part of the chloride channel complex, which is observed in different cells and thought to be affected in cystic fibrosis?, *Biol. Chem. Hoppe Seyler*, 371, 1047, 1990.
180. **Thinnes, F. P., Babel, D., Hein, A., Jurgens, L., Konig, U., Schmid, A., and Hilschmann, N.,** A flip-flop model of the chloride channel complex explains the dysregulation of the chloride flow in the plasmalemma of cells in cystic fibrosis, *Klin. Wochenschr.*, 69, 283, 1992.
181. **Breuer, W.,** Reconstitution of a kidney chloride channel and its identification by covalent labelling, *Biochim. Biophys. Acta*, 1022, 229, 1990.
182. **Zidovetzky, R., Yarden, Y., Schlessinger, J., and Jovin, T. M.,** Rotational diffusion of epidermal growth factor complexed to cell surface receptors reflects rapid microaggregation and endocytosis of occupied receptors, *Proc. Natl. Acad. Sci. U.S.A.*, 78, 6981, 1981.
183. **Edidin, M.,** Function by association? MHC antigens and membrane receptor complexes, *Immunol. Today*, 9, 218, 1988.
184. **Zidovetzky, R., Bartholdi, M., Arndt-Jovin, D. J., and Jovin, T. M.,** Rotational dynamics of the Fc receptor for immunoglobulin E on histamine-releasing rat basophylic leukemia cells, *Biochemistry*, 25, 4397, 1986.
185. **Edidin, M. and Wei, T.,** Diffusion rates of cell surface antigens of mouse-human heterokaryons. II. Effect of membrane potential on lateral diffusion, *J. Cell Biol.*, 75, 483, 1977.
186. **Zidovetzky, R., Yarden, Y., Schlessinger, J., and Jovin, T. M.,** Microaggregation of hormone-occupied epidermal growth factor receptors on plasma membrane preparations, *EMBO J.*, 5, 247, 1986.
187. **Szöllősi, J., Damjanovich, S., Balázs, M., Nagy, P., Trón, L., Fulwyler, M. J., and Brodsky, F. M.,** Physical association between MHC class I and class II molecules detected on the cell surface by flow cytometric energy transfer, *J. Immunol.*, 143, 208, 1989.

188. **van Belzen, N., Spaargaren, M., Verkleij, A. J., and Boonstra, J.**, Interaction of EGF receptors with the cytoskeleton is related to receptor clustering, *J. Cell. Physiol.*, 145, 365, 1990.
189. **Glant, T. and Mikecz, K.**, Antigenic profiles of human, bovine and canine articular chondrocytes, *Cell. Tissue Res.*, 244, 359, 1986.
190. **Schumacher, M.**, Rapid membrane effects of steroid hormones: an emerging concept in neuroendocrinology, *Trends Neurosci.*, 13, 359, 1990.
191. **Spät, A., Enyedi, P., Hajnoczky, G., and Hunyadi, L.**, Generation and role of calcium signal in adrenal glomerulosa cells, *Exp. Physiol.*, 76, 859, 1991.
192. **Borel, J. F.**, Pharmacology of cyclosporin (Sandimmune) IV. Pharmacological properties in vivo, *Pharmacol. Rev.*, 41, 259, 1989.
193. **Borel, J. F., Feurer, C., Gubler, M. U., and Stahelin, H.**, Biological effects of cyclosporin A: a new antilymphocytic agent, *Agents Actions*, 6, 468, 1976.
194. **Wenger, R. M.**, Synthesis of cyclosporin and analogues: structural and conformational requirements for immunosuppressive activity, *Prog. Allergy*, 38, 46, 1986.
195. **Hiestand, P. C., Gunn, H. C., Gale, J. M., Ryffel, B., and Borel, J. F.**, Comparison of the pharmacological profiles of cyclosporin, (Nva^2)-cyclosporin and (Val^2)-dihydro-cyclosporin, *Immunology*, 55, 249, 1985.
196. **Kino, T., Hatanaka, H., Hashimoto, M., Nishiyama, M., Goto, T., Okuhara, M., Kohsaka, M., Aoki, M., and Imanaka, H.**, FK-506, a novel immunosuppressant isolated from a Streptomyces. I. Fermentation, isolation and physico-chemical and biological characterization, *J. Antibiot.*, 40, 1249, 1987.
197. **Kino, T., Hatanaka, H., Miyata, S., Inamura, N., Nishiyama, M., Yajima, T., Goto, T., Okuhara, M., Kohsaka, M., Aoki, H., and Ochiai, T.**, FK-506, a novel immunosuppressant isolated from Streptomyces. II. Immunosuppressive effect of FK-506 in vitro, *J. Antibiot.*, 40, 1256, 1987.
198. **Kino, T., Inamura, N., Sakai, F., Nakahara, K., Goto, T., Okuhara, M., Kohsaka, M., Aoki, H., and Ochiai, T.**, Effect of FK-506 on human mixed lymphocyte reaction in vitro, *Transplant. Proc.*, 19(Suppl. 6), 36, 1987.
199. **Krönke, M., Leonard, W. J., Depper, J. M., Arya, S. K., Wong-Staal, F., Gallo, R. C., Waldmann, T. A., and Greene, W. C.**, Cyclosporin A inhibits T-cell growth factor gene expression at the level of mRNA transcription, *Proc. Natl. Acad. Sci. U.S.A.*, 81, 5214, 1984.
200. **Elliot, J. F., Lin, Y., Mizel, S. B., Bleackley, R. C., Harnish, D. G., and Paetkau, V.**, Induction of interleukin-2 messenger RNA inhibited by cyclosporin A, *Science*, 226, 1439, 1984.
201. **Handschumacher, R. E., Harding, M. W., Rice, J., Drugge, R. J., and Speicher, R. W.**, Cyclophilin: a specific cytosolic binding protein for CsA, *Science*, 544, 1984.
202. **Gergely, P. and Aszalós, A.**, Cyclosporin A on the anti CD-3 antibody and con-A induced activation and membrane potential of human T lymphocytes, *Immunopharmacology*, 20, 191, 1990.
203. **DiPadova, F. E.**, Pharmacology of cyclosporine (Sandimmune) V. Pharmacological effects on immune function: in vitro studies, *Pharmacol. Rev.*, 41, 373, 1989.
204. **Aszalós, A.**, Cyclosporin elicits a non-responsive state and a shift in K^+ fluxes in the early phase of activation of human lymphocytes with anti-CD3, *Eur. J. Pharmacol.*, 205, 125, 1991.
205. **Harding, M. W. and Handschumacher, R. E.**, Cyclophylin is a primary target molecule for cyclosporin A. Structural and functional implications, *Transplantation*, 46, 298, 1988.
206. **Harding, M. W., Galat, A., Uehling, D. E., and Schreiber, S. L.**, A receptor for the immunosuppressant FK-506 is a cis-trans peptidyl-prolyl isomerase, *Nature*, 341, 758, 1989.
207. **McKeon, F.**, When worlds collide: immunosuppressants meet protein phosphatases, *Cell*, 66, 823, 1991.
208. **Graziadei, L., Riabowdl, K., and Bar-Sagi, D.**, Co-capping of ras-protein with surface immunoglobulins in B lymphocytes, *Nature*, 347, 393, 1990.
209. **Singer, S. and Nicolson, G. L.**, The fluid mosaic model of the structure of cell membranes. Cell membranes are viewed as two-dimensional solutions of oriented globular proteins and lipids, *Science*, 175, 720, 1972.
210. **Damjanovich, S., Trón, L., Szöllősi, J., Zidovetzki, R., Vaz, W. C. L., Regaterio, F., Arndt-Jovin, D. J., and Jovin, T. M.**, Distribution and mobility of murine histocompatibility H2K^k antigens in the cytoplasmic membrane, *Proc. Natl. Acad. Sci. U.S.A.*, 80, 5985, 1983.
211. **Damjanovich, S., Trón, L., Szöllősi, J., Mátyus, L., and Szabó, G., Jr.**, Dynamic properties of the murine histocompatibility H-2K^k antigen in cytoplasmic membrane, *Mol. Immunol.*, 21, 1151, 1984.

212. **Szöllősi, J., Trón, L., Damjanovich, S., Helliwell, S. H., Arndt-Jovin, D. J., and Jovin, T. M.,** Energy transfer measurements on the cell surface. Distribution of H2K^k antigens, in *Quo Vadis? Flow Cytometry and Monoclonal Antibodies for Therapy Monitoring*, Gros, P., Jansen, F. K., Poncelet, P., and Roncucci, R., Eds., Sanofi Group, Montpellier, France, 1984, 33.
213. **Szöllősi, J., Trón, L., Damjanovich, S., Helliwell, S. H., Arndt-Jovin, D. J., and Jovin, T. M.,** Fluorescence energy transfer measurements on cell surfaces: a critical comparison of steady state fluorimetric and flow cytometric methods, *Cytometry*, 5, 210, 1984.
214. **Szöllősi, J., Mátyus, L., Trón, L., Balázs, M., Ember, I., Fulwyler, M. J., and Damjanovich, S.,** Flow cytometric measurements of fluorescence energy transfer using single laser excitation, *Cytometry*, 8, 120, 1987.
215. **Szöllősi, J., Damjanovich, S., Goldman, C. K., Fulwyler, M. J., Aszalós, A., Goldstein, G., Rao, P., Talle, M. A., and Waldmann, T. A.,** Flow cytometric resonance energy transfer measurements support the association of a 95-kD peptide termed T27 with the 55 kD Tac peptide, *Proc. Natl. Acad. Sci. U.S.A.*, 84, 7246, 1987.
216. **Szöllősi, J., Feuerstein, B. G., Hyun, W., Das, K. M., and Marton, L. J.,** Attachment of A172 human glioblastoma cells affects calcium signaling: a comparison of image cytometry, flow cytometry and spectrofluorimetry, *Cytometry*, 12, 707, 1992.
217. **Trón, L., Szöllősi, J., Damjanovich, S., Helliwell, S. H., Arndt-Jovin, D. J., and Jovin, T. M.,** Flow cytometric measurements of fluorescence resonance energy transfer on cell surfaces. Quantitative evaluation of the transfer efficiency on a cell-by-cell basis, *Biophys. J.*, 45, 939, 1984.
218. **Trón, L., Szöllősi, J., and Damjanovich, S.,** Proximity measurements of cell surface proteins by fluorescence energy transfer, *Immunol. Lett.*, 16, 1, 1987.
219. **Mátyus, L. and Edidin, M.,** Flow cytometry and cell sorting, in *Topics in Fluorescence Spectroscopy*, Vol. 1, Lakowicz, J. R., Ed., Plenum Press, New York, 1991, 411.
220. **Edidin, M., Aszalós, A., Damjanovich, S., and Waldmann, T. M.,** Lateral diffusion measurements give evidence for association of the Tac peptide of IL-2 receptor with T27 peptide in the plasma membrane of HUT-102-B2 cells, *J. Immunol.*, 141, 1205, 1988.
221. **Mátyus, L.,** Fluorescence resonance energy transfer measurements on cell surfaces. A spectroscopic tool for determining protein interactions, *J. Photochem. Photobiol.*, 12, 323, 1992.
222. **Burton, J., Goldman, C. K., Rao, P., Moos, M., and Waldmann, T. A.,** Association of intercellular adhesion molecule 1 with the multichain high affinity interleukin 2 receptor, *Proc. Natl. Acad. Sci. U.S.A.*, 87, 7329, 1990.
223. **De la Hera, A., Muller, U., Olsson, C., Isaaz, S., and Tunnaclife, A.,** Structure of the T cell antigen receptor (TCR): Two CD3 ε subunits in a functional TCR/CD3 complex, *J. Exp. Med.*, 173, 7, 1991.
224. **Jovin, T. M.,** Fluorescence polarization and energy transfer: theory and application, in *Flow Cytometry and Sorting*, 1st ed., Melamed, M. R., Mullaney, P. F., and Mendelsohn, M. L., Eds., John Wiley & Sons, New York, 1979, 137.
225. **Jovin, T. M. and Arndt-Jovin, D. J.,** FRET microscopy: digital imaging of fluorescence resonance energy transfer. Application in cell biology, in *Cell Structure and Function by Microspectrofluorimetry*, Kohen, E., Ploem, J. S., and Hirschberg, J. G., Eds., Academic Press, Orlando, FL, 1989, 99.
226. **Jovin, T. M. and Arndt-Jovin, D. J.,** Luminescence digital imaging microscopy, *Annu. Rev. Biophys. Biophys. Chem.*, 18, 271, 1989.
227. **Jovin, T. M., Arndt-Jovin, D. J., Marriott, G., Clegg, R. M., Robert-Nicoud, M., and Schormann, T.,** Distance, wavelength and time: the versatile 3rd dimension in light emission microscopy, in *Optical Microscopy for Biology*, 1st ed., Herman, B. and Jacobson, K., Eds., John Wiley & Sons, New York, 1990, 575.
228. **Richard, E. A. and Miller, C.,** Steady-state coupling of ion-channel conformation to a membrane ion gradient, *Science*, 247, 1208, 1990.
229. **Fromherz, P.,** Spatio-temporal patterns in the fluid-mosaic model of membranes, *Biochim. Biophys. Acta*, 944, 108, 1988.
230. **Fromherz, P.,** Self-organization of the fluid mosaic of charged channel proteins in membranes, *Proc. Natl. Acad. Sci. U.S.A.*, 85, 6353, 1988.
231. **Fromherz, P.,** Dissipative structures of ion channels in the fluid mosaic model of a membrane cable, *Ber. Bunsenges. Phys. Chem.*, 92, 1010, 1988.
232. **Saffman, P. G. and Delbrueck, M.,** Brownian motion in biological membranes, *Proc. Natl. Acad. Sci. U.S.A.*, 72, 3111, 1975.

233. **Saffman, P. G.,** Brownian motion in thin sheets of viscous fluid, *J. Fluid Mech.*, 73, 593, 1976.
234. **Schreiber, A. B., Hoebeke, J., Vray, B., and Strossberg, D.,** Evidence for reversible microclustering of lentil lectin membrane receptors on HeLa cells, *FEBS Lett.*, 111, 303, 1980.
235. **Schreiber, A. B., Hebeke, J., Vray, B., and Strossberg, D.,** Resonance energy studies of the mechanism of microclustering of lentil lectin membrane receptors on HeLa cells, *Exp. Cell Res.*, 132, 273, 1981.
236. **Frye, L. D. and Edidin, M.,** The rapid intermixing of cell surface antigens after formation of mouse-human heterokaryons, *J. Cell Sci.*, 7, 319, 1970.
237. **Edidin, M., Kuo, S. C., and Sheetz, M. P.,** Lateral movements of membrane glycoproteins restricted by dynamic cytoplasmic barriers, *Science*, 254, 1379, 1991.
238. **Quinn, P. J.,** The fluidity of cell membrane and its regulation, *Prog. Biophys. Mol. Biol.*, 38, 1, 1981.
239. **Pasta, R. W., Venable, R. M., and Karplus, M.,** Model for the structure of the lipid bilayer, *Proc. Natl. Acad. Sci. U.S.A.*, 88, 892, 1991.
240. **Lakos, Zs., Somogyi, B., Balázs, M., Matkó, J., and Damjanovich, S.,** The effect of transmembrane potential on the dynamic behaviour of cell membrane, *Biochim. Biophys. Acta*, 1023, 41, 1990.
241. **Lelkes, P.,** Potential dependent rigidity changes in lipid membrane vesicles, *Biochem. Biophys. Res. Commun.*, 90, 656, 1979.
242. **Weiss, R. E., Roberts, W. M., Stühmer, W., and Almers, W.,** Mobility of voltage-dependent ion channels and lectin receptors in the sarcolemma of frog skeletal muscle, *J. Gen. Physiol.*, 87, 955, 1986.
243. **Martonosi, A. N., Ed.,** *Membranes and Transport*, Vol. 1 and 2, Plenum Press, New York, 1982.
244. **Martonosi, A. N., Jóna, I., Molnár, E., Seidler, N. W., Buchet, R., and Varga, S.,** Emerging views on the structure and dynamics of the Ca^{2+}-ATP-ase in sarcoplasmic reticulum, *FEBS Lett.*, 268, 365, 1990.
245. **Förster, T.,** Zwischenmolekulare Energiewanderung und fluoreszenz, *Ann. Physik*, 2, 55, 1948.
246. **Szabó, G., Jr., Pine, S. P., Weaver, J. L., Kasari, M., and Aszalós, A.,** Epitope mapping by photobleaching fluorescence energy transfer measurements using a laser scanning microscope system, *Biophys. J.*, 61, 661, 1992.
247. **Damjanovich, S. and Somogyi, B.,** Molecular enzyme model based on oriented energy transfer, *J. Theor. Biol.*, 41, 567, 1973.
248. **Damjanovich, S., Somogyi, B., and Welch, G. R.,** Protein fluctuation and enzyme activity, *J. Theor. Biol.*, 105, 25, 1981.
249. **Somogyi, B. and Damjanovich, S.,** Relationship between the lifetime of an enzyme substrate complex and the properties of the molecular environment, *J. Theor. Biol.*, 48, 393, 1975.
250. **Somogyi, B. and Damjanovich, S.,** A microenvironmental approach to enzyme dynamics, in *The Fluctuating Enzyme*, Welch, G. R., Ed., John Wiley & Sons, New York, 1986, 341.
251. **Somogyi, B., Welch, G. R., and Damjanovich, S.,** The dynamic basis of energy transduction in enzymes, *Biochim. Biophys. Acta*, 768, 81, 1984.
252. **Welch, G. R., Somogyi, B., and Damjanovich, S.,** The role of protein fluctuation in enzyme action: a review, *Prog. Biophys. Mol. Biol.*, 39, 109, 1982.
253. **Burkli, A. and Cherry, R. J.,** Rotational motion and flexibility of Ca^{2+}, Mg^{2+}-dependent ATP-ase in sarcoplasmic reticulum membrane, *Biochemistry*, 20, 138, 1981.
254. **Beece, D., Eisenstein, L., Frauenfelder, H., Good, D., Marden, M. C., Reinisch, L., Reynolds, A. H., Sorensen, L. B., and Yue, K. T.,** Solvent viscosity and protein dynamics, *Biochemistry*, 19, 5147, 1980.
255. **Cone, C. D., Jr.,** Variation of transmembrane potential level as a basic mechanism of mitotic control, *Oncology*, 24, 438, 1970.
256. **Cohen, L. B.,** Changes in neuron structure during action potential propagation and synaptic transmission, *Physiol. Rev.*, 53, 573, 1973.
257. **Lee, S. C. and Deutsch, C.,** Temperature dependence of K+ channel properties in human T lymphocytes, *Biophys. J.*, 57, 49, 1990.
258. **Matkó, J., Trón, L., Balázs, M., Hevessy, J., Somogyi, B., and Damjanovich, S.,** Correlation between activity and dynamics of protein matrix of phosphorylase B, *Biochemistry*, 19, 5782, 1980.
259. **Jóna, I. and Martonosi, A.,** The effects of membrane potential and lanthanides on the conformation of the Ca^{2+} transport ATPase in sarcoplasmic reticulum, *Biochem. J.*, 234, 363, 1986.
260. **Tsong, T. Y. and Astumian, R. D.,** Electroconformational coupling and membrane protein function, *Prog. Biophys. Mol. Biol.*, 50, 1, 1987.

261. **Boyse, E. A., Old, L. J., and Stockert, E.,** An approach to the mapping of antigens on the cell surface, *Genetics*, 60, 886, 1968.
262. **Flaherty, L. and Zimmerman, D.,** Surface mapping of mouse thymocytes, *Proc. Natl. Acad. Sci. U.S.A.*, 76, 1990, 1979.
263. **Vaz, W. L. C., Kaufman, K., and Nikisch, A.,** Use of energy transfer to assay the association of proteins with lipid membranes, *Anal. Biochem.*, 83, 385, 1977.
264. **Vaz, W. L. C., Derzko, Z. I., and Jacobson, K. A.,** Photobleaching measurements of lateral diffusion of lipids and proteins in artificial phospholipid bilayer membranes, in *Cell Surface Reviews*, Vol. 8, Poste, G. and Nicolson, G., Eds., Elsevier, Amsterdam, 1982, 83.
265. **Vaz, W. L. C., Goodsaid-Zalduondo, F., and Jacobson, K.,** Lateral diffusion of lipids and proteins in bilayer membranes, *FEBS Lett.*, 174, 199, 1984.
266. **Vaz, W. L. C., Clegg, R. M., and Hallman, D.,** Translational diffusion of lipids in liquid crystalline phase phosphatidylcholine multibilayers. A comparison of experiment with theory, *Biochemistry*, 24, 781, 1985.
267. **Clegg, R. M. and Vaz, W. L. C.,** Translational diffusion of proteins and lipids in artificial bilayer membranes. A comparison of experiment with theory, in *Progress in Protein-Lipid Interactions*, Vol. 1, Watts, A. and DePont, J. J., Eds., Elsevier, Amsterdam, 1985, 173.
268. **Borochov, H. and Schinitzky, M.,** Vertical displacement of membrane proteins mediated by changes in microviscosity, *Proc. Natl. Acad. Sci. U.S.A.*, 73, 4526, 1976.
269. **Boonstra, J., Nelemans, S., Feijen, A., Bierman, A., Van Zoelen, E. J., Van der Saag, P. T., and DeLaat, S. W.,** Effect of fatty acids on plasma membrane lipid dynamics and cation permeability in neuroblastoma cells, *Biochim. Biophys. Acta*, 692, 321, 1982.
270. **Bretscher, M. S.,** Directed lipid flow in cell membranes, *Nature*, 260, 21, 1976.
271. **Geppert, T. D. and Lipsky, P. E.,** Association of various T cell-surface molecules with the cytosceleton. Effect of cross-linking and activation, *J. Immunol.*, 146, 3298, 1991.
272. **Aszalós, A., Damjanovich, S., and Gottesman, M.,** Depolymerization of microtubules alters membrane potential and affects the motional freedom of membrane proteins, *Biochemistry*, 25, 5804, 1986.
273. **Fernandez, S. M. and Berlin, R. D.,** Cell surface distribution of lectin receptors determined by resonance energy transfer, *Nature*, 264, 411, 1976.
274. **Blatt, E. and Jovin, T. A.,** Rotational dynamics of biological macromolecules, in *Photophysical and Photochemical Tools in Polymer Science*, 1st ed., Wnnick, E. A., Ed., D. Reidel, Dordrecht, The Netherlands, 1986, 351.
275. **Blatt, E. and Sawyer, W. H.,** Depth-dependent fluorescence quenching in micelles and membranes, *Biochim. Biophys. Acta*, 822, 43, 1985.
276. **Cheng, K. H., Wiedmer, T., and Sims, P. J.,** Fluorescence resonance energy transfer study of the associative state of membrane-bound complexes of complement proteins C5b-8, *J. Immunol.*, 135, 459, 1985.
277. **Cherry, R. J.,** Rotational and lateral diffusion of membrane proteins, *Biochim. Biophys. Acta*, 559, 289, 1979.
278. **Cherry, R. J.,** Transient dichroism and rotational diffusion of macromolecules: application to membrane proteins, in *Spectroscopy and Dynamics of Molecular Biological Systems*, 1st ed., Bayley, P. M. and Dale, R. E., Eds., Academic Press, London, 1985, 79.
279. **Cherry, R. J.,** Effect of mellitin and membrane potential on the mobility of band 3 proteins in human erythrocyte membranes, in *Dynamics of Biochemical Systems*, 1st ed., Damjanovich, S., Keleti, T., and Trón, L., Eds., Elsevier, Amsterdam, 1986, 487.
280. **Dale, R. E., Eisinger, J., and Blumberg, W. E.,** The orientational freedom of molecular probes. The orientation factor in intramolecular energy transfer, *Biophys. J.*, 26, 161, 1979.
281. **Dale, R. E., Novros, J., Roth, S., Edidin, M., and Brand, L.,** Application of Foerster long-range excitation energy transfer to the determination of distributions of fluorescently-labelled concanavalin A receptor complexes at the surfaces of yeast and of normal and malignant fibroblasts, in *Fluorescent Probes*, 1st ed., Beddard, G. S. and West, M. A., Eds., Academic Press, London, 1981, 159.
282. **Damjanovich, S., Bahr, W., and Jovin, T. M.,** The functional and fluorescence properties of Escherichia coli RNA polymerase reacted with fluorescamine, *Eur. J. Biochem.*, 72, 559, 1977.
283. **Eckert, R., Paschke, D., and Hülser, D. F.,** Patch clamp techniques for the characterization of membrane channels, in *Physical Characterization of Biological Cells*, Schütt, W., Klinkmann, H., Lamprecht, I., and Wilson, T., Eds., Verlag Gesundheit GmbH, Berlin, 1991, 353.

284. **Eliot, T., Cerundolo, V., Elvin, J., and Townsend, A.,** A peptide-induced conformational change of the class I heavy chain, *Nature*, 351, 402, 1991.
285. **Fishman, H. M.,** Relaxations, fluctuations and ion transfer across membranes, *Q. Rev. Biophys.*, 46, 127, 1985.
286. **Grinvald, A., Hildesheim, R., Farber, I. C., and Anglister, L.,** Improved fluorescent probes for the measurement of rapid changes in membrane potential, *Biophys. J.*, 39, 301, 1982.
287. **Grynkiewicz, G., Poenie, M., and Tsien, R. Y.,** A new generation of Ca^{2+} indicators with greatly improved fluorescence properties, *J. Biol. Chem.*, 260, 3440, 1985.
288. **Gupta, R. K., Salzberg, B. M., Grinvald, A., Cohen, L. B., Kamino, K., Lesher, S., Boyle, M. B., Waggoner, A. S., and Wang, C. H.,** Improvements in optical methods for measuring rapid changes in membrane potential, *J. Membr. Biol.*, 58, 123, 1981.
289. **Johnstone, R. M., Laris, P. C., and Eddy, A. A.,** The use of fluorescent dyes to measure membrane potentials: a critique, *J. Cell Physiol.*, 112, 298, 1982.
290. **Kaplan, J. H. and Somlyo, A. P.,** Flash photolysis of caged compounds: new tools for cellular physiology, *Trends Neurosci.*, 12, 54, 1989.
291. **Kubitscheck, U., Kircheis, M., Schweitzer-Stenner, R., Dreybrodt, W., Jovin, T. M., and Pecht, I.,** Fluorescence resonance energy transfer on single living cells. Application to binding of monovalent haptens to cell-bound immunoglobulin E, *Biophys. J.*, 60, 307, 1991.
292. **Penner, R. and Neher, E.,** The patch clamp technique in the study of secretion, *Trends Neurosci.*, 12, 159, 1989.
293. **Pershadsingh, H. A., Johnstone, R. M., and Larish, P. C.,** Influence of (DL)-propanolol and Ca^{2+} on membrane potential and amino acid transport in Ehrlich ascites tumour cells, *Biochim. Biophys. Acta*, 509, 360, 1978.
294. **Restall, C. J., Dale, R. E., Murray, E. K., Gilbert, C. W., and Chapman, D.,** Rotational diffusion of calcium-dependent adenosine-triphosphatase in sarcoplasmic reticulum: a detailed study, *Biochemistry*, 23, 6765, 1984.
295. **Riegler, R. and Ehrenberg, M.,** Fluorescence relaxation spectroscopy in the analysis of macromolecular structure and motion, *Q. Rev. Biophys.*, 9, 1, 1976.
296. **Rigler, R.,** Progress in intensity correlation spectroscopy and analysis of structure and dynamics of biopolymers, in *Spectroscopy and Dynamics of Molecular Biological Systems*, Bailey, P. M. and Dale, R. E., Eds., Academic Press, London, 1985, 35.
297. **Smith, T. C.,** The use of fluorescence dyes to measure membrane potential: a response, *J. Cell. Physiol.*, 122, 302, 1982.
298. **Averdunk, R. and Lauf, P. K.,** Effect of mitogens on Na^+/K^+ transport, ^3H-ouabain binding and adenosine triphosphatase activity in lymphocytes, *Exp. Cell Res.*, 93, 331, 1975.
299. **Birx, D. L., Berger, M., and Fleisher, T. A.,** The interference of T cell activation by Ca^{2+} channel blocking agents, *J. Immunol.*, 133, 2904, 1984.
300. **Blumenthal, R., Klausner, R. D., and Weinstein, J. N.,** Voltage dependent translocation of the asialoglycoprotein receptor across lipid membranes, *Nature*, 288, 333, 1980.
301. **Brodie, C. and Samson, S. R.,** Nerve growth factor and fibroblast growth factor influence post-fusion expression of Na^+ channels in cultured rat skeletal muscle, *J. Cell. Physiol.*, 144, 492, 1990.
302. **Castle, N. A., Haylett, D. G., and Jenkinson, D. H.,** Toxins in the characterization of potassium channels, *Trends Neurosci.*, 12, 59, 1989.
303. **Chambord, J. C., Paris, S., L'Allemain, G., and Pouyssegur, J.,** Two growth factor signalling pathways in fibroblasts distinguished by pertussis toxin, *Nature*, 326, 800, 1987.
304. **Choquet, D. and Korn, H.,** Dual effects of serotonin on a voltage gated conductance in lymphocytes, *Proc. Natl. Acad. Sci. U.S.A.*, 85, 4557, 1988.
305. **Choquet, D., Sarthou, P., Primi, D., Cazenave, A., and Korn, H.,** Cyclic AMP-modulated K^+ channels in murine B cells and their precursors, *Science*, 325, 1211, 1987.
306. **Corda, D., Pasternak, C., and Schinitzky, M.,** Increase in lipid microviscosity of unilamellar vesicles upon creation of transmembrane potential, *J. Membr. Biol.*, 65, 235, 1982.
307. **Corin, A. F., Matayoshi, E. D., and Jovin, T. M.,** Triplet state spectroscopy for investigating diffusion and chemical kinetics, in *Spectroscopy and Dynamics of Molecular Biological Systems*, Bailey, P. M. and Dale, R. E., Eds., Academic Press, London, 1985, 53.
308. **Derkach, V., Suprement, A., and North, R. A.,** $5HT_3$ receptors are membrane ion channels, *Nature*, 339, 706, 1989.

309. **Deutsch, C., Holian, A., Holian, S. K., Daniele, R. O., and Wilson, D. F.,** Transmembrane electrical and pH gradient across human erythrocytes and human peripheral lymphocytes, *J. Cell. Comp. Physiol.*, 99, 79, 1979.
310. **Deutsch, C. J. and Price, M.,** Role of extracellular Na^+ and K^+ in lymphocyte activation, *J. Cell. Comp. Physiol.*, 113, 73, 1982.
311. **Doherty, P. A. and Snider, M. D.,** Effects of hypertonic and sodium free medium on transport of a membrane glycoprotein along the secretary pathway in cultured mammalian cells, *J. Cell. Physiol.*, 146, 34, 1991.
312. **Dunne, M. J. and Petersen, O. H.,** Potassium selective ion channel in insulin secreting cells: physiology, pharmacology and their role in stimulus secretion coupling, *Biochim. Biophys. Acta*, 1071, 67, 1991.
313. **Felber, S. M. and Brand, M. D.,** Factors determining the plasma membrane potential of lymphocytes, *Biochem. J.*, 204, 577, 1982.
314. **Felber, S. M. and Brand, M. D.,** Early plasma membrane potential changes during stimulation of lymphocytes by concanavalin A, *Biochem. J.*, 210, 885, 1983.
315. **Felber, S. M. and Brand, M. D.,** Concanavalin A causes an increase in sodium permeability and intracellular sodium content of pig lymphocytes, *Biochem. J.*, 210, 893, 1983.
316. **Grimley, P. M. and Aszalós, A.,** Early membrane depolarization by alpha interferon: biologic correlation with antiproliferative signal, *Biochem. Biophys. Res. Commun.*, 146, 300, 1987.
317. **Grunicke, H. H.,** The cell membrane as a target for cancer chemotherapy, *Eur. J. Cancer*, 27, 281, 1991.
318. **Guy, R. H. and Conti, F.,** Pursuing the structure and function of voltage gated channels, *Trends Neurosci.*, 13, 201, 1990.
319. **Hardle, R. C.,** A histamine-activated chloride channel involved in neurotransmission at a photoreceptor synapse, *Nature*, 339, 704, 1989.
320. **Harris, D. T., Kozumbo, W. J., Cerutti, P., and Cerottini, J. C.,** Molecular mechanisms involved in T cell activation. I. Evidence for independent signal-transducing pathways in lymphokine production vs. proliferation in cloned cytotoxic T lymphocytes, *J. Immunol.*, 138, 600, 1987.
321. **Ip, S. H. and Cooper, R. A.,** Decreased membrane fluidity during differentiation of human promyelocytic leukemia cells in culture, *Blood*, 56, 227, 1980.
322. **Johnson, S. M. and Kramers, M.,** Membrane viscosity differences in normal and leukemic human lymphocytes, *Biochem. Biophys. Res. Commun.*, 80, 451, 1978.
323. **Kiefer, H., Blume, A. J., and Kaback, H. R.,** Membrane potential changes during mitogenic stimulation of mouse spleen lymphocytes, *Proc. Natl. Acad. Sci. U.S.A.*, 77, 2200, 1980.
324. **Kim, D., Lewis, D. L., Graziadei, L., Neer, E. J., Bar-Sagi, D., and Chapman, D. E.,** G-protein $\beta\gamma$-subunits activate the cardiac muscarinic K^+ channel via phospholipase A_2, *Nature*, 337, 557, 1989.
325. **Mason, M. J. and Grinstein, S.,** Effect of cytoplasmic acidification on the membrane potential of T lymphocytes: role of trace methods, *J. Membr. Biol.*, 116, 139, 1990.
326. **Mastro, A. M. and Smith, M. S.,** Ca^{2+} dependent activation of lymphocytes by ionophore, A23187 and a phorbol ester tumor promoter, *J. Cell. Physiol.*, 116, 51, 1983.
327. **McCloskey, M. A. and Cahalan, M. D.,** G-protein control of potassium channel activity in a mast cell line, *J. Gen. Physiol.*, 95, 205, 1990.
328. **Meldolesi, J. and Pozzan, T.,** Pathways of Ca^{2+} influx at the plasma membrane: voltage-, receptor-, and second messenger-operated channels, *Exp. Cell Res.*, 171, 271, 1987.
329. **Monroe, J. G. and Cambier, J. C.,** B cell activation. I. Anti-immunoglobulin induced receptor crosslinking results in a decrease in the plasma membrane potential of murine B lymphocytes, *J. Exp. Med.*, 157, 2073, 1983.
330. **Monroe, J. G. and Cambier, J. C.,** B cell activation. II. Receptor cross-linking by thymus-dependent and thymus-independent antigens induces a rapid decrease in the plasma membrane potential of antigen binding B lymphocytes, *J. Immunol.*, 131, 2641, 1983.
331. **Monroe, J. G. and Cambier, J. C.,** B cell activation. III. B cell plasma membrane depolarization and hyper Ia antigen expression induced by receptor immunoglobulin cross-linking are coupled, *J. Exp. Med.*, 158, 1598, 1983.
332. **Monroe, J. G., Niedel, J. E., and Cambier, J. C.,** B cell activation. IV. Induction of cell membrane depolarization and hyper Ia expression by phorbol ester suggest a role for protein kinase C in murine B lymphocyte activation, *J. Immunol.*, 132, 1472, 1984.

333. **Morris, A. P., Gallacher, D. V., Irvine, R. F., and Petersen, O. H.,** Synergism of inositol trisphosphate and tetratrisphosphate in activating Ca^{2+} dependent K^+ channels, *Nature*, 330, 653, 1987.
334. **Nicolson, G. L.,** Transmembrane control of the receptors on normal and tumor cells. I. Cytoplasmic influence over cell surface components, *Biochim. Biophys. Acta*, 457, 57, 1976.
335. **Panet, R., Amir, I., Snyder, D., Zonenshein, L., Atlan, H., Laskov, R., and Panet, A.,** Effect of Na^+ flux inhibitors on induction of c-fos, c-myc and ODC (ornithin decarboxilase) genes during cell cycle, *J. Cell. Physiol.*, 140, 161, 1989.
336. **Pershadsingh, H. A., Stubbs, E. B., Jr., Noteboom, W. D., Vorbeck, M. L., and Martin, A. P.,** Influence of Ca^{2+} on the plasma membrane potential and electrogenic uptake of glycine by myeloma cells. Involvement of a Ca^{2+} activated K^+ channel, *Biochim. Biophys. Acta*, 821, 445, 1985.
337. **Sternweis, P. C. and Pang, I. H.,** The G protein-channel connection, *Trends Neurosci.*, 13, 122, 1990.
338. **Tatham, P. E. and Delves, P. J.,** Flow cytometric detection of membrane potential changes in murine lymphocytes induced by concanavalin A, *Biochem. J.*, 221, 137, 1984.
339. **Flanagan, M. W., Corthesy, B., Bram, R. J., and Crabtree, G. R.,** Nuclear association of a T-cell transcription factor blocked by FK-506 and cyclosporin A, *Nature*, 352, 803, 1991.
340. **Trauble, H. and Eibl, H.,** Electrostatic effects on lipid phase transitions: membrane structure and ionic environment, *Proc. Natl. Acad. Sci. U.S.A.*, 71, 214, 1974.
341. **van Blitterswijk, W. J., Emmelot, P., Hilkman, H. A., Oomen-Meulemans, A. P., and Inbar, M.,** Differences in lipid fluidity among isolated plasma membranes of normal and leukemic lymphocytes and membranes exfoliated from their cell surface, *Biochim. Biophys. Acta*, 467, 309, 1977.
342. **Trón, L., Aszalós, A., Balázs, M., Mulhern, S. A., Szöllősi, J., and Damjanovich, S.,** On the biophysics of transmembrane signalling, *Mol. Immunol.*, 25, 1075, 1988.
343. **Chen, L. A., Dale, R. E., Roth, S., and Brand, L.,** Nanosecond time-dependent fluorescence depolarization of diphenyl-hexatriene in dimyristoillecithin vesicles and the determination of "microviscosity", *J. Biol. Chem.*, 252, 2163, 1977.
344. **Jenis, D. M., Stepanowski, A. L., Blair, O. C., Burger, D. E., and Sartorelli, A. C.,** Lectin receptor proximity on HL-60 leukemia cells determined by fluorescence energy transfer using flow cytometry, *J. Cell. Physiol.*, 121, 501, 1984.
345. **Liegler, T., Szöllősi, J., Hyun, W., and Goodenow, R. S.,** Proximity measurements between H-2 antigens and the insulin receptor by fluorescence energy transfer: evidence that a close association does not influence insulin binding, *Proc. Natl. Acad. Sci. U.S.A.*, 88, 6755, 1991.
346. **Harel-Bellan, A., Ferris, D. K., Vinocour, M., Holt, J. T., and Farrar, W. L.,** Specific inhibition of c-myc protein biosynthesis using an antisense synthetic deoxyoligonucleotide in human T lymphocytes, *J. Immunol.*, 141, 2431, 1988.
347. **Gorvel, J. P., Miskal, Z., Liegey, F., Rigal, A., and Maroux, S.,** Conformational change of rabbit aminopeptidase N into erythrocyte plasma membrane domains analyzed by flow cytometry fluorescence energy transfer, *J. Cell. Biol.*, 108, 2193, 1984.
348. **Sharon, M., Gnarra, J. R., Baniyash, M., and Leonard, W. J.,** Possible association between IL-2 receptors and class I HLA molecules on T cells, *J. Immunol.*, 141, 3512, 1988.
349. **Mittler, R. S., Goldman, S. J., Spitalny, G. L., and Burakoff, S. J.,** T-cell receptor-CD4 physical association in a murine T-cell hybridoma: induction by antigen receptor ligation, *Proc. Natl. Acad. Sci. U.S.A.*, 86, 8531, 1989.
350. **Szegedi, Gy., Suranyi, P., and Damjanovich, S.,** Antigen density on the surface of mononuclear cells in SLE, *Immunol. Lett.*, 15, 243, 1987.
351. **Balázs, M., Szöllősi, J., Lee, W. C., Haugland, R. P., Guzikowski, A. P., Fulwyler, M. J., Damjanovich, S., Feuerstein, B. G., and Pershadsingh, H. A.,** Fluorescent tetradecanoylphorbol acetate: a novel probe of phorbol ester binding domains, *J. Cell. Biochem.*, 46, 266, 1991.
352. **Smith, K. A.,** The two-chain structure of high-affinity IL-2 receptor, *Immunol. Today*, 8, 11, 1987.
353. **Smith, K. A.,** Interleukin-2 inception, impact and implication, *Science*, 240, 1169, 1988.
354. **Busse, R., Fichtner, H., Lückkoff, A., and Kohlhard, M.,** Hyperpolarization and increased free calcium in acetylcholine-stimulated endothelial cells, *Am. J. Physiol.*, 255, H965, 1988.
355. **Fischer, P. B., Flamm, M., Schachter, D., and Weinstein, I. B.,** Tumor promoters induce membrane changes detected by fluorescence polarization, *Biochem. Biophys. Res. Commun.*, 86, 1063, 1979.
356. **L'Allemain, G., Franchi, A., Cragoe, E., Jr., and Pouyssegur, J.,** Blockade of the Na^+/H^+ antiport abolishes growth factor induced DNA synthesis in fibroblasts, *J. Biol. Chem.*, 259, 4313, 1984.

357. **Gáspár, R., Krasznai, Z., Márián, T., Trón, L., Recchioni, R., Falasca, M., Moroni, F., Pieri, C., and Damjanovich, S.**, Bretylium-induced voltage-gated sodium current in human lymphocytes, *Biochim. Biophys. Acta*, 1137, 143, 1992.
358. **Oettgen, H. C., Terhorst, C., Cantley, L. C., and Rosoff, P. M.**, Stimulation of the T3-T cell receptor complex induces a membrane potential sensitive calcium influx, *Cell*, 40, 583, 1985.
359. **Didier, A., Aussel, C., Polassy, C., and Fehlmann, M.**, IL-1 signalling for IL-2 production in T cells involves a rise in phosphatidylserine synthesis, *J. Immunol.*, 141, 3078, 1988.
360. **Eardley, D. D. and Koshland, M. E.**, Glycosil-phosphatidyl-inositol: a candidate system for interleukin-2 signal transduction, *Science*, 251, 78, 1991.
361. **Kozumbo, W., Harris, D. T., Gromkowski, S., Cerottini, J.-C., and Cerutti, P. A.**, Molecular mechanism involved in T cell activation. II. The phosphatidyl-inositol signal-transducing mechanism mediates antigen induced lymphokine production but not IL-2-induced proliferation in cloned cytotoxic T lymphocytes, *J. Immunol.*, 138, 606, 1987.
362. **Mitchell, L. C., Davis, L. S., and Lipsky, P. E.**, Promotion of human T lymphocyte proliferation by IL-4, *J. Immunol.*, 142, 1548, 1989.
363. **Tosibio, M. L., Gutrierrez-Ramos, J. C., Pezzi, L., Marcos, M. A. R., and Martinez, C.**, Interleukin-2-dependent proliferation in T cell development, *Nature*, 342, 82, 1989.
364. **Yoshimasa, T., Sibley, D. R., Bouvier, M., Lefkowitz, R. J., and Caron, M. G.**, Crosstalk between cellular signalling pathways suggested by phrobol-ester induced adenylate cyclase phosphorylation, *Nature*, 327, 67, 1987.
365. **Wuthrich, K., von Freyberg, B., Weber, C., Wider, G., Traber, R., Widmer, G., and Braun, W.**, Receptor induced conformation of the immunosuppressant cyclosporin A, *Science*, 254, 953, 1991.
366. **Kallen, J., Spitzfaden, C., Zurini, M. G. M., Wider, G., Widmer, H., Wutrich, K., and Walkinshaw, M. D.**, Structure of human cyclophilin and its binding site for cyclosporin A determined by X-ray crystallography and NMR spectroscopy, *Nature*, 353, 276, 1991.
367. **Bierer, B. E., Mattial, P. S., Standaert, R. F., Herzenberg, L. A., Burakoff, S. J., Crabtree, G., and Schreiber, S. L.**, Two distinct signal transmission pathways in T lymphocytes are inhibited by complexes formed between an immunophilin and either FK 506 or rapamycin, *Proc. Natl. Acad. Sci. U.S.A.*, 87, 9231, 1990.
368. **Hess, A. D. and Colombani, P. M.**, Mechanism of the action of cyclosporin: role of calmodulin, cyclophylin and other cyclosporin binding proteins, *Transplant. Proc. XVIII*, Suppl. 5, 219, 1986.
369. **Redelman, D.**, CsA does not inhibit the PHA stimulate increase in intracellular Ca^{2+} concentration but inhibits the increase in E rosette receptor (CD-2) expression and appearance of IL-2R-s (CD-25), *Cytometry*, 9, 156, 1990.

INDEX

G

H

I

J

K

L

M

N

O

P

T

U

V

W